AF328069

HYDRODYNAMICS AROUND CYLINDRICAL STRUCTURES

ADVANCED SERIES ON OCEAN ENGINEERING

Series Editor-in-Chief
Philip L- F Liu (*Cornell University*)

Advanced Series on Ocean Engineering – Volume 12

HYDRODYNAMICS AROUND CYLINDRICAL STRUCTURES

B. MUTLU SUMER
JØRGEN FREDSØE

Department of Hydrodynamics and Water Resources (ISVA)
Technical University of Denmark

Published by

World Scientific Publishing Co. Pte. Ltd.

P O Box 128, Farrer Road, Singapore 912805

USA office: Suite 1B, 1060 Main Street, River Edge, NJ 07661

UK office: 57 Shelton Street, Covent Garden, London WC2H 9HE

British Library Cataloguing-in-Publication Data
A catalogue record for this book is available from the British Library.

First published 1997
Reprinted 1999

HYDRODYNAMICS AROUND CYLINDRICAL STRUCTURES

ISBN 981-02-2898-8
ISBN 981-02-3056-7 (pbk)

Printed in Singapore.

Preface

Flow around a circular cylinder is a classical topic within hydrodynamics. Since the rapid expansion of the offshore industry in the sixties, the knowledge of this kind of flow has also attracted considerable attention from many mechanical and civil engineers working in the offshore field.

The purpose of the present book is

- To give a detailed, updated description of the flow pattern around cylindrical structures (including pipelines) in the presence of waves and/or current.

- To describe the impact (lift and drag forces) of the flow on the structure.

- And finally to describe the possible vibration patterns for cylindrical structures. This part will also describe the flow around a vibrating cylinder and the resulting forces.

The scope does not deviate very much from the book by Sarpkaya and Isaacson (1980) entitled "Mechanics of Wave Forces on Offshore Structures". However, while Sarpkaya and Isaacson devoted around 50% of the book to the drag-dominated regime and around 50% to diffraction, the present book concentrates mainly on the drag-dominated regime. A small chapter on diffraction is included for the sake of completeness. The reason for our concentration on the drag-dominated regime (large KC-numbers) is that it is in this field the most progress and development have taken place during the last almost 20 years since Sarpkaya and Isaacson's book. In the drag-dominated regime, flow separation, vortex shedding, and turbulence have a large impact on the resulting forces. Good understanding of this impact has been gained by detailed experimental investigations, and much has been achieved, also in the way of the numerical modelling, especially during the last 5-10 years, when the computer capacity has exploded.

In the book the theoretical and the experimental development is described. In order also to make the book usable as a text book, some classical flow solutions are included in the book, mainly as examples.

Acknowledgement:

The writers would like to express their appreciation of the very good scientific climate in the area offshore research in Denmark. In our country the hydrodynamic offshore research was introduced by professor Lundgren at our institute in the beginning of the seventies. In the late seventies and in the eighties the research was mainly concentrated in the Offshore Department at the Danish Hydraulic Institute. Significant contributions to the understanding of pipeline hydrodynamics were here obtained by Vagner Jacobsen and Mads Bryndum, two colleagues whose support has been of inestimable importance to us.

In 1984 a special grant from the university made it possible to ask one of the authors (Mutlu Sumer) to join the Danish group on offshore engineering so that he could convey his experience on fluid forces acting on small sediment particles to larger structures. This has been followed up by many grants from the Danish Technical Council (STVF), first through the FTU-programme and next through the frame-programme "Marine Technique" (1991-97). The present book is an integrated output from all these efforts and grants. The book has been typewritten by Hildur Juncker and the drawings have been prepared by Liselotte Norup, Eva Vermehren, Erling Poder, and Nega Beraki. Our librarian Kirsten Djørup has corrected and improved our written English.

Credits

The authors and World Scientific Publishing Co Pte Ltd gratefully acknowledge the courtesy of the organizations who granted permission to use illustrations and other information in this book.

Fig. 3.4:
Reprinted from H. Honji: "Streaked flow around an oscillating circular cylinder". J. Fluid Mech., 107:509-520, 1982, with kind permission from Cambridge University Press, Publishing Division, The Edinburgh Building, Shaftesbury Road, Cambridge CB2 2RU, UK.

Fig. 3.7:
Reprinted from C.H.K. Williamson: "Sinusoidal flow relative to circular cylinders". J. Fluid Mech., 155:141-174, 1985, with kind permission from Cambridge University Press, Publishing Division, The Edinburgh Building, Shaftesbury Road, Cambridge CB2 2RU, UK.

Figs. 4.51-4.53:
Reprinted from E.-S. Chan, H.-F. Cheong and B.-C. Tan: "Laboratory study of plunging wave impacts on vertical cylinders". Coastal Engineering, 25:87-107, 1995, with kind permission from Elsevier Science, Sara Burgerhartstraat 25, 1055 KV Amsterdam, The Netherlands.

Fig. 5.4b:
Reprinted from J.E. Fromm and F.H. Harlow: "Numerical solution of the problem of vortex street development". The Physics of Fluids, 6(7):975-982, 1963, with kind permission from American Institute of Physics, Office of Rights and Permissions, 500 Sunnyside Blvd., Woodbury, NY 11797, USA.

Fig. 5.9:
Reprinted from P. Justesen: "A numerical study of oscillating flow around a circular cylinder". J. Fluid Mech., 222:157-196, 1991, with kind permission from Cambridge University Press, Publishing Division, The Edinburgh Building, Shaftesbury Road, Cambridge CB2 2RU, UK.

Fig. 5.14:
Reprinted from T. Sarpkaya, C. Putzig, D. Gordon, X. Wang and C. Dalton: "Vortex trajectories around a circular cylinder in oscillatory plus mean flow". J. Offshore Mech. and Arctic Engineering, 114:291-298, 1992, with kind permission from Production Coordinator, Technical Publishing Department, ASME International, 345 East 47th Street, New York, NY 10017-2392, USA.

Fig. 5.26:
Reprinted from P.K. Stansby and P.A. Smith: "Viscous forces on a circular cylinder in orbital flow at low Keulegan-Carpenter numbers". J. Fluid Mech., 229:159-171, with kind permission from Cambridge University Press, Publishing Division, The Edinburgh Building, Shaftesbury Road, Cambridge CB2 2RU, UK.

Fig. 8.50:
Reprinted from R. King: "A review of vortex shedding research and its application". Ocean Engineering, 4:141-172, 1977, with kind permission from Elsevier Science Ltd., The Boulevard, Langford Lane, Kidlington OX5 1GB, UK.

List of symbols

The main symbols used in the book are listed below. In some cases, the same symbol was used for more than one quantity. This is to maintain generally accepted conventions in different areas of fluid mechanics. In most cases, however, their use is restricted to a single chapter, as indicated in the following list.

Main symbols

A	amplitude of vibrations
A	cross-sectional area of body (Chapter 4)
$A_{\max}$	maximum value of vibration amplitude
a	amplitude of oscillatory flow, or amplitude of horizontal component of orbital motion
a	acceleration (Chapter 4)
a	distance between discrete vortices in an infinite row of vortices (Chapter 5)
a	amplitude of surface elevation (Chapter 7)
b	amplitude of vertical component of orbital motion
C	concentration or passive quantity (or temperature)
C_D	drag coefficient
C_D'	oscillating component of drag coefficient
C_L	lift coefficient
C_L'	oscillating component of lift coefficient
C_{LA}	lift coefficient corresponding to F_{yA}
C_{LT}	lift coefficient corresponding to F_{yT}
C_{Ld}, C_{Lm}	lift force coefficients (drag and inertia components, respectively)
$C_{L\,\max}$	lift coefficient corresponding to $F_{L\,\max}$
C_{Lrms}	lift coefficient corresponding to F_{Lrms}
C_{Trms}	force coefficient corresponding to F_{Trms}

C_M	inertia coefficient
C_m	hydrodynamic-mass coefficient
C_{mc}	hydrodynamic-mass coefficient in current
C_s	force coefficient corresponding to force f
c	viscous damping coefficient
c	wave celerity (Chapter 4, Appendix III)
c_p	pressure coefficient
D	cylinder diameter (or pipeline diameter)
$D(f,\theta)$	directional spectrum
E	ellipticity of orbital motion
E	elasticity modulus (Chapter 11)
$\overline{E}$	mean wave energy
E_T	total energy
E_d	energy dissipated in one cycle of vibrations
e	gap between cylinder and wall, or clearance between pipeline and seabed
F	Morison force per unit length of structure
F	external force
F_D	drag force per unit length of structure
F_D'	oscillating component of drag force per unit length of structure
F_K	Froude-Krylov force per unit height of vertical structure
$F_{K,tot}$	total Froude-Krylov force on vertical structure
F_L	lift force per unit length of structure
F_L'	oscillating component of lift force per unit length of structure
$F_{L\,\max}$	maximum value of lift force per unit length of structure
F_{Lrms}	root-mean-square value of lift force per unit length of structure
F_N	force component normal to structure, per unit length of structure
F_T	total (resultant) force per unit length of structure
F_{Trms}	root-mean-square value of total (resultant) force per unit length of structure
F_d	damping force
F_f	friction drag per unit length of structure
F_p	form drag per unit length of structure
F_p, F_m	predicted and measured in-line forces, respectively (Chapter 4)
F_{rms}	root-mean-square value of in-line force per unit length of structure
F_x, F_y	force components in Cartesian coordinate system
$F_{x,tot}$	total force on vertical cylinder
F_y	lift force per unit length of structure
F_{yA}	maximum value of lift force away from wall per unit length of structure
F_{yT}	maximum value of lift force towards wall per unit length of structure
F_z	lift force per unit length of structure
F_0	force due to potential flow per unit length of cylinder
f	frequency, frequency of vibrations

f	impact force on vertical cylinder due to breaking waves (Chapter 4)
f_L	fundamental lift frequency
f_n	undamped natural frequency (or natural frequency)
f_{nc}	natural frequency in current
f_t	frequency of transition waves
f_v	vortex-shedding frequency
f_w	frequency of oscillatory flow, frequency of waves
f_x	frequency of in-line vibrations
f_y	frequency of cross-flow vibrations in forced vibration experiments
f_0	peak frequency
g	acceleration due to gravity
H	wave height
H_m	maximum wave height
H_{rms}	root-mean-square value of wave height
H_s	significant wave height
$H_{1/3}$	significant wave height $(= H_s)$
h	water depth
h	distance between two infinite rows of vortices (Chapter 5)
I	inertia moment
I_u	turbulence intensity
i	imaginary unit
Im	imaginary part
K	diffusion coefficient (or thermal conductivity)
K_s	stability parameter
KC	Keulegan-Carpenter number
KC_r	Keulegan-Carpenter number for random oscillatory flow
k_s	Nikuradse's equivalent sand roughness
k	cylinder roughness (Chapter 4)
k	spring constant (Chapters 8-11)
k	wave number
k_r, k_i	real and imaginary parts of wave number k
L	correlation length
L	wave length (Chapter 6, Appendix III)
M	mass ratio
M	overturning moment (Chapter 6)
m	mass of body, per unit length of structure unless otherwise is stated
m'	hydrodynamic mass, per unit length of structure unless otherwise is stated
m'_c	hydrodynamic mass in current, per unit length of structure unless otherwise is stated
m_n	nth moment of spectrum
N	normalized vibration frequency in oscillatory flows or in waves f/f_w $(=$ number of vibrations per flow cycle$)$
$N(z)$	tension (Chapter 11)

N_L	normalized lift frequency, $f_L/f_w (=$ number of oscillations in lift per flow cycle)
n	normal direction
P	pressure force
Pr	probability of occurrence
p	pressure
p	probability density function (Chapter 7)
p'	fluctuating pressure
p_0	hydrostatic pressure
p^+	excess pressure
q	spectral width parameter
q_0	speed
R	autocovariance function (Chapter 7)
R	correlation
Re	Reynolds number
Re_r	Reynolds number for random oscillatory flow
r, θ	polar coordinates
r, θ	spherical coordinates (in axisymmetric flow) (Chapter 5)
r_0	cylinder radius
r_0	sphere radius (Chapter 5)
St	Strouhal number
$S(f)$	spectrum function of surface elevation (wave spectrum)
$S_a(f)$	spectrum function of acceleration
$S_{Fx}(f)$	force spectrum
$S_U(f)$	spectrum function of velocity
$S_\eta(f)$	spectrum function of surface elevation (wave spectrum)
T	period of oscillatory flow, period of waves
T_R	return period
T_c	mean crest period
T_s	significant wave period
T_v	vortex-shedding period
T_w	period of oscillatory flow, period of waves
T_z	mean zero-upcrossing period
$\overline{T}$	mean period
T_0	peak period
t	time
U	outer flow velocity
U_N	flow velocity component normal to cylinder
U_{Trms}	root-mean-square value of resultant velocity
U_c	current velocity
U_f	wall shear stress velocity
U_m	maximum value of oscillatory-flow velocity, maximum value of horizontal component of orbital velocity
U_{rms}	root-mean-square value of horizontal velocity

U_w	wind speed
u	flow velocity in boundary layer
u, v, w	velocity components in Cartesian coordinates
u', v'	infinitesimal disturbances introduced in velocity components
$\mathbf{u}$	velocity vector
V	volume of body
V_m	maximum value of vertical component of orbital velocity
V_r	reduced velocity
V_{rms}	root-mean-square value of vertical velocity
v	speed
v_r, v_θ	velocity components in polar coordinates, or spherical coordinates (axisymmetric)
W_0, W_1	complex potential
w	complex potential
x	streamwise distance, or horizontal distance
x_d	"dynamic" motion
x_f	forced motion
x, y	Cartesian coordinates
y	distance from wall
x, y	x- and y-displacements of structure (Chapter 8-11)
z	vertical coordinate measured from mean water level upwards (Chapter 6, Appendix III)
z	spanwise separation distance, or spanwise distance
z	complex coordinate, $z = x + iy = re^{i\theta}$ (Chapter 5)
β	ratio of Reynolds number to Keulegan-Carpenter number
Γ	circulation
Γ_i	vortex strength, corresponding to ith vortex
δ	boundary layer thickness
δ	goodness-of-fit parameter (Chapter 4)
δ	phase difference between incident wave and force (Chapter 6)
δ	logarithmic decrement (Chapter 8)
δ^*	displacement thickness of boundary layer
δt	time increment
ε	spectral width parameter
ε_p	1 for $p = 0$; 2 for $p \geq 1$
ζ	total damping
ζ_f	fluid damping
ζ_s	structural damping
η	surface elevation
θ	polar coordinate or spherical coordinate
θ	wave direction (Chapter 7)
κ	strength of individual vortices in an infinite row
λ	wave length of wavy trajectory of cylinder towed in still fluid
μ	dynamic viscosity

ν	kinematic viscosity
ρ	fluid density
σ_U	standard deviation of flow velocity
σ_η	standard deviation of quantity η
τ	shear stress
τ	normalized wave period (Chapter 7)
τ_0	wall shear stress
τ_w	wall shear stress (Chapter 4)
ϕ	angular coordinate
ϕ	phase difference between cylinder vibration and flow velocity (Chapter 3)
ϕ	potential function (Chapters 4, 6 and Appendix III)
ϕ_i	potential function for incident waves
ϕ_s	potential function for scattered (reflected plus diffracted) waves (Chapter 6)
ϕ_s	separation angle
φ	phase delay
ψ	stream function
ψ'	infinitesimal disturbance in stream function
ω	angular frequency, also angular frequency of external force (for a vibrating system)
ω	vorticity defined by $\omega = \partial v/\partial x - \partial u/\partial y$ (Chapter 5)
ω_d	damped natural angular frequency
ω_{dv}	angular frequency of damped free vibrations
ω_n	undamped natural angular frequency
ω_r, ω_i	real and imaginary parts of angular frequency ω
ω_v	angular frequency of undamped free vibrations
overbar	time average
overdot	differentiation with respect to time

Contents

7. Forces on a cylinder in irregular waves

8. Flow-induced vibrations of a free cylinder in steady currents

9. Flow-induced vibrations of a free cylinder in waves

10. Vibrations of marine pipelines

11. Mathematical modelling of flow-induced vibrations

Chapter 1. Flow around a cylinder in steady current

1.1 Regimes of flow around a smooth, circular cylinder

The non-dimensional quantities describing the flow around a smooth circular cylinder depend on the cylinder Reynolds number

$$Re = \frac{DU}{\nu} \tag{1.1}$$

in which D is the diameter of the cylinder, U is the flow velocity. and ν is the kinematic viscosity. The flow undergoes tremendous changes as the Reynolds number is increased from zero. The flow regimes experienced with increasing Re are summarized in Fig. 1.1. Fig. 1.2, on the other hand, gives the definition sketch regarding the two different flow regions referred to in Fig. 1.1, namely the wake and the boundary layer. While the wake extends over a distance which is comparable with the cylinder diameter, D, the boundary layer extends over a very small thickness. δ. which is normally small compared with D. The boundary layer thickness, in the case of laminar boundary layer, for example, is (Schlichting. 1979)

	Description	Re range
a)	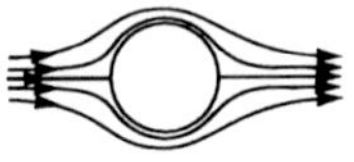 No separation. Creeping flow	$Re < 5$
b)	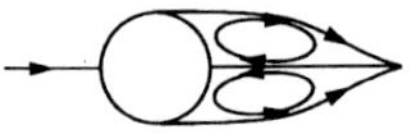 A fixed pair of symmetric vortices	$5 < Re < 40$
c)	 Laminar vortex street	$40 < Re < 200$
d)	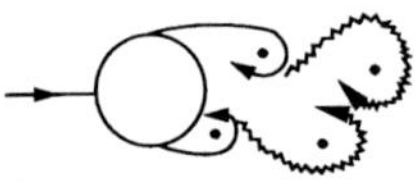 Transition to turbulence in the wake	$200 < Re < 300$
e)	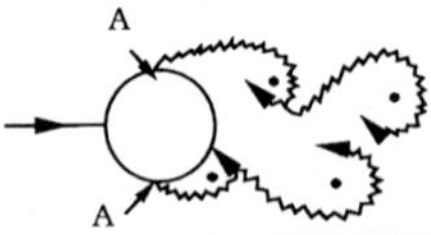 Wake completely turbulent. A:Laminar boundary layer separation	$300 < Re < 3 \times 10^5$ Subcritical
f)	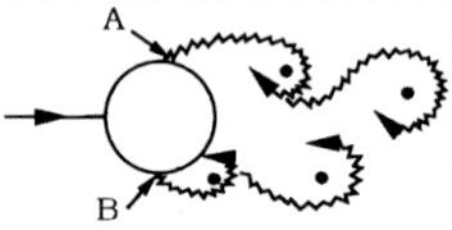 A:Laminar boundary layer separation B:Turbulent boundary layer separation;but boundary layer laminar	$3 \times 10^5 < Re < 3.5 \times 10^5$ Critical (Lower transition)
g)	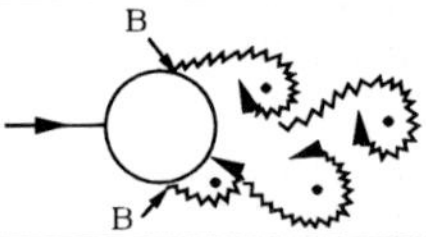 B: Turbulent boundary layer separation;the boundary layer partly laminar partly turbulent	$3.5 \times 10^5 < Re < 1.5 \times 10^6$ Supercritical
h)	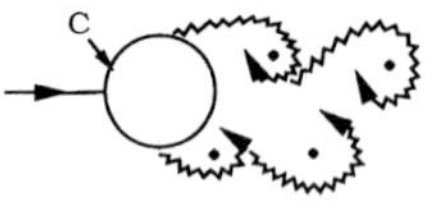 C: Boundary layer completely turbulent at one side	$1.5 \times 10^6 < Re < 4 \times 10^6$ Upper transition
i)	 C: Boundary layer completely turbulent at two sides	$4 \times 10^6 < Re$ Transcritical

Figure 1.1 Regimes of flow around a smooth, circular cylinder in steady current.

$$\frac{\delta}{D} = O\left(\frac{1}{\sqrt{Re}}\right) \tag{1.2}$$

and it is seen that $\delta/D << 1$ for Re larger than $O(100)$, say.

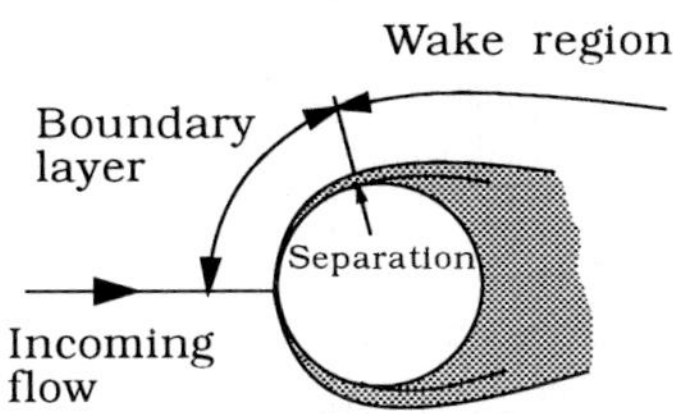

Figure 1.2 Definition sketch.

Now, returning to Fig. 1.1, for very small values of Re no separation occurs. The separation first appears when Re becomes 5 (Figs. 1.1a).

For the range of the Reynolds number $5 < Re < 40$, a fixed pair of vortices forms in the wake of the cylinder (Fig. 1.1 b). The length of this vortex formation increases with Re (Batchelor, 1967).

When the Reynolds number is further increased, the wake becomes unstable, which would eventually give birth to the phenomenon called vortex shedding in which vortices are shed alternately at either side of the cylinder at a certain frequency. Consequently, the wake has an appearance of a vortex street (see Fig. 1.3d–f).

For the range of the Reynolds number $40 < Re < 200$ the vortex street is laminar (Fig. 1.1c). The shedding is essentially two-dimensional, i.e., it does not vary in the spanwise direction (Williamson, 1989).

With a further increase in Re, however, transition to turbulence occurs in the wake region (Fig. 1.1d). The region of transition to turbulence moves towards the cylinder, as Re is increased in the range $200 < Re < 300$ (Bloor, 1964). Bloor (1964) reports that at $Re = 400$, the vortices, once formed, are turbulent. Observations show that the two-dimensional feature of the vortex shedding observed in the range $40 < Re < 200$ becomes distinctly three-dimensional in this range (Gerrard, 1978 and Williamson, 1988); the vortices are shed in cells in the spanwise direction. (It may be noted that this feature of vortex shedding prevails for all the other Reynolds number regimes $Re > 300$. This topic will be studied in some detail in Section 1.2.2 in the context of correlation length).

For $Re > 300$, the wake is completely turbulent. The boundary layer over the cylinder surface remains laminar, however, for increasing Re over a very wide

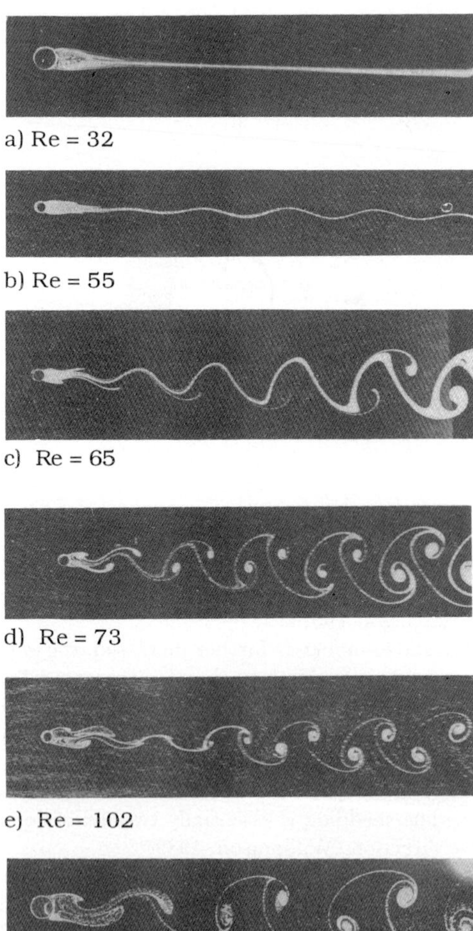

Figure 1.3 Appearance of vortex shedding behind a circular cylinder in a stream of oil (from Homann, 1936) with increasing Re.

range of Re, namely $300 < Re < 3 \times 10^5$. This regime is known as the subcritical flow regime (Fig. 1.1e).

With a further increase in Re, transition to turbulence occurs in the boundary layer itself. The transition first takes place at the point where the boundary layer separates, and then the region of transition to turbulence moves upstream over the cylinder surface towards the stagnation point as Re is increased (Figs. 1.1f – 1.1i).

In the narrow Re band $3 \times 10^5 < Re < 3.5 \times 10^5$ (Fig. 1.1f) the boundary layer becomes turbulent at the separation point, but this occurs only at one side of the cylinder. So the boundary layer separation is turbulent at one side of the cylinder and laminar at the other side. This flow regime is called the critical (or the lower transition) flow regime. The flow asymmetry causes a non-zero mean lift on the cylinder, as seen from Fig. 1.4.

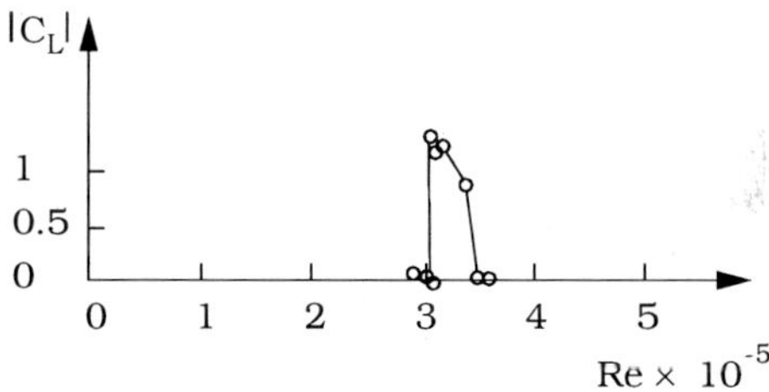

Figure 1.4   Non-zero mean lift in the critical-flow regime ($3 \times 10^5 < Re < 3.5 \times 10^5$). Schewe (1983).

The side at which the separation is turbulent switches from one side to the other occasionally (Schewe, 1983). Therefore, the lift changes direction, as the one-sided transition to turbulence changes side, shifting from one side to the other (Schewe, 1983).

The next Reynolds number regime is the so-called supercritical flow regime where $3.5 \times 10^5 < Re < 1.5 \times 10^6$ (Fig. 1.1g). In this regime, the boundary layer separation is turbulent on both sides of the cylinder. However, transition to turbulence in the boundary layer has not been completed yet; the region of transition to turbulence is located somewhere between the stagnation point and the separation point.

The boundary layer on one side becomes fully turbulent when Re reaches the value of about 1.5×10^6. So, in this flow regime, the boundary layer is completely turbulent on one side of the cylinder and partly laminar and partly turbulent on the other side. This type of flow regime, called the upper-transition flow regime, prevails over the range of Re, $1.5 \times 10^6 < Re < 4.5 \times 10^6$ (Fig. 1.1h).

Finally, when Re is increased so that $Re > 4.5 \times 10^6$, the boundary layer over the cylinder surface is virtually turbulent everywhere. This flow regime is called the transcritical flow regime.

Regarding the terminology in relation to the described flow regimes and also the ranges of Re in which they occur, there seems to be no general consensus among various authors (Farell, 1981). The preceding classification and the description are mainly based on Roshko's (1961) and Schewe's (1983) works. Roshko's work covered the Reynolds number range from 10^6 to 10^7, which revealed the existence of the upper transition and the transcritical regimes, while Schewe's work, covering the range $2.3 \times 10^4 < Re < 7.1 \times 10^6$, clarified further details of the flow regimes from the lower transition to the transcritical flow regimes.

1.2 Vortex shedding

The most important feature of the flow regimes described in the previous section is the vortex-shedding phenomenon, which is common to all the flow regimes for $Re > 40$ (Fig. 1.1). For these values of Re, the boundary layer over the cylinder surface will separate due to the adverse pressure gradient imposed by the divergent geometry of the flow environment at the rear side of the cylinder. As a result of this, a shear layer is formed, as sketched in Fig. 1.5.

As seen from Fig. 1.6, the boundary layer formed along the cylinder contains a significant amount of vorticity. This vorticity is fed into the shear layer formed downstream of the separation point and causes the shear layer to roll up into a vortex with a sign identical to that of the incoming vorticity. (Vortex A in Fig. 1.5).

Likewise, a vortex, rotating in the opposite direction, is formed at the other side of the cylinder (Vortex B).

Mechanism of vortex shedding

It has been mentioned in the previous section that the pair formed by these two vortices is actually unstable when exposed to the small disturbances for Reynolds numbers $Re > 40$. Consequently, one vortex will grow larger than the other if $Re > 40$. Further development of the events leading to vortex shedding has been described by Gerrard (1966) in the following way.

The larger vortex (Vortex A in Fig. 1.7a) presumably becomes strong enough to draw the opposing vortex (Vortex B) across the wake, as sketched in Fig. 1.7a. The vorticity in Vortex A is in the clockwise direction (Fig. 1.5b), while that in Vortex B is in the anti-clockwise direction. The approach of vorticity of

a)

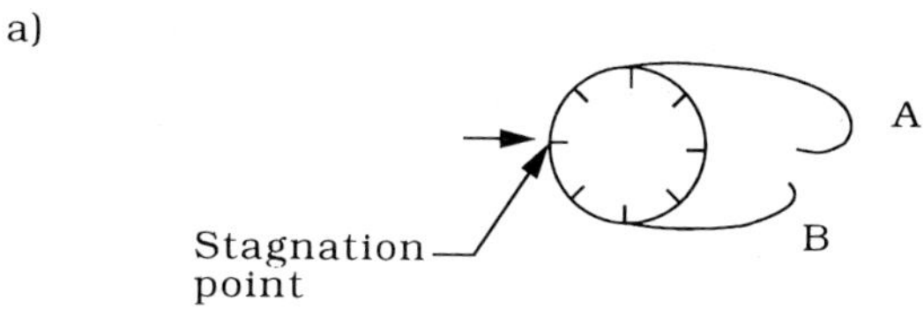

b)

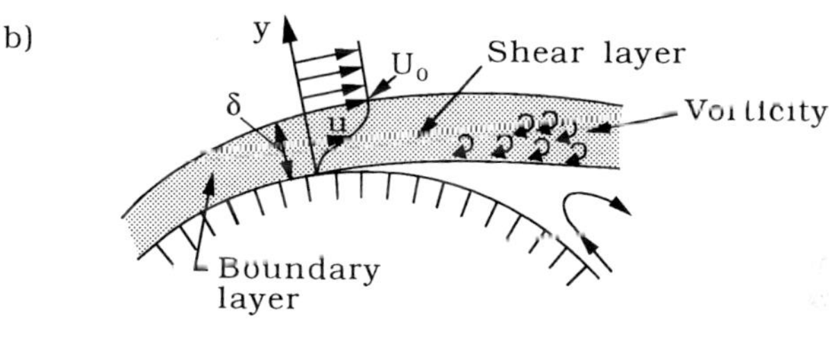

Detailed picture of
flow near separation

Figure 1.5 The shear layer. The shear layers on both sides roll up to form
the lee-wake vortices, Vortices A and B.

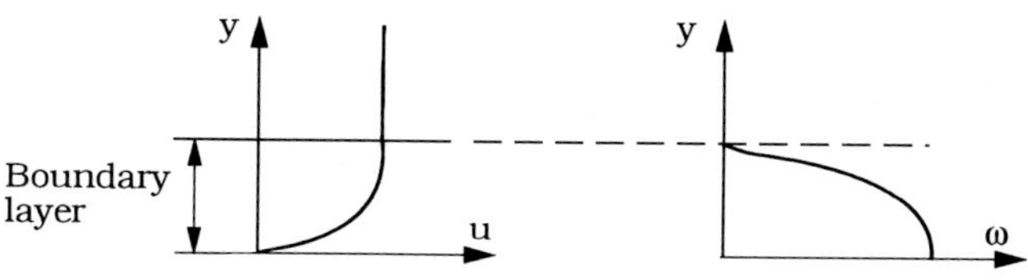

Figure 1.6 Distribution of velocity and vorticity in the boundary layer. ω
is the vorticity, namely $\omega = \frac{1}{2}\frac{\partial u}{\partial y}$.

the opposite sign will then cut off further supply of vorticity to Vortex A from its boundary layer. This is the instant where Vortex A is shed. Being a free vortex, Vortex A is then convected downstream by the flow.

Following the shedding of Vortex A, a new vortex will be formed at the same side of the cylinder, namely Vortex C (Fig. 1.7b). Vortex B will now play the same role as Vortex A, namely it will grow in size and strength so that it will draw Vortex C across the wake (Fig. 1.7b). This will lead to the shedding of Vortex B. This process will continue each time a new vortex is shed at one side of the cylinder where the shedding will continue to occur in an alternate manner between the sides of the cylinder.

a)

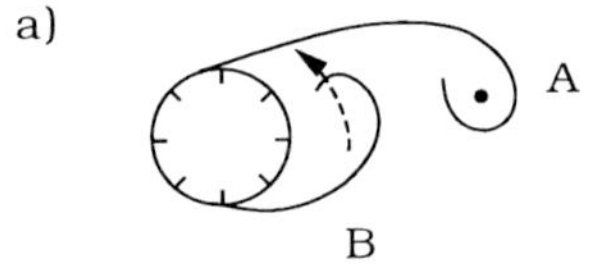

b)

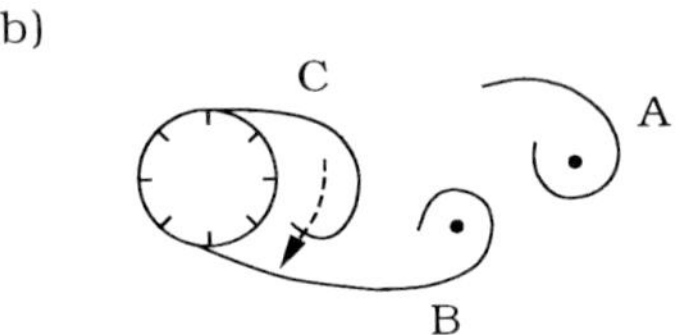

Figure 1.7 (a): Prior to shedding of Vortex A, Vortex B is being drawn across the wake. (b): Prior to shedding of Vortex B, Vortex C is being drawn across the wake.

The sequence of photographs given in Fig. 1.8 illustrates the time development of the process during the course of shedding process.

One implication of the foregoing discussion is that the vortex shedding occurs only when the two shear layers interact with each other. If this interaction is inhibited in one way or another, for example by putting a splitter plate at the downstream side of the cylinder between the two shear layers, the shedding would be prevented, and therefore no vortex shedding would occur in this case. Also, as another example, if the cylinder is placed close to a wall, the wall-side shear layer will not develop as strongly as the opposing shear layer; this will presumably lead to a weak interaction between the shear layers, or to practically no interaction if the cylinder is placed very close to the wall. In such situations, the vortex shed-

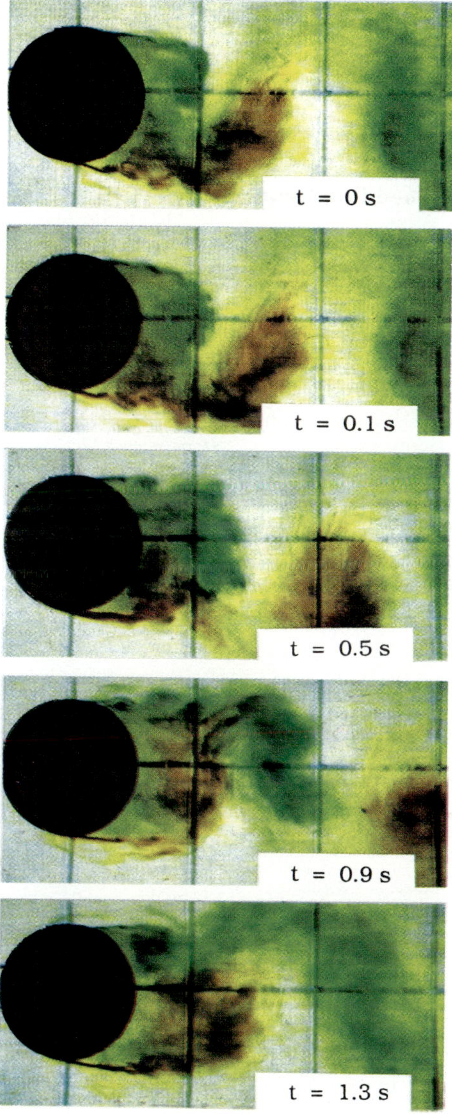

Figure 1.8 Time development of vortex shedding during approximately two-third of the shedding period. $Re = 7 \times 10^3$.

ding is suppressed. The effect of close proximity of a wall on the vortex shedding
will be examined in some detail later in the next section.

1.2.1 Vortex-shedding frequency

The vortex-shedding frequency, when normalized with the flow velocity U
and the cylinder diameter D, can on dimensional grounds be seen to be a function
of the Reynolds number:

$$St = St(Re) \tag{1.3}$$

in which

$$St = \frac{f_v D}{U} \tag{1.4}$$

and f_v is the vortex-shedding frequency. The normalized vortex-shedding frequency, namely St, is called the Strouhal number. Fig. 1.9 illustrates how the
Strouhal number varies with Re, while Fig. 1.10 gives the power spectra corresponding to Schewe's (1983) data shown in Fig. 1.9.

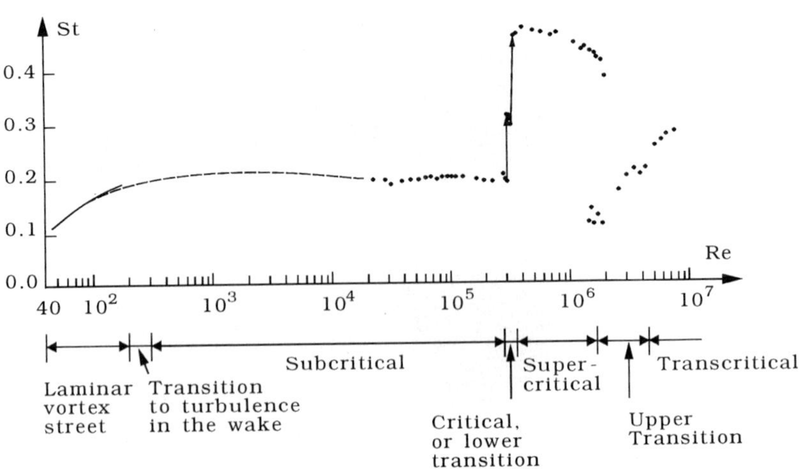

Figure 1.9 Strouhal number for a smooth circular cylinder. Experimental
data from: Solid curve: Williamson (1989). Dashed curve:
Roshko (1961). Dots: Schewe (1983).

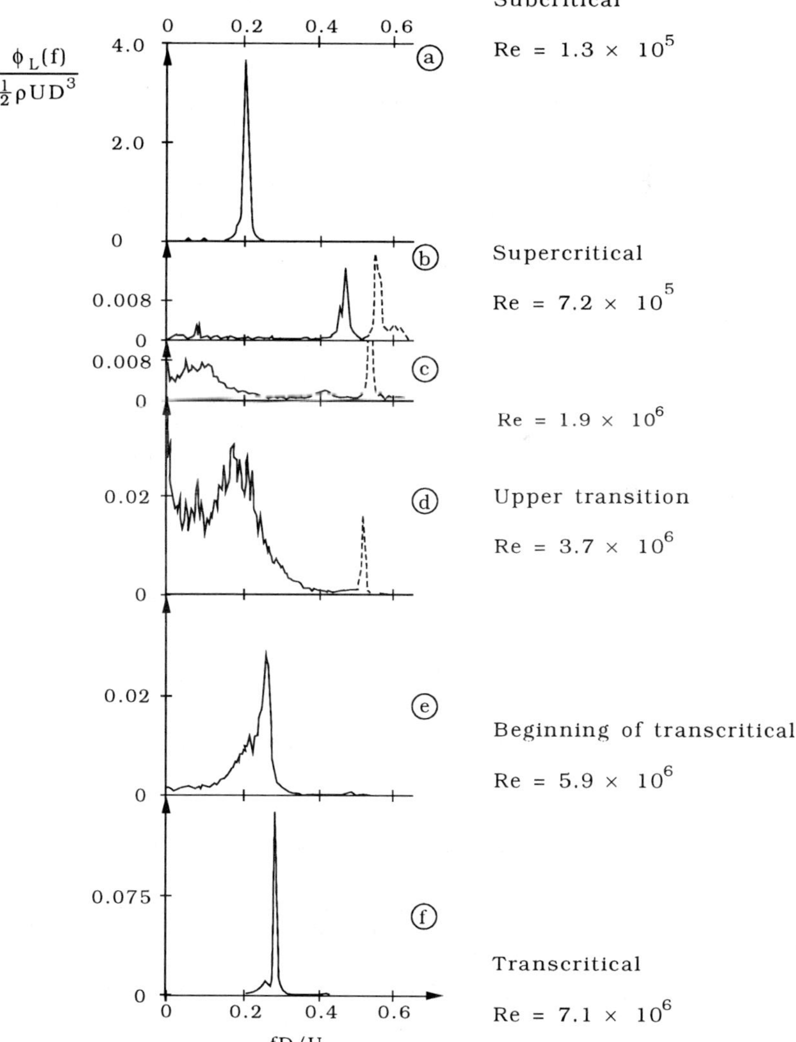

Figure 1.10 Power spectra of the lift oscillations corresponding to Schewe's data in Fig. 1.9 (Schewe, 1983).

The vortex shedding first appears at $Re = 40$. From Fig. 1.9, the shedding frequency St is approximately 0.1 at this Re. It then gradually increases as Re is increased and attains a value of about 0.2 at $Re \cong 300$, the lower end of the subcritical flow regime. From this Re number onwards throughout the subcritical range St remains practically constant (namely, at the value of 0.2).

The narrow-band spectrum with the sharply defined dominant frequency in Fig. 1.10a indicates that vortex shedding in the subcritical range occurs in a well-defined, regular fashion.

As seen from Fig. 1.9, the Strouhal frequency experiences a sudden jump at $Re = 3 - 3.5 \times 10^5$, namely in the critical Re number range, where St increases from 0.2 to a value of about 0.45. This high value of St is maintained over a rather large part of the supercritical Re range, subsequently it decreases slightly with increasing Reynolds number.

The large increase in St in the supercritical-flow range is explained as follows: in the supercritical flow regime, the boundary layer on both sides of the cylinder is turbulent at the separation points. This results in a delay in the boundary-layer separation where the separation points move downstream, as sketched in Fig. 1.11. This means that the vortices (now being closer to each other) would interact at a faster rate than in the subcritical flow regime, which would obviously lead to higher values of the Strouhal number.

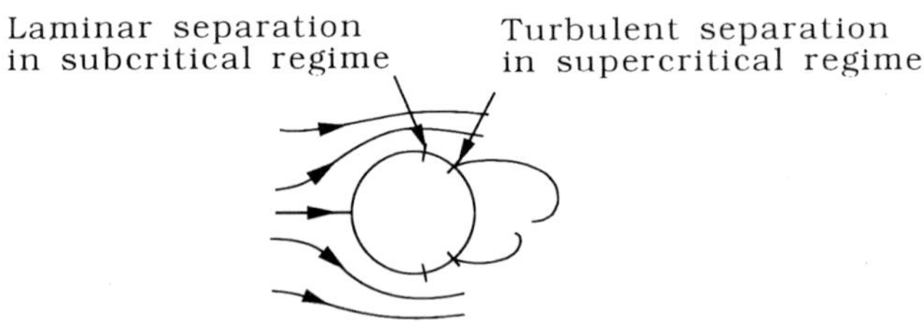

Figure 1.11 Sketch showing positions of separation points at different separation regimes.

The power spectrum (Fig. 1.10b) at $Re = 7.2 \times 10^5$, a Reynolds number which is representative for the supercritical range, indicates that in this Re range, too, the shedding occurs in a well-defined, orderly fashion, since the power spectrum appears to be a narrow-band spectrum with a sharply defined, dominant peak. The fact that the magnitude of the spectrum itself is extremely small (cf. Figs. 1.10a and 1.10b) indicates, however, that the shed vortices are not as strong as they are in the subcritical flow regime. An immediate consequence of this, as will be shown later, is that the lift force induced by the vortex shedding is relatively weak in this Re range.

The Strouhal number experiences yet another discontinuity when Re reaches the value of about 1.5×10^6. At this Reynolds number, transition to turbulence in one of the boundary layers has been completed (Fig. 1.1h). So, the boundary layer at one side of the cylinder is completely turbulent and that at the other side of the cylinder is partly laminar and partly turbulent, an asymmetric situation with regard to the formation of the lee-wake vortices. This situation prevails over the whole upper transition region (Fig. 1.1h). Now, the asymmetry in the formation of the lee-wake vortices inhibits the interaction of these vortices partially, resulting in an irregular, disorderly vortex shedding. This can be seen clearly from the broad-band spectra in Figs. 1.10c and d.

The regular vortex shedding is re-established, however, (see the narrow-band power spectra in Fig. 1.10e and f), when Re is increased to values larger than approximately 4.5×10^6, namely the transcritical flow regime where the Strouhal number takes the value of $0.25 - 0.30$ (Fig. 1.9).

Effect of surface roughness

For rough cylinders the normalized shedding frequency, namely the Strouhal number, should be a function of both Re and the relative roughness

$$St = St(Re, \; k_s/D) \tag{1.5}$$

in which k_s is the Nikuradse's equivalent sand roughness of the cylinder surface.

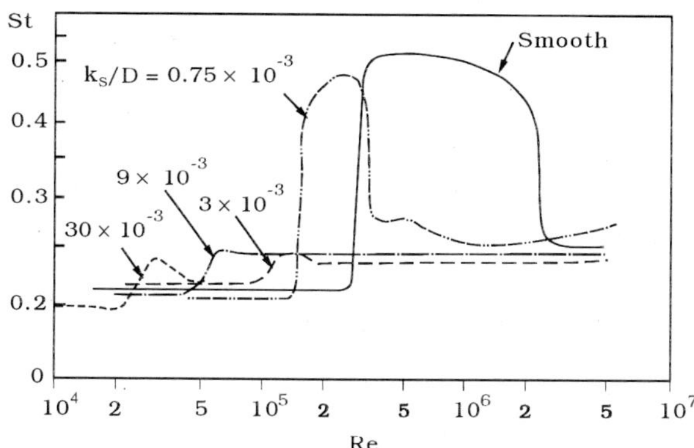

Figure 1.12 Effect of surface roughness on vortex-shedding frequency. Strouhal number against Reynolds number. Circular cylinder. Achenbach and Heinecke (1981).

Fig. 1.12 illustrates the effect of the relative roughness on the Strouhal number where the experimentally obtained St values for various values of k_s/D are plotted against Re (Achenbach and Heinecke, 1981). Clearly, the effect is significant. From the figure, it is apparent that, for rough cylinders with $k_s/D > 3\times10^{-3}$, the critical (the lower transition), the supercritical and the upper transition flow regimes merge into one narrow region in the St-Re plane, and the flow regime switches directly to transcritical over this narrow Re range, and this occurs at very low values of Re number. (The figure indicates for example that, at Re 0.3×10^5 for $k_s/D = 30 \times 10^{-3}$ and at $Re \cong 1.5 \times 10^5$ for $k_s/D = 3 \times 10^{-3}$). This result is in fact anticipated, as it is well known that transition to turbulence occurs much earlier (i.e., at much smaller values of Reynolds number) over rough walls.

Example 1.1:   Nikuradse's equivalent sand roughness

In practice there exists an extremely wide variety of surface roughnesses, from small protrusions existing in the texture of the surface itself to extremely large roughnesses in the form of marine growth such as mussels and acorn barnacles, etc..

Therefore, normally it is not an easy task to relate the roughness of the surface to some typical scale of the roughness elements, partly because the elements are quite unevenly distributed. (On a loose sand bed, for example, the roughness is measured to be 2-3 times the grain diameter). To tackle this problem, the concept "Nikuradse's equivalent sand roughness" has been introduced. The idea is to relate any kind of roughness to the Nikuradse roughness so that comparison can be made on the same basis. Very systematic and careful measurements on rough pipes were carried out by Nikuradse (1933), who used circular pipes. Sand with known grain size was glued on the pipe wall inside the pipe. By measuring the flow resistance and velocity profiles, Nikuradse obtained the following velocity distribution law

$$\frac{u}{U_f} = 5.75 \log_{10} \frac{y}{k_s} + 8.5 \tag{1.6}$$

which can be put in the following form

$$\frac{u}{U_f} = \frac{1}{\kappa} \ln \frac{30y}{k_s} \tag{1.7}$$

in which u is the streamwise velocity, U_f is the wall shear-stress velocity, κ is the Karman constant ($\cong 0.4$), y is the distance from the wall and k_s is the height of the sand roughness that Nikuradse used in his experiments (a detailed account of the subject is given by Schlichting (1979)). To judge about the roughness of a particular surface, the usual practice is first to measure the velocity distribution above the surface in consideration and then, based on this measured velocity distribution $u(y)$, to determine k_s, the Nikuradse's equivalent sand roughness of the surface, from Eq. 1.7.

Effect of cross-sectional shape

Fig. 1.13 shows the Strouhal-number data compiled by Blevins (1977) for various non-circular cross sections, while Fig. 1.14 presents the Strouhal numbers for a variety of profile shapes compiled by ASCE Task Committee (1961). Modi, Wiland, Dikshit and Yokomizo (1992) give a detailed account of flow and vortex shedding around elliptic cross-section cylinders.

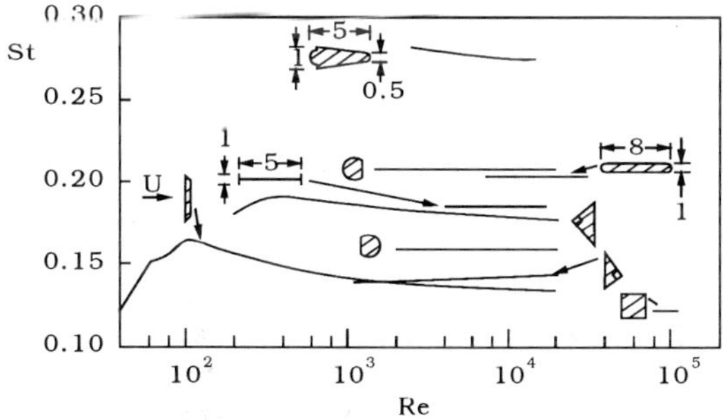

Figure 1.13 Effect of cross-sectional shape on vortex-shedding frequency. Strouhal number against Reynolds number. Blevins (1977).

As far as the large Reynolds numbers are concerned ($Re \gtrsim 10^5$), the vortex formation process is relatively uninfluenced by the Reynolds number for the cross sections with fixed separation points such as rectangular cylinders. So, the Strouhal number may not undergo large changes with increasing Re for such cross-sectional shapes, in contrast to what occurs in the case of circular cylinders.

Effect of incoming turbulence

Quite often, the approach flow is turbulent. For example, a cylinder placed on the sea bottom would feel the approach-flow turbulence which is generated within the bottom boundary layer. The turbulence in the approach flow is also an influencing factor with regard to the vortex shedding. The effect of turbulence

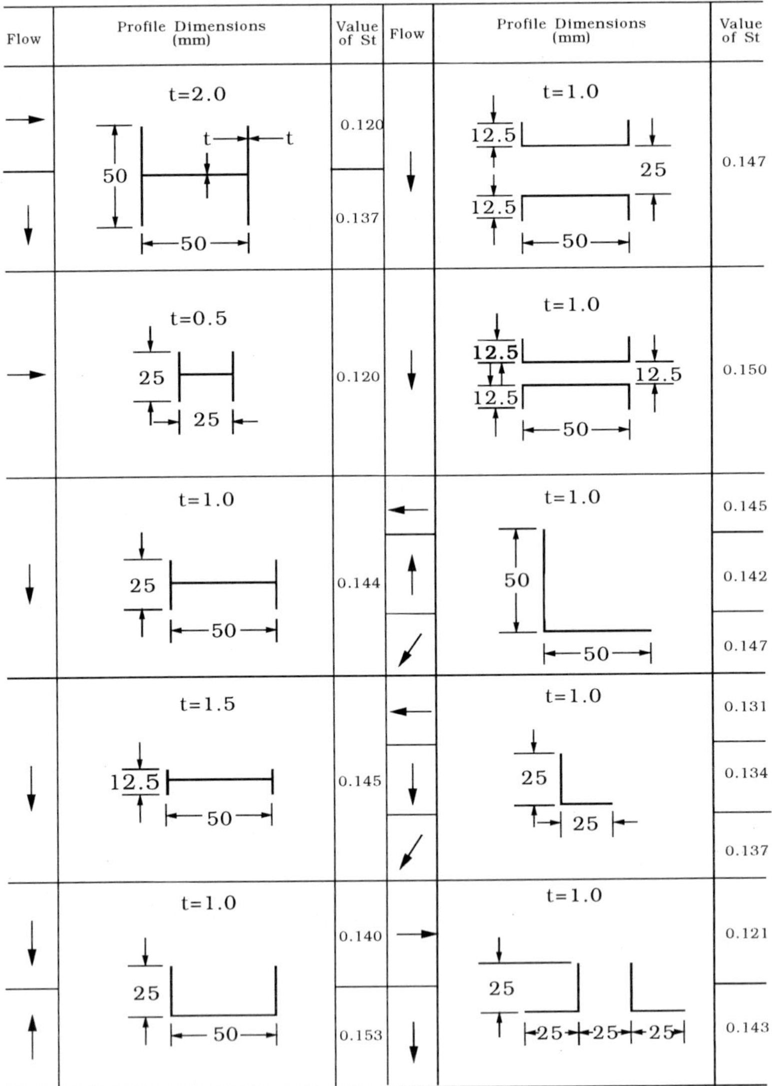

Figure 1.14 Effect of cross-sectional shape on Strouhal number. Strouhal numbers for profile shapes. ASCE Task Committee (1961).

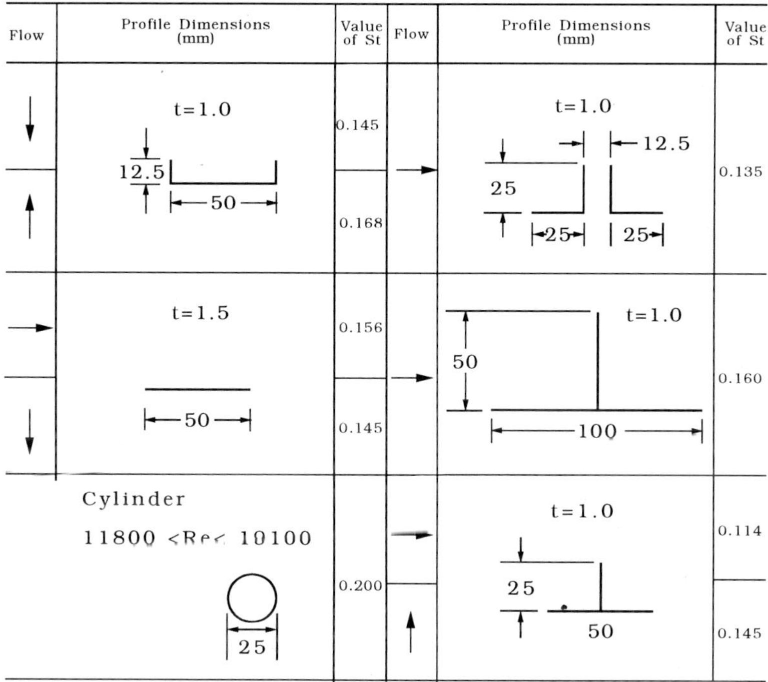

Figure 1.14 (continued.)

on the vortex shedding has been studied by various authors, for example by Cheung and Melbourne (1983), Kwok (1986) and Norberg and Sundén (1987) among others. Fig. 1.15 presents the Strouhal number data obtained by Cheung and Melbourne for various levels of turbulence in their experimental tunnel. Here, I_u is the turbulence intensity defined by

$$I_u = \frac{\sqrt{\overline{u'^2}}}{\overline{u}} \tag{1.8}$$

in which $\sqrt{\overline{u'^2}}$ is the root-mean-square value of the velocity fluctuations and $\overline{u}$ is the mean value of the velocity.

The variation of St with the Reynolds number changes considerably with the level of turbulence in the approach flow. The effect of turbulence is rather similar to that of cylinder roughness. The critical, the supercritical, and the upper transition flow regimes seem to merge into one transitional region.

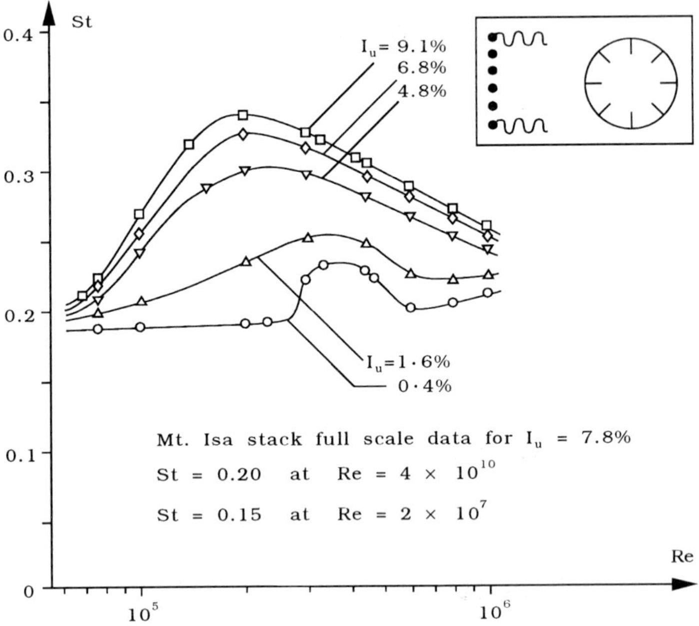

Figure 1.15 Effect of turbulence in the approach flow on vortex-shedding frequency. Strouhal numbers as a function of Reynolds number for different turbulence intensities. I_u is the level of turbulence (Eq. 1.8). Cheung and Melbourne (1983).

It appears from the figure that the lower end of this transition range shifts towards the smaller and smaller Reynolds numbers with the increased level of turbulence. This is obviously due to the earlier transition to turbulence in the cylinder boundary layer with increasing incoming turbulence intensity.

Effect of shear in the incoming flow

The shear in the approach flow is also an influencing factor in the vortex shedding process. The shear could be present in the approach flow in two ways: it could be present in the spanwise direction along the length of the cylinder (Fig. 1.16a), or in the cross-flow direction (Fig. 1.16b). The characteristics of shear flow around bluff bodies including the non-circular cross-sections have been reviewed by Griffin (1985a and b). In the case when the shear is present in the spanwise direction (Fig. 1.16a), the vortex shedding takes place in spanwise cells, with a frequency constant over each cell. Fig. 1.17 clearly shows this; it is seen that the shedding occurs in four cells, each with a different frequency. When the Strouhal number is based on the local velocity (the dashed lines in the figure), the data are grouped around the Strouhal number of about 0.25.

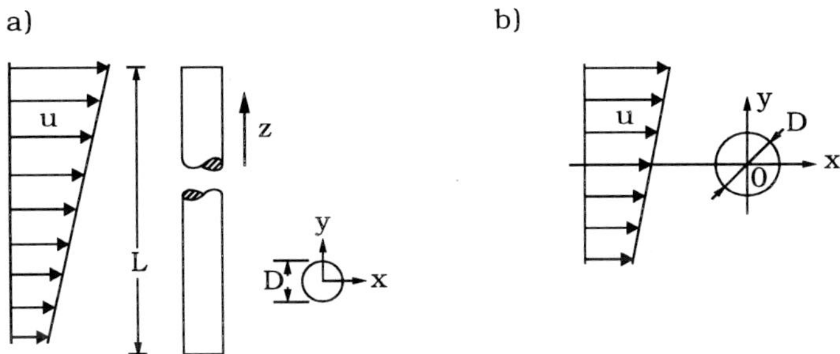

Figure 1.16 Two kinds of shear in the approach flow. a: Shear is in the spanwise direction. b: Shear is in the cross-flow direction.

Regarding the length of cellular structures, research shows that the length of cells is correlated with the degree of the shear. The general trend is that the cell length decreases with increasing shear (Griffin, 1985a).

When the shear takes place in the cross-stream direction (the conditions in the spanwise direction being uniform), the shedding is only slightly influenced for small and moderate values of the shear steepness s which is defined by

$$s = \frac{D}{U_c}\frac{du}{dy} \tag{1.9}$$

For large values of s, however, the shedding is influenced somewhat substantially (Kiya, Tamura and Arie, 1980). Fig. 1.18 shows the Strouhal number plotted against the Reynolds number for three different values of s. As is seen for $s = 0.2$, the Strouhal number is increased substantially relative to the uniform-flow case ($s = 0$).

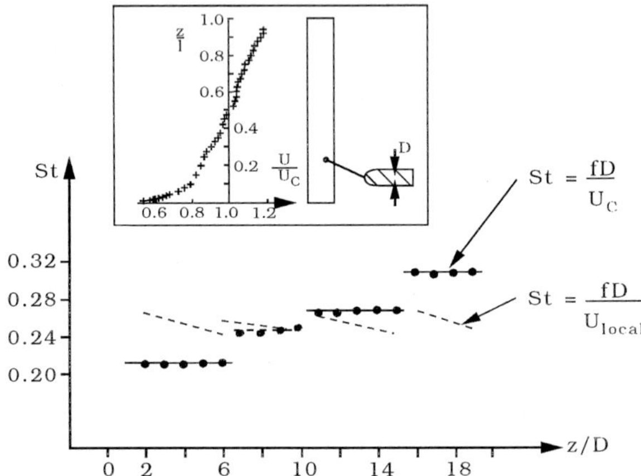

Figure 1.17 Effect of shear in the approach flow on vortex-shedding frequency. Shear in the spanwise direction. Circles: Strouhal number based on the centre-line velocity U_c. Dashed lines: Strouhal number based on the local velocity, U_{local}. $Re = 2.8 \times 10^4$. The shear steepness: $s = 0.025$. Maull and Young (1973).

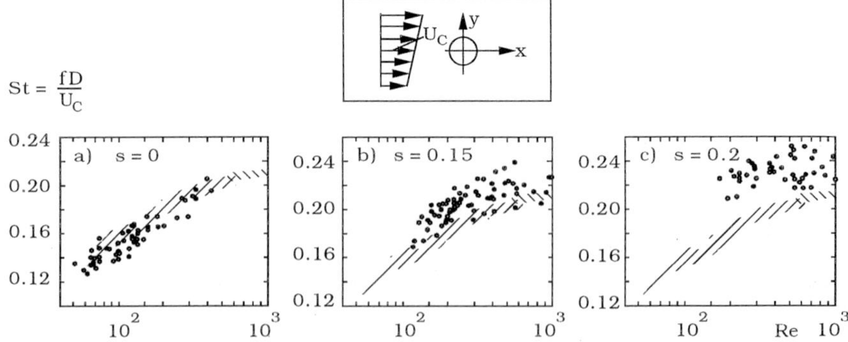

Figure 1.18 Effect of shear in the approach flow frequency. Shear in cross-flow direction. The Strouhal number against the Reynolds number for three different values of the shear steepness s. Hatched band: Uniform-flow results. Circles: Shear-flow results. Kiya et al. (1980).

Effect of wall proximity

This topic is of direct relevance with regard to pipelines. When a pipeline is placed on an erodible sea bed, scour may occur below the pipe due to flow action. This may lead to suspended spans of the pipeline where the pipe is suspended above the bed with a small gap, usually in the range from O(0.1D) to O(1D). Therefore it is important to know what kind of changes take place in the flow around and in the forces on such a pipe.

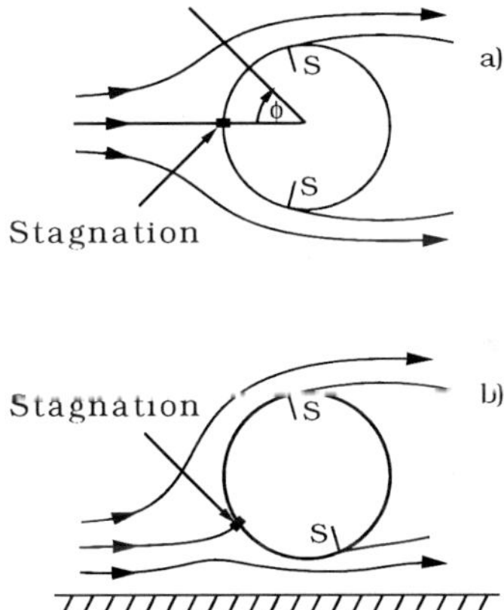

Figure 1.19 Flow around a) a free cylinder, b) a near-wall cylinder. $S =$ separation points.

When a cylinder is placed near a wall, a number of changes occur in the flow around the cylinder. These changes are summarized as follows:

1) Vortex shedding is suppressed for the gap-ratio values smaller than about $e/D = 0.3$, as will be seen later in the section. Here, e is the gap between the cylinder and the wall.

2) The stagnation point moves to a lower angular position as sketched in Fig. 1.19. This can be seen clearly from the pressure measurements of Fig. 2.20a and Fig. 2.20b where the mean pressure distributions around the cylinder are given for three different values of the gap ratio. While the stagnation point is located at about $\phi = 0°$ when $e/D = 1$, it moves to the angular position of about $\phi = -40°$ when the gap ratio is reduced to $e/D = 0.1$.

3) Also, the angular position of the separation points changes. The separation point at the free-stream side of the cylinder moves upstream and that at the wall side moves downstream, as shown in the sketch given in Fig. 1.19. The

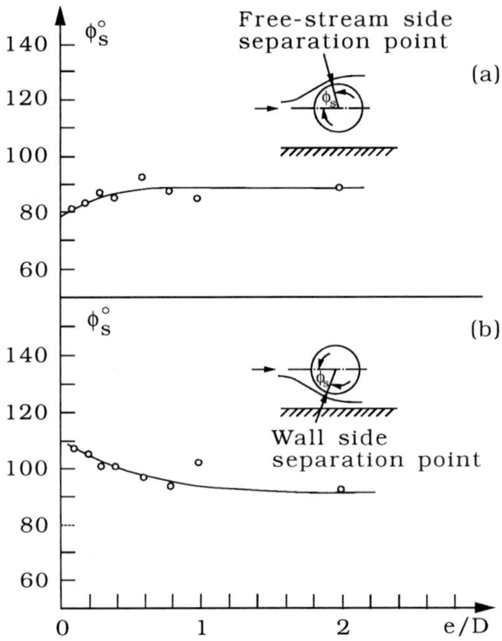

Figure 1.20 Angle of separation as a function of the gap ratio. (a): At the free-stream side of the cylinder and (b): At the wall side of the cylinder. $Re = 6 \times 10^3$. Jensen and Sumer (1986).

separation angle measured for a cylinder with $Re = 6 \times 10^3$ is shown in Fig. 1.20; the figure indicates that for example for $e/D = 0.1$ the separation angle at the free-stream side is $\phi \cong 80°$, while it is $\phi \cong -110°$ at the wall side for the same gap ratio.

4) Finally, the suction is larger on the free-stream side of the cylinder than on the wall-side of the cylinder, as is clearly seen in Fig. 2.20b and c. When the cylinder is placed away from the wall, however (Fig. 2.20a) this effect disappears and the symmetry is restored.

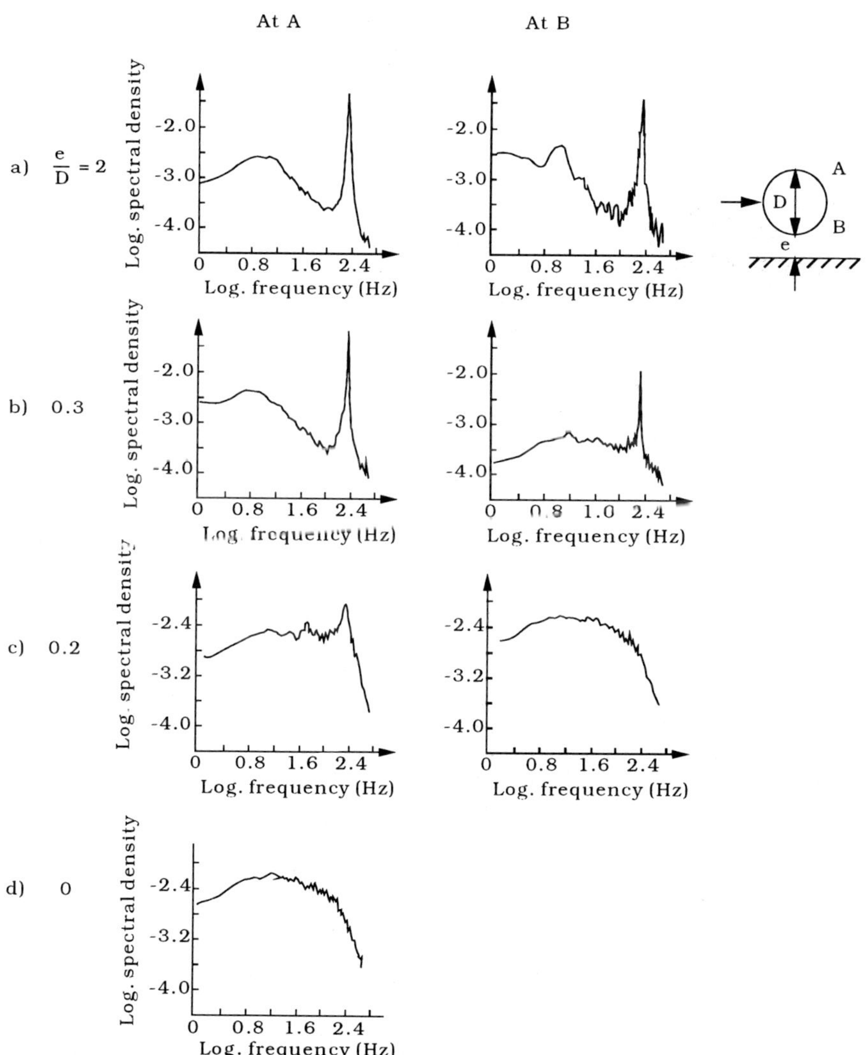

Figure 1.21 Effect of wall proximity on vortex shedding. Power spectra of the hot-wire signal received from the wake. Bearman and Zdravkovich (1978).

Vortex shedding may be suppressed for a cylinder which is placed close to a wall. Fig. 1.21 presents power spectra of the hot-wire signals received from both sides of the wake of a cylinder placed at different distances from a wall (Bearman and Zdravkovich, 1978). As is clearly seen, regular vortex shedding, identified by the sharply defined, dominant peaks in the power spectra, persists only for values of the gap-to-diameter ratio e/D down to about 0.3. This result, recognized first by Bearman and Zdravkovich, was later confirmed by the measurements of Grass, Raven, Stuart and Bray (1984). The photographs shown in Fig. 1.22 demonstrate the supression of vortex shedding for gap ratios e/D below 0.3.

The suppression of vortex shedding is linked with the asymmetry in the development of the vortices on the two sides of the cylinder. The free-stream-side vortex grows larger and stronger than the wall-side vortex. Therefore the interaction of the two vortices is largely inhibited (or, for small e/D, totally inhibited), resulting in partial or complete suppression of the regular vortex shedding.

Regarding the effect of wall proximity on the vortex-shedding frequency for the range of e/D where the vortex shedding exists, measurements show that the shedding frequency tends to increase (yet slightly) with decreasing gap ratio. In Fig. 1.23 are plotted the results of two studies, namely Grass et al. (1984) and Raven, Stuart, Bray and Littlejohns (1985). Grass et al.'s experiments were done in a laboratory channel with both smooth and rough beds. The surface of the test cylinder was smooth. Their results collapse onto a common curve when plotted in the normalized form presented in the figure where St_0 is the Strouhal number for a wall-free cylinder. The data points of Raven et al.'s study, on the other hand, were obtained in an experimental program conducted in the Severn Estuary (UK) where a full-scale pipeline (50.8 cm in diameter with a surface roughness of $k/D = 8.5 \times 10^{-3}$) was used. In both studies, St is defined by the velocity at the top of the cylinder. There are other data available such as Bearman and Zdravkovich (1978) and Angrilli, Bergamaschi and Cossalter (1982). While Bearman and Zdravkovich's measurements indicate that the shedding frequency practically does not change over the range $0.3 \leq e/D \leq 3$, Angrilli et al.'s measurements show that there is a systematic (yet, slight) increase in the shedding frequency with decreasing gap ratio in their measurement range $0.5 \leq e/D \leq 6$ (they report a 10% increase in the shedding frequency at $e/D = 0.5$).

It is apparent from the existing data that the vortex-shedding frequency is insensitive to the gap ratio, although there seems to be a tendency that it increases slightly with decreasing gap ratio. This slight increase in the Strouhal frequency may be attributed to the fact that the presence of the wall causes the wall-side vortex to be formed closer to the free-stream-side vortex. As a result of this, the two vortices interact at a faster rate, leading to a higher St frequency.

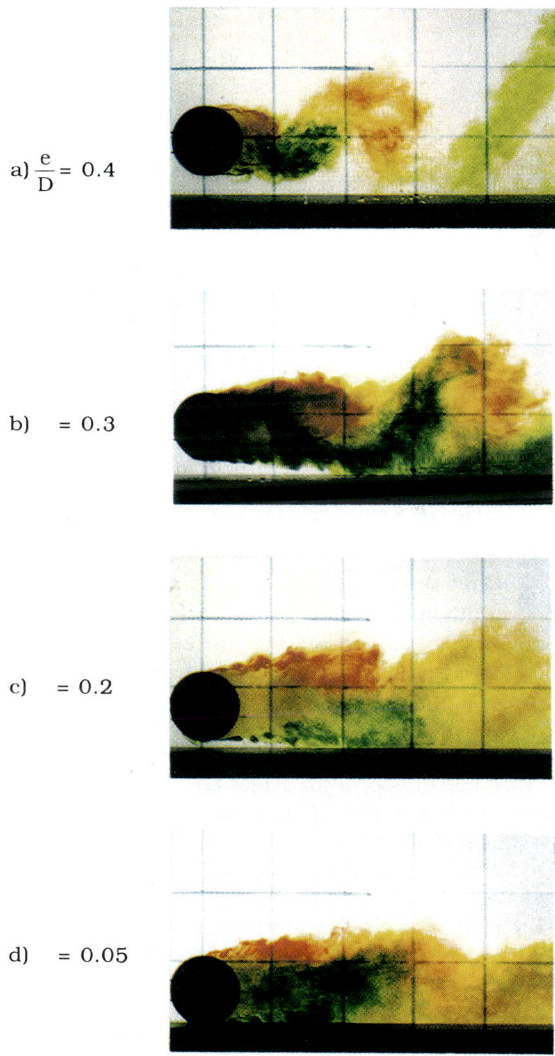

Figure 1.22 Effect of wall proximity on vortex shedding. Flow in the wake of a near-wall cylinder. Shedding is apparent for $e/D = 0.4$ and 0.3 but suppressed for $e/D = 0.2$ and 0.05. $Re = 7 \times 10^3$.

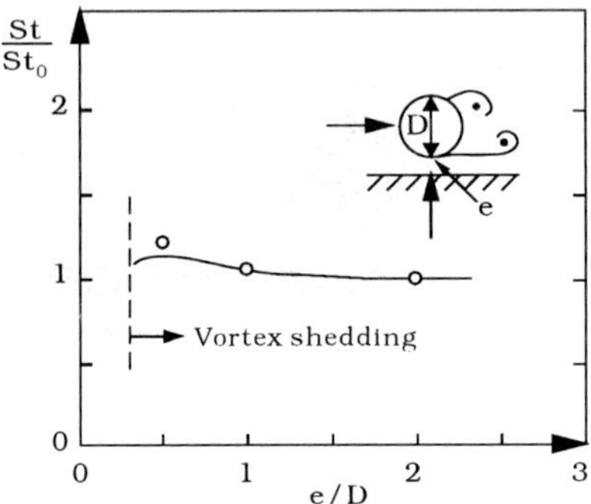

Figure 1.23 Effect of wall proximity on vortex shedding frequency. Normalized Strouhal number as a function of gap ratio. St_o is the Strouhal number for wall-free cylinder. Circles: Raven et al. (1985). Solid curve: Grass et al. (1984).

Jensen, Sumer, Jensen and Fredsøe (1990) investigated the flow around a pipeline (placed initially on a flat bed) at five characteristic stages of the scour process which take place underneath the pipeline. Each stage was characterized in the experiments by a special, frozen scoured bed profile, which was an exact copy of the measured bed profile of an actual scour test. The investigated scour profiles and the corresponding mean flow field are shown in Fig. 1.24. It was observed that no vortex shedding occurred for the first two stages, namely stages I and II, while vortex shedding did occur for stages III - V. Fig. 1.25 depicts the shedding frequency corresponding to the different stages.

The variation of the Strouhal number, which goes from as high a value as 0.36 for Stage III to an equilibrium value of 0.17 in Stage V, can be explained by the geometry of the downstream scour profile as follows.

For profiles III and IV, the steep slope of the upstream part of the dune behind the cylinder forces the shear layer originating from the lower edge of the cylinder to bend upwards, thus causing the associated lower vortex to interact with the upper one prematurely, leading to a premature vortex shedding. The result of this is a higher vortex shedding frequency and a very narrow formation region. The flow visualization study carried out in the same experiments (Jensen et al., 1990) confirmed the existence of this narrow region.

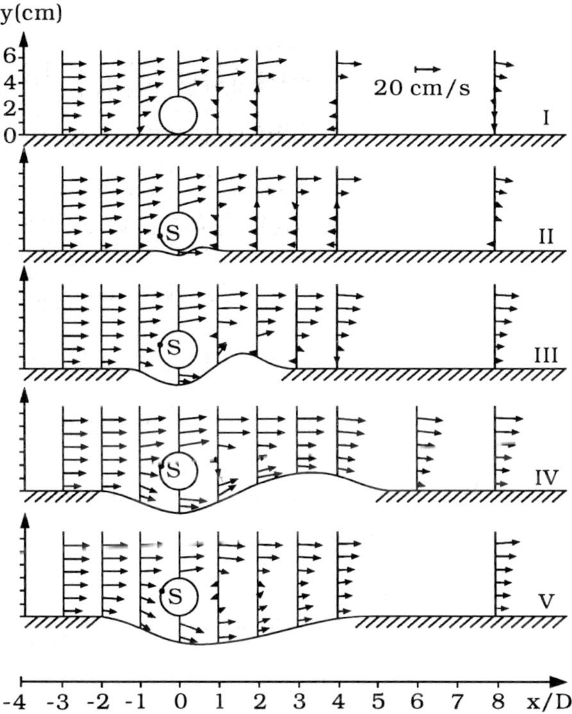

Figure 1.24 Vector plot of the mean velocities, $S =$ the approximate position of the stagnation point. Jensen et al. (1990).

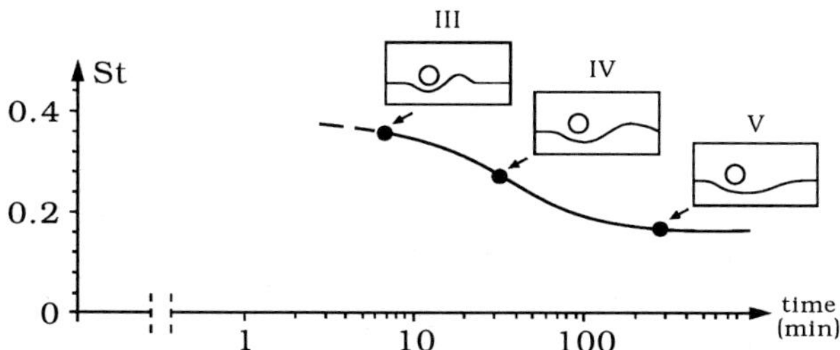

Figure 1.25 Time development of Strouhal number during the scour process below a pipeline. Jensen et al. (1990).

1.2.2 Correlation length

As has been mentioned in Section 1.1, vortex shedding in the turbulent wake regime (i.e. $Re \gtrsim 200$) occurs in cells along the length of the cylinder.

These spanwise cell structures are visualized in Fig. 1.26 which shows the time evolution of the shedding process in plan view.

The cells are quite clear from the photographs in Fig. 1.26. Shedding does not occur uniformly along the length of the cylinder, but rather in cells (designated by A, B and C in Fig. 1.26). It can also be recognized from the pictures in Fig. 1.26 that the cells along the length of the cylinder are out of phase. Consequently, the maximum resultant force acting on the cylinder over its total length may be smaller than the force acting on the cylinder over the length of a single cell.

The average length of the cells may be termed the correlation length. The precise determination of the correlation length requires experimental determination of the spanwise variation of the correlation coefficient of some unsteady quantity related to vortex shedding, such as fluctuating surface pressure, or a fluctuating velocity just outside the shear layer at separation.

The correlation coefficient is defined by

$$R(z) = \frac{\overline{p'(\zeta)\, p'(\zeta + z)}}{\sqrt{\overline{p'^{\,2}(\zeta)}}\sqrt{\overline{p'^{\,2}(\zeta + z)}}} \tag{1.10}$$

in which ζ is the spanwise distance, z is the spanwise separation between two measurement points, and p' is the fluctuating part of the unsteady quantity in consideration. The overbar denotes the time averaging. The correlation length L, on the other hand, is defined by the integral

$$L = \int_0^{\infty} R(z)\,dz \tag{1.11}$$

Fig. 1.27 gives a typical example of the correlation coefficient obtained in a wind tunnel with a cylinder 7.6 cm in diameter and 91.4 cm in length with large streamlined end plates (Novak and Tanaka, 1977). The Reynolds number was 1.9×10^4. The measured quantity was the surface pressure at an angle $60°$ to the main stream direction. The correlation length corresponding to the correlation coefficient, given in Fig. 1.27, on the other hand is found to be $L/D \cong 3$ from Eq. 1.11.

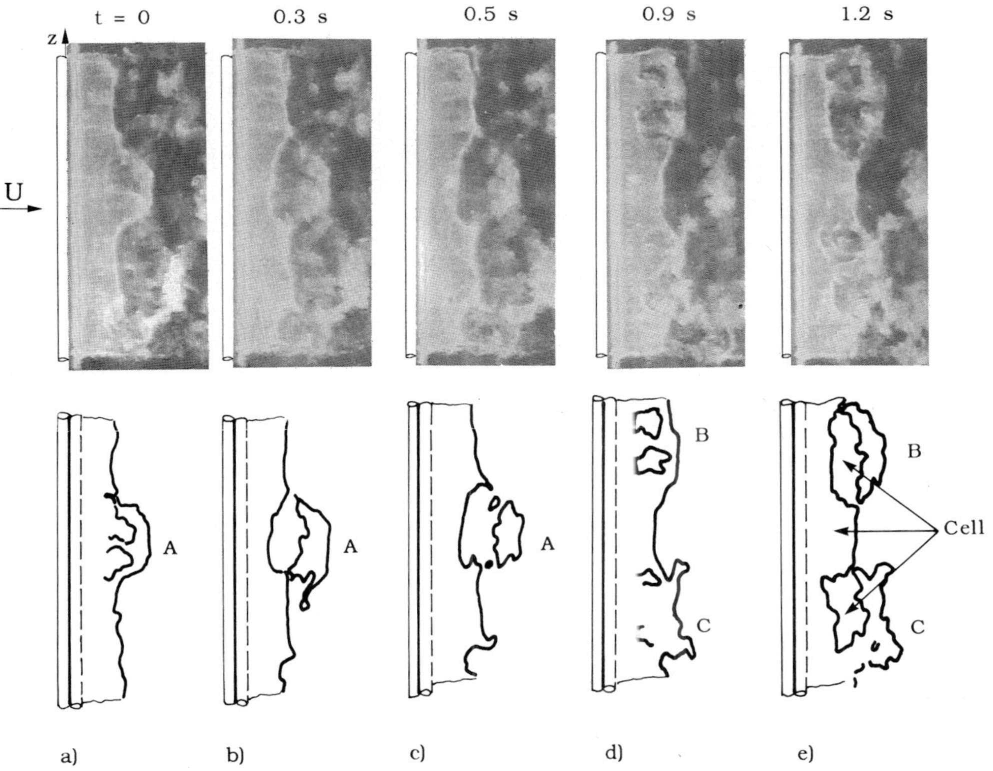

Figure 1.26 Photographs, illustrating the time evolution of spanwise cell structure. Cylinder smooth. $Re = 6 \times 10^3$.

For a smooth cylinder, the correlation length changes with the Reynolds number. Table 1.1 presents the correlation-length data compiled by King (1977).

Table 1.1 Correlation lengths and Reynolds numbers of smooth cylinders.

Reynolds number	Correlation length	Source
$40 < Re < 150$	(15–20)D	Gerlach and Dodge (1970)
$150 < Re < 10^5$	(2–3)D	Gerlach and Dodge (1970)
$1.1 \times 10^4 < Re < 4.5 \times 10^4$	(3–6)D	El-Baroudi (1960)
$\geq 10^5$	0.5D	Gerlach and Dodge (1970)
2×10^5	1.56D	Humphreys (1960)

The table shows that the correlation length is (15–20)D for $40 < Re < 150$ but experiences a sudden drop to (2–3)D at $Re = 150$. The latter Re number is quite close to the Reynolds number (see Fig. 1.1d), at which the laminar vortex shedding regime disappears. Regarding the finite (although large) values of the correlation length in the range $40 < Re < 150$, the correlation length in this flow regime should theoretically be infinite, since the vortex regime in this range is actually two-dimensional. However, purely two-dimensional shedding cannot be achieved in practice due to the existing end conditions. A slight divergence from the purely two-dimensional shedding, in the form of the so-called oblique shedding (see for example Williamson, 1989), may result in finite correlation lengths.

Other factors also affect the correlation. The correlation increases considerably when the cylinder is *oscillated* in the cross-flow direction. Fig. 1.28 presents the correlation coefficient data obtained by Novak and Tanaka (1977) for several values of the double-amplitude-diameter ratio $2A/D$ where A is the amplitude of cross-flow vibrations of the cylinder. The figure shows that the correlation coefficient increases tremendously with the amplitude of oscillations. Similar results were obtained by Toebes (1969) who measured the correlation coefficient of fluctuating velocity in the wake region near the cylinder. Fig. 1.29 presents the variation of the correlation length as a function of the amplitude-to-diameter ratio (curve *a* in Fig. 1.29). Clearly, the correlation length increases extensively with increasing the amplitude of oscillations.

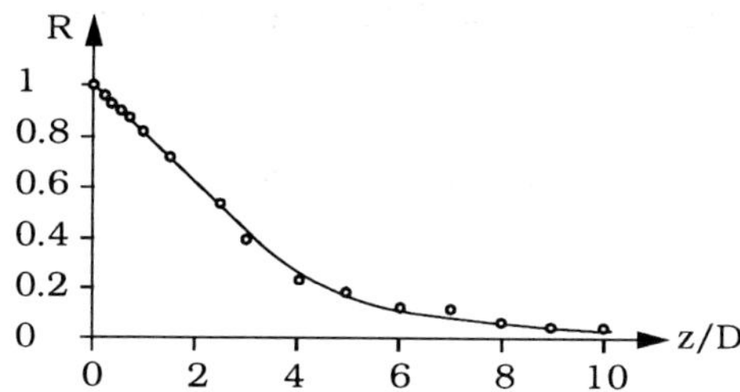

Figure 1.27 Correlation coefficient of surface pressure fluctuations as function of the spanwise separation distance z. Cylinder smooth. $Re = 1.9 \times 10^4$. Pressure transducers are located at 60° to the main stream direction. Novak and Tanaka (1977).

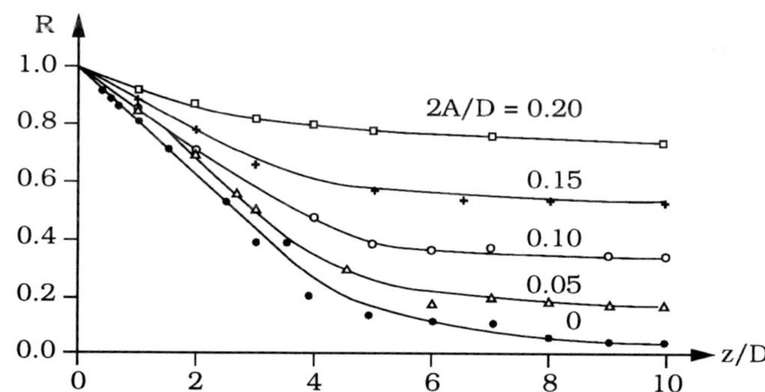

Figure 1.28 Effect of cross-flow vibration of cylinder on correlation coefficient of surface pressure fluctuations. Cylinder smooth. $Re = 1.9 \times 10^4$. Pressure transducers are located at 60° to the main stream direction. A is the amplitude of the cross-flow vibrations of cylinder. Novak and Tanaka (1977).

Turbulence in the approaching flow is also a significant factor for the correlation length, as is seen from Fig. 1.29. The turbulence in the tests presented in this figure was generated by a coarse grid in the experimental tunnel used in Novak and Tanaka's (1977) study. The figure indicates that the presence of turbulence in the approaching flow generally reduces the correlation length. It is interesting to note that with $2A/D = 0.2$, while the correlation length increases from about 3 diameters to 43 diameters for a turbulence-free, smooth flow, the increase is not so dramatic when some turbulence is introduced into the flow; the correlation length increases to only about 10 diameters in this latter situation.

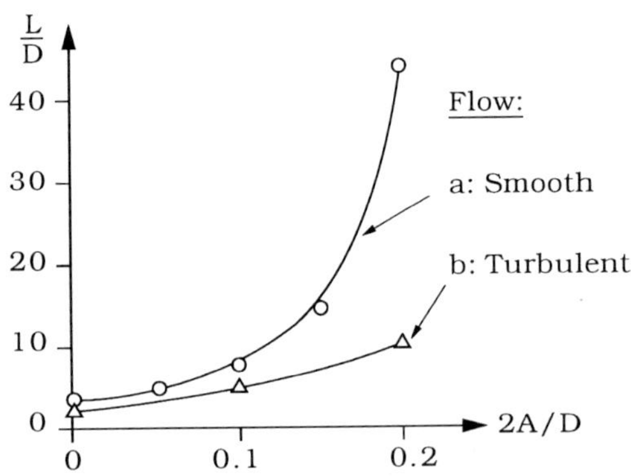

Figure 1.29 Correlation length. Cylinder smooth. $Re = 1.9 \times 10^4$. Pressure transducers are located at $60°$ to the main stream direction. A is the amplitude of cross-flow vibrations of the cylinder. Turbulence in the tunnel was generated by a coarse grid, and its intensity, $I_u = 11\%$. Novak and Tanaka (1977).

The subject has been most recently studied by Szepessy and Bearman (1992). These authors studied the effect of the aspect ratio (namely the cylinder length-to-diameter ratio) on vortex shedding by using moveable end plates. They found that the vortex-induced lift showed a maximum for an aspect ratio of 1, where the lift could be almost twice the value for very large aspect ratios. This increase of the lift amplitude was found to be accompanied by enhanced spanwise correlation of the flow.

Finally, it may be noted that Ribeiro (1992) gives a comprehensive review of the literature on oscillating lift on circular cylinders in cross-flow.

REFERENCES

Achenbach, E. and Heinecke E. (1981): On vortex shedding from smooth and rough cylinders in the range of Reynolds numbers 6×10^3 to 5×10^6. J. Fluid Mech., 109:239-251.

Angrilli, F., Bergamaschi, S. and Cossalter, V. (1982): Investigation of wall-induced modifications to vortex shedding from a circular cylinder. Trans. of the ASME, J. Fluids Engrg., 104:518-522.

ASCE Task Committee on Wind Forces (1961): Wind forces on structures. Trans. ASCE, 126:1124-1198.

Batchelor, G.K. (1967): An Introduction to Fluid Dynamics. Cambridge University Press.

Bearman, P.W. and Zdravkovich, M.M. (1978): Flow around a circular cylinder near a plane boundary. J. Fluid Mech., 89(1):33 48.

Blevins, R.D. (1977): Flow-induced Vibrations. Van Nostrand.

Bloor, M.S. (1964): The transition to turbulence in the wake of a circular cylinder. J. Fluid Mech., 19:290-304.

Cheung, J.C.K. and Melbourne, W.H. (1983): Turbulence effects on some aerodynamic parameters of a circular cylinder at supercritical Reynolds numbers. J. of Wind Engineering and Industrial Aerodynamics, 14:399-410.

El-Baroudi, M.Y. (1960): Measurement of Two-Point Correlations of Velocity near a Circular Cylinder Shedding a Karman Vortex Street. University of Toronto, UTIAS, TN31.

Farell, C. (1981): Flow around fixed circular cylinders: Fluctuating loads. Proc. of ASCE, Engineering Mech. Division, 107:EM3:565-588. Also see the closure of the paper. Journal of Engineering Mechanics, ASCE, 109:1153-1156, 1983.

Gerlach, C.R. and Dodge, F.T. (1970): An engineering approach to tube flow-induced vibrations. Proc. Conf. on Flow-Induced Vibrations in Reactor System Components, Argonne National Laboratory, pp. 205-225.

Gerrard, J.H. (1966): The mechanics of the formation region of vortices behind bluff bodies. J. Fluid Mech., 25:401-413.

Gerrard, J.H. (1978): The wakes of cylindrical bluff bodies at low Reynolds number. Phil. Transactions of the Royal Soc. London, Series A, 288(A1354):351-382.

Grass, A.J., Raven, P.W.J., Stuart, R.J. and Bray, J.A. (1984): The influence of boundary layer velocity gradients and bed proximity on vortex shedding from free spanning pipelines. Trans. ASME, J. of Energy Res. Technology, 106:70-78.

Griffin, O.M. (1985a): Vortex shedding from bluff bodies in a shear flow: A Review. Trans. ASME, J. Fluids Eng., 107:298-306.

Griffin, O.M. (1985b): The effect of current shear on vortex shedding. Proc. Int. Symp. on Separated Flow Around Marine Structures. The Norwegian Inst. of Technology, Trondheim, Norway, June 26-28, 1985, pp. 91-110.

Homann, F. (1936): Einfluss grosser Zähigkeit bei Strömung um Zylinder. Forschung auf dem Gebiete des Ingenieurwesen, 7(1):1-10.

Humphreys, J.S. (1960): On a circular cylinder in a steady wind at transition Reynolds numbers. J. Fluid Mech., 9:603-612.

Jensen, B.L. and Sumer, B.M. (1986): Boundary layer over a cylinder placed near a wall. Progress Report No. 64, Inst. of Hydrodynamics and Hydraulic Engineering, ISVA, Techn. Univ. Denmark, pp. 31-39.

Jensen, B.L., Sumer, B.M., Jensen, H.R. and Fredsøe, J. (1990): Flow around and forces on a pipeline near a scoured bed in steady current. Trans. of the ASME, J. of Offshore Mech. and Arctic Engrg., 112:206-213.

King, R. (1977): A review of vortex shedding research and its application. Ocean Engineering, 4:141-171.

Kiya, M., Tamura, H. and Arie, M. (1980): Vortex shedding from a circular cylinder in moderate-Reynolds-number shear flow. J. Fluid Mech., 141:721-735.

Kwok, K.C.S. (1986): Turbulence effect on flow around circular cylinder. J. Engineering Mechanics, ASCE, 112(11):1181-1197.

Maull, D.J. and Young, R.A. (1973): Vortex shedding from bluff bodies in a shear flow. J. Fluid Mech., 60:401-409.

Modi, V.J., Wiland, E., Dikshit, A.K. and Yokomizo, T. (1992): On the fluid dynamics of elliptic cylinders. Proc. 2nd Int. Offshore and Polar Engrg. Conf., San Francisco, CA, 14-19 June 1992, III:595-614.

Nikuradse, J. (1933): Strömungsgesetze in rauhen Rohren. Forsch. Arb.Ing.-Wes. No. 361.

Norberg, C. and Sundén, B. (1987): Turbulence and Reynolds number effects on the flow and fluid forces on a single cylinder in cross flow. Jour. Fluids and Structures, 1:337-357.

Novak, M. and Tanaka, H. (1977): Pressure correlations on a vibrating cylinder. Proc. 4th Int. Conf. on Wind Effects on Buildings and Structures, Heathrow, U.K., Ed. by K.J. Eaton. Cambridge Univ. Press, pp. 227-232.

Raven, P.W.J., Stuart, R.J., Bray, J.A. and Littlejohns, P.S. (1985): Full-scale dynamic testing of submarine pipeline spans. 17th Annual Offshore Technology Conference, Houston, Texas, May 6-9., paper No. 5005, 3:395-404.

Ribeiro, J.L.D. (1992): Fluctuating lift and its spanwise correlation on a circular cylinder in a smooth and in a turbulent flow: a critical review. Jour. of Wind Engrg. and Indust. Aerodynamics, 40:179-198.

Roshko, A. (1961): Experiments on the flow past a circular cylinder at very high Reynolds number. J. Fluid Mech., 10:345-356.

Schewe, G. (1983): On the force fluctuations acting on a circular cylinder in cross-flow from subcritical up to transcritical Reynolds numbers. J. Fluid Mech., 133:265-285.

Schlichting, G. (1979): Boundary Layer Theory. 7.ed. McGraw-Hill Book Company.

Szepessy, S. and Bearman, P.W. (1992): Aspect ratio and end plate effects on vortex shedding from a circular cylinder. J. Fluid Mech., 234:191-217.

Toebes, G.H. (1969): The unsteady flow and wake near an oscillating cylinder. Trans. ASME J. Basic Eng., 91:493-502.

Williamson, C.H.K. (1988): The existence of two stages in the transition to three-dimensionality of a cylinder wake. Phys. Fluids, 31(11):3165-3168.

Williamson, C.H.K. (1989): Oblique and parallel modes of vortex shedding in the wake of a circular cylinder at low Reynolds number. J. Fluid Mech., 206:579-627.

Chapter 2. Forces on a cylinder in steady current

The flow around the cylinder described in Chapter 1 will exert a resultant force on the cylinder. There are two contributions to this force, one from the pressure and the other from the friction.

The in-line component of the mean resultant force due to pressure (the in-line mean *pressure* force) per unit length of the cylinder is given by

$$\overline{F}_p = \int_0^{2\pi} \overline{p}\cos(\phi)r_0 d\phi, \tag{2.1}$$

(see Fig. 2.1 for the definition sketch), while that due to friction (the in-line mean *friction* force) is given by

$$\overline{F}_f = \int_0^{2\pi} \overline{\tau}_0 \sin(\phi)r_0 d\phi \tag{2.2}$$

in which $\overline{p}$ is the pressure and $\overline{\tau}_0$ is the wall shear stress on the cylinder surface, and the overbar denotes time-averaging.

The total in-line force, the so-called **mean drag**, is the sum of these two forces:

$$\overline{F}_D = \overline{F}_p + \overline{F}_f \tag{2.3}$$

$\overline{F}_p$ is termed the **form drag** and $\overline{F}_f$ the **friction drag**.

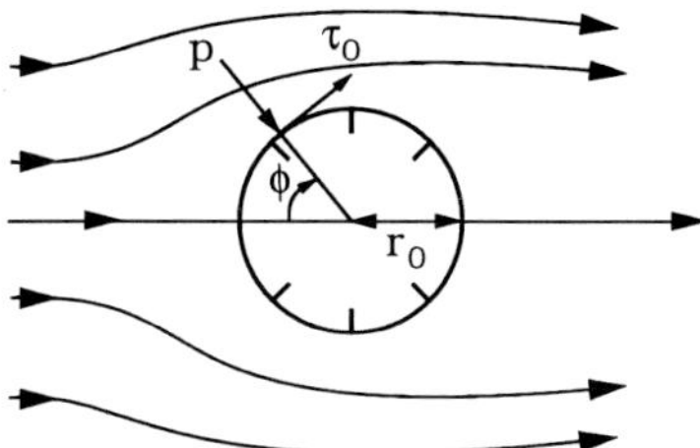

Figure 2.1 Definition sketch.

Regarding the cross-flow component of the mean resultant force, this force will be nil due to symmetry in the flow. However, the instantaneous cross-flow force on the cylinder, i.e., the instantaneous **lift force**, is non-zero and its value can be rather large, as will be seen in the next sections.

2.1 Drag and lift

As has been discussed in Chapter 1, the regime of flow around a circular cylinder varies as the Reynolds number is changed (Fig. 1.1). Also, effects such as the surface roughness, the cross-sectional shape, the incoming turbulence, and the shear in the incoming flow influence the flow. However, except for very small Reynolds numbers ($Re \lesssim 40$), there is one feature of the flow which is common to all the flow regimes, namely the vortex shedding.

As a consequence of the vortex-shedding phenomenon, the pressure distribution around the cylinder undergoes a periodic change as the shedding process progresses, resulting in a periodic variation in the force components on the cylinder.

Fig. 2.2 shows a sequence of flow pictures of the wake together with the measured pressure distributions and the corresponding force components, which are calculated by integrating the pressure distributions over the cylinder surface (the time span covered in the figure is slightly larger than one period of vortex shedding). Fig. 2.3, on the other hand, depicts the force traces corresponding to the same experiment as in the previous figure.

The preceding figures show the following two important features: first, the force acting on the cylinder in the in-line direction (the drag force) does change periodically in time oscillating around mean drag, and secondly, although the incoming flow is completely symmetric with respect to the cylinder axis, there exists a non-zero force component (with a zero mean, however) on the cylinder

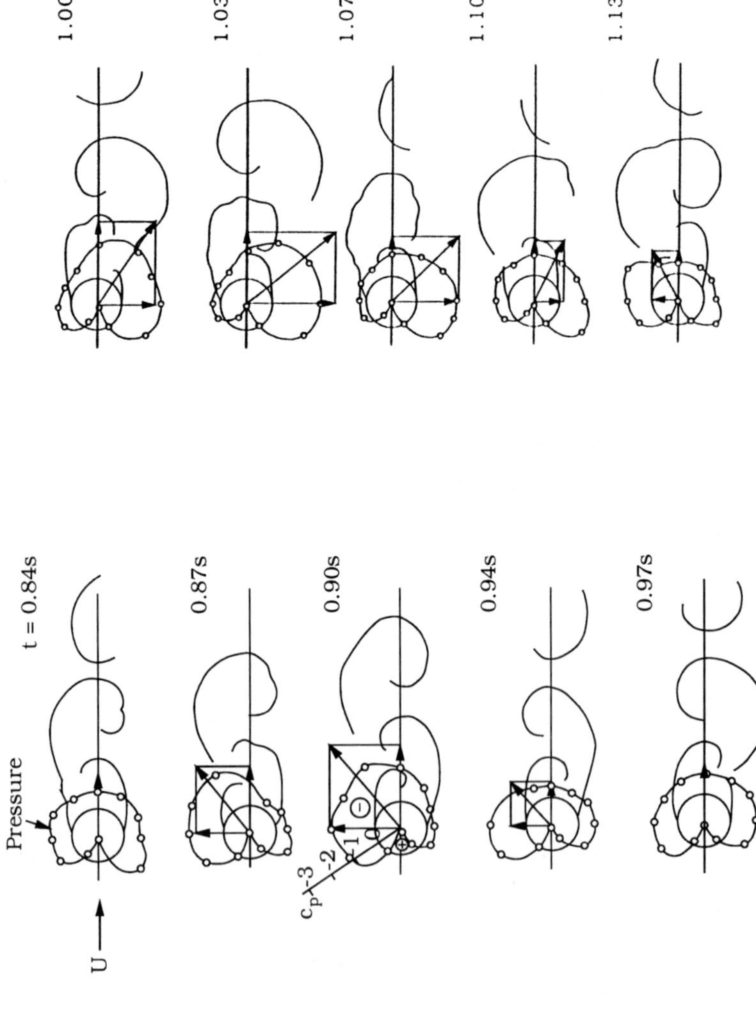

Figure 2.2 Time development of pressure distribution and the force components, as the vortex shedding progresses. $Re = 1.1 \times 10^5$, $D = 8$ cm and $U = 1.53$ m/s. $c_p = (p - p_0)/(\frac{1}{2}\rho U^2)$. Drescher (1956).

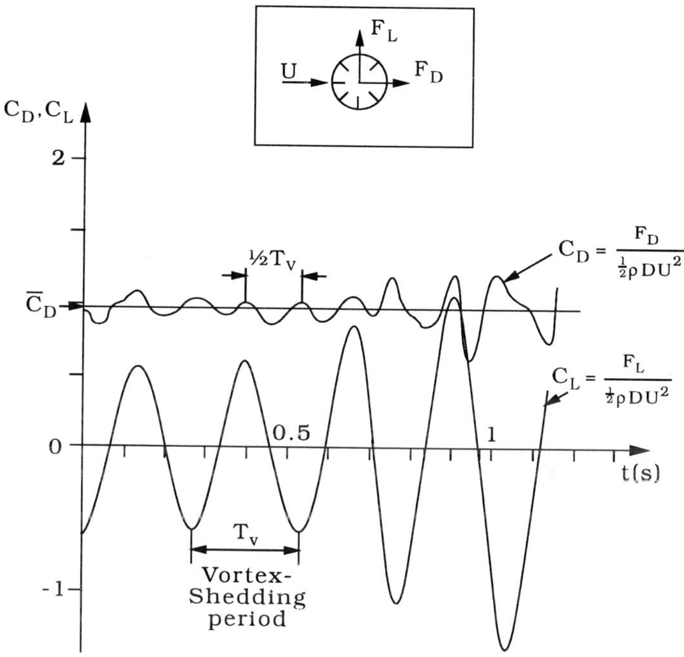

Figure 2.3 Drag and lift force traces obtained from the measured pressure distributions in the previous figure. $C_D = F_D/(\frac{1}{2}\rho DU^2)$ and $C_L = F_L/(\frac{1}{2}\rho DU^2)$. Drescher (1956).

in the transverse direction (the lift force), and this, too, varies periodically with time.

In the following paragraphs we will first concentrate our attention on the mean drag, then we will focus on the oscillating components of the forces, namely the oscillating drag force and the oscillating lift force.

2.2 Mean drag

Form drag and friction drag

Fig. 2.4 shows the relative contribution to the total mean drag force from friction as function of the Re-number. The figure clearly shows that, for the range of Re numbers normally encountered in practice, namely $Re \gtrsim 10^4$, the contribution of the friction drag to the total drag force is less than 2 - 3%. So the friction drag can be omitted in most of the cases, and the total mean drag can be assumed to be composed of only one component, namely the form drag

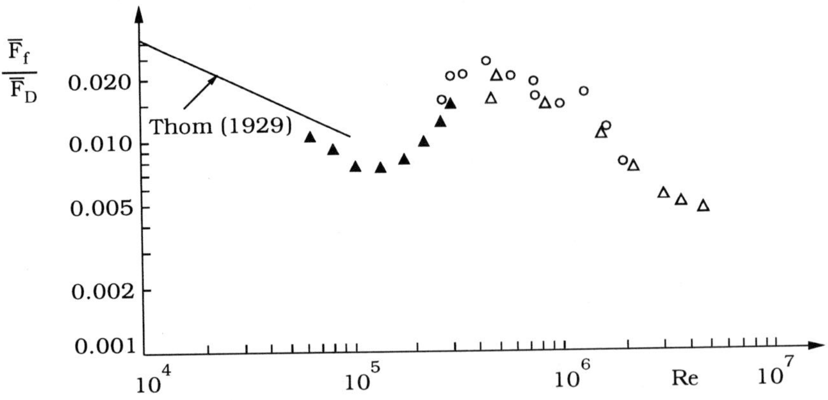

Figure 2.4 Relative contribution of the friction force to the total drag for circular cylinder. Achenbach (1968).

$$\overline{F}_D \cong \int_0^{2\pi} \overline{p}\cos(\phi)r_0\,d\phi \tag{2.4}$$

Fig. 2.5a depicts several measured pressure distributions for different values of Re, while Fig. 2.5b presents the corresponding wall shear stress distributions. Fig. 2.5a contains also the pressure distribution obtained from the potential flow theory, which is given by

$$\overline{p} - p_0 = \frac{1}{2}\rho U^2 (1 - 4\sin^2\phi) \tag{2.5}$$

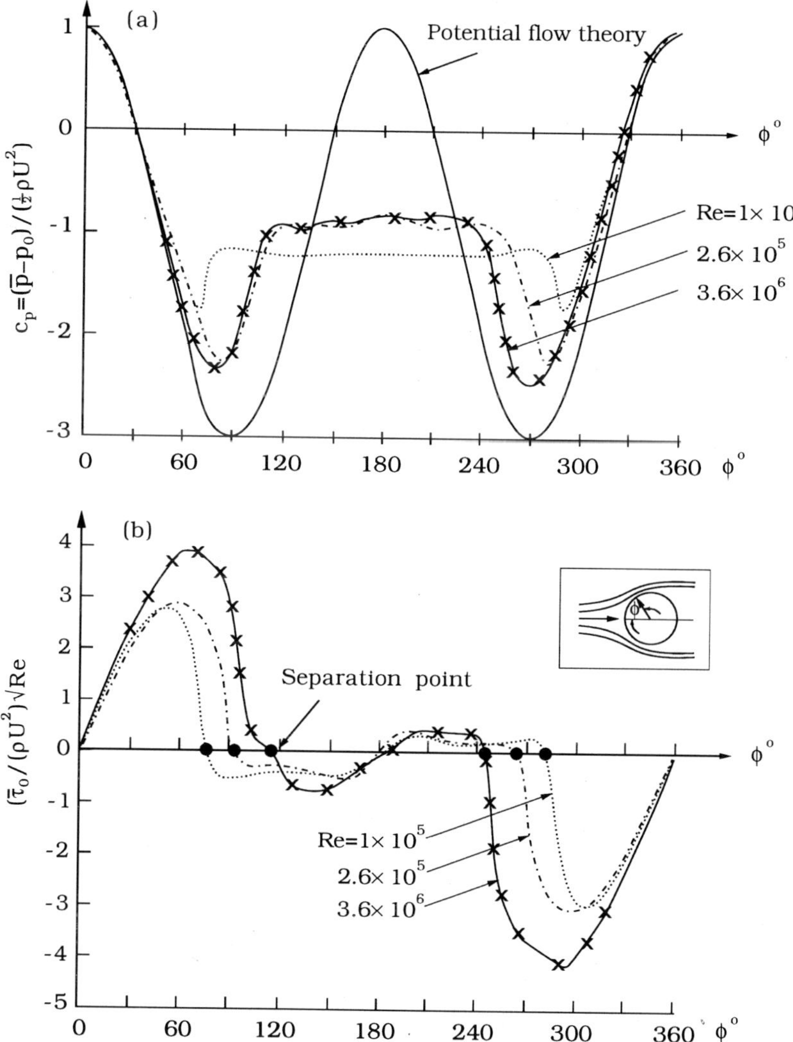

Figure 2.5 Pressure distribution and wall shear stress distribution at different Re numbers for a smooth cylinder. Achenbach (1968).

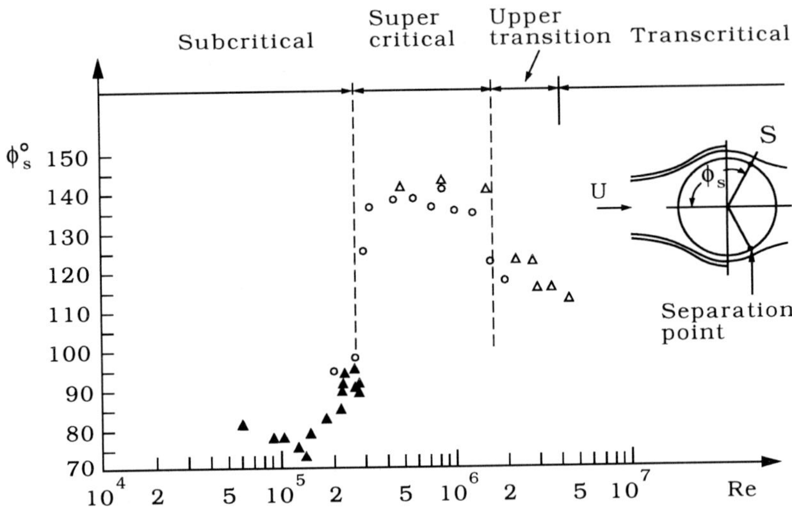

Figure 2.6 Position of the separation point as a function of the Reynolds number for circular cylinder. Achenbach (1968).

in which p_0 is the hydrostatic pressure. Fig. 2.6 gives the position of separation points as a function of Re.

The main characteristic of the measured pressure distributions is that the pressure at the rear side of the cylinder (i.e., in the wake region) is always negative (in contrast to what the potential-flow theory gives). This is due to separation. Fig. 2.5a further indicates that the pressure on the cylinder remains practically constant across the cylinder wake. This is because the flow in the wake region is extremely weak as compared to the outer-flow region.

Drag coefficient

The general expression for the drag force is from Eqs. 2.1–2.3 given by

$$\overline{F}_D = \int_0^{2\pi} \left(\overline{p}\cos(\phi) + \overline{\tau}_0 \sin(\phi) \right) r_0 \, d\phi \tag{2.6}$$

This equation can be written in the following form

$$\frac{\overline{F}_D}{\frac{1}{2}\rho D U^2} = \int_0^{2\pi} \left[\left(\frac{\overline{p} - p_0}{\rho U^2} \right) \cos(\phi) + \left(\frac{\overline{\tau}_0}{\rho U^2} \right) \sin(\phi) \right] d\phi \tag{2.7}$$

in which $D = 2r_0$, the cylinder diameter. The right-hand-side of the equation is a function of the Re number, since both the pressure term and the wall shear stress term are functions of the Re number for a smooth cylinder (Fig. 2.5). Therefore Eq. 2.7 may be written in the following simple form

$$\frac{\overline{F}_D}{\frac{1}{2}\rho DU^2} = \overline{C}_D \tag{2.8}$$

$\overline{C}_D$ is called the mean drag coefficient, or in short, the drag coefficient, and is a function of Re.

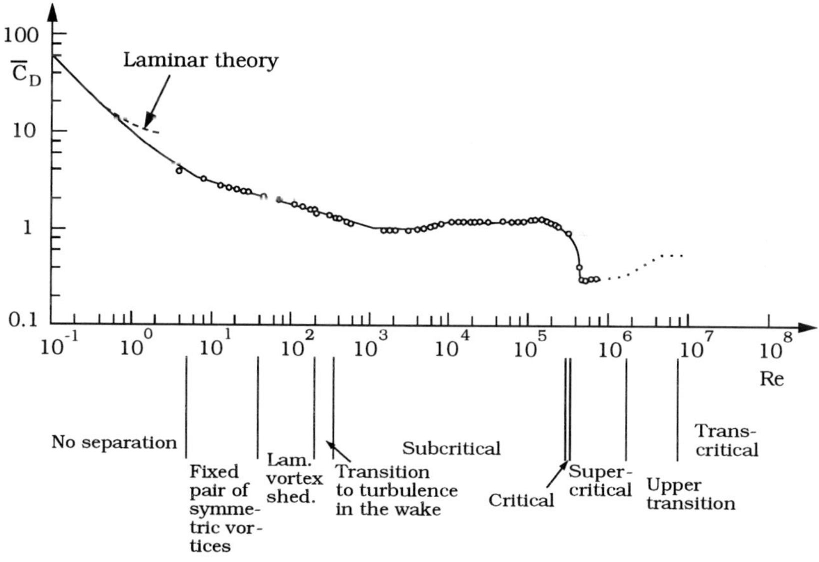

Figure 2.7 Drag coefficient for a smooth circular cylinder as a function of the Reynolds number. Dashed curve: The Oseen-Lamb laminar theory (Eq. 5.41). Measurements by Wieselsberger for $40 < Re < 5 \times 10^5$ and Schewe (1983) for $Re > 10^5$. The diagram minus Schewe's data was taken from Schlichting (1979).

Fig. 2.7 presents the experimental data together with the result of the laminar theory, illustrating the variation of $\overline{C}_D$ with respect to the Re number, while Fig. 2.8 depicts the close up picture of this variation in the most interesting

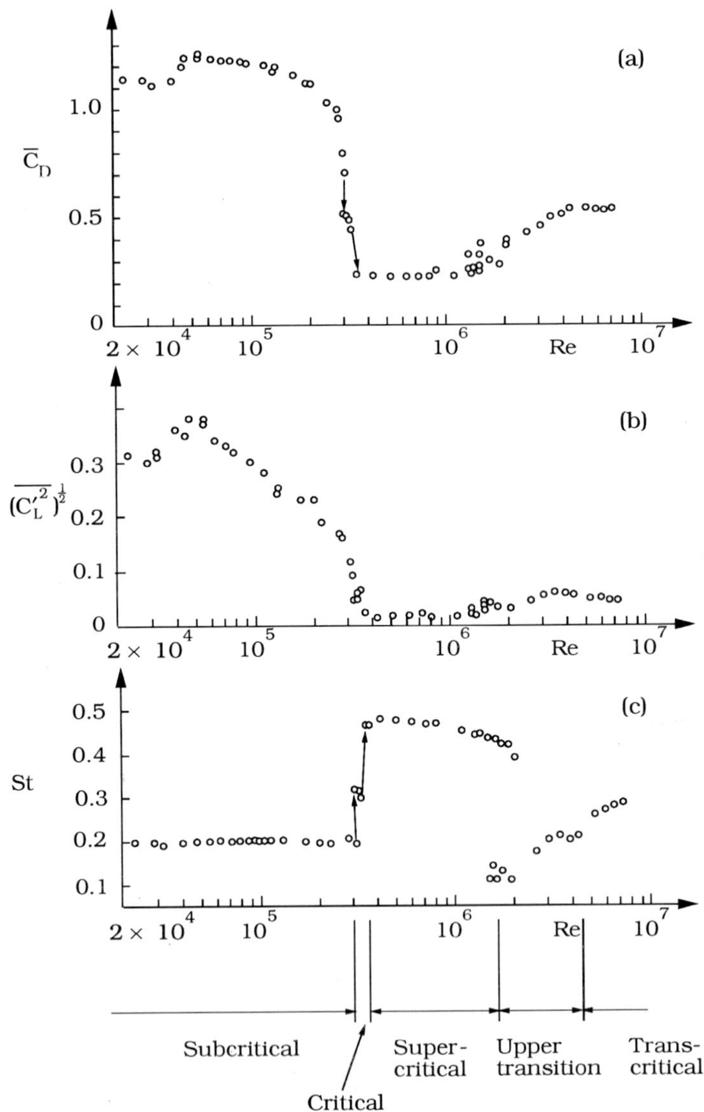

Figure 2.8 Drag coefficient, r.m.s. of the lift oscillations and Strouhal number as function of Re for a smooth circular cylinder. Schewe (1983).

range of Re numbers, namely $Re \stackrel{>}{\sim} 10^4$. The latter figure also contains information about the oscillating lift force and the Strouhal number, which are maintained in the figure for the sake of completeness. The lift force data will be discussed later in the section dealing with the oscillating forces.

As seen from Fig. 2.7, $\overline{C}_D$ decreases monotonously with Re until Re reaches the value of about 300. However, from this Re number onwards, $\overline{C}_D$ assumes a practically constant value, namely 1.2, throughout the subcritical Re range ($300 < Re < 3 \times 10^5$). When Re attains the value of 3×10^5, a dramatic change occurs in $\overline{C}_D$; the drag coefficient decreases abruptly and assumes a much lower value, about 0.25, in the neighbouring Re range, the supercritical Re range, $3.5 \times 10^5 < Re < 1.5 \times 10^6$ (Fig. 2.8a). This phenomenon, namely the drastic fall in $\overline{C}_D$, is called the **drag crisis**.

The drag crisis can best be explained by reference to the pressure diagrams given in Fig. 2.5. Note that the friction drag can be disregarded in the analysis because it constitutes only a very small fraction of the total drag.

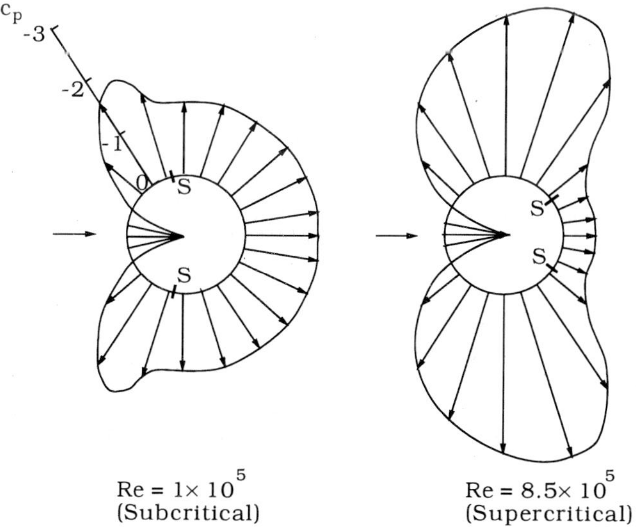

Figure 2.9 Pressure distributions. $c_p = (\overline{p} - p_0)/(\frac{1}{2}\rho U^2)$. S denotes the separation points. Achenbach (1968).

Two of the diagrams, namely the one for $Re = 1 \times 10^5$ (a representative Re number for subcritical flow regime) and that for $Re = 8.5 \times 10^5$ (a representative Re number for supercritical flow regime) are reproduced in Fig. 2.9. From the figure, it is evident that the drag should be smaller in the supercritical flow regime than in the subcritical flow regime. Clearly, the key point here is that the separation point moves from $\phi_s = 78°$ ($Re = 1 \times 10^5$, the laminar separation) to $\phi_s = 140°$ ($Re = 8.5 \times 10^5$, the turbulent separation), when the flow regime is changed from subcritical to supercritical (Fig. 2.6), resulting in an extremely narrow wake with substantially smaller negative pressure, which would presumably lead to a considerable reduction in the drag.

Returning to Figs. 2.7 and 2.8 it is seen that the drag coefficient increases as the flow regime is changed from supercritical to upper-transition, and then $\overline{C}_D$ attains a constant value of about 0.5, as Re is increased further to transcritical values, namely $Re > 4.5 \times 10^6$. Again, the change in $\overline{C}_D$ for these higher flow regimes can be explained by reference to the pressure distributions given in Fig. 2.5 along with the information about the separation angle given in Fig. 2.6.

Effect of surface roughness

In the case of rough cylinders, the mean drag, as in the case of smooth cylinders, can be assumed to be composed of only one component, namely the form drag; indeed, Achenbach's (1971) measurements demonstrate that the contribution of the friction drag to the total drag does not exceed 2–3%, thus can be omitted in most of the cases (Fig. 2.10).

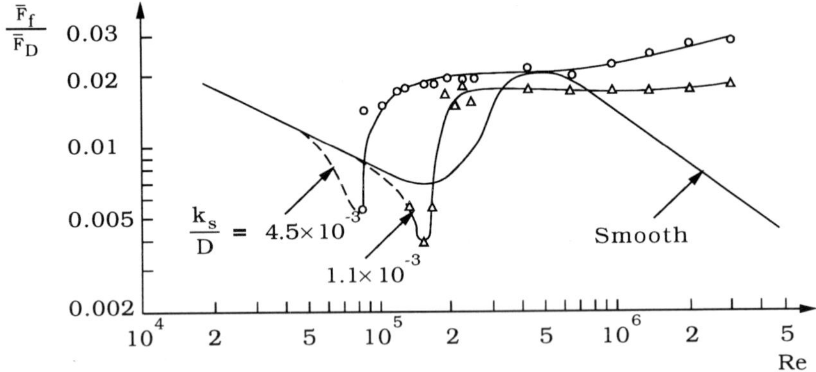

Figure 2.10 Relative contribution of the friction force to the total drag. Effect of cylinder roughness. Achenbach (1971).

The drag coefficient, $\overline{C}_D$, now becomes not only a function of Re number but also a function of the roughness parameter k_s/D

$$\overline{C}_D = \overline{C}_D \left(Re, \frac{k_s}{D} \right) \tag{2.9}$$

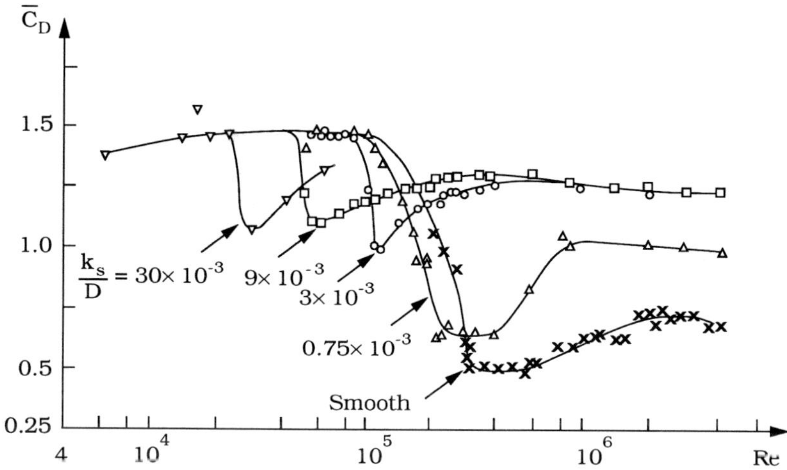

Figure 2.11 Drag coefficient of a circular cylinder at various surface roughness parameters k_s/D. Achenbach and Heinecke (1981).

in which k_s is the Nikuradse equivalent sand roughness.

Fig. 2.11 depicts $\overline{C}_D$ plotted as a function of these parameters. The way in which $\overline{C}_D$ varies with Re for a given k_s/D is sketched in Fig. 2.12.

As seen from the figures, the Reynolds-number ranges observed for the smooth-cylinder case still exist. However, two of the high Re number ranges, namely the supercritical range and the upper transition range seem to merge into one single range as the roughness is increased. Furthermore, the following observations can be made from the figure:

1) For small Re numbers (i.e., the subcritical Re numbers), $\overline{C}_D$ takes the value obtained in the case of smooth cylinders, namely 1.4, irrespective of the cylinder roughness.

2) The $\overline{C}_D$-versus-Re curve shifts towards the lower end of the Re-number range indicated in the figure, as the cylinder roughness is increased. Clearly, this behaviour is related to the early transition to turbulence in the boundary layer with increasing roughness.

3) The drag crisis, which is characterized by a marked depression in the $\overline{C}_D$ curve, is not as extensive as it is in the smooth-cylinder case: while $\overline{C}_D$ falls from 1.4 to a value of about 0.5 in the case of smooth cylinder, it falls from 1.4 only to a value of about 1.1 in the case of rough cylinder with $k_s/D = 30 \times 10^{-3}$. This is directly linked with the angular location of the separation points. Fig. 2.13 compares the latter quantity for cylinders with different roughnesses. It is seen that, in the supercritical range, while ϕ_s is equal to 140° in the case of a smooth

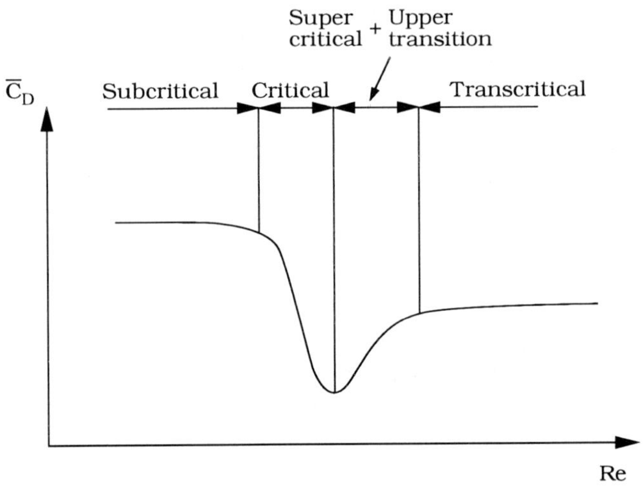

Figure 2.12 General form of $\overline{C}_D = \overline{C}_D(Re)$ curve for a rough cylinder.

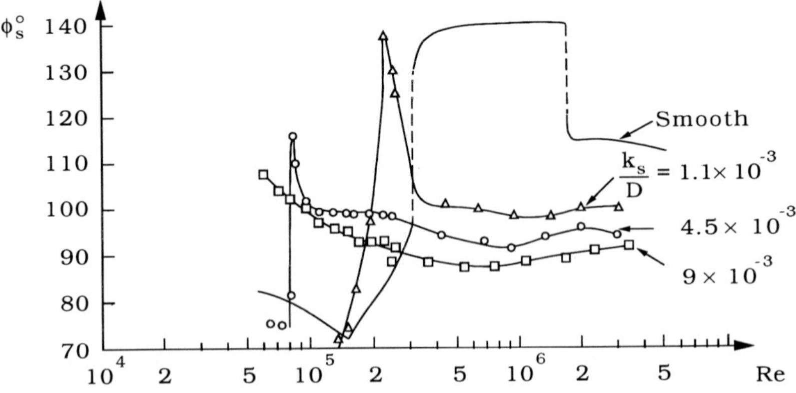

Figure 2.13 Circular cylinder. Angular position of boundary-layer separation at various roughness parameters. Achenbach (1971).

cylinder, it is only 115° for the case of a rough cylinder with $k_s/D = 4.5 \times 10^{-3}$. (This is because of the relatively weaker momentum exchange near the wall in the case of rough wall due to the larger boundary-layer thickness). Therefore, the picture given in Fig. 2.9b for the smooth-cylinder situation (where $\phi_s = 140°$) will not be the same for the rough cylinder ($\phi_s = 115°$). As a matter of fact, the pressure-distribution picture for the rough cylinder in consideration ($\phi_s = 115°$) must lie somewhere between the picture given in Fig. 2.9a and that given in Fig. 2.9b, which implies that the fall in the mean drag due to the drag crisis in this case will not be as extensive as in the case of a smooth cylinder, as clearly indicated in Fig. 2.11.

Regarding the transcritical Re numbers in Fig. 2.11, the transcritical range covers smaller and smaller Re numbers as the roughness is increased. Also, the $\overline{C}_D$ coefficient in the transcritical range takes higher and higher values with increasing roughness, see Table 2.1. Clearly, this behaviour is closely linked with the behaviour of the cylinder boundary layer. Finally, Fig. 2.14 gives the drag coefficient as a function of cylinder roughness for the transcritical Re-number range.

Table 2.1 Transcritical Re number range for various values of the relative roughness. Data from Fig. 2.11.

k_s/D Cylinder roughness	Transcritical Reynolds number range
0	$Re > (3-4) \times 10^6$
0.75×10^{-3}	$Re > 9 \times 10^5$
3×10^{-3}	$Re > 5 \times 10^5$
9×10^{-3}	$Re > 3 \times 10^5$
30×10^{-3}	$Re > (1-2) \times 10^5$

The reader is referred to the following work for further details of the effect of the cylinder roughness on the mean drag: Achenbach (1968, 1971) and Güven, Patel and Farell (1975 and 1977), Güven, Farell and Patel (1980), Shih, Wang, Coles and Roshko (1993) among others.

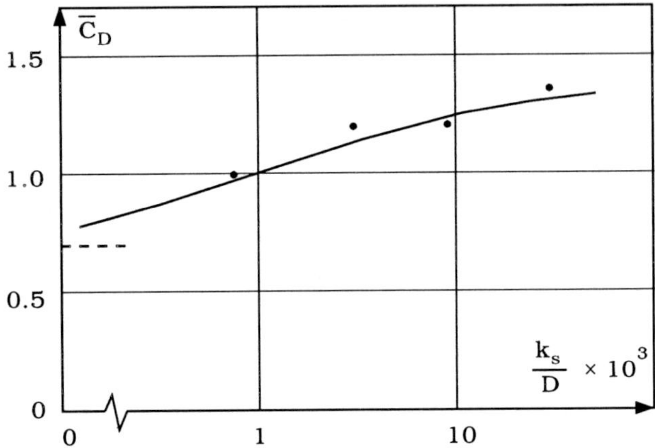

Figure 2.14 Drag coefficient for rough cylinders in the *transcritical Re-number* range (Table 2.1). Data from Fig. 2.11.

2.3 Oscillating drag and lift

A cylinder which is exposed to a steady flow experiences oscillating forces if $Re > 40$, where the wake flow becomes time-dependent (Section 1.1). The origin of the oscillating forces is the vortex shedding. As already discussed in Section 1.1, the key point is that the pressure distribution around the cylinder undergoes a periodic change as the vortex shedding progresses, resulting in a periodic variation in the force (Figs. 2.2 and 2.3). A close inspection of Fig. 2.2 reveals that the upward lift is associated with the growth of the vortex at the lower edge of the cylinder ($t = 0.87 \text{ - } 0.94$ s), while the downward lift is associated with that at the upper edge of the cylinder ($t = 1.03 \text{ - } 1.10$ s). Also, it is readily seen that both vortices give a temporary increase in the drag.

As seen from Fig. 2.3, the lift force on the cylinder oscillates at the vortex-shedding frequency, $f_v(= 1/T_v)$, while the drag force oscillates at a frequency which is twice the vortex-shedding frequency. Fig. 2.3 further indicates that the amplitude of the oscillations is not a constant set of value. As is seen, it varies from one period to the other. It may even happen that some periods are missed. Nevertheless, the magnitude of the oscillations can be characterized by their statistical properties such as the root-mean-square (*r.m.s.*) value of the oscillations. Fig. 2.15 gives the oscillating-force data compiled by Hallam, Heaf and Wootton

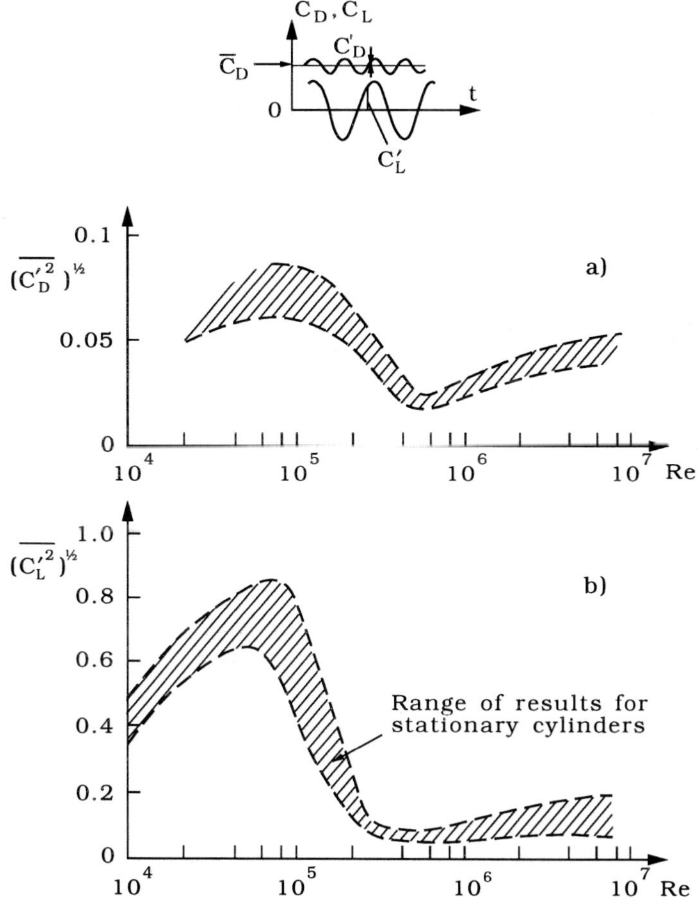

Figure 2.15 *R.m.s.*-values of drag and lift oscillations. $C'_D = F'_D/$
$\left(\frac{1}{2}\rho DU^2\right)$ and $C'_L = F'_L/\left(\frac{1}{2}\rho DU^2\right)$. Hallam et al. (1977).

(1977), regarding the magnitude of the oscillations in the force coefficients where C'_D and C'_L are defined by the following equations

$$F'_D = \frac{1}{2}\rho C'_D DU^2 \tag{2.10}$$

$$F'_L = \frac{1}{2}\rho C'_L DU^2 \tag{2.11}$$

in which F_D' is the oscillating part of the drag force

$$F_D' = F_D - \overline{F}_D \ , \qquad (2.12)$$

and F_L' is the oscillating lift force

$$F_L' = F_L - \overline{F}_L = F_L - 0 = F_L \ , \qquad (2.13)$$

$(\overline{C_D'^2})^{1/2}$ and $(\overline{C_L'^2})^{1/2}$ are the r.m.s. values of the oscillations C_D' and C_L', respectively. The magnitude of the oscillating forces is a function of Re, which can be seen very clearly from Fig. 2.8, where C_L' data from a single set of experiments are shown along with the C_D and the St-number variations obtained in the same work. It is evident that the $r.m.s.$-value of C_L' experiences a dramatic change in the same way as in the case of C_D and St in the critical flow regime, and then it attains an extremely low value in the supercritical flow regime. This point has already been mentioned in Section 1.2.1 in connection with the frequency of vortex shedding with reference to the power spectra of the lift oscillations illustrated in Fig. 1.10 (cf. Fig. 1.10a and 1.10b, and note the difference in the scales of the vertical axes of the two figures). The main reason behind this large reduction in the $r.m.s.$-value of C_L' is that, in the supercritical flow regime, the interaction between the vortices in the wake is considerably weaker, partly because the boundary layer separates at an extremely large angular position (Fig.2.6) meaning that the vortices are much closer to each other in this flow regime, and partly because the boundary-layer separation is turbulent (Fig. 1.1).

2.4 Effect of cross-sectional shape on force coefficients

The shape of the cross-section has a large influence on the resulting force. A detailed table giving the variation in the force coefficient with various shapes of cross-sections is given in Appendix I.

There are two points which need to be elaborated here. One is the Reynolds number dependence in the case of cross-sectional shapes with sharp edges. In this case, practically no Reynolds number dependence should be expected since the separation point is fixed at the sharp corners of the cross section. So, no change in force coefficients is expected with Re number for these cross-sections in contrast to what occurs in the case of circular cross-sections.

Secondly, non-circular cross-sections may be subject to steady lift at a certain angle of attack. This is due to the asymmetry of the flow with respect to the principle axis of the cross-sectional area. A similar kind of steady lift has been observed even for circular cylinders in the critical flow regime (Schewe, 1983) where the asymmetry occurs due to the one-sided transition to turbulence (Section 1.1).

Fig. 2.16 presents the force coefficient regarding this steady lift for different cross-sections.

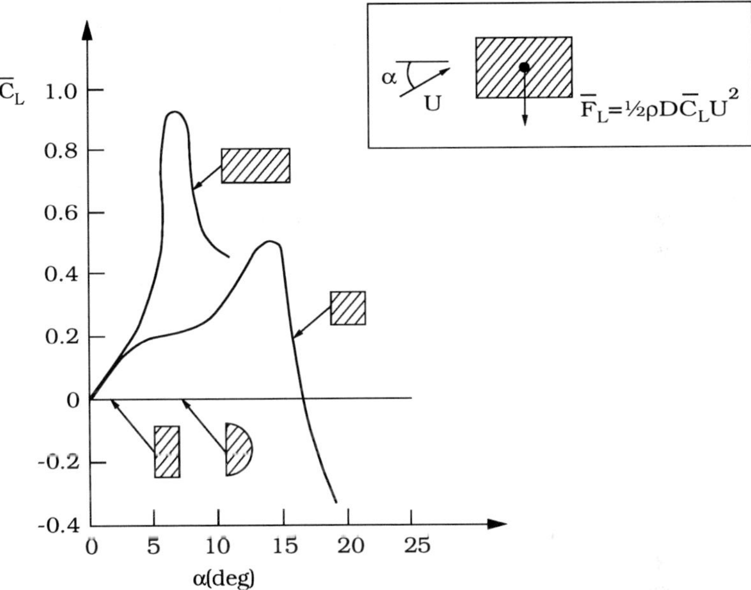

Figure 2.16 Steady lift force coefficients, $Re = 33,000$ to $66,000$. Parkinson and Brooks (1961).

2.5   Effect of incoming turbulence on force coefficients

The turbulence in the approaching flow may affect the force coefficients, Cheung and Melbourne (1983), Kwok (1986), and Norberg and Sundén (1987). The effect is summarized in Fig. 2.17 based on the data presented in Cheung and Melbourne (1983). The dashed lines in the figure correspond to the case where the turbulence level is very small, and therefore the flow in this case may be considered smooth.

The figures clearly show that the force coefficients are affected quite considerably by the incoming turbulence. Increasing the turbulence level from almost smooth flow (the dashed curves) to larger and larger values acts in the same way as increasing the cylinder roughness (cf. Fig. 2.17a and Fig. 2.11). As has been discussed in the context of the effect of roughness, the increased level of incoming turbulence will directly influence the cylinder boundary layer and hence its separation. This will obviously lead to changes in the force and therefore in the force cocfficients.

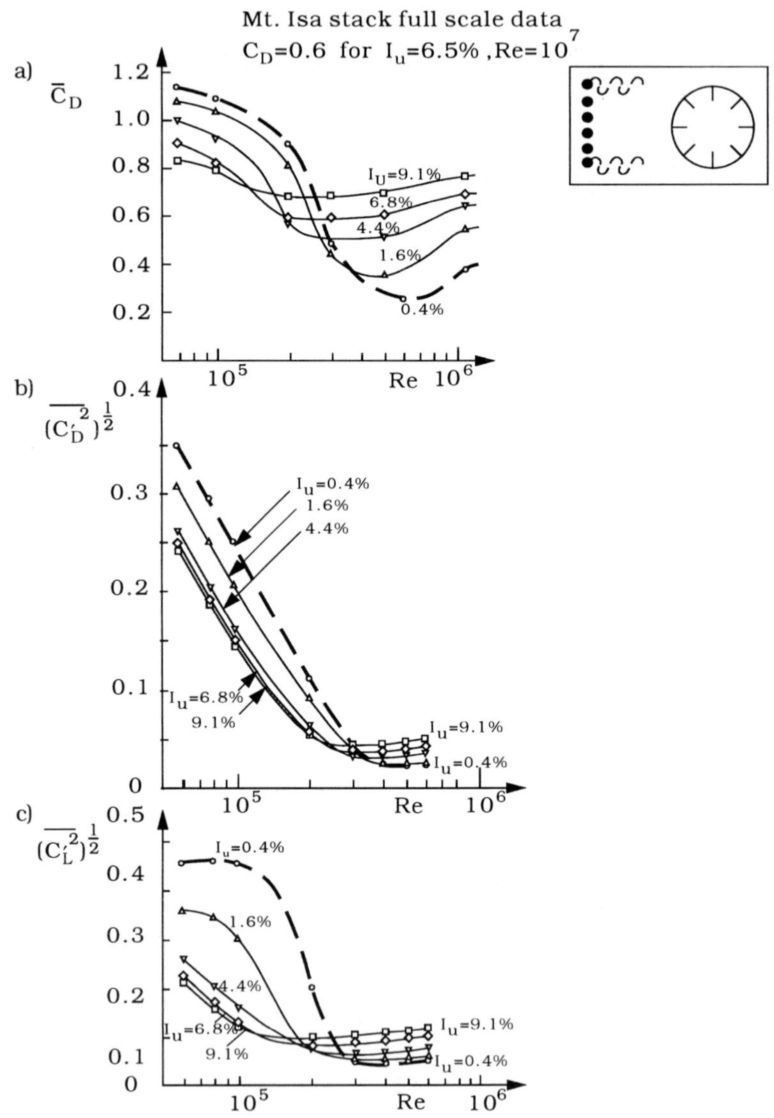

Figure 2.17 Effect of turbulence on the force coefficients. I_u is defined in
Eq. 1.8. Cheung and Melbourne (1983).

2.6 Effect of angle of attack on force coefficients

When a cylinder is placed at an angle to the flow (Fig. 2.18), forces on the cylinder may change. Experiments show, however, that in most of the cases the so-called independence or cross-flow principle is applicable (Hoerner, 1965). Namely, the component of the force normal to the cylinder may be calculated from

$$F_N = \frac{1}{2}\rho C_D D\, U_N^2 \qquad (2.14)$$

in which U_N is the velocity component normal to the cylinder axis. The drag coefficient in the preceding equation can be taken as that obtained for a cylinder normal to the flow. So, C_D *is independent of the angle of attack,* θ.

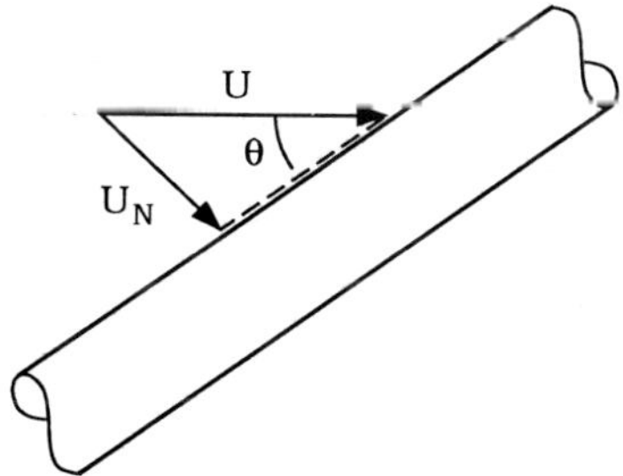

Figure 2.18 Definition sketch. Angle of attack of flow, θ, is different from 90°.

It may be argued that the flow sees an elliptical cross-section in the case of an oblique attack, and therefore separation may be delayed, resulting in a value of C_D different from that obtained for a cylinder normal to the flow. Observations show, however, that, although the approaching flow is at an angle, the streamlines in the neighbourhood of the cylinder are bent in such a way that the actual flow past the cylinder is at an angle of about $\theta = 90°$ (Fig. 2.19). Therefore, the position of the separation point practically does not change, meaning that C_D should be independent of θ. Kozakiewicz, Fredsøe and Sumer (1995), based on their flow-visualization experiments, give the critical value of θ approximately 35°. For $\theta \lesssim 35°$, the streamlines do not bend, implying that, for such small values of θ, C_D is no longer independent of θ, and therefore the independence principle will be violated.

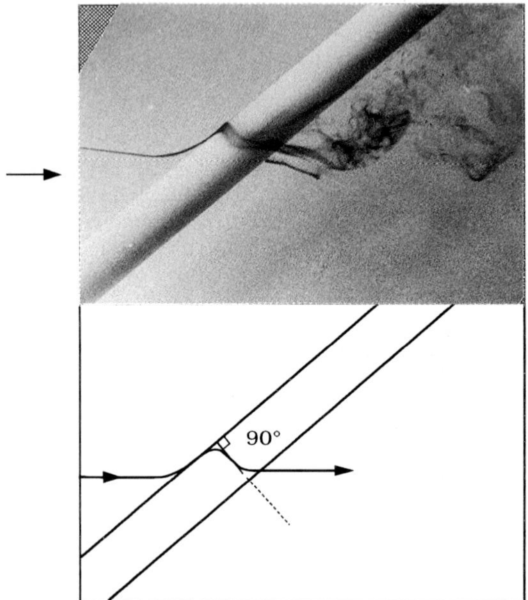

Figure 2.19 Visualization of flow past a circular cylinder in the case of oblique attack (θ being different from 90°). Kozakiewicz et al. (1995).

Regarding the lift, Kozakiewicz et al. (1995) report that the independence principle is valid also for the lift force for the tested range of θ for their force measurements, namely $45° \leq \theta \leq 90°$. They further report that the vortex shedding frequency (obtained from the lift-force spectra) is close to the value calculated from the Strouhal relationship. The lift force power spectrum becomes broader, however, as θ is decreased.

Kozakiewicz et al.'s (1995) study covers also the case of a near-bottom cylinder (the pipeline problem) with the gap between the cylinder and the bottom being 0.1 D in one case and nil in the other. Apparently, the independence principle is valid also for the near-bottom-cylinder situation for the tested range of $\theta(45° \leq \theta \leq 90°)$.

Finally, it may be noted that, although, theoretically, the independence principle is justified only in the subcritical range of Re, it has been proved to hold true also in the postcritical flows (Norton, Heideman and Mallard, 1981). However, there is evidence (Bursnall and Loftin, 1951) that for the transcritical values of Re the independence principle may not be applied.

2.7 Forces on a cylinder near a wall

The changes in the flow caused by the wall proximity is discussed in Section 1.2.1; these changes will obviously influence the forces acting on the cylinder.

This section will describe the effect of wall proximity on the forces on a cylinder placed near (or on) a wall. The following aspects of the problem will be examined: the drag force, the lift force, the oscillating components of the drag and the lift, and finally the forces on a pipeline placed in/over a scour trench.

Drag force on a cylinder near a plane wall

Fig. 2.20 depicts the pressure distributions around a cylinder placed at three different distances from a plane wall (Bearman and Zdravkovich, 1978). Fig. 2.21, on the other hand, presents the experimental data on the drag coefficient from the works by Kiya (1968), Roshko, Steinolffron and Chattoorgoon (1975), Zdravkovich (1985) and Jensen, Sumer, Jensen and Fredsøe (1990). The drag coefficient is defined in the same way as in Eq. 2.8.

The general trend is that the drag coefficient decreases with decreasing gap ratio near the wall. This result is consistent with the pressure distributions given in Fig. 2.20.

The differences between the various experiments in the figure may be attributed to the change in the Reynolds number.

One characteristic point in the variation of $\overline{C}_D$ with respect to e/D is that, as seen from the figure, $\overline{C}_D$ increases in a monotonous manner with increasing e/D up to a certain value of e/D, and then it remains reasonably constant for further increase in e/D (Fig. 2.22). This behaviour has been linked by Zdravkovich (1985) to the thickness of the boundary layer of the approaching flow: the flat portion of the curve occurs for such large gap ratios that the cylinder is embedded fully in the potential flow region. At lower gap ratios the cylinder is embedded partly in the potential flow region and partly in the boundary layer of the incoming flow. The curves belonging to Zdravkovich's (1985) data in Fig. 2.21 with two different values of δ/D, namely $\delta/D = 0.5$ and $\delta/D = 1$ where $\delta =$ the thickness of the boundary layer in the approaching flow, demonstrates this characteristic behaviour.

Lift force on a cylinder near a plane wall

The mean flow around a near-wall cylinder is not symmetric, therefore a non-zero mean lift must exist (in contrast to the case of a free cylinder). Fig. 2.20 shows that, while the mean pressure distribution around the cylinder is almost symmetric when $e/D = 1$, meaning that practically no lift exists, this symmetry

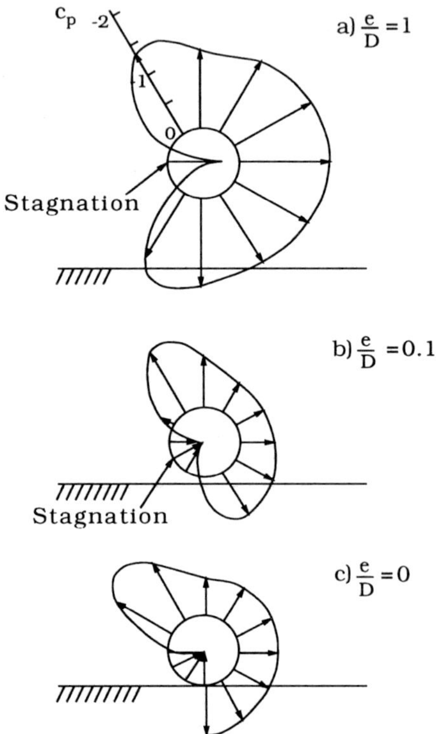

Figure 2.20 Pressure distributions on a cylinder near a wall as a function of gap ratio e/D. $c_p = (p - p_0)/(\frac{1}{2}\rho U^2)$ where p_0 is the hydrostatic pressure. Bearman and Zdravkovich (1978).

clearly disappears for the gap ratios $e/D = 0.1$ and 0, resulting in a non-zero mean lift on the cylinder. This lift, as seen from the figure, is directed away from the wall.

The variation of the lift force with respect to the gap ratio can best be described by reference to the simple case, the shear-free flow situation, depicted in Fig. 2.23. In the figure are plotted Fredsøe, Sumer, Andersen and Hansen's (1985) experimental data, Fredsøe and Hansen's (1987) modified potential-flow solution and also the potential-flow solution for a wall-mounted cylinder (see, for example, Yamamoto, Nath and Slotta (1974) for the latter). The shear-free flow in Fredsøe et al.'s study was achieved by towing the cylinder in still water. The $\overline{C}_L$ coefficient

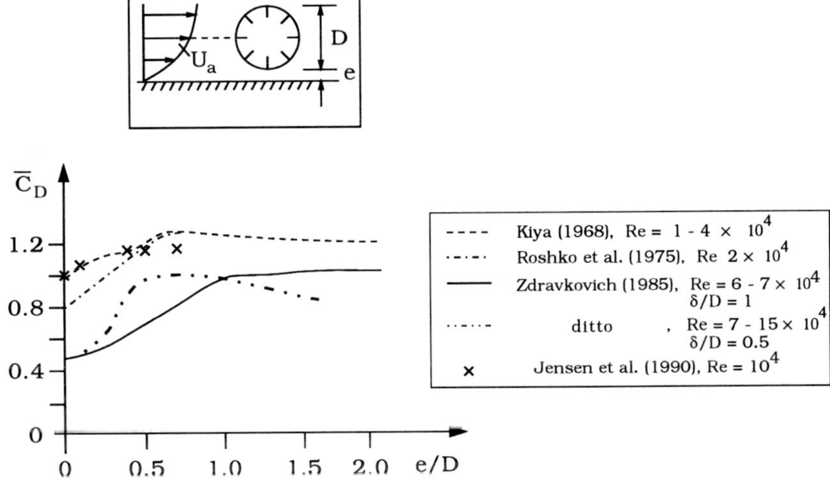

Figure 2.21 Drag coefficient for a cylinder near a plane wall, $\overline{C}_D = \overline{F}_D/(\frac{1}{2}\rho U_a^2 D)$. In the figure δ is the boundary-layer thickness of the approaching flow.

plotted in the figure is defined by

$$\overline{F}_L = \frac{1}{2}\rho \overline{C}_L D U^2 \tag{2.15}$$

where $\overline{F}_L$ is the mean lift force on the cylinder, and the positive lift means that it is directed away from the wall.

The figure indicates that while the lift is fairly small for gap ratios such as $e/D = 0.2 - 0.3$, it increases tremendously as the gap ratio is decreased. This is because, as mentioned previously, 1) the stagnation point moves to lower and lower angular positions, as the gap is decreased (Fig. 2.24); also, 2) the suction on the free-stream side of the cylinder becomes larger and larger with decreasing gaps. The combined action of these two effects result in larger and larger lift forces, as the cylinder is moved towards the wall.

Regarding the potential-flow solution plotted in Fig. 2.23, the potential flow solution for a wall-mounted cylinder was given by von Müller (1929) in closed form as $F_L = \rho U^2 D\pi(\pi^2 + 3)/18$, which gives a lift force directed away from the wall with a lift coefficient $C_L = 4.49$, as seen in the figure. When the cylinder is placed a small distance away from the wall, however, the potential flow solution gives a negative lift, Yamamoto et al. (1974), Fredsøe and Hansen (1987). Fredsøe and

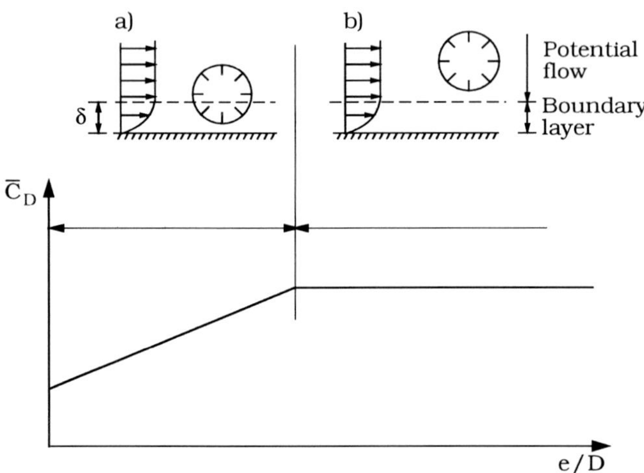

Figure 2.22 Schematic variation of drag coefficient with the gap ratio.

Hansen modified the potential flow solution by superposing a vortex body around the cylinder onto the existing potential flow such that the velocity at the top and at the bottom of the cylinder becomes equal, in accordance with the experimental observation which is referred to in the same study. Fredsøe and Hansen's modified potential-flow solution, as is seen from Fig. 2.23, agrees quite satisfactorily with the experimental results.

When a shear is introduced in the approaching flow, the variation of the lift force with respect to the gap ratio changes considerably very close to the wall, as seen in Fig. 2.25, where $\overline{C}_L$ is defined by Eq. 2.15 with U replaced by U_a, the undisturbed flow velocity at the level of the cylinder axis. The shear-flow data plotted in this figure were obtained in an experiment conducted at practically the same Reynolds number, employing the same test cylinder as in Fig. 2.23. The only difference between the two tests is that in the shear-free flow experiments the cylinder was towed in still water, while in the shear-flow experiments the cylinder was kept stationary and subject to the boundary-layer flow established in an open channel with a smooth bottom.

Clearly, the difference observed in Fig. 2.25 in the $\overline{C}_L$ versus e/D behaviour is due to the shear in the approaching flow. The lift undergoes a substantial drop for very small gap ratios. Fredsøe and Hansen (1987) links this drop to the change in the stagnation pressure in the following way: First they show that the stagnation point does not move significantly by the introduction of the shear. So the direction of pressure force is much the same in both cases. The major difference is that the stagnation pressure is reduced considerably with the introduction of the shear,

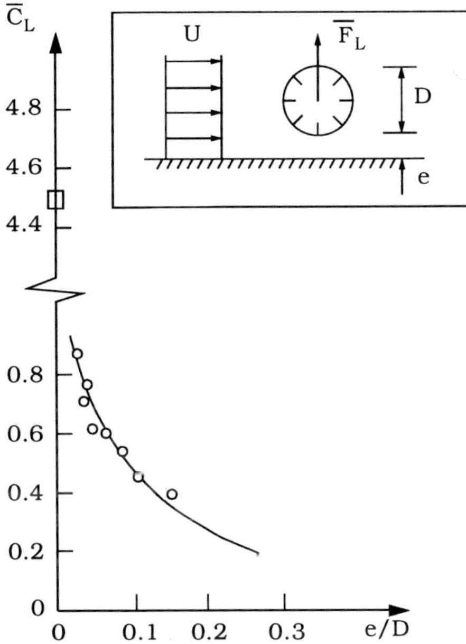

Figure 2.23 Lift force for a cylinder in a shear-free flow $\overline{C}_L = \overline{F}_L/(\frac{1}{2}\rho U^2 D)$. Circles: Experiments, $10^4 < Re < 3\times10^4$ (Fredsøe et al., 1985). Solid curve: Fredsøe and Hansen's (1987) modified potential-flow solution. Square: Potential-flow solution (see for example Yamamoto et al., 1974).

as sketched in Fig. 2.26; while the stagnation pressure in the shear-free flow, implementing the Bernoulli equation and taking the far-field pressure, is equal to

$$p = \frac{1}{2}\rho U^2 \, , \tag{2.16}$$

the same quantity in the case of shear flow, to a first approximation, is

$$p = \frac{1}{2}\rho U_s^2 \tag{2.17}$$

where U_s is the far-field flow velocity associated with the stagnation streamline.

Clearly, the pressure in Eq. 2.17 is much smaller than that in Eq. 2.16 (Fig. 2.26). This reduction in the stagnation pressure, while keeping the direction of

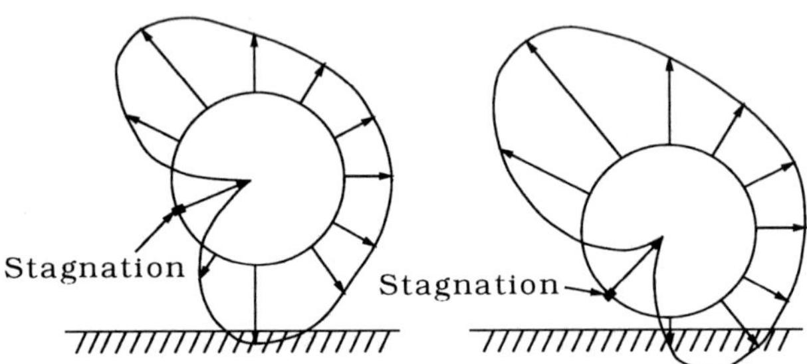

Figure 2.24 Sketches showing the changes in the stagnation point and the pressure distribution, as the cylinder is moved towards the wall: The stagnation point moves to lower and lower angular positions, and the suction on the free-stream side of the cylinder becomes larger and larger than that on the wall side.

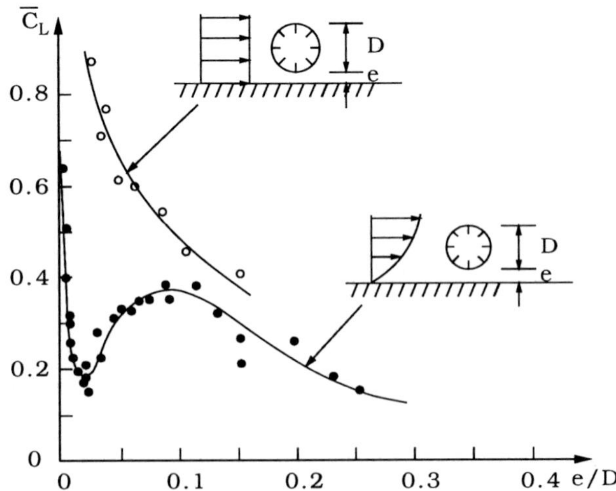

Figure 2.25 Comparison of $\overline{C}_L$ in shear-free and shear flows, $10^3 < Re < 3 \times 10^4$. The boundary-layer thickness to diameter ratio $\delta/D = 5$. In the shear flow case $\overline{C}_L$ is defined by $\overline{F}_L = \frac{1}{2}\rho\overline{C}_L D U_a^2$ where U_a is the undisturbed velocity at the axis of the cylinder. Fredsøe et al. (1985).

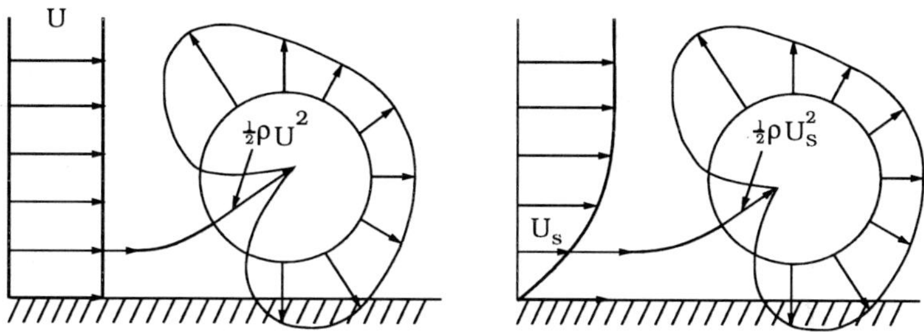

Figure 2.26 Comparison of shear-free and shear flows. Stagnation pressure
decreases considerably in the shear-flow case.

pressure forces unchanged, presumably causes the lift to be reduced substantially
in the case of shear flow.

When the cylinder is moved extremely close to the wall, however, more and
more fluid will be diverted to pass over the cylinder, which will lead to larger and
larger suction pressure on the free-stream side of the cylinder. Indeed, when the
cylinder is sitting on the wall, the suction pressure on the cylinder surface will be
the largest (Fig. 2.20c). This effect may restore the lift force in the shear-flow
case for very small gap values, as is implied by Fig. 2.25.

Fig. 2.27 presents data regarding the lift on a cylinder in a shear flow
obtained at different Reynolds numbers.

Oscillating drag and lift on a cylinder near a plane wall

The vortex-induced, oscillating lift and drag will cease to exist in the case
when the gap ratio is smaller than about 0.3, simply because the vortex shedding
is suppressed for these gap ratios (Section 1.2.1).

Although the shedding exists for gap ratios larger than 0.3, it will, however,
be influenced by the close proximity of the wall when e/D is not very large.
Therefore the oscillating forces will be affected, too, by the close proximity of the
wall. Fig. 2.28 illustrates this influence regarding the $r.m.s.$-value of the oscillating
lift force. The figure shows that the oscillating lift becomes weaker and weaker, as
the gap ratio is decreased. Note that the C_L' coefficient here is defined in the same
way as in Eq. 2.11 provided that U is replaced by the velocity U_a, the undisturbed
flow velocity at the level of the cylinder axis.

Finally, Fig. 2.29 compares the vortex-shedding induced oscillating lift with
the mean lift caused by the wall proximity. The C_L coefficient plotted in the figure
representing the vortex-induced oscillating lift is the lift coefficient associated with
the maximum value of the oscillating lift force. As is seen from the figure, the
wall-induced lift and the vortex-induced lift appear to be in the same order of

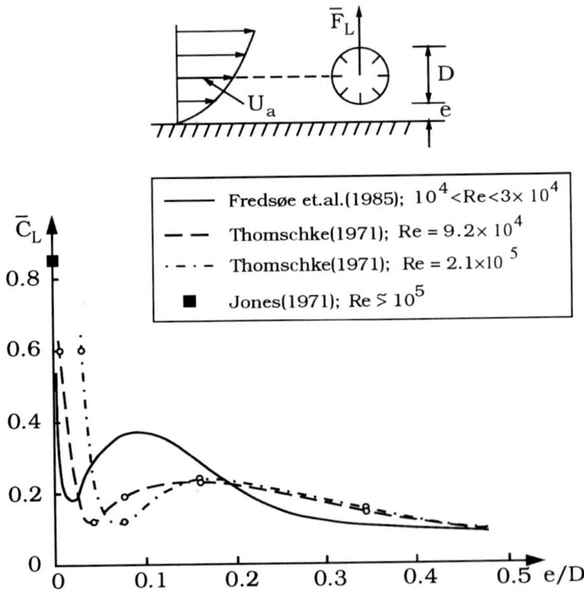

Figure 2.27 Lift force on a near-wall cylinder in a shear flow.
$$\overline{C}_L = \overline{F}_L / (\tfrac{1}{2}\rho U_a^2 D).$$

magnitude in the neighbourhood of $e/D = 0.3$. With decreasing values of e/D, however, the wall-induced lift increases quite substantially. The figure further indicates that, with e/D larger than 0.3 up to 0.4 - 0.5, the two effects, namely the wall-induced steady lift force and the vortex-induced oscillating lift force, may be present concurrently, meaning that, while the cylinder undergoes a steady lift, it will also be subject to an oscillating lift force induced by vortex shedding.

Forces on a pipeline in/over a scour trench

As mentioned in Section 1.2.1, when a pipeline is placed on an erodible bed, scour may occur below the pipe due to flow action, leading to suspended spans of the pipeline. Jensen et al. (1990) investigated the flow around and forces on a pipeline (placed initially on a flat bed) at five characteristic stages of the scour process. The results regarding the flow description have been given in Section 1.2.1 under the heading "Effect of wall proximity" (Figs. 1.24 and 1.25). Fig. 2.30 gives the force coefficients obtained in the same study. The force coefficients are defined, based on the undisturbed velocity at the axis of the pipe. As mentioned in the flow

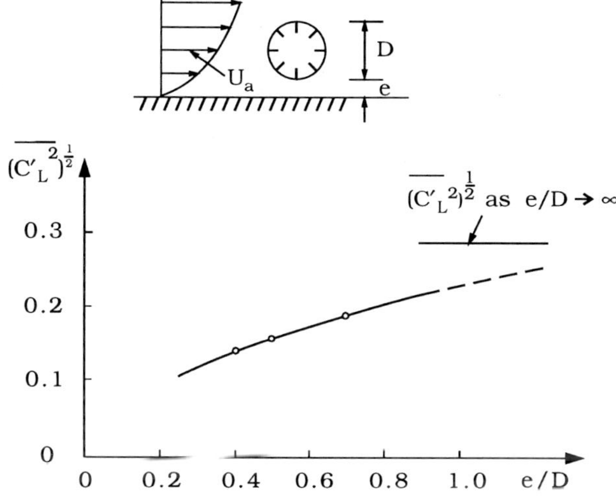

Figure 2.28 *R.m.s.*-value of oscillating lift coefficient. $C'_L = F'_L/(\frac{1}{2}\rho U_a^2 D)$. $Re = 10^4$. Circles: Jensen et al. (1990). Asymptotic value for $e/D = \infty$ from Schewe (1983).

description, each profile corresponds to a particular instant in the course of the scour process from which the profiles are taken. It is interesting to note that $\overline{C}_D$ and $\overline{C}_L$ reach their equilibrium values at rather early stages of the scour process. It is also interesting to observe that the pipe experiences a negative lift force as soon as the tunnel erosion (Stage II) comes into action. It is seen that this lift force remains negative throughout the scour process.

The negative lift in Stage II can be attributed to the strong suction below and behind the cylinder caused by the gap flow, which is also the cause of the relatively high value of $\overline{C}_D$ obtained for Stage II. As for Stage V, the negative lift can be explained by the position of the stagnation point and the angle of attack of the approaching flow. This angle can in Fig. 1.24 be found to be around 10–15 degrees, which fits well with the angle of the resultant force vector with respect to the horizontal.

The phenomenon, namely the "premature" vortex shedding, which causes the high Strouhal numbers in the initial stages of the scour process (Stages III and IV in Fig. 1.25), is also the main cause of the variation in the mean double amplitude of the fluctuating lift force: the larger the strength of the vortices shed, the larger the fluctuating lift force. Since the vortices shed from the pipe become stronger and stronger as the scour progresses, the fluctuating lift force should

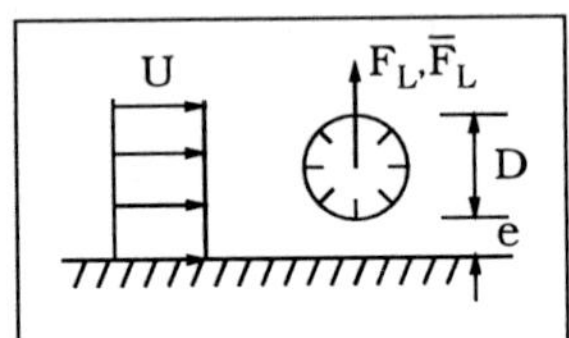

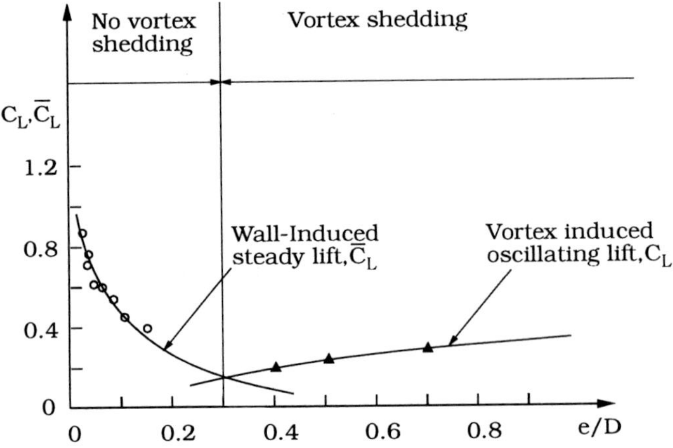

Figure 2.29 Force coefficients of the mean lift force $\left(\overline{C}_L\right)$ and the oscillating lift force $\left(C_L\right)$ on cylinder as a function of the gap ratio. The coefficient C_L is based on the amplitude of the oscillating lift force.

correspondingly increase, as indicated by Fig. 2.30c.

Figure 2.31 compares the results presented in Fig. 2.30 with those obtained with a plane bed in the same study. The plane-bed counterpart of each scour profile is selected on the basis of equal non-dimensional clearance between the pipe and the bed (i.e., equal to e/D, see Fig. 2.31).

As seen from the figure, $\overline{C}_D$ is not affected much, whether the bed is a plane bed or a scoured one. As for the mean lift coefficient $\overline{C}_L$, the difference between a plane bed and a scoured bed is that the pipe experiences a negative lift force in the case of a scoured bed, while it experiences a positive one when the bed is plane (Fig. 2.25).

As for the fluctuating lift force C_L' there is practically no difference between a plane bed and a scoured bed for large values of e/D. However, this is not the case for small values of e/D, where the effect of upstream slope of the dune behind

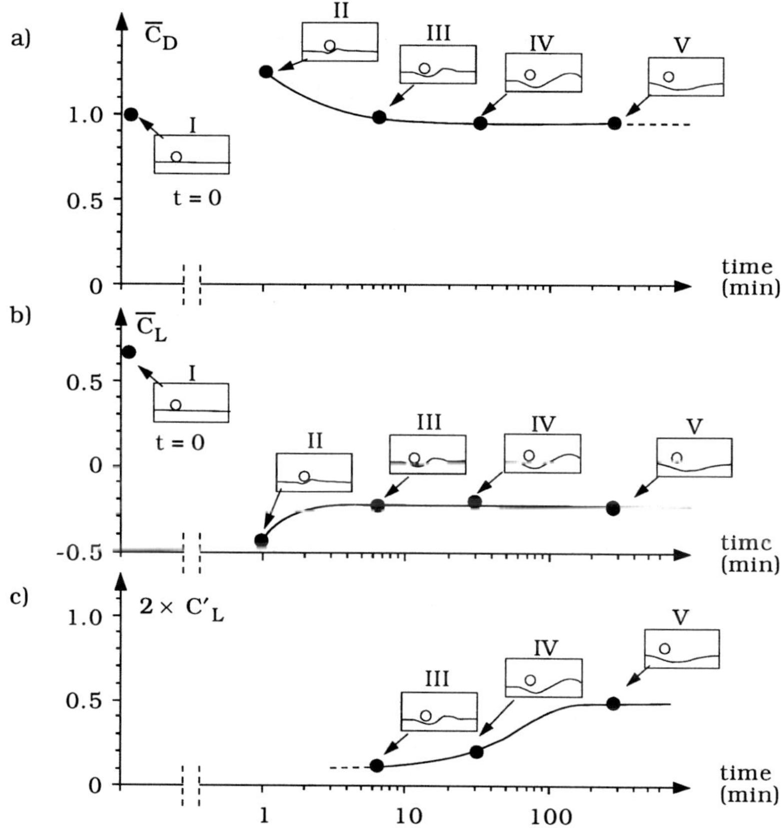

Figure 2.30 Time development of force during the scour process below a pipeline. (a) mean drag coefficient; (b) mean lift coefficient; (c) mean double amplitude of the fluctuating lift force. Jensen et al. (1990).

the pipeline is felt very strongly in the vortex-shedding process, as explained in the flow description in conjunction with Fig. 1.25 in Section 1.2.1.

Stansby and Starr (1992) report the results of measurements of drag on a pipe undergoing a gradual sinking, as the scour process progresses in a live, sand bed. According to Stansby and Starr, the drag coefficient is reduced from $\overline{C}_D = 1$ for a pipe sitting on the bed to $\overline{C}_D = 0.3$–0.4 when the pipe sinked in the sand to a level of about $e/D = -0.6$. This is obviously due to the fact that the pipe is protected against the flow, as it is buried in the sand bed.

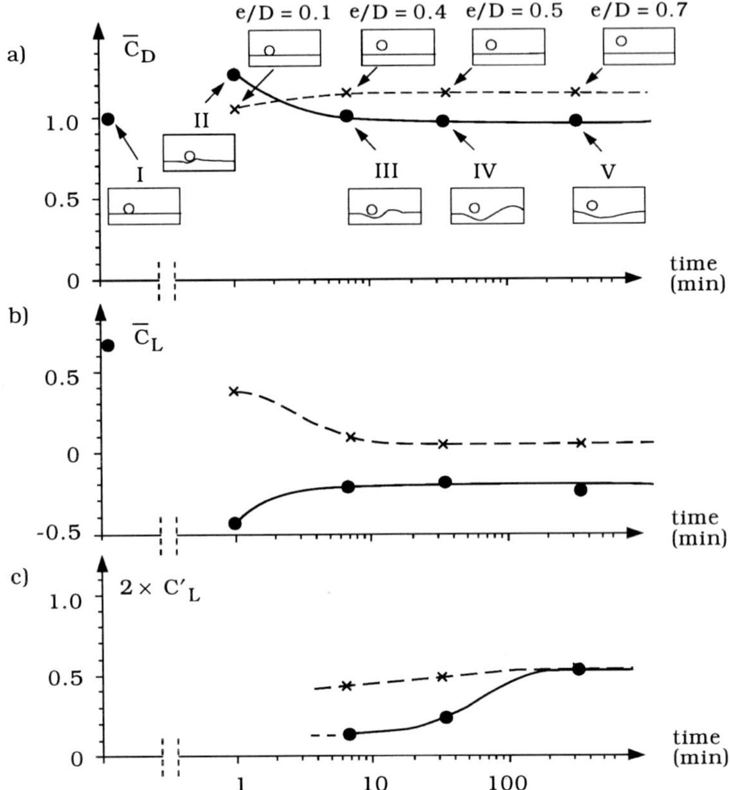

Figure 2.31 Comparison of forces between the cases of a cylinder over a scoured and a plane bed. (a) mean drag coefficient; (b) mean lift coefficient; (c) mean double amplitude of the fluctuating lift force. Jensen et al. (1990).

When the pipelines are placed in a trench hole, the forces are reduced considerably (Fig. 2.32). As seen, both the drag and the lift are reduced by a factor 5–10, depending on the position of the pipe in the trench hole. This is because the pipe is protected against the main body of the flow by the trench *(sheltering effect)*. Jensen and Mogensen (1982) report that in the case of a trench hole the same size as that in Fig. 2.32 but with a much steeper slope (namely 1:1), the reduction in the forces is even much larger.

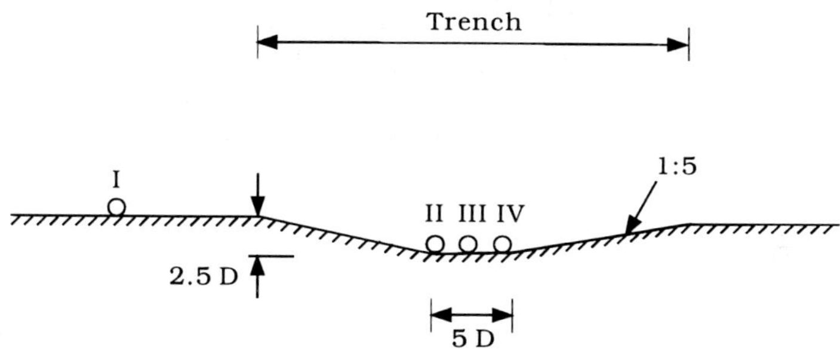

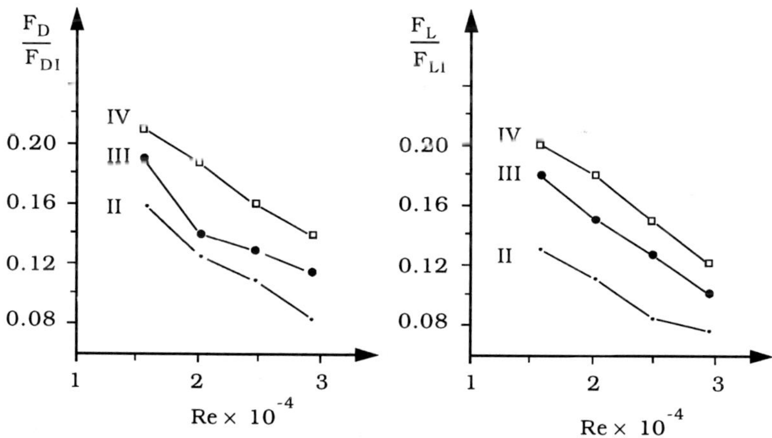

Figure 2.32 Relative drag and lift forces on a pipeline placed in a trench for several positions (Positions II, III and IV). F_{DI} and F_{LI} are the corresponding forces on the same pipeline sitting on a flat bed (Position I). Jensen and Mogensen (1982).

REFERENCES

Achenbach, E. (1968): Distribution of local pressure and skin friction around a circular cylinder in cross-flow up to $Re = 5 \times 10^5$. J. Fluid Mech., 34(4):625-639.

Achenbach, E. (1971): Influence of surface roughness on the cross-flow around a circular cylinder. J. Fluid Mech., 46:321-335.

Achenbach, E. and Heinecke E. (1981): On vortex shedding from smooth and rough cylinders in the range of Reynolds numbers 6×10^3 to 5×10^6. J. Fluid Mech., 109:239-251.

Bearman, P.W. and Zdravkovich, M.M. (1978): Flow around a circular cylinder near a plane boundary. J. Fluid Mech., 89(1):33-48.

Bursnall, W.J. and Loftin, L.K. (1951): Experimental Investigation of the Pressure Distribution about a Yawed Circular Cylinder in the Critical Reynolds Number Range. NACA, Technical Note 2463.

Cheung, J.C.K. and Melbourne, W.H. (1983): Turbulence effects on some aerodynamic parameters of a circular cylinder at supercritical Reynolds numbers. J. of Wind Engineering and Industrial Aerodynamics, 14:399-410.

Drescher, H. (1956): Messung der auf querangeströmte Zylinder ausgeübten zeitlich veränderten Drücke. Z. f. Flugwiss, 4(112):17-21.

Fredsøe, J. and Hansen, E.A. (1987): Lift forces on pipelines in steady flow. J. Waterway, Port, Coastal and Ocean Engineering, ASCE, 113(2):139-155.

Fredsøe, J., Sumer, B.M., Andersen, J. and Hansen, E.A. (1985): Transverse vibrations of a cylinder very close to a plane wall. Proc. 4th Symposium on Offshore Mechanics and Arctic Engineering, OMAE, Dallas, TX, 1:601-609. Also, Trans. of the ASME, J. Offshore Mechanics and Arctic Engineering, 109:52-60.

Güven, O., Patel, V.C. and Farell, C. (1975): Surface roughness effects on the mean flow past circular cylinders. Iowa Inst. Hydraulic Res., Rep. No. 175.

Güven, O., Patel, V.C. and Farell, C. (1977): A model for high-Reynolds-number flow past rough-walled circular cylinders. Trans. ASME, J. Fluids Engrg., 99:486-494.

Güven, O., Farell, C. and Patel, V.C. (1980): Surface-roughness effects on the mean flow past circular cylinders. J. Fluid Mech., 98(4):673-701.

Hallam, M.G., Heaf, N.J. and Wootton, L.R. (1977): Dynamics of Marine Structures. CIRIA Underwater Engineering Group, Report UR8, Atkins Research and Development, London, U.K.

Hoerner, S.F. (1965): Fluid-Dynamic Drag. Practical Information on Aerodynamic Drag and Hydrodynamic Resistance. Published by the Author. Obtainable from ISVA.

Jensen, R. and Mogensen, B. (1982): Hydrodynamic forces on pipelines placed in a trench under steady current conditions. Progress Report No. 57, Inst. of Hydrodynamics and Hydraulic Engineering, ISVA, Techn. Univ. Denmark, pp. 43-50.

Jensen, B.L., Sumer, B.M., Jensen, H.R. and Fredsøe, J. (1990): Flow around and forces on a pipeline near a scoured bed in steady current. Trans. of the ASME, J. of Offshore Mech. and Arctic Engrg., 112:206-213.

Jones, W.T. (1971): Forces on submarine pipelines from steady currents. Paper presented at the Petroleum Mechanical Engineering with Underwater Technology Conf., Sept. 19-23, 1971, Houston, Texas, Underwater Technology Div., ASME.

Kiya, M. (1968): Study on the turbulent shear flow past a circular cylinder. Bulletin Faculty of Engrg., Hokkaido University, 50:1-100.

Kozakiewicz, A., Fredsøe, J. and Sumer, B.M. (1995): Forces on pipelines in oblique attack. Steady current and waves. Proc. 5th Int. Offshore and Polar Engineering Conf., The Hague, Netherlands, June 11-16, 1995, II:174-183.

Kwok, K.C.S. (1986): Turbulence effect on flow around circular cylinder. J. Engineering Mechanics, ASCE, 112(11):1181-1197.

Müller, W. von (1929): Systeme von Doppelquellen in der ebenen Strömung, insbesondere die Strömung um zwei Kreiszylinder. Zeitschrift für angewandte Mathematik und Mechanik, 9(3):200-213.

Norberg, C. and Sundén, B. (1987): Turbulence and Reynolds number effects on the flow and fluid forces on a single cylinder in cross flow. Jour. Fluids and Structures, 1.337-357.

Norton, D.J., Heideman, J.C. and Mallard, W.W. (1981): Wind tests of inclined circular cylinders. Proc. 13th Annual OTC in Houston, TX, May 4-7, OTC 4122, pp. 67-70.

Parkinson, G.V. and Brooks, N.P.H. (1961): On the aeroelastic instability of bluff cylinders. J. Appl. Mech., 28:252-258.

Roshko, A., Steinolffron, A. and Chattoorgoon, V. (1975): Flow forces on a cylinder near a wall or near another cylinder. Proc. 2nd US Conf. Wind Engrg. Research, Fort Collins, Co., Paper IV-15.

Schewe, G. (1983): On the force fluctuations acting on a circular cylinder in cross-flow from subcritical up to transcritical Reynolds numbers. J. Fluid Mech., 133:265-285.

Schlichting, G. (1979): Boundary Layer Theory. 7.ed. McGraw-Hill Book Company.

Shih, W.C.L., Wang, C., Coles, D. and Roshko, A. (1993): Experiments on flow past rough circular cylinders at large Reynolds numbers. J. Wind Engrg. and Industrial Aerodynamics, 49:351-368.

Stansby, P.K. and Starr, P. (1992): On a horizontal cylinder resting on a sand bed under waves and current. Int. J. Offshore and Polar Engrg., 2(4):262-266.

Thom, A. (1929): An investigation of fluid flow in two dimensions. Aero. Res. Counc. London, R. and M. No. 1194, pp. 166-183.

Thomschke, H. (1971): Experimentelle Untersuchung der stationären Umströmung von Kugel und Zylinder in Wandnähe. Fakultät für Maschinenbau der Universität Karlsruhe, Karlsruhe, West Germany.

Yamamoto, T., Nath, J.H. and Slotta, L.S. (1974): Wave forces on cylinders near plane boundary. J. Waterway, Port, Coastal Ocean Div., ASCE, 100(4):345-360.

Zdravkovich, M.M. (1985): Forces on a circular cylinder near a plane wall. Applied Ocean Research, 7:197-201.

Chapter 3. Flow around a cylinder in oscillatory flows

As shown in Chapter 1, the hydrodynamic quantities describing the flow around a smooth, circular cylinder in steady currents depend on the Reynolds number. In the case where the cylinder is exposed to an oscillatory flow an additional parameter - the so-called Keulegan-Carpenter number - appears. The Keulegan-Carpenter number - the KC number - is defined by

$$KC = \frac{U_m T_w}{D} \qquad (3.1)$$

in which U_m is the maximum velocity and T_w is the period of the oscillatory flow. If the flow is sinusoidal with the velocity given by

$$U = U_m \sin(\omega t) \qquad (3.2)$$

then the maximum velocity will be

$$U_m = a\omega = \frac{2\pi a}{T_w} \qquad (3.3)$$

where a is the amplitude of the motion. For the sinusoidal case the KC number will therefore be identical to

$$KC = \frac{2\pi a}{D} \qquad (3.4)$$

The quantity ω in Eq. 3.2 is the angular frequency of the motion

$$\omega = 2\pi f_w = \frac{2\pi}{T_w} \tag{3.5}$$

in which f_w is the frequency.

The physical meaning of the KC number can probably be best explained by reference to Eq. 3.4. The numerator on the right-hand-side of the equation is proportional to the stroke of the motion, namely $2a$, while the denominator, the diameter of the cylinder D, represents the width of the cylinder (Fig. 3.1). Small KC numbers therefore mean that the orbital motion of the water particles is small relative to the total width of the cylinder. When KC is very small, separation behind the cylinder may not even occur.

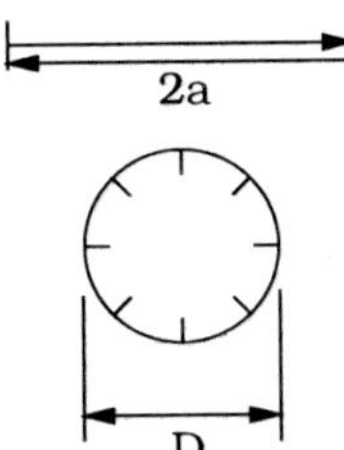

Figure 3.1 Definition sketch.

Large KC numbers, on the other hand, mean that the water particles travel quite large distances relative to the total width of the cylinder, resulting in separation and probably vortex shedding. For very large KC numbers ($KC \rightarrow \infty$), we may expect that the flow for each half period of the motion resembles that experienced in a steady current.

3.1 Flow regimes as a function of KC number

Fig. 3.2 summarizes the changes that occur in the flow as the Keulegan-Carpenter number is increased from zero. The picture presented in the figure is for $Re = 10^3$ in which Re is defined as

$$Re = \frac{DU_m}{\nu} \tag{3.6}$$

(As Re is changed, the flow regimes shown in Fig. 3.2 may also change, accompanied by possible changes at the upper and lower limits of the indicated KC

a) 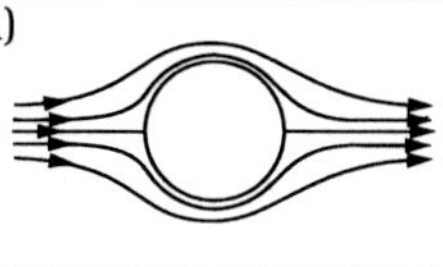	No separation. Creeping (laminar) flow.	KC < 1.1
b) 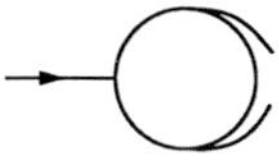	Separation with Honji vortices. See Figs. 3.3 and 3.4	1.1 < KC < 1.6
c) 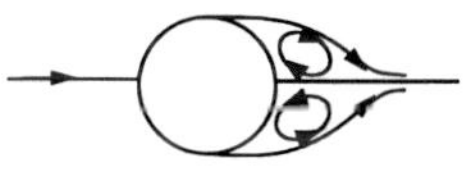	A pair of symmetric vortices	1.6 < KC < 2.1
d) 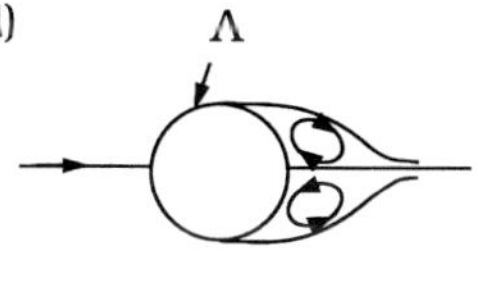	A pair of symmetric vortices. Turbulence over the cylinder surface (A).	2.1 < KC < 4
e) 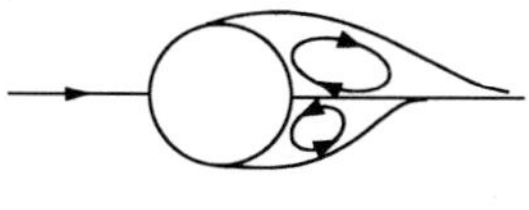	A pair of asymmetric vortices	4 < KC < 7
f) 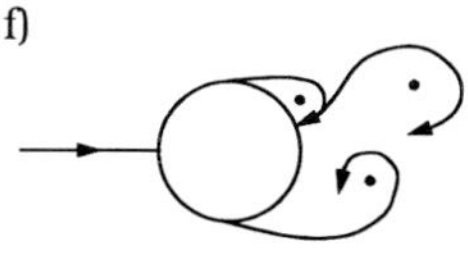	Vortex shedding	7 < KC Shedding regimes

Figure 3.2 Regimes of flow around a smooth, circular cylinder in oscillatory flow. $Re = 10^3$. Source for $KC < 4$ is Sarpkaya (1986a) and for $KC > 4$ Williamson (1985). Limits of the KC intervals may change as a function of Re (see Figs. 3.15 and 3.16).

ranges. We shall concentrate our attention first on the KC dependence, however. The influence of Re will be discussed in Section 3.3).

As seen from Fig. 3.2, for very small values of KC, no separation occurs, as expected. The separation first appears when KC is increased to 1.1; this occurs in the form of the so-called **Honji instability** (Figs. 3.3 and 3.4). When this KC number is reached, the purely two-dimensional flow over the cylinder surface breaks into a three-dimensional flow pattern where equally-spaced, regular streaks are formed over the cylinder surface, as sketched in Fig. 3.3. These streaks can be made visible by flow-visualization techniques. Observations show that the marked fluid particles, which were originally on the surface of the cylinder, would always end up in these narrow, streaky flow zones. The observations also show that these streaks eventually are subject to separation in every half period prior to the flow reversal, each separated streak being in the form of a mushroom-shape vortex (Figs.3.3 and 3.4). This phenomenon was first reported by Honji (1981) and later by Sarpkaya (1986a). Subsequently, Hall (1984) carried out a linear stability analysis and showed that the oscillatory viscous flow becomes unstable to axially periodic vortices (i.e. Vortices B in Fig. 3.3) above a critical KC number for a given Re, validating the experimentally observed flow instability.

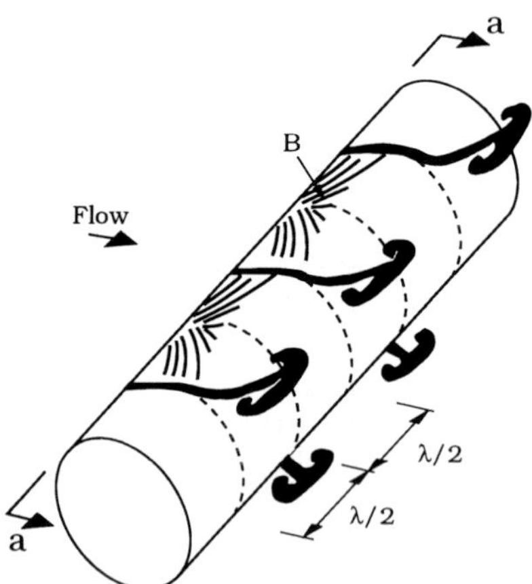

Figure 3.3 Honji streaks, which are subject to separation in the form of mushroom-shape vortices; see the photograph in Fig. 3.4, viewed in a-a, for the separated, mushroom-shape vortices.

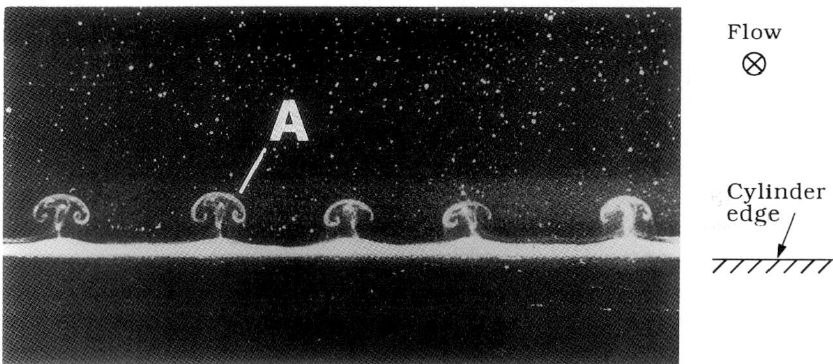

Figure 3.4 Separated mushroom-shape vortices (A) viewed in a-a indicated in Fig. 3.3. Oscillatory flow is in the direction perpendicular to this page. From Honji (1981) with permission – see Credits.

The flow regime where separation takes place in the form of Honji instability occurs in a narrow KC interval, namely $1.1 < KC < 1.6$ (Fig. 3.2b). With a further increase of KC number, however, separation begins to occur in the form of a pair of symmetric, ordinary, attached vortices as indicated in Fig. 3.2c and d. This regime covers the KC range $1.6 < KC < 4$ with the subrange $2.1 < KC < 4$ where turbulence is observed over the cylinder (Sarpkaya, 1986a). It must be remembered that the limits for the indicated KC ranges in the figure are those for $Re = 1000$.

When KC is increased even further, the symmetry between the two attached vortices breaks down. (The vortices are still attached, and no shedding occurs, however). This regime prevails over the KC range $4 < KC < 7$ (Fig. 3.2e). The significance of this regime is that the lift force is no longer nil, and this is due to the asymmetry in the formation of the attached vortices.

Fig. 3.5 illustrates the time evolution of vortex motions as the flow progresses for the regimes where separation occurs in the form of a pair of symmetrically attached vortices, namely for the KC range $1.6 < KC < 4$ (Fig. 3.2c and d). The arrows in the figure refer to the cylinder motion in an otherwise still fluid. As seen from the figure, the vortices which form behind the cylinder (Vortex M) are washed over the cylinder by the end of the previous half period and form a pair of vortex pairs with the newly formed vortices (Vortex N) which would eventually move away from the cylinder due to the self-induced velocity fields of the vortex pairs.

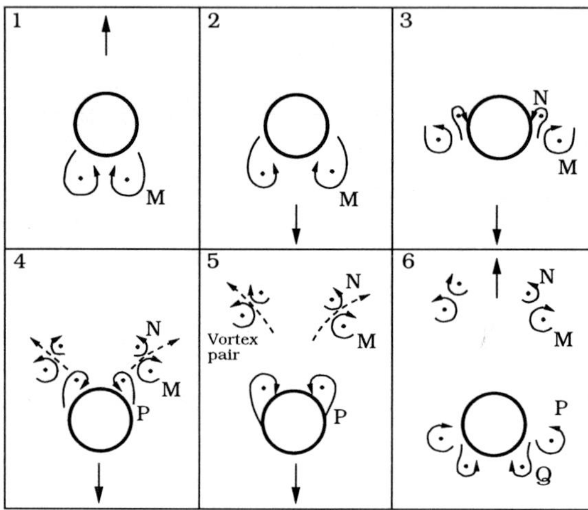

Figure 3.5 $1.6 < KC < 4$. $Re = 10^3$. Evolution of vortex motions for the regime with a pair of separation vortices (Fig. 3.2c-d). Arrows refer to cylinder motion. The vortices are viewed from a fixed camera. Williamson (1985).

Returning to Fig. 3.2, with a further increase of the Keulegan-Carpenter number, we come to the so-called **vortex-shedding regimes** ($KC > 7$) (Fig. 3.2f). The following section will focus on these flow regimes.

3.2 Vortex-shedding regimes

The vortex-shedding regimes have been investigated extensively by, among others, Bearman, Graham and Singh (1979), Singh (1979), Grass and Kemp (1979), Bearman and Graham (1979), Bearman, Graham, Naylor and Obasaju (1981) and more recently by Williamson (1985). These works have shed considerable light on the understanding of the complex behaviour of vortex motions in various regimes. Based on the previous research and his own work, Williamson (1985) has described the vortex trajectory patterns in quite a systematic manner. The following description is mainly based on Williamson (1985).

In the vortex-shedding regimes the vortex shedding occurs during the course of each half period of the oscillatory motion. There are several such regimes, each of which has a different vortex flow pattern, observed for different ranges of the

7<KC<15. Single-Pair regime

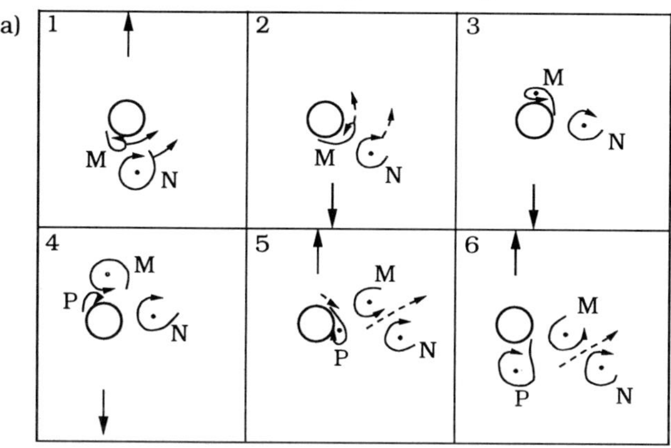

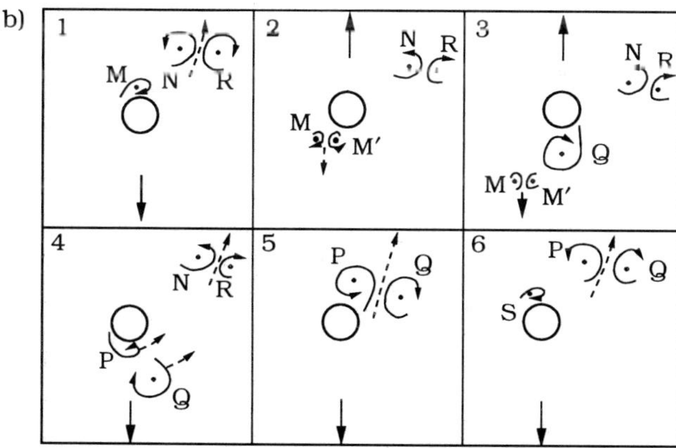

Figure 3.6 a) $7 < KC < 13$. The arrows refer to cylinder motion, but the vortices are viewed from a reference frame which moves with the cylinder. The wake consists of a series of vortices convecting out to one side of the cylinder in the form of a street (the transverse vortex street).

b) $13 < KC < 15$. The wake consists of a series of pairs convecting away each cycle at around 45° to the flow oscillation direction, and on one side of the cylinder only. Both in (a) and in (b), there is always one pair of vortices which convect away from the cylinder. Williamson (1985).

KC number. These KC ranges are $7 < KC < 15$, $15 < KC < 24$, $24 < KC < 32$, $32 < KC < 40$, etc.

$7 < \mathbf{KC} < 15$ (single-pair regime)

Fig. 3.6 illustrates the time development of vortex motions in this regime. The major portion of the KC range, namely $7 < KC < 13$ (Fig. 3.6a), is known as the transverse-vortex-street regime.

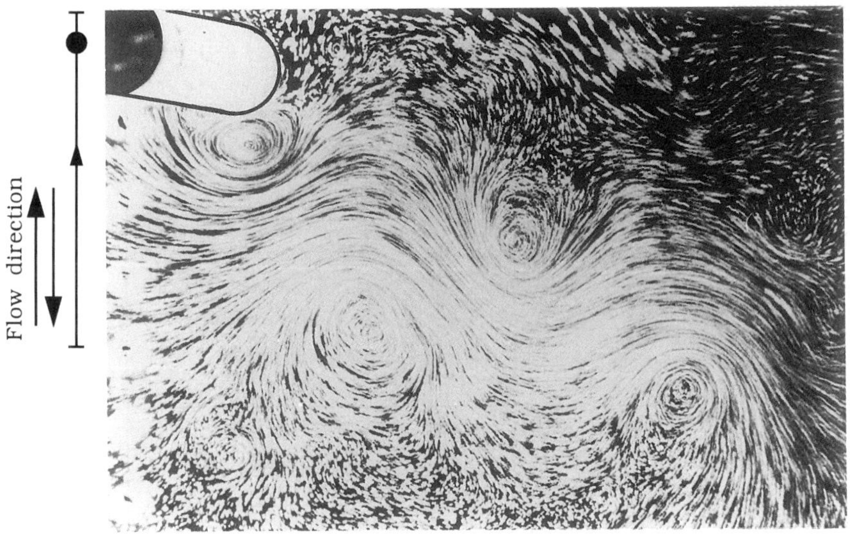

Figure 3.7 Transverse-street wake for $KC = 12$. In this photograph the cylinder is moving up, and is near the end of a half cycle. Due to the induced velocities of the main vortices, one of which is shed in each half cycle, the trail of vortices convects away at around 90° to the oscillation direction in the form of a street. In this case the street travels to the right. Williamson (1985) with permission - see Credits.

Fig. 3.6a, Frame 1, indicates that Vortex N has just been shed and there is a growing vortex (Vortex M) at the other side of the cylinder. When the flow reverses (Fig. 3.6a, Frame 2), both vortices are washed over the cylinder. As the half period progresses, Vortex M itself is shed and, being a free vortex, it forms a vortex pair with Vortex N (Fig. 3.6a, Frame 4). The vortex pair M +

N will then move away from the cylinder under its self-induced velocity field. As implied in the preceding, the concept "pairing" here means that two vortices, of opposite sign, come together and each is convected by the velocity field of the other. It is evident from the figure that there will be one vortex pair convecting away from the cylinder at the end of each full period. This would apparently lead to a **transverse vortex street** (i.e., a vortex street in the direction perpendicular to the flow direction), as depicted in Fig. 3.7: in this figure, the vortex street is formed at the lower side of the cylinder. Observations show, however, that the vortex street changes sides occasionally. The position of the vortex street relative to the cylinder may be important from the point of view of the lift force acting on the cylinder. Due to the asymmetry, a non-zero mean lift must exist in this flow regime. When the vortex street changes side, then the direction of this lift force will change correspondingly.

$15 < KC < 24$. Double-Pair regime

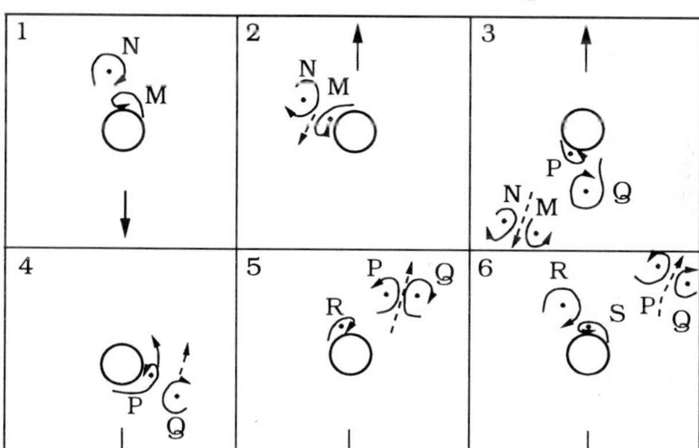

Figure 3.8 $15 < KC < 24$. The arrows refer to cylinder motion, but the vortices are viewed from a reference frame which moves with the cylinder. The wake is the result of two vortices being shed in each half cycle. Two trails of vortex pairs convect away from the cylinder in opposite directions and from opposite sides of the cylinder (for example vortices N + M and P + Q). Williamson (1985).

Regarding the second portion of the KC range, namely $13 < KC < 15$ (Fig. 3.6b), the pattern of vortex motions changes somewhat in this range of KC number; the pairs now convect away at around 45° to the flow direction, and this occurs at one side only.

From both Fig. 3.6a and Fig. 3.6b it is seen that there is always one vortex pair convecting away from the cylinder in one period of the motion; this is, for example in Fig. 3.6a, the pair M + N, while in Fig. 3.6b it is N + R in the first period and P + Q in the following period. For this reason Williamson (1985) calls this regime ($7 < KC < 15$) the single-pair regime.

$15 < KC < 24$ (double-pair regime) and further KC regimes

Fig. 3.8 gives the time development of vortex motions in the case when $15 < KC < 24$, while Fig. 3.9 gives that in the case when $24 < KC < 32$. The detailed descriptions are given in the figure captions. However, it is readily seen that there are two vortex pairs convecting away from the cylinder in the former case, while there are three vortex pairs convecting away from the cylinder in the latter case.

24< KC< 32. Three-Pairs regime

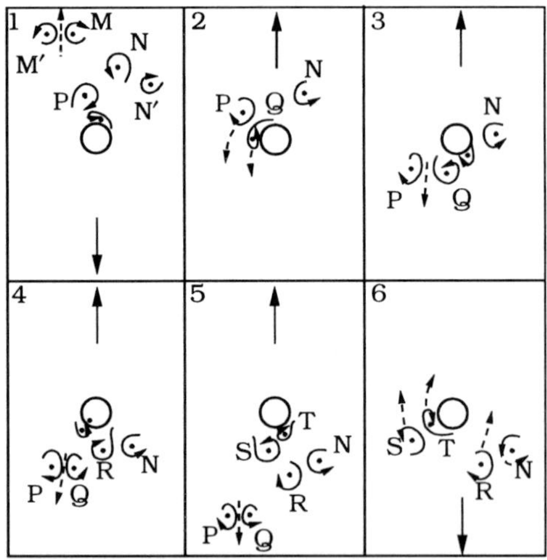

Figure 3.9 $24 < KC < 32$. The arrows refer to cylinder motion, but the vortices are viewed from a reference frame which moves with the cylinder. The wake is the result of three vortices being shed in a half cycle, and comprises three vortex pairings in a cycle (for example vortices P + Q, N + R and S + T). Williamson (1985).

For further KC regimes, the number of vortex pairs will be increased by one each time the KC regime is changed to a higher one; the number of vortex pairs which are convecting away from the cylinder will be four in the case when $32 < KC < 40$ and five in the case when $40 < KC < 48$ and so on. This means that there will be two more vortex sheddings in one full period each time the KC range is changed to a higher regime. This result is a direct consequence of the Strouhal law in oscillatory flows, as shown in Example 3.1.

Example 3.1:

Consider the oscillatory flow given in Fig. 3.10a. Its KC number is

$$KC = \frac{(2a)\pi}{D} \tag{3.7}$$

Now, suppose that we increase the KC number by $\Delta(KC) = 8$ so that the number of vortices shed for one full period is increased by 2, or for one half period by 1 (Fig. 3.10b). In this new situation, KC number will be

$$KC + \Delta(KC) = \frac{(2a + \ell)\pi}{D} \tag{3.8}$$

in which ℓ is the increase in the double-amplitude of the motion. Since $\Delta(KC)$ is 8, the length ℓ from Eqs. 3.7 and 3.8 will then be

$$\ell = \frac{8}{\pi}D \tag{3.9}$$

Given the fact that the increase in the number of vortex sheddings in one half period is 1, the size of ℓ should then be just enough to accommodate one complete vortex shedding (Fig. 3.10b). In other words, the time period during which the cylinder travels over the length ℓ should be identical to half of the vortex-shedding period, $(1/2)T_v$:

$$\ell = \left(\frac{1}{2}T_v\right)U \tag{3.10}$$

in which U is the average velocity of the cylinder during this travel. From Eqs. 3.9 and 3.10, the frequency of the vortex shedding,

$$f_v = \frac{1}{T_v} \tag{3.11}$$

will then be:

$$\frac{f_v D}{U} = \frac{1}{2}\frac{\pi}{8} = 0.20 \tag{3.12}$$

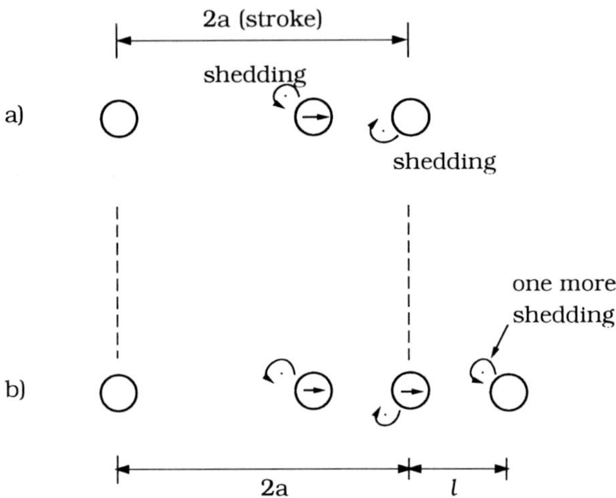

Figure 3.10 Definition sketch. In (b): KC number is increased such that the new KC number is in the next, higher KC regime.

As is seen, this is nothing but the Strouhal law with the normalized frequency being 0.20. So, as a conclusion it may be stated that the observed increase in the number of vortices shed, namely 2 in one full period when KC range is changed to a higher regime, is a direct consequence of the familiar Strouhal law.

Vortex-shedding frequency and lift frequency

In contrast to steady currents, the concept "frequency of vortex shedding" is not quite straightforward in oscillatory flows, particularly for lower KC regimes such as the single-pair regime and the double-pair regime. This is mainly due to the presence of flow reversals. The subject can probably be best explained by reference to Figs. 3.11 and 3.12. These figures depict time series of the lift force acting on a cylinder and the corresponding motion of vortices, which are reproduced from Figs. 3.6a and 3.8. (The force time series have been obtained simultaneously with the flow visualizations of vortex motions so that a direct relation between the lift variation and the motion of vortices could be established, Williamson (1985)).

In Fig. 3.11, each negative peak (marked A and C) is caused by the growth and shedding of a vortex (such as N, Frame 1, and M, Frame 4) during each half period, in exactly the same fashion as in steady currents (see Fig. 2.2 and related

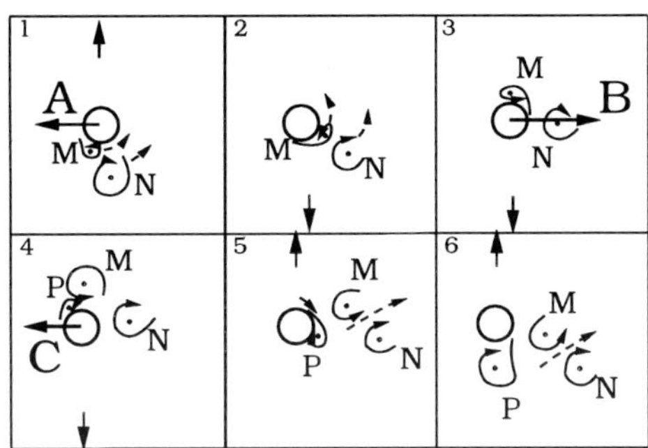

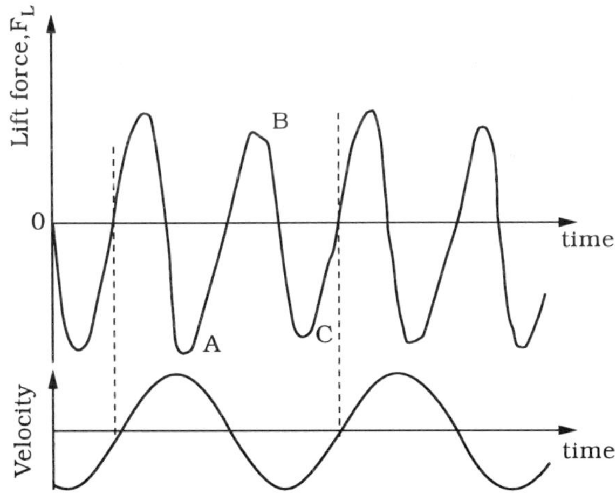

Figure 3.11 $KC = 11$. Lift-force time series obtained simultaneously in the same experiment as the flow visualization study of vortex motions depicted in Fig. 3.6a, which are reproduced here for convenience. The vertical arrows refer to cylinder motion. In the lift-force time series, the peaks marked A and C are caused by the growth and shedding of Vortex N and M (Frames 1 and 4) respectively, while the peak marked B is caused by the return of Vortex N towards the cylinder just after the flow reversal (Frame 3). Williamson (1985).

discussion, Section 2.3). The positive peaks, on the other hand, (for example that marked B) are induced by the return of the most recently shed vortex towards the cylinder just after flow reversal such as N in Frame 3. (The fact that the cylinder experiences a positive lift force when there is a vortex moving over the cylinder in the fashion as in Frame 3 was shown also by the theoretical work of Maull and Milliner (1978)). As is seen, not all the peaks in the lift force time series are induced by the vortex shedding.

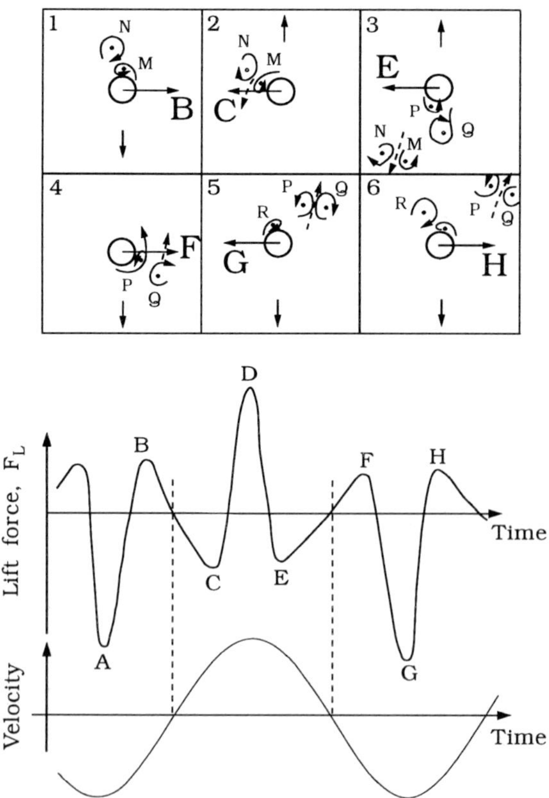

Figure 3.12 $15 < KC < 24$. Lift-force time series obtained simultaneously in the same experiment as the flow visualization study of vortex motions depicted in Fig. 3.8, which are reproduced here for convenience. Williamson (1985).

When closely examined, Fig. 3.12 will also indicate that the peaks marked A, B, D, E, G and H are caused by vortex shedding, while those marked C and F are induced by the return of the most recently shed vortex towards the cylinder just after the flow reversal, as described in the previous example.

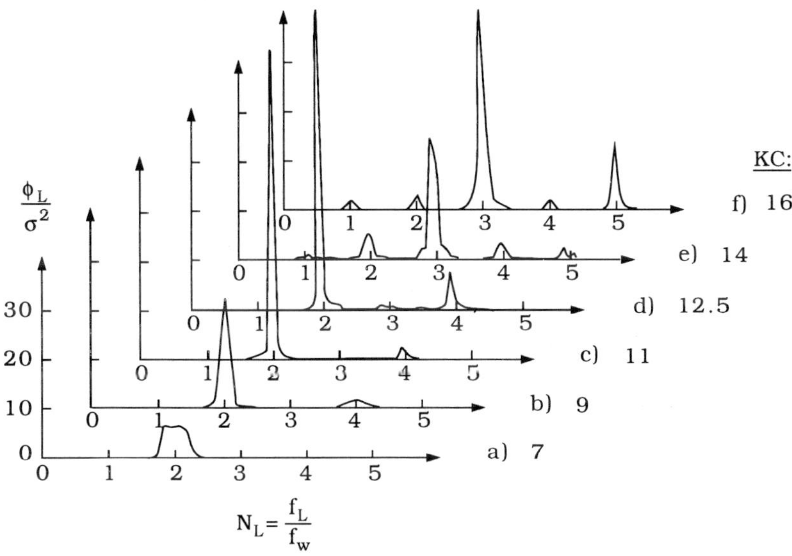

Figure 3.13 Power spectra of lift. The quantity σ^2 is the variance of the lift fluctuations. $Re = 5 \times 10^5$. Justesen (1989).

As a rule, we may say that the peak in the lift force which occurs just after the flow reversal is related to the return of the most recently shed vortex to the cylinder, while the rest of the peaks in the lift variation is associated with the vortex shedding. So, it is evident that, in oscillatory flows, the lift-force frequency is not identical to the vortex-shedding frequency.

One way of determining the lift frequency is to obtain the power spectrum of the lift force and identify the dominant frequency. This frequency is called the **fundamental lift frequency**. Fig. 3.13 gives an example; a sequence of power spectra obtained for different values of KC number in an experiment where Re is maintained constant at $Re = 5 \times 10^5$ are given. Here, ϕ_L and σ^2 are the power spectrum and the variance of the lift force, respectively. As seen, the fundamental lift frequency normalized by the oscillatory-flow frequency, namely

$$N_L = \frac{f_L}{f_w} \tag{3.13}$$

is 2 (that is, two oscillations in the lift force per flow cycle) for $KC = 7$; 9; 11; and 12.5, while it switches to 3 at the value of KC somewhere between $KC = 12.5$ and 14 and is maintained at 3 (that is, three oscillations in the lift force per flow cycle) for $KC = 14$ and 16. The actual time series of the lift force corresponding to the spectrum for $KC = 16$ in the previous figure is given in Fig. 3.14, to illustrate further the relation between the actual lift-force time series and the corresponding spectrum.

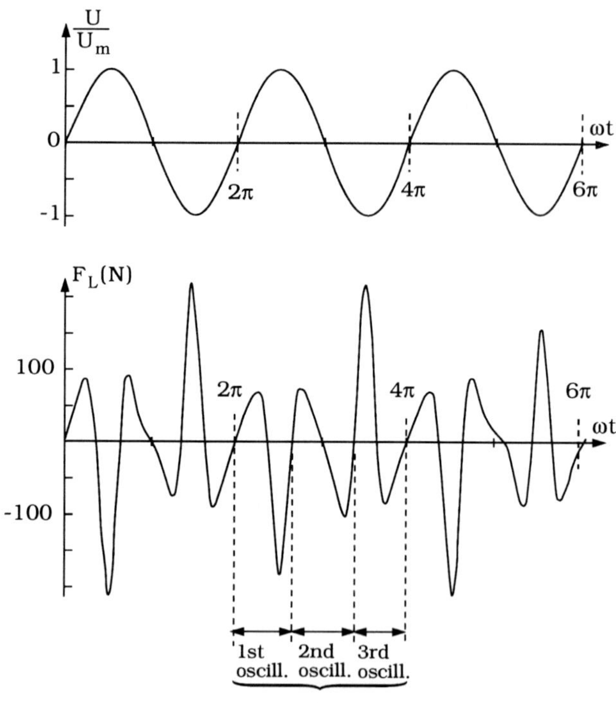

Figure 3.14 Time series for the lift force corresponding to the spectrum for $KC = 16$ in Fig. 2.13. $Re = 5 \times 10^5$. Justesen (1989).

Williamson's work (1985), where the ratio of Re to KC was maintained constant at $\beta = Re/KC = 255$ in one series of the tests and at 730 in the other, has indicated that the fundamental lift frequency increases with increasing KC, as shown in Table 3.1.

In these experiments, the KC number at which N_L switches from 2 to 3 is $KC = 15$, in contrast to the observation made in Fig. 3.13 where N_L switches from 2 to 3 at KC of about 13. This slight difference with regard to the KC number is related to the Reynolds number dependence.

Table 3.1 Fundamental lift frequencies observed in the experiments of Williamson (1985).

KC regime	KC range	Reynolds number Re	Normalized fundamental lift frequency (= the number of oscillations in the lift per flow cycle) $N_L = \frac{f_L}{f_w}$
Single pair	$7 < KC < 15$	$1.8 - 3.8 \times 10^3$	2
Double pair	$15 < KC < 24$	$3.8 - 6.1 \times 10^3$	3
Three pairs	$24 < KC < 32$	$6.1 - 8.2 \times 10^3$	4
Four pairs	$32 < KC < 40$	$8.2 - 10 \times 10^3$	5

3.3 Effect of Reynolds number on flow regimes

The detailed picture of the flow regimes as functions of both the KC number and the Re number is given in Figs. 3.15 and 3.16. Fig. 3.15 describes the role of Re for small KC numbers ($KC < 3$). The figure illustrates how the boundaries between the different flow regimes, as described in Fig. 3.2, vary as a function of Re. Furthermore, the following points may be made with regard to Fig. 3.15:

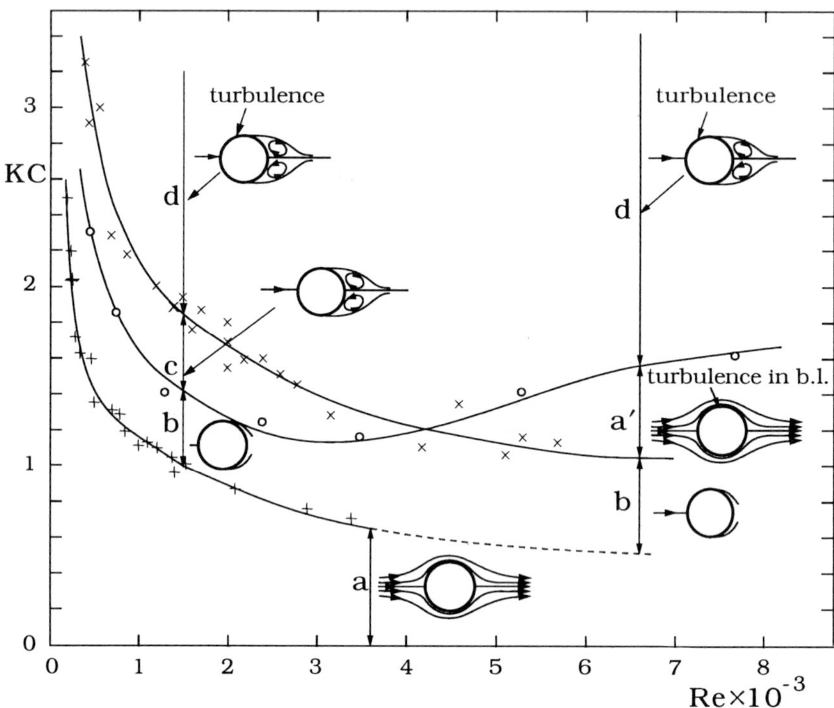

Figure 3.15 Regimes of flow around a smooth, circular cylinder in oscillatory flow for small KC numbers ($KC < 3$). (For large KC numbers, see Fig. 3.16). Explanation of various flow regimes in this figure: a: No separation. Creeping flow. a': No separation. Boundary layer is turbulent. b: Separation with Honji vortices (Fig. 3.3). c: A pair of symmetric vortices. d: A pair of symmetric vortices, but turbulence over the cylinder surface. Data: Circles from Sarpkaya (1986a); crosses for $Re < 1000$ from Honji (1981) and crosses for $Re > 1000$ from Sarpkaya (1986a). The diagram is adapted from Sarpkaya (1986a).

1) The curves which represent the inception of separation in Fig. 3.15 must be expected to approach asymptotically to the line $Re = 5$, as $KC \to \infty$ (steady current), to reconcile with the steady current case depicted in Fig. 1.1.

2) For large Re numbers (larger than about 4×10^3), the non-separated flow regime may re-appear with an increase in the KC number, after the Honji type

separation has taken place (Fig. 3.15, Region a′). This is linked with the transition to turbulence in the boundary layer. Once the flow in the boundary layer becomes turbulent, this will delay separation and therefore the non-separated flow regime will be re-established. However, in this case, the non-separated flow will be no longer a purely viscous, creeping type of flow, but rather a non-separated flow with turbulence over the cylinder surface. The transition to separated flow, on the other hand, occurs directly with the formation of a pair of symmetric vortices (Region d, in Fig. 3.15).

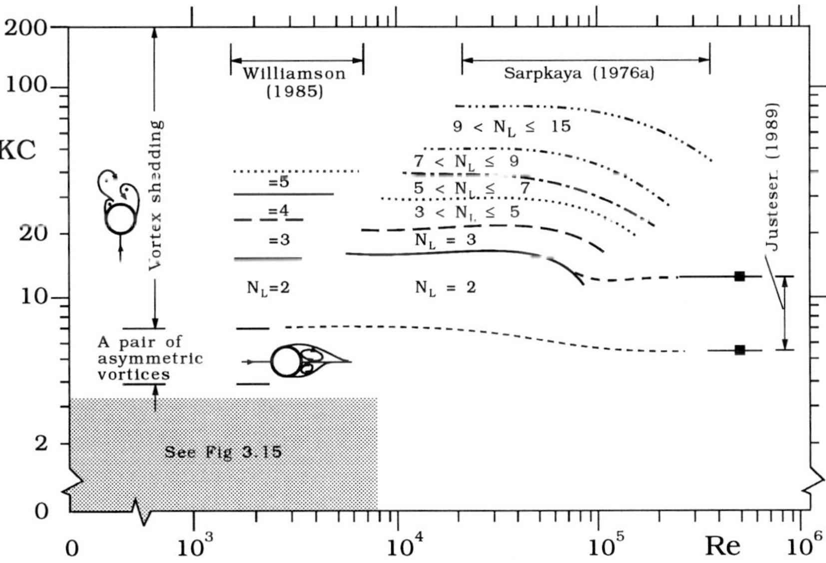

Figure 3.16 Vortex-shedding regimes around a smooth circular cylinder in oscillatory flow. Data: Lines, Sarpkaya (1976a) and Williamson (1985) and; squares from Justesen (1989). The quantity N_L is the number of oscillations in the lift force per flow cycle: $N_L = f_L/f_w$ in which f_L is the fundamental lift frequency and f_w is the frequency of oscillatory flow.

Regarding the effect of Re for larger KC numbers ($KC > 3$) depicted in Fig. 3.16, the presently available data are not very extensive. It is evident that no detailed account of various upper Reynolds-number regimes, known from the steady-current research (such as the lower transition, the supercritical, the upper transition and the transcritical regimes), is existent. Nevertheless, Sarpkaya's

(1976a) extensive data covering a wide range of KC for lower Re regimes along with Williamson's (1985) and Justesen's (1989) data may indicate what happens with increasing the Reynolds number.

Regarding the vortex-shedding regimes, it is evident from the figure that the curves begin to bend down, as Re approaches to the value 10^5, meaning that in this region the normalized lift frequency N_L increases with increasing Re. This is consistent with the corresponding result in steady currents, namely that the shedding frequency increases with increasing Re at 3.5×10^5 when the flow is switched from subcritical to supercritical through the critical (lower transition) flow regime (Fig. 1.9).

Finally, it may be mentioned that Tatsumo and Bearman (1990) presented the results of a detailed flow visualization study of flow at low KC numbers and low $\beta(= Re/KC)$ numbers.

3.4   Effect of wall proximity on flow regimes

The influence of wall proximity on the flow around and forces on a cylinder has already been discussed in the context of steady currents (Sections 1.2.1 and 2.7). As has been seen, several changes occur in the flow around the cylinder when the cylinder is placed near a wall, such as the break-up of symmetry in the flow, the suppression of vortex shedding, etc..

The purpose of the present section is to examine the effect of wall proximity on the regimes of flow around a cylinder exposed to an oscillatory flow. The analysis is mainly based on the work of Sumer, Jensen and Fredsøe (1991) where a flow visualization study of vortex motions around a smooth cylinder was carried out along with force measurements. The Re-range of the flow-visualization experiments was $10^3 - 10^4$, while that of the force measurements was $0.4 \times 10^5 - 1.1 \times 10^5$.

Flow regimes

$4 \leq \mathbf{KC} < 7$

Fig. 3.17 illustrates how the vortices evolve during the course of the oscillatory motion for $KC = 4$ for three different values of the gap-to-diameter ratio e/D, e being the gap between the cylinder and the wall. The symmetry observed in the formation and also in the motion of the vortices (Fig. 3.17a) is no longer present when $e/D = 0.1$ (Fig. 3.17b). This is also clear from the lift-force traces given in Fig. 3.18 where almost no lift force is exerted on the cylinder for $e/D = 2$, while a non-zero lift exists for $e/D = 0.1$. Here C_L is the lift coefficient defined by

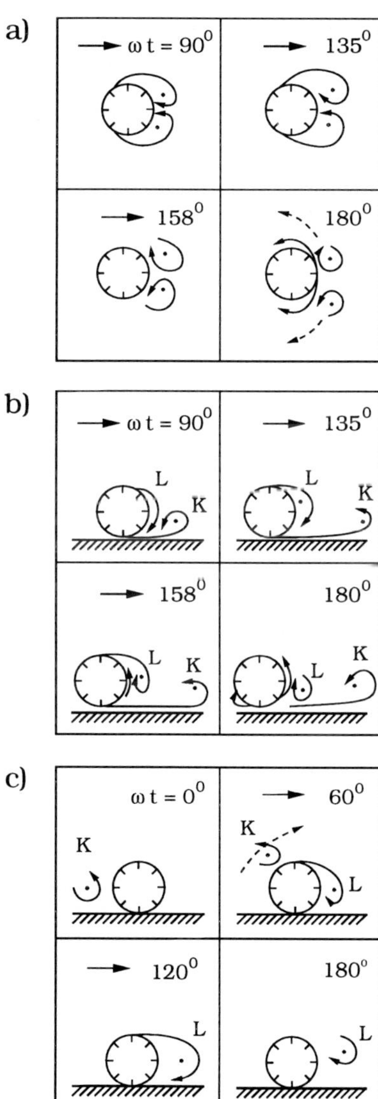

Figure 3.17 Evolution of vortex motions. $KC = 4$. Gap-to-diameter-ratio values: (a) $e/D = 2$, (b) $e/D = 0.1$, (c) $e/D = 0$. Sumer et al. (1991).

$$F_y = \frac{1}{2}\rho C_L D U_m^2 \qquad (3.14)$$

The vortex regime is quite simple for the wall-mounted cylinder (Fig. 3.17c): a vortex grows behind the cylinder each half-period, and is washed over the cylinder as the next half-period progresses. Jacobsen, Bryndum and Fredsøe (1984) give a detailed account of the latter where the motion of the lee-wake vortex over the cylinder is linked to the maximum pressure gradient in the outer flow. The lift-force trace is presented in Fig. 3.18c. The peaks in the lift force are associated with the occurrences where the vortices (Vortex K, Vortex L,... in Fig. 3.17c) are washed over the cylinder.

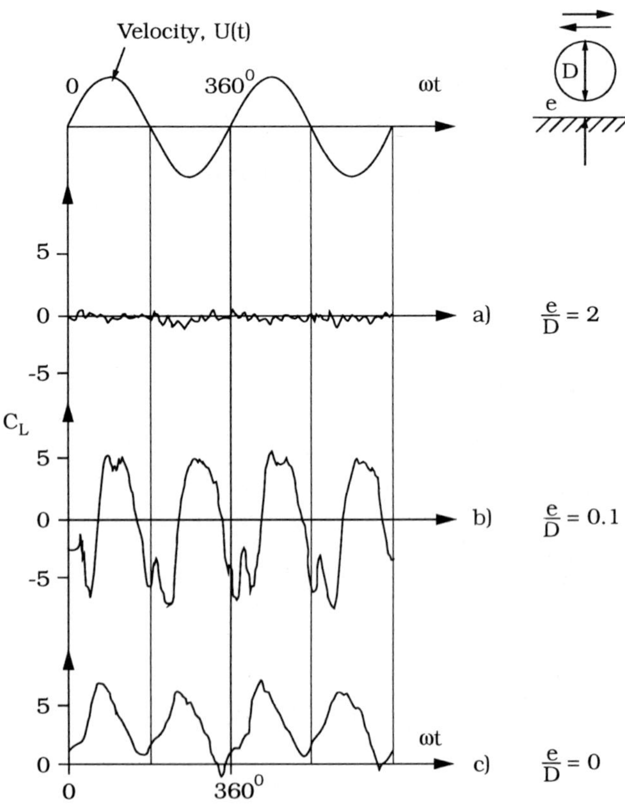

Figure 3.18 Lift-force traces. $KC = 4$. Sumer et al. (1991).

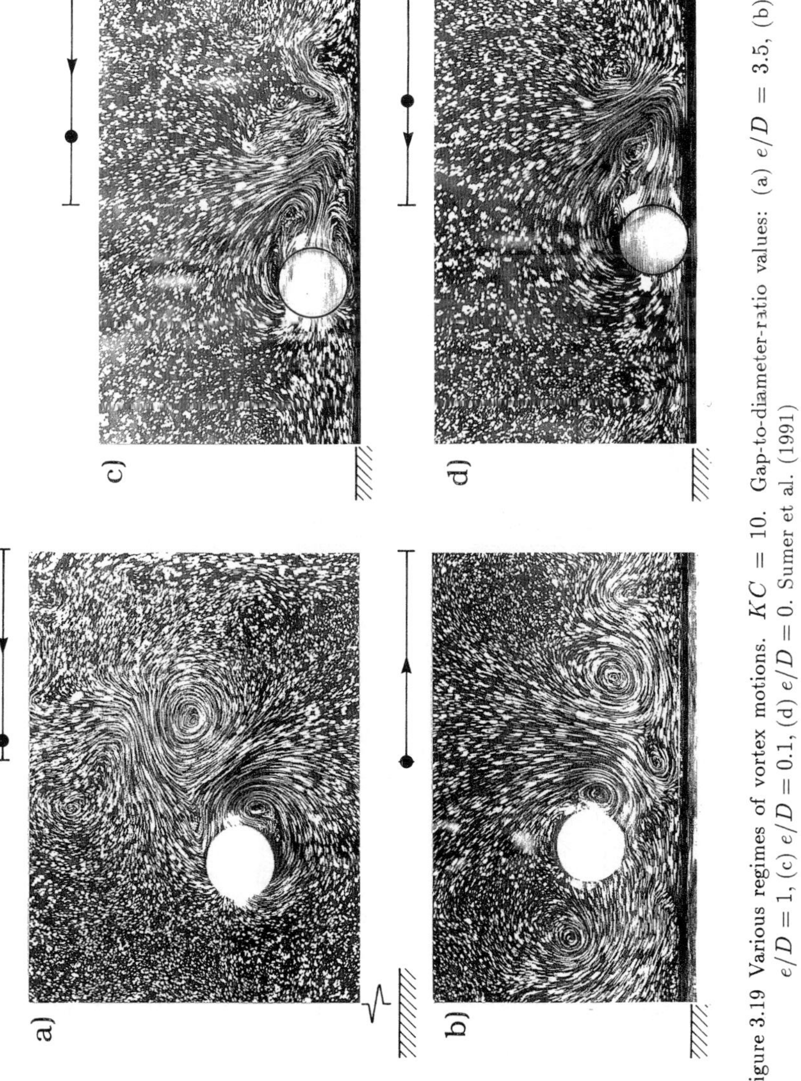

Figure 3.19 Various regimes of vortex motions. $KC = 10$. Gap-to-diameter-ratio values: (a) $e/D = 3.5$, (b) $e/D = 1$, (c) $e/D = 0.1$, (d) $e/D = 0$. Sumer et al. (1991)

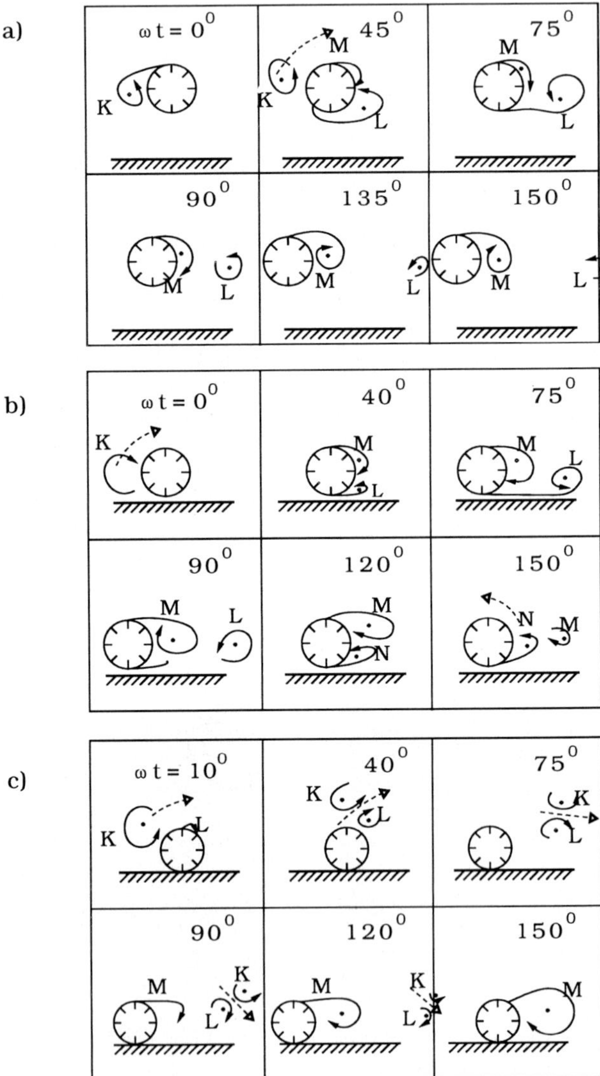

Figure 3.20 Evolution of vortex motions in the range $7 < KC < 15$. In the tests presented here $KC = 10$. Gap-to-diameter-ratio values: (a) $e/D = 1$, (b) $e/D = 0.1$, (c) $e/D = 0$. Sumer et al. (1991).

$7 < \mathbf{KC} < 15$

One of the interesting features of this KC regime for a wall-free cylinder is the formation of the transverse vortex street where the shed vortices form a vortex street perpendicular to the flow direction (Figs. 3.6a and 3.7). Sumer et al.'s (1991) work shows that *the transverse street regime disappears when the gap between the cylinder and the wall becomes less than about 1.7–1.8 times the cylinder diameter*. Figs. 3.19a and 3.19b illustrate two different vortex flow regimes, one with a gap ratio above this critical value (the transverse street regime) and the other below it, where the transverse vortex street is replaced by a wake region which lies parallel to the flow oscillation direction.

(a) $\mathbf{e/D} = \mathbf{1}$. Fig. 3.20a illustrates the time development of vortex motions during one half-period of the motion, while Fig. 3.21b presents the corresponding lift-force trace. Fig. 3.20a indicates that there is only one vortex shed (Vortex L) during one half-period of the motion. Fig. 3.21b shows how the lift force evolves during the course of the motion. The negative peak (B in Fig. 3.21b) is caused by the development of Vortex K (Fig. 3.20a, $\omega t = 0° - 45°$) (see Maull and Millincr (1978) for the relation between the vortex motion and the forces). As Vortex K is washed over the cylinder, the cylinder experiences a positive lift force, and the development of Vortex L also exerts a positive lift force (C in Fig. 3.21b). As Vortex L moves away from the cylinder ($\omega t = 135° - 150°$), the positive lift exerted on the cylinder by Vortex L is diminished.

(b) $\mathbf{e/D} = \mathbf{0.1}$. The main difference between this case and the previous one is that here the wall-side vortex (Vortex N) grows quite substantially. It is this latter vortex which is washed over the cylinder, whereas in the former case it was the free-stream-side vortex (Vortex M).

The positive peak in the lift force (D in Fig. 3.21c) is caused by the development of Vortex L. The negative peak in the lift force (E in Fig. 3.21c) is caused by the development of Vortex N combined with the high velocities in the gap induced by the flow reversal.

(c) $\mathbf{e/D} = \mathbf{0}$. In this case, the vortex which develops behind the cylinder in the previous half-period (Vortex K in Fig. 3.20c) and the vortex which is newly created (Vortex L in Fig. 3.20c) form a vortex pair. This pair is then set into motion owing to its self-induced velocity field, and thus steadily moves away from the cylinder in the downstream direction (see Fig. 3.20c, $\omega t = 40° - 120°$). Following the removal of Vortex L, a new vortex (Vortex M) begins to develop behind the cylinder.

The visualization results show that the way in which the vortex flow regime develops for the wall-mounted cylinder ($e/D = 0$) remains the same, irrespective of the range of KC. It should be noted, however, that the individual events such as the formation of the vortex pair etc. may occur at different phase (ωt) values for different KC ranges.

The peaks in the lift-force trace are caused by the passage of Vortex K over the cylinder.

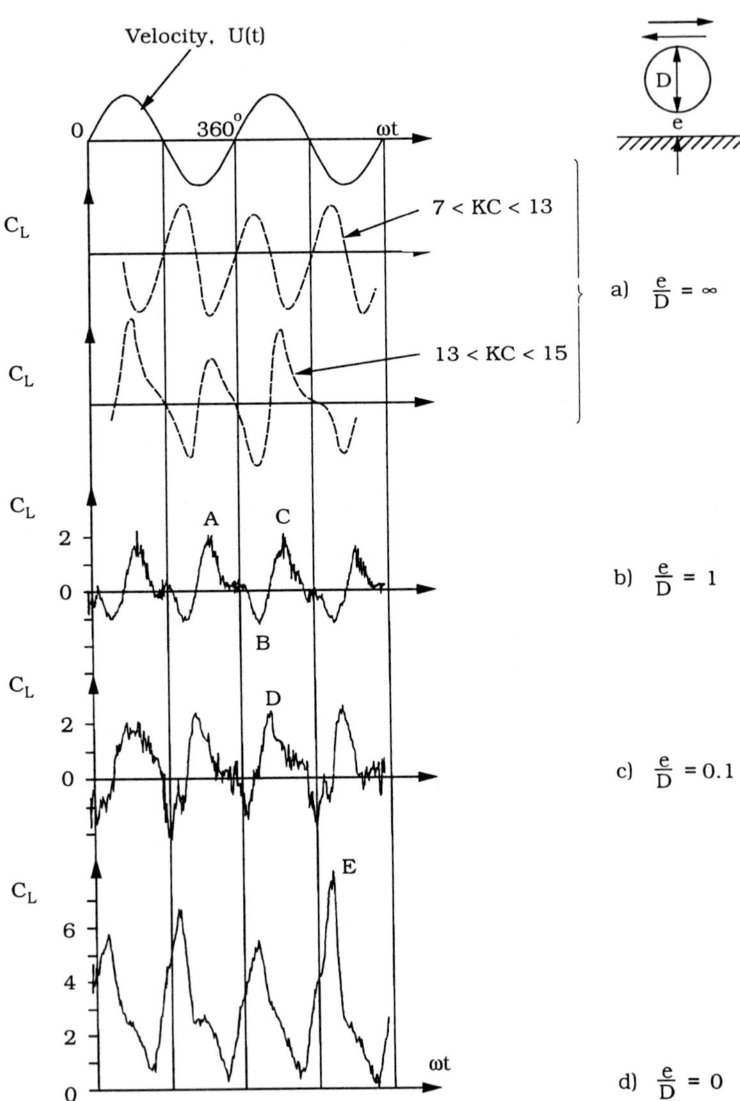

Figure 3.21 Lift-force traces in the range $7 < KC < 15$. Positive lift is directed away from the wall. The wall-free cylinder traces (a), $e/D = \infty$, are taken from Williamson (1985). For the tests presented here $KC = 10$. Sumer et al. (1991).

$15 < \mathbf{KC} < 24$ and further KC regimes

First, the KC regime $15 < KC < 24$ will be considered.

(a) $\mathbf{e/D = 1}$. In this KC regime for wall-free cylinders there is no symmetry between the half-periods, as far as the vortex motions are concerned (Figs. 3.8 and 3.12), and this also applies to the present case where $e/D = 1$, as seen from Fig. 3.22a; the vortex which is washed over the cylinder alternates between the wall side and the free-stream side each half-period. The lift-force variation (Fig. 3.23b) supports this asymmetric flow picture.

(b) $\mathbf{e/D = 0.1}$. Here, the flow is asymmetry; it is always the wall-side vortex (Vortex P, Fig. 3.22b) which is washed over the cylinder before the flow reverses to start a new half-period.

The lift force is directed away from the wall most of the time (Fig. 3.23c). Furthermore, it contains distinct, short-duration peaks in its variation with time (F, G in Fig. 3.23c). The flow-visualization tests show that these peaks are associated with the vortex shedding at the wall side of the cylinder: such peaks occur whenever there is a growing vortex on that side of the cylinder (Fig. 3.22b: $\omega t = 50° - 60°$ and $\omega t = 80° - 93°$).

Fig. 3.24 represents the lift force traces separately for the interval $0.05 \leq e/D \leq 0.4$. For values of the gap ratio smaller than approximately 0.3, the lift force becomes asymmetric, being directed away from the wall for most of the time, containing the previously mentioned distinct short-duration peaks. These peaks are present even for the gap ratio $e/D = 0.05$. These short-duration peaks indicate that the vortex shedding is maintained even for very small gap ratios such as $e/D = 0.1$, in contrast to what occurs in steady currents where the vortex shedding is maintained for values of gap ratio down to only about $e/D = 0.3$ (Section 1.2.1, Fig. 1.21). This aspect of the problem will be discussed in greater detail later in this section.

(c) $\mathbf{e/D = 0}$. It is apparent from Fig. 3.22c that the manner in which the vortex flow regime develops is exactly the same as in the range $7 < KC < 15$ (cf. Figs. 3.20c and 3.22c). However, the streamwise distance that the vortex pair travels is now relatively larger.

The lift force (Fig. 3.23d) varies with respect to time in the same way as in Fig. 3.21d where $7 < KC < 15$. However, the peaks in the present case occur relatively earlier than those in Fig. 3.21d.

The visualization tests of Sumer et al. (1991) indicate that, as in Williamson (1985), the flow patterns for the KC regimes beyond $KC = 24$ differ only in the number of vortices shed with no basic changes in the actual flow patterns.

Vortex shedding

Whether vortex shedding will be suppressed for small values of the gap ratio can be detected from the flow-visualization films as well as from the lift-force traces. The results of such an analysis are plotted in Fig. 3.25. From the figure, the following observations can be made.

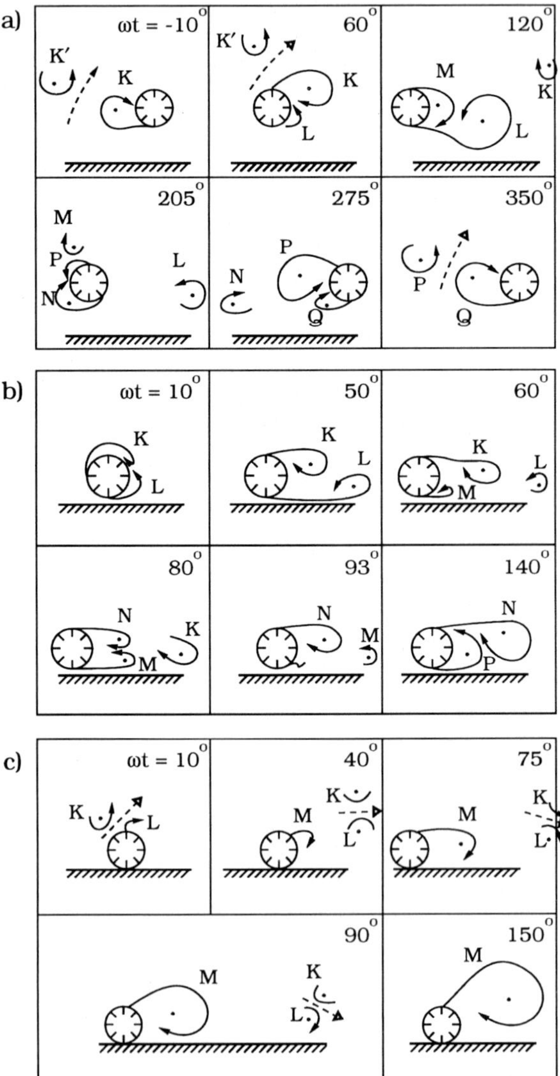

Figure 3.22 Evolution of vortex motions in the range $15 < KC < 24$. In the tests presented here $KC = 20$. Gap-to-diameter-ratio values: (a) $e/D = 1$, (b) $e/D = 0.1$, (c) $e/D = 0$. Sumer et al. (1991).

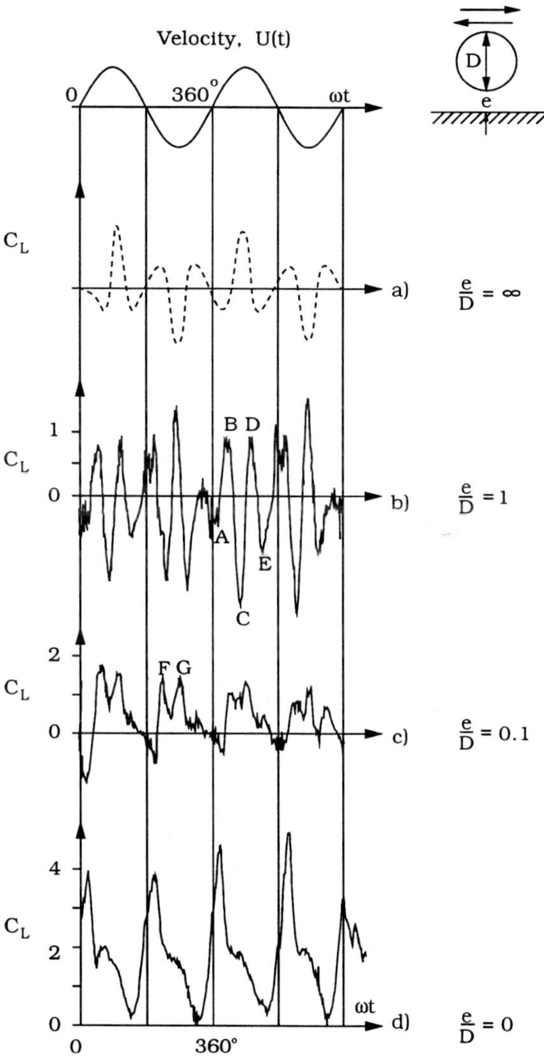

Figure 3.23 Lift-force traces in the range $15 < KC < 24$. Positive lift is directed away from the wall. The wall-free cylinder $(e/D = \infty)$ trace (a) is taken from Williamson (1985), see Fig. 3.12. In the tests presented here $KC = 20$. Sumer et al. (1991).

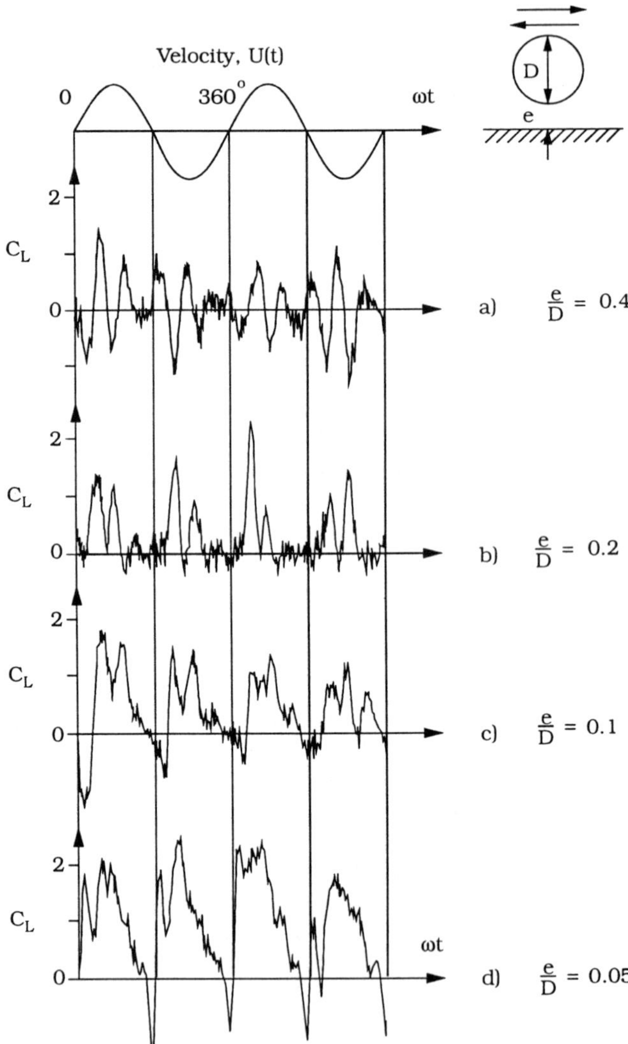

Figure 3.24 Lift-force traces for the ranges $0.05 \leq e/D \leq 0.4$ and $15 < KC < 24$. Positive lift is directed away from the wall. In the tests presented here $KC = 20$. (a) $e/D = 0.4$, (b) $e/D = 0.2$, (c) $e/D = 0.1$, (d) $e/D = 0.05$. Sumer et al. (1991).

1) For large values of KC, it appears that the gap ratio below which the vortex shedding is suppressed approaches the critical value $e/D \approx 0.25$ deduced from the work by Bearman and Zdravkovich (1978) and by Grass et al. (1984) for steady currents, (Section 1.2.1).

2) Although the borderline between the two regions in the figure, namely the vortex-shedding region and the region where the vortex shedding is suppressed, is not expected to be a clean-cut curve, there is a clear tendency that the vortex shedding is maintained for smaller and smaller values of the gap ratio as KC is decreased.

Vortex shedding is maintained even for very small gap ratios such as $e/D = 0.1$ for $KC = 10 - 20$, as shown in the photograph in Fig. 3.19c. Likewise, Fig. 3.24c implies that shedding occurs for that value of the gap ratio, as the short-duration peaks in the lift-force time series are associated with vortex shedding. The reason why vortex shedding is maintained for such small gap ratios is because the water discharge at the wall side of the cylinder is much larger in oscillatory flow at small KC than in steady currents due to the large pressure gradient from the wave.

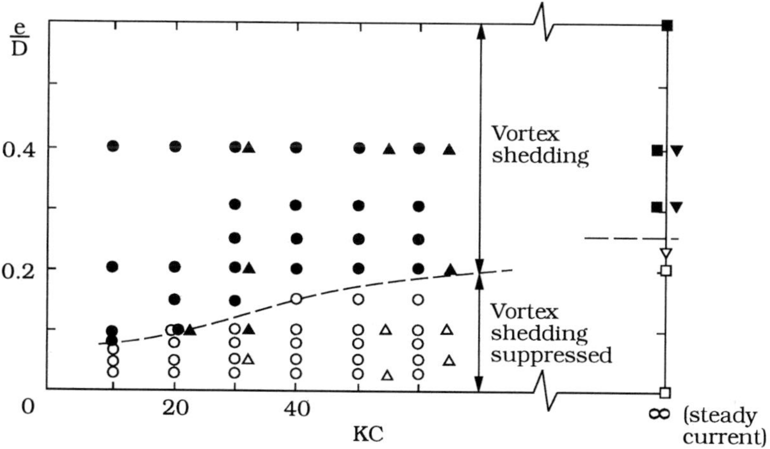

Figure 3.25 Diagram showing where the vortex shedding is suppressed in the $(e/D, KC)$-plane. Open symbols: vortex shedding is suppressed. Filled symbols: vortex shedding exists. o, △, experiments of Sumer et al. (1991). (o from flow visualization, △ from lift-force traces); □, Bearman and Zdravkovich (1978); ▽, Grass et al. (1984).

The frequency of vortex shedding can be defined by an average frequency based on the number of the short-duration peaks in the lift force over a certain period, as sketched in Fig. 3.26. The figure depicts the Strouhal number, based on this frequency and the maximum flow velocity

$$St = \frac{f_v D}{U_m} \tag{3.15}$$

as a function of the gap ratio. The shedding frequency actually varies over the cycle. The f_v-value used in the definition of St in the preceding equation is averaged over a sufficiently long period of time. Fig. 3.27 presents the same data in the normalized form St/St_0 where St_0 is the value of St attained for large values of e/D. Also plotted in Fig. 3.27 are the results of two studies conducted in steady currents, namely Grass et al. (1984) and Raven et al. (1985). The details regarding these two latter studies have already been mentioned in the previous chapter (see Fig. 1.23 and the related text). From Figs. 3.26 and 3.27 the following conclusions can be drawn.

1) For a given e/D, St increases (albeit slightly) with decreasing KC (Fig. 3.26).

2) The measurements collapse remarkably well on a single curve when plotted in the normalized form St/St_0 versus e/D (Fig. 3.27), where the influence of the close proximity of the wall on St can be seen even more clearly.

3) It is apparent that St increases as the gap ratio decreases. The increase in St frequency can be considerable (by as much as 50%) when the cylinder is placed very near the wall ($e/D = 0.1 - 0.2$). This is because the presence of the wall causes the wall-side vortex to be formed closer to the free-stream-side vortex. As a result of this, the two vortices interact at a faster rate, leading to a higher St frequency.

Finally, Sumer et al.'s (1991) work indicates that there is almost no noticeable difference between the shedding frequency obtained in their smooth-cylinder experiments and that obtained in their supplementary experiments with a rough cylinder (the cylinder roughness in the latter experiments is about $k_s/D = 10^{-2}$).

3.5 Correlation length

It has been seen that vortex shedding around a cylinder occurs in cells along the length of the cylinder (Section 1.2.2), and that the spanwise correlation coefficient is one quantity which gives information about the length of these cells. The studies concerning the effect of Re number, the effect of cylinder vibration, and the effect of turbulence in the incoming flow on correlation in steady currents have been reviewed in Section 1.2.2. In the present section, we will focus on the correlation measurements made for cylinders exposed to oscillatory flows.

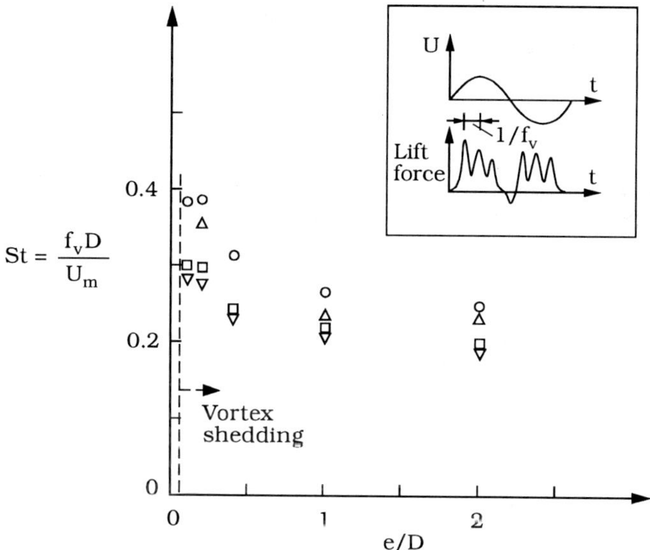

Figure 3.26 Strouhal number versus gap ratio. o, $KC = 20$; $\triangle$, $KC = 30$; $\square$, $KC = 55$; $\triangledown$, $KC = 65$. Sumer et al. (1991).

These measurements have been made by Obasaju, Bearman and Graham (1988), Kozakiewicz, Sumer and Fredsøe (1992) and Sumer, Fredsøe and Jensen (1994).

Obasaju et al.'s (1988) study has clearly demonstrated that the correlation is strongly dependent on the Keulegan-Carpenter number. Fig. 3.28 depicts their results, z being the spanwise separation (see Eq. 1.10). In the study of Obasaju et al., the correlation measurements were made by measuring the pressure differential, i.e. the difference between the pressures on the diametrically opposite points at the top and bottom of the cylinder. Fig. 3.28 indicates that the correlation coefficient takes very large values when KC is small, while it takes the lowest value when KC is at about 22. Obasaju et al. (1988) give a detailed account of the behaviour of the correlation coefficient as a function of the KC number. They link the low correlation measured at $KC = 22$ to the fact that $KC = 22$ lies at the boundary between the two KC-regimes, $15 < KC < 24$ and $24 < KC < 30$, while they argue that the correlation is measured to be high at $KC = 10$ because $KC = 10$ lies in the center of the KC-regime $7 < KC < 15$ (see also Bearman, 1985).

Fig. 3.29 illustrates the time evolution of the correlation coefficient for a given value of the spanwise separation distance, namely $z/D = 1.8$, as the flow

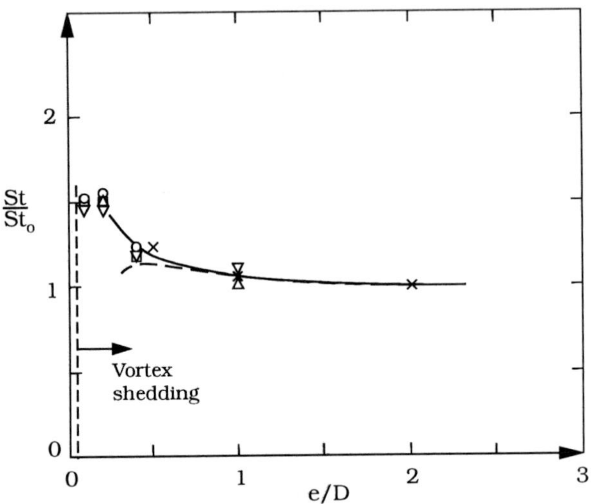

Figure 3.27 Normalized Strouhal number as function of gap ratio. o, $KC = 20$; $\triangle$, $KC = 30$; $\square$, $KC = 55$; ∇, $KC = 65$; $\times$, steady current (Raven et al., 1985); - -, steady current (Grass et al., 1984). Sumer et al. (1991).

progresses. Here $KC = 65$, and the figure is taken from Kozakiewicz et al.'s (1992) study where the cylinder was placed at a distance from a plane wall with the gap-ratio $e/D = 1.5$, sufficiently away from the wall so that the wall effects could be considered insignificant.

The correlation coefficient is calculated from the signals received from the pressure transducers mounted along the length of the cylinder using the following equations, Eqs. 3.16 and 3.20):

$$R(z, \omega t) = \frac{\overline{p'(\zeta, \omega t)p'(\zeta + z, \omega t)}}{[\overline{p'^2(\zeta, \omega t)}]^{1/2}[\overline{p'^2(\zeta + z, \omega t)}]^{1/2}} \tag{3.16}$$

in which ζ is the spanwise distance, z is the spanwise separation between two pressure transducers, and p' is the fluctuation in pressure defined by

$$p' = p - \bar{p} \tag{3.17}$$

the pressure transducers being at the free-stream-side of the cylinder.

The overbar in the preceding equations denotes ensemble averaging:

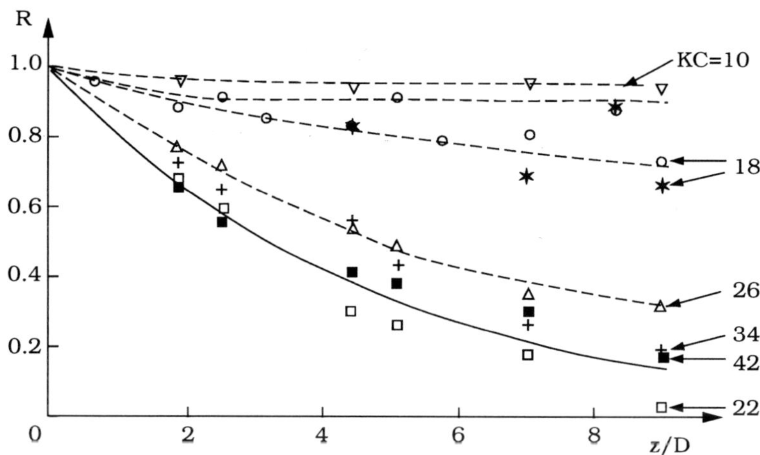

Figure 3.28 Average values of correlation coefficients versus spanwise separation. (a) ▽, $KC - 10$; ○, 18; *, 18; ⊔, 22; △, 20; +, 34; ■, 42. Note $\beta(= Re/KC) = 683$ except for the case denoted by * where $\beta = 1597$. Obasaju et al. (1988).

$$\bar{p} = \frac{1}{M} \sum_{j=1}^{M} p[\zeta, \omega(t + (j-1)T)] \tag{3.18}$$

$$\overline{p'^2} = \frac{1}{M} \sum_{j=1}^{M} \{p'[\zeta, \omega(t + (j-1)T)]\}^2 \tag{3.19}$$

$$\overline{p'(\zeta, \omega t)p'(\zeta + z, \omega t)} = \frac{1}{M} \sum_{j=1}^{M} p'[\zeta, \omega(t+(j-1)T]p'[\zeta+z, \omega(t+(j-1)T] \tag{3.20}$$

in which T is the period of the oscillatory flow, and M is the total number of flow cycles sampled.

Fig. 3.29 shows that the correlation coefficient increases towards the end of every half period, and attains its maximum at the phase $\omega t \cong 165°$, about 15° before the outer flow reverses. This phase value corresponds to the instant where the flow at the measurement points comes to a standstill, as can be traced from the pressure traces given in Kozakiewicz et al. (1992). As the flow progresses from this point onwards, however, the correlation gradually decreases and assumes its

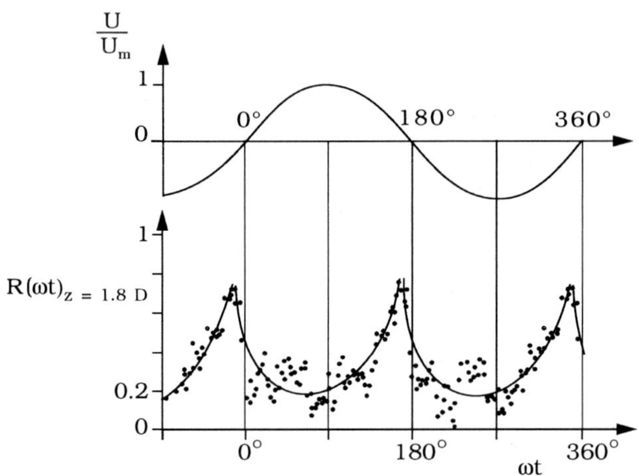

Figure 3.29 Correlation coefficient as a function of phase ωt. $KC = 65$, $Re = 6.8 \times 10^4$, $e/D = 1.5$ (sufficiently large for the wall effects to be considered insignificant), $z/D = 1.8$. Kozakiewicz et al. (1992).

minimum value for some period of time. Then it increases again towards the end of the next half period.

Fig. 3.30 shows three video sequences at the phase values $\omega t = 113°$, $165°$ and $180°$. The flow picture in Fig. 3.30b shows that the shear layer marked by the hydrogen bubble has rolled up into its vortex (A in Fig. 3.30b) and is standing motionless. As time progresses from this point onwards, however, this vortex begins to move in the reverse direction and is washed over the cylinder as a coherent entity along the length of the cylinder (Fig. 3.30c). Now, comparison of Fig. 3.30a with Fig. 3.30b indicates that while spanwise cell structures can easily be identified in the former ($\omega t = 113°$), no such structure is apparent in Fig. 3.30b ($\omega t = 165°$), meaning that the spanwise correlation should be distinctly larger in the latter than in the former case. The same is also true for $\omega t = 180°$ where, again, large correlations should be expected. This is indeed the case found in the preceding in relation to Fig. 3.29.

Effect of wall proximity on correlation

Kozakiewicz et al.'s (1992) study covers also the near-wall cylinder case. Fig. 3.31 shows the correlation coefficients for four different test data with $e/D = 2.3$, 1.5, 0.1 and 0 where e is the gap between the wall and the cylinder.

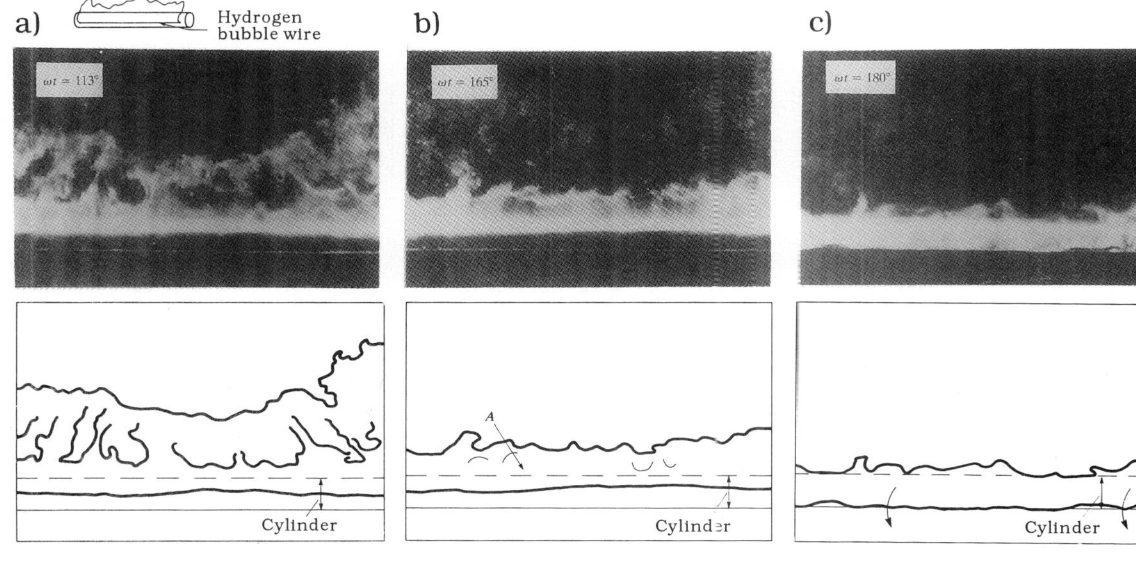

Figure 3.30 Hydrogen-bubble flow visualization sequence of pictures showing the time development of the spanwise cell structures for a stationary cylinder. $D = 2$ cm, $KC = 40$, $Re = 2 \times 10^3$. Kozakiewicz et al. (1992). The cylinder is located well away from a wall, namely the gap-to-diameter ratio $e/D = 3$; therefore, the effect of wall proximity could be considered insignificant.

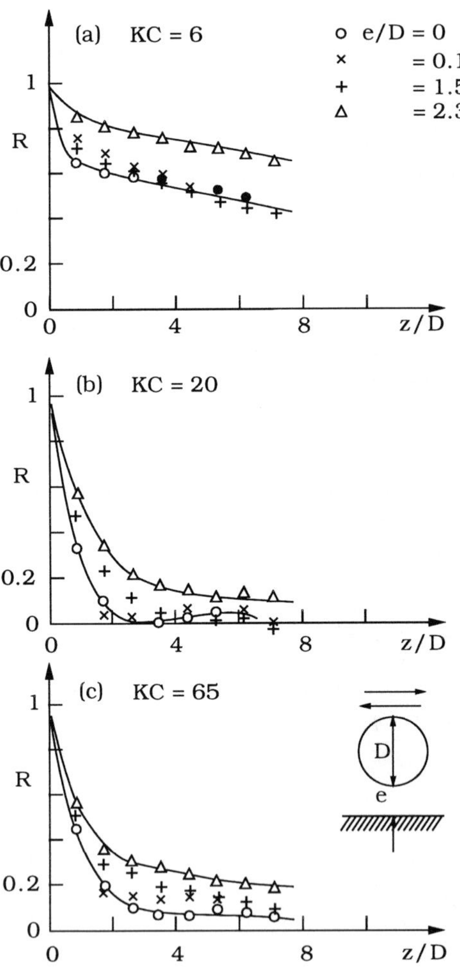

Figure 3.31 Period-averaged correlation coefficient. Wall proximity effect regarding the pressure fluctuations. See Fig. 3.32 for the wall proximity effect regarding the correlation of the *lift force*. Kozakiewicz et al. (1992).

The correlation coefficients presented in Fig. 3.31 are the period-averaged correlation coefficient, which is defined by

$$R(z) = \frac{1}{2\pi} \int_0^{2\pi} R(z,\ \omega t)\ d(\omega t) \tag{3.21}$$

The general trend in Fig. 3.31 is that the correlation coefficient decreases with decreasing gap ratio. However, caution must be exercised in interpreting the results in the figure. While R for $e/D = 2.3$ and 1.5 can be regarded as the correlation coefficient also for the lift force on the cylinder (since the fluctuations p' for which R is calculated are caused by the vortex shedding), this is not the case for $e/D = 0.1$ and 0. First of all, for $e/D = 0$, the vortex shedding is totally absent (Fig. 1.21), and the fluctuations in the measured pressure, p', in this case degenerate from those induced by the highly organized vortex-shedding phenomenon ($e/D = 2.3$ and 1.5) to those due to disorganized turbulence. So, the correlation, R, for this case, namely $e/D = 0$, only give information about the length scale in the spanwise direction of this turbulence.

For $e/D = 0.1$, on the other hand, the vortex shedding may be maintained particularly for small KC numbers (see Fig. 3.25). However, the lift in this case consists of two parts, a low frequency portion which is caused by the close proximity of the wall and the superimposed high-frequency fluctuations which are caused by vortex shedding (Fig. 3.23c). As such, the correlation, R, calculated on the basis of fluctuations, p', which are associated with the vortex shedding only, cannot be regarded as the correlation coefficient also for the lift force for the case of $e/D = 0.1$.

Regarding the correlation of the lift force itself, Kozakiewicz et al. (1992) did some indicative experiments for the wall-mounted cylinder situation with the pressure transducers positioned on the flow side of the cylinder. Clearly, with this arrangement the pressure time-series can be substituted in place of the lift force ones, as far as the correlation calculations are concerned. Regarding the lift force itself, the lift in this case ($e/D = 0$) is not caused by the pressure fluctuations (as opposed to what occurs in the case of a wall-free cylinder, Fig. 3.23a), but rather by the contraction of the streamlines near the flow side of the cylinder as well as by the movement of the lee-wake vortex over the cylinder, which results in the observed peak in the lift force prior to the flow reversal in each half-cycle of the motion (Fig. 3.23d). Hence, the correlation in connection with the lift force in this case cannot be calculated by Eq. 3.16 (which is based on the pressure fluctuations rather than on the pressure itself); instead, the usual time-averaging should be employed, i.e. the correlation is calculated by Eq. 1.10.

Fig. 3.32 presents the spanwise correlation coefficients obtained for the wall-mounted cylinder, where the results for $e/D = 2.3$ of Fig. 3.31 are replotted to facilitate comparison. The correlations in these diagrams are now all associated with the lift force; therefore comparison can be made on the same basis. The figure indicates that, as expected, the correlation increases tremendously as the gap ratio changes from 2.3 (the wall-free cylinder) to nil (the wall-mounted cylinder).

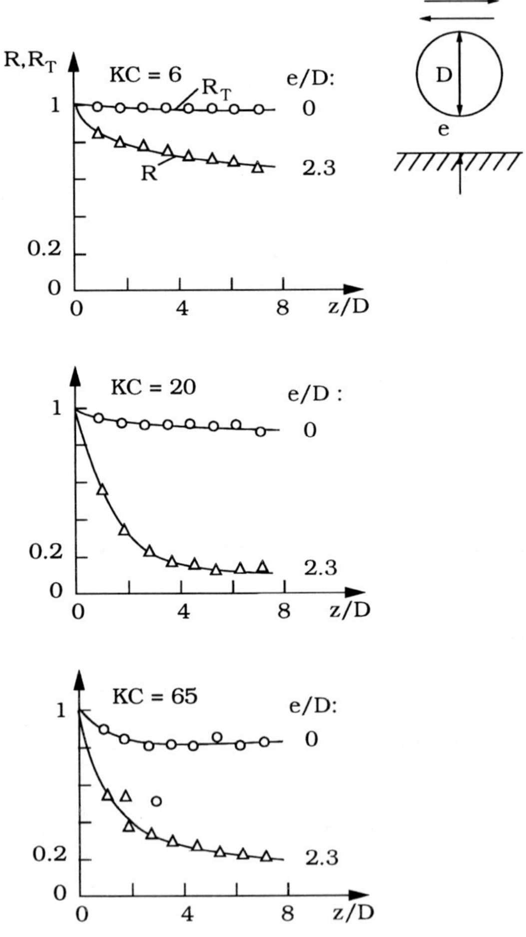

Figure 3.32 Correlation coefficient for the *lift force* on cylinder, showing wall proximity effect. R_T for the wall-mounted cylinder is computed direct from pressure signals employing time-averaging according to Eq. 1.10. Kozakiewicz et al. (1992).

Effect of vibrations on correlation

This section focuses on the effect of vibrations on the correlation when the cylinder is vibrated in a direction perpendicular to the flow only. Fig. 3.33 presents the correlation coefficients as functions of the double-amplitude-to-diameter ratio for three KC numbers, Kozakiewicz et al. (1992). In the study of Kozakiewicz et al., the vibrations were not free, but rather forced vibrations. Also, the cylinder vibrations were synchronized with the outer oscillatory-flow motion. The results of Fig. 3.33 may be compared with the corresponding results of Novak and Tanaka (1977) obtained for steady currents (Fig. 1.28). Note that in Novak and Tanaka's study the cylinder is vibrated with a frequency equal to its vortex-shedding frequency, which is identical to the fundamental lift frequency. Likewise, in the study presented in Fig. 3.33, the cylinder is vibrated with a frequency equal to the fundamental lift frequency. If this frequency is denoted by f_L and the frequency of the oscillatory flow by f_w, then $N_L = f_L/f_w$ will become the number of oscillations in the lift force for one cycle of the flow as discussed in Section 3.2 (see Eq. 3.13). In Kozakiewicz et al.'s study N_L was set equal to 13 for $KC = 65$, to 4 for $KC = 20$, and to 2 for $KC = 6$. Note that these figures are in accordance with Sarpkaya's (1976a) stationary-cylinder lift-force frequency results (Fig. 3.16) and also with Sumer and Fredsøe's (1988) results with regard to the cross flow vibration frequency of a flexibly-mounted cylinder subject to an oscillatory flow.

Returning to Fig. 3.33, the following conclusions can be deduced from the figure:

1) A constant increase in the correlation coefficient with increasing amplitudes takes place up to the values of $2A/D$ of about 0.2 for $KC = 6$ and up to $2A/D = 0.3$ for $KC = 20$ and 65. This can be seen even more clearly from Fig. 3.34 where the correlation coefficient at the spanwise distance $z = D$ is plotted as a function of $2A/D$. The way in which the correlation coefficient increases with increasing amplitude-to-diameter ratio is in accord with the steady current results (Fig. 3.34d). However, this increase is not as large as in steady currents.

2) The correlation decreases, however, for further increase in the value of $2A/D$. This may be attributed to the change in the flow regime with increasing $2A/D$ (this change in the flow regime with increasing $2A/D$ has been demonstrated by Williamson and Roshko (1988) for a cylinder exposed to a steady current). No pressure correlation data are available for the steady-current situation for values of $2A/D$ larger than 0.25. Therefore, no comparison could be made as far as such high values of $2A/D$ are concerned. There are, however, correlation measurements (Ramberg and Griffin, 1976) for $2A/D$ values as large as 0.7, where the correlation coefficient is based on wake velocity signals; these measurements indicate that the correlation coefficient increases in a monotonous manner with increasing amplitudes.

In a subsequent study, Sumer et al. (1994) measured the correlation on a freely-vibrating cylinder. Their results indicated that the correlation increases monotonously with increasing amplitude of vibrations (Fig. 3.35). The observed

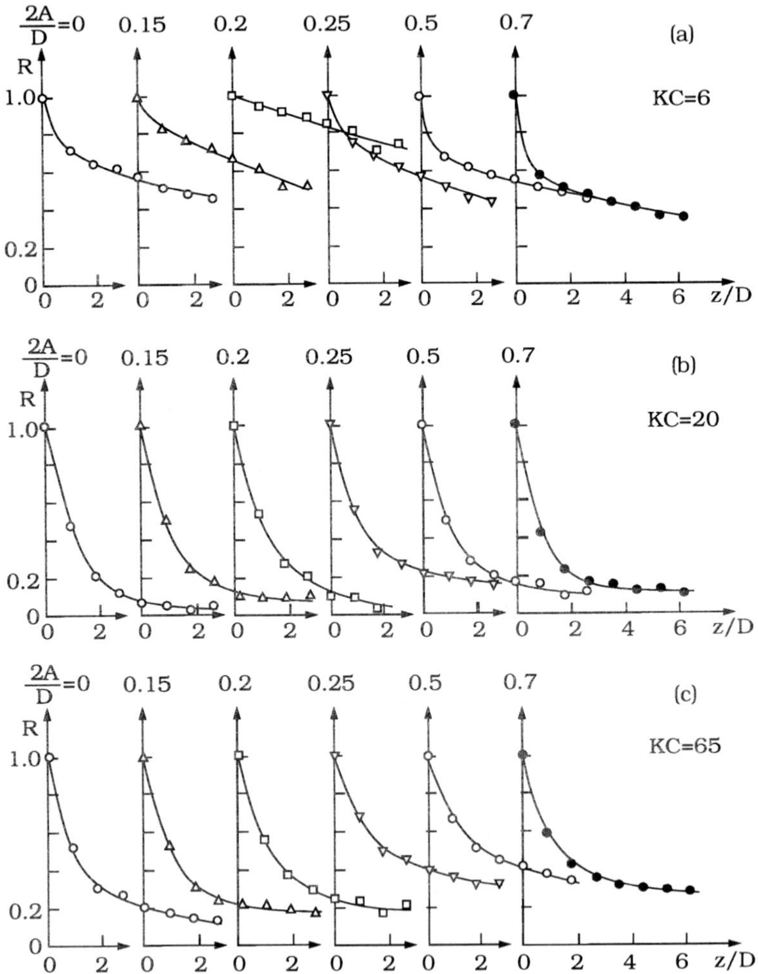

Figure 3.33 Period-averaged correlation coefficient for vibrating cylinder for $e/D = 1.5$. (a) $N_L = 2$ and $Re = 3.4 \times 10^4$ for $KC = 6$; (b) $N_L = 4$ and $Re = 6.8 \times 10^4$ for $KC = 20$ and (c) $N_L = 13$ and $Re = 6.8 \times 10^4$ for $KC = 65$. Vibrations are forced vibrations and N_L being the normalized fundamental lift frequences (Eq. 3.13). Kozakiewicz et al. (1992).

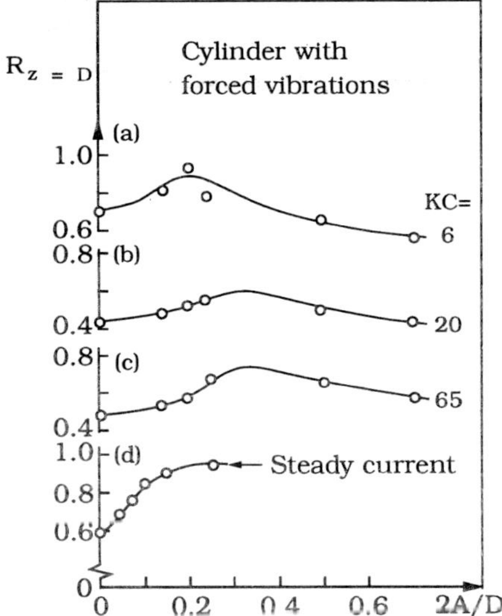

Figure 3.34 (a), (b) and (c): Period-averaged correlation cofficient with respect to vibration amplitudes for different KC numbers; (d): Steady-current data (Howell and Novak, 1979), $e/D = \infty$ and $Re = 7.5 \times 10^4$. Vibrations are forced vibrations. Kozakiewicz et al. (1992).

difference between the variation of correlation coefficients in the case of forced vibrations (Fig. 3.34) and that in the case of self-induced vibrations (the freely-vibrating-cylinder case, Fig. 3.35) is attributed to the change in the phase between the cylinder vibration and the flow velocity:

In the tests of Kozakiewicz et al. (Fig. 3.34), the cylinder motion is synchronized with the outer, oscillatory-flow motion such that the instants corresponding to the zero upcrossings in the outer-flow velocity time series coincide with the zero downcrossings in the cylinder-vibration time series. In the tests of Sumer et al. (Fig. 3.35), however, the vibrations are self-induced, and apparently the phase between the cylinder vibration and the flow velocity, ϕ, is not constant, but rather a function of the reduced velocity (Fig. 3.36). Obviously, any change in the quantity ϕ may influence the end result considerably. This may explain the disagreement between the results of Kozakiewicz et al.'s study (Fig. 3.34) and those of Sumer et al.'s study (Fig. 3.35).

Fig. 3.37 illustrates how the vibration frequency influences the correlation coefficient. Here $N_L - 13$ is the number of vibrations in one cycle of the oscillatory

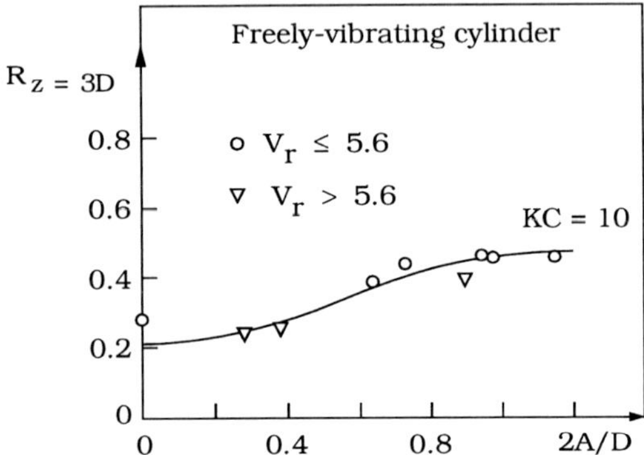

Figure 3.35 Period-averaged correlation coefficient with respect to vibra-
tion amplitudes. V_r is the reduced velocity defined by $V_r = U_m/(D f_n)$ in which f_n is the natural frequency of the flexibly-
mounted cylinder. Vibrations are not forced, but rather self-
induced vibrations. Sumer et al. (1994).

flow, and it corresponds to the fundamental lift frequency corresponding to a
stationary cylinder. As is seen, R decreases as the vibration frequency moves
away from the fundamental lift-force frequency. This result is in agreement with
the corresponding result obtained in Toebes' (1969) study for the steady-current
situation.

3.6 Streaming

In the case of unseparated flow around the cylinder, a constant, secondary
flow in the form of recirculating cells emerge around the cylinder (Fig. 3.38).
This is called streaming. A simple explanation for the emergence of this steady
streaming may be given as follows.

The flow velocity experienced at any point near the surface of the cylinder
(Point A, say, in Fig. 3.38) is asymmetric with respect to two consecutive half
periods of the flow. Namely, the velocity is relatively larger when the flow is in
the direction of converging surface geometry than that when the flow is in the
opposite direction, as sketched in Fig. 3.39 (this is due to the difference in the
response of the cylinder boundary layer in the two half periods, namely in the
converging half period and in the diverging half period). This asymmetry in the
velocity results in a non-zero mean velocity in the direction towards the top in the

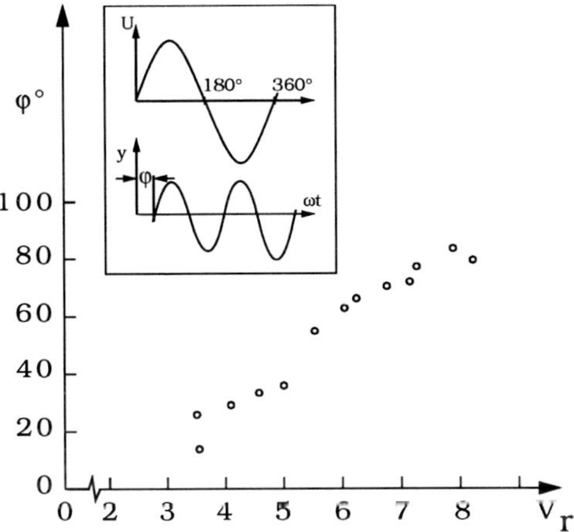

Figure 3.36 Phase difference between the cylinder vibration and the flow velocity in the tests presented in the previous figure. Sumer et al. (1994).

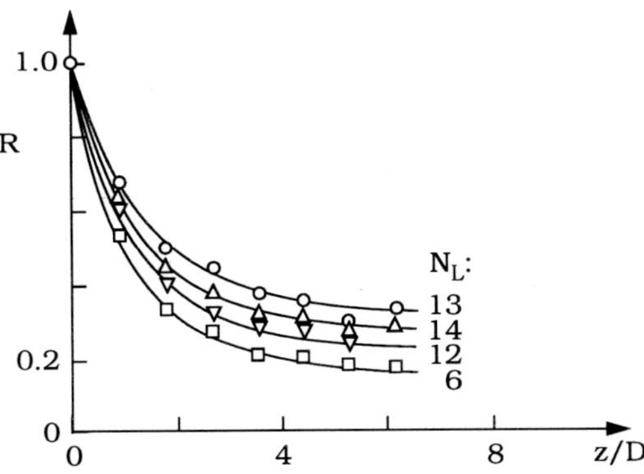

Figure 3.37 Effect of vibration frequency on period-averaged correlation coefficients for $KC = 65$, $e/D = 1.5$ and $2A/D = 0.25$. Vibrations are forced vibrations. Kozakiewicz et al. (1992).

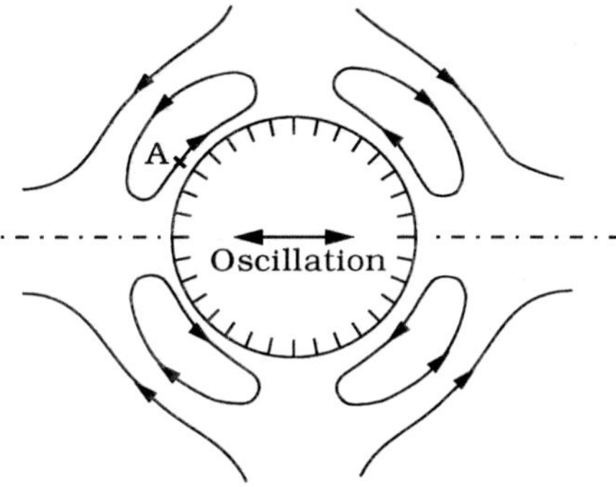

Figure 3.38 Steady streaming around a cylinder which is subject to an oscillatory, unseparated flow.

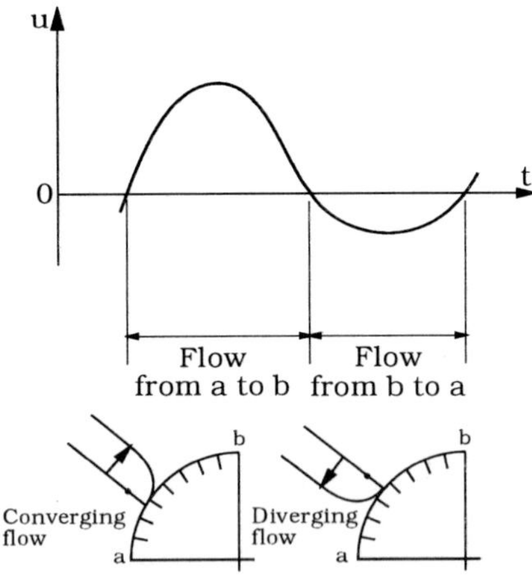

Figure 3.39 Asymmetry in two consecutive half periods in the velocity at a point near the cylinder surface that results in a steady streaming towards the top of the cylinder.

upper half of the cylinder and towards the bottom in the lower half of the cylinder. This presumably leads to the recirculating flow pattern shown in Fig. 3.38.

The streaming has been the subject of an extensive research with regard to its application in the field of acoustics (see Schlichting (1979, p.428) and Wang (1968)). It may be important also in the field of offshore engineering in conjunction with the sediment motion and the related deposition and scour processes around very large, bottom-seated marine structures which are subject to waves.

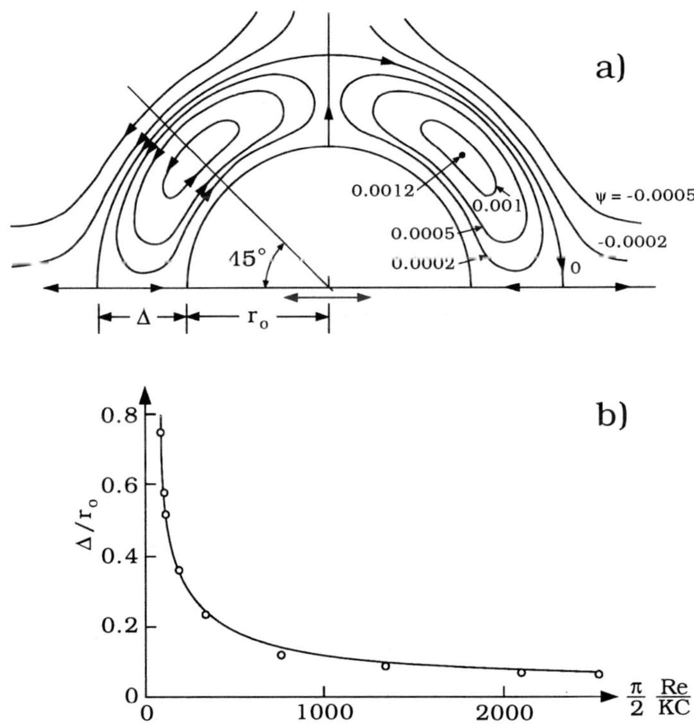

Figure 3.40 (a): The steady streaming caused by an oscillating circular cylinder. $Re = 2$, $KC = 3 \times 10^{-2}$. (b): The thickness of recirculating cells. ○, experiment, (Holtsmark et al., 1954); —, theory by Wang (1968).

Wang (1968) developed an analytical theory for very small Re numbers (creeping flow) and KC numbers. Wang's results compare very well with the experiments. In the study, analytical expressions were obtained for the stream function and the drag coefficient. Fig. 3.40a shows the flow picture obtained by

Wang for $Re = 2$ and $KC = 3 \times 10^{-2}$, while Fig. 3.40b depicts the variation of the thickness of the recirculating cells as a function of Re and KC numbers.

For large Re numbers, apparently no study is available in the literature. Therefore it is difficult to make an assessment of the thickness of the recirculating cells and the magnitude of the streaming. However, the results of Sumer, Laursen and Fredsøe's study (1993) on oscillatory flow in a convergent/divergent tunnel, where the Reynolds number was rather large (indeed, so large that the boundary layer was turbulent) suggest that the thickness of the recirculating cell may be in the order of magnitude of the boundary-layer thickness and the magnitude of the streaming may be in the order of magnitude of $O(0.1U_m)$.

In a recent study (Badr, Dennis, Kocabiyik and Nguyen, 1995), the solution of N.-S. equations was achieved for $Re = 10^3$ and $KC = 2$ and 4. The time-averaged flow field over one period obtained by the authors revealed the presence of the steady streaming pattern (depicted in Fig. 3.38) even in the case of separated flow.

REFERENCES

Badr, H.M., Dennis, S.C.R., Kocabiyik, S. and Nguyen, P. (1995): Viscous oscillatory flow about a circular cylinder at small to moderate Strouhal number. J. Fluid Mech., 303:215-232.

Bearman, P.W. (1985): Vortex trajectories in oscillatory flow. Proc. Int. Symp. on Separated Flow Around Marine Structures. The Norwegian Inst. of Technology, Trondheim, Norway, June 26-28, 1985, p. 133-153.

Bearman, P.W. and Graham, J.M.R. (1979): Hydrodynamic forces on cylindrical bodies in oscillatory flow. Proc. 2nd Int. Conf. on the Behaviour of Offshore Structures, London, 1:309-322.

Bearman, P.W. and Zdravkovich, M.M. (1978): Flow around a circular cylinder near a plane boundary. J. Fluid Mech., 89:33-48.

Bearman, P.W., Graham, J.M.R., Naylor, P. and Obasaju, E.D. (1981): The role of vortices in oscillatory flow about bluff cylinders. Proc. Int. Symp. on Hydrodyn. in Ocean Engr., Trondheim, Norway, 1:621-643.

Bearman, P.W., Graham, J.M.R. and Singh, S. (1979): Forces on cylinders in harmonically oscillating flow. Proc. Symp. on Mechanics of Wave Induced Forces on Cylinders, Bristol, ed. T.L. Shaw, Pitman, pp. 437-449.

Grass, A.J. and Kemp, P.H. (1979): Flow visualization studies of oscillatory flow past smooth and rough circular cylinders. Proc. Symp. on Mechanics of Wave-Induced Forces on Cylinders, Bristol, ed. T.L. Shaw, Pitman, pp. 406-420.

Grass, A.J., Raven, P.W.J., Stuart, R.J. and Bray, J.A. (1984): The influence of boundary layer velocity gradients and bed proximity on vortex shedding from free spanning pipelines. Trans. ASME, J. Energy Resour. Tech., 106:70-78.

Hall, P. (1984): On the stability of the unsteady boundary layer on a cylinder oscillating transversely in a viscous fluid. J. Fluid Mech., 146:347-367.

Holtsmark, J., Johnsen, I., Sikkeland, I. and Skavlem, S. (1954): Boundary layer flow near a cylindrical obstacle in an oscillating incompressible fluid. J. Acoust. Soc. Am., 26:26-39.

Honji, H. (1981): Streaked flow around an oscillating circular cylinder. J. Fluid Mech., 107:509-520.

Howell, J.F. and Novak, M. (1979): Vortex shedding from a circular cylinder in turbulent flow. Proc. 5th Int. Conf. on Wind Engrg., Paper V-11.

Jacobsen, V., Bryndum, M.B. and Fredsøe, J. (1984): Determination of flow kinematics close to marine pipelines and their use in stability calculations. In Proc. 16th Annual Offshore Technology Conf. Paper OTC 4833.

Justesen, P. (1989): Hydrodynamic forces on large cylinders in oscillatory flow. J. Waterway, Port, Coastal and Ocean Engineering, ASCE, 115(4):497-514.

Kozakiewicz, A., Sumer, B.M. and Fredsøe, J. (1992): Spanwise correlation on a vibrating cylinder near a wall in oscillatory flows. J. Fluids and Structures, 6:371-392.

Maull, D.J. and Milliner, M.C. (1978): Sinusoidal flow past a circular cylinder. Coastal Engineering, 2:149-168.

Novak, M. and Tanaka, H. (1977): Pressure correlations on a vibrating cylinder. Proc. 4th Int. Conf. on Wind Effects on Buildings and Structures, Heathrow, U.K., Cambridge Univ. Press, pp. 227-232.

Obasaju, E.D., Bearman, P.W. and Graham, J.M.R. (1988): A study of forces, circulation and vortex patterns around a circular cylinder in oscillating flow. J. Fluid Mech., 196:467-494.

Ramberg, S.E. and Griffin, O.M. (1976): Velocity correlation and vortex spacing in the wake of a vibrating cable. Trans. ASME, J. Fluids Eng., 98:10-18.

Raven, P.W.C., Stuart, R.J. and Littlejohns, P.S. (1985): Full-scale dynamic testing of submarine pipeline spans. 17th Annual Offshore Technology Conf., Houston, TX, May 6-9, Paper 5005, 3:395-404.

Sarpkaya, T. (1976a): In-line and transverse forces on smooth and sand-roughened cylinders in oscillatory flow at high Reynolds numbers. Naval Postgraduate School, Monterey, CA, Tech. Rep. NPS-69SL76062.

Sarpkaya, T. (1986a): Force on a circular cylinder in viscous oscillatory flow at low Keulegan-Carpenter numbers. J. Fluid Mech., 165:61-71.

Schlichting, H. (1979): Boundary-Layer Theory. 7. ed., McGraw-Hill Book Co.

Singh, S. (1979): Forces on bodies in oscillatory flow. Ph.D. thesis, Univ. London.

Sumer, B.M. and Fredsøe, J. (1988): Transverse vibrations of an elastically mounted cylinder exposed to an oscillating flow. Jour. Offshore Mechanics and Arctic Engineering, ASME, 110:387-394.

Sumer, B.M., Jensen, B.L. and Fredsøe, J. (1991): Effect of a plane boundary on oscillatory flow around a circular cylinder. J. Fluid Mech., 225:271-300.

Sumer, B.M., Fredsøe, J. and Jensen, K. (1994): A note on spanwise correlation on a freely vibrating cylinder in oscillatory flow. Jour. Fluids and Structures, 8:231-238.

Sumer, B.M., Laursen, T.S. and Fredsøe, J. (1993): Wave boundary layers in a convergent tunnel. Coastal Engineering, 20:3/4:317-342.

Tatsumo, M. and Bearman, P.W. (1990): A visual study of the flow around an oscillating circular cylinder at low Keulegan-Carpenter numbers and low Stokes numbers. J. Fluid Mech., 211:157-182.

Toebes, G.H. (1969): The unsteady flow and wake near an oscillating cylinder. ASME, Journal of Basic Engineering, 91:493-502.

Wang, C.Y. (1968): On high-frequency oscillatory viscous flows. J. Fluid Mech., 32:55-68.

Williamson, C.H.K. (1985): Sinusoidal flow relative to circular cylinders. J. Fluid Mech., Vol. 155, pp. 141-174.

Williamson, C.H.K. and Roshko, A. (1988): Vortex formation in the wake of an oscillating cylinder. Jour. of Fluids and Structures, 2:355-381.

Chapter 4. Forces on a cylinder in regular waves

Similar to steady currents, a cylinder subject to an oscillatory flow may experience two kinds of forces: the in-line force and the lift force (Fig. 4.1). In the following, first, the in-line force on a smooth, circular cylinder will be considered and subsequently the attention will be directed to the lift force. The remainder of the chapter will focus on the influence on the force components of the following effects: surface roughness, angle of attack, co-existing current and orbital motion (real waves).

4.1 In-line force in oscillatory flow

In steady currents, the force acting on a cylinder in the in-line direction is given by

$$F = \frac{1}{2}\rho C_D DU |U| \tag{4.1}$$

where F is the force per unit length of the cylinder and C_D is the drag coefficient. (Note that the velocity-squared term in Eq. 2.8, namely U^2, is written in the present context in the form of $U|U|$ to ensure that the drag force is always in the direction of velocity). In the case of oscillatory flows, however, there will be two additional contributions to the total in-line force:

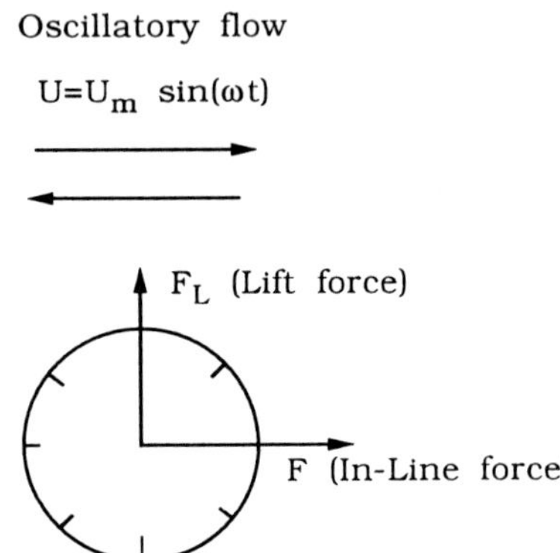

Figure 4.1 Definition sketch.

$$F = \frac{1}{2}\rho C_D D U |U| + m' \dot{U} + \rho V \dot{U} \tag{4.2}$$

in which $m' \dot{U}$ is called the *hydrodynamic-mass force* while $\rho V \dot{U}$ is called the *Froude-Krylov force* where m' is the hydrodynamic mass and V is the volume of the cylinder, which for a unit length of the cylinder reduces to A, the cross-sectional area of the cylinder. The following paragraphs give a detailed account of these two forces.

4.1.1 Hydrodynamic mass

The hydrodynamic mass can be illustrated by reference to the following example. Suppose that a thin, infinitely long plate with the width b is immersed in still water and that it is impulsively moved from rest (Fig. 4.2). When the plate is moved in its own plane, it will experience almost no resistance, considering that the frictional effects are negligible due to the very small thickness of the plate. Whereas, when it is moved in a direction perpendicular to its plane, there will be a tremendous resistance against the movement.

The reason why this resistance is so large is that it is not only the plate but also the fluid in the immediate neighbourhood of the plate, which has to be accelerated in this case due to the pressure from the plate.

The hydrodynamic mass is defined as the mass of the fluid around the body which is accelerated with the movement of the body due to the action of pressure.

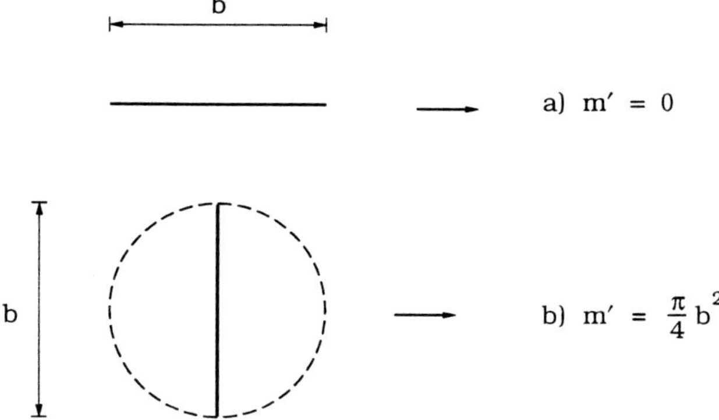

Figure 4.2 Movement of an infinitely long plate in an otherwise still fluid. a)
Movement of the plate in its own plane and b) that in a direction
perpendicular to its own plane.

If the hydrodynamic mass is denoted by m', the force to accelerate the total mass,
namely the mass of the body, m, and the hydrodynamic mass, m', may be written
as

$$F = (m + m')a \qquad (4.3)$$

where a is the acceleration.

Usually, the hydrodynamic mass is calculated by neglecting frictional effects,
i.e. the flow is calculated by expressing fluid force equilibrium between pressure
and inertia. Hereby the flow field introduced by accelerating the body through
the fluid can be calculated using potential flow theory.

The procedure to calculate the hydrodynamic mass for a body placed in
a still water can now be summarized as follows. 1) Accelerate the body in the
water; (this acceleration will create a pressure gradient around the body resulting
in the hydrodynamic-mass force); 2) calculate the flow field around the body; 3)
calculate the pressure on the surface of the body based on the flow information in
the previous step; and finally 4) determine the force on the body from the pressure
information. In the following we shall implement this procedure to calculate the
hydrodynamic mass for a free circular cylinder.

Example 4.1: Hydrodynamic mass for a circular cylinder

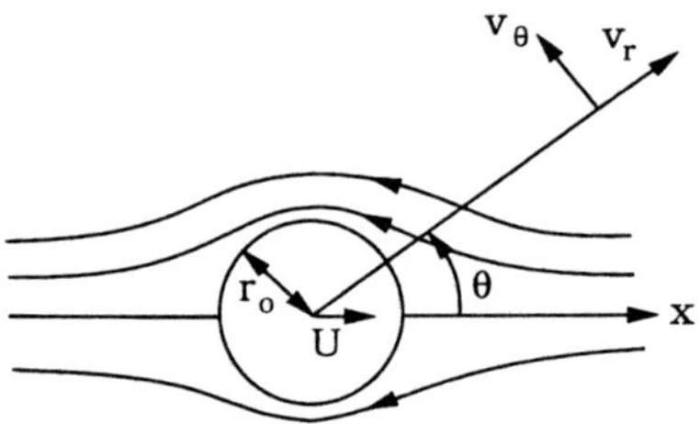

Figure 4.3 Potential flow around an accelerated cylinder, moving with velocity U in an otherwise still fluid.

When a cylinder is held stationary and the fluid moves with a velocity U in the negative direction of the x-axis, the velocity potential is given by (Milne-Thomson, 1962, Section 6.22):

$$\phi = U(r + \frac{r_0^2}{r})cos\theta \tag{4.4}$$

If we superimpose on the whole system a velocity U in the positive direction of the x-axis, the cylinder will move forward with velocity U and the fluid will be at rest at infinity (Fig. 4.3), so that ϕ is given by (Milne-Thomson, 1962, Section 9.20):

$$\phi = U\frac{r_0^2}{r}\cos\theta \tag{4.5}$$

The velocity components v_r and v_θ will then be calculated as follows

$$v_\theta = -\frac{1}{r}\frac{\partial\phi}{\partial\theta} = U\frac{r_0^2}{r^2}\sin\theta \tag{4.6}$$

$$v_r = -\frac{\partial\phi}{\partial r} = U\frac{r_0^2}{r^2}\cos\theta \tag{4.7}$$

The pressure around the cylinder can now be calculated, employing the general Bernoulli equation (Milne-Thomson, 1962, Section 3.60):

$$\frac{p}{\rho} + \frac{1}{2}v^2 - \frac{\partial \phi}{\partial t} = \text{constant} \tag{4.8}$$

in which v is the speed

$$v^2 = v_r^2 + v_\theta^2 \tag{4.9}$$

On the cylinder surface v^2 will be

$$v^2 = U^2(\sin^2\theta + \cos^2\theta) = U^2 \tag{4.10}$$

therefore the pressure on the cylinder surface from Eq. 4.8 can be written as

$$\frac{p}{\rho} = \frac{\partial \phi}{\partial t} + \text{constant} \tag{4.11}$$

where the constant term includes also $\frac{1}{2}U^2$, as the latter does not vary with the independent variables r and θ. This term, as a matter of fact, is not significant as it does not contribute to the resulting force. So, dropping the constant, the pressure on the cylinder surface may be written as

$$p = \rho\frac{\partial \phi}{\partial t} = \rho\frac{\partial}{\partial t}\left(U\frac{r_0^2}{r_0}\cos\theta\right) = \rho r_0 \cos\theta\frac{\partial U}{\partial t}$$

or

$$p = \rho r_0 a \cos\theta \tag{4.12}$$

in which a is the acceleration, i.e. $a = \partial U/\partial t$.

The resultant force can then be calculated by integrating the pressure around the cylinder

$$P = -\int_0^{2\pi} p\cos\theta(r_0\, d\theta) \tag{4.13}$$

The vertical component of the force will be automatically zero due to symmetry. So the resultant force will be

$$P = -a\rho r_0^2 \int_0^{2\pi} \cos^2\theta\; d\theta$$

or

$$P = -\rho r_0^2 a\pi \tag{4.14}$$

In other words, the force required to accelerate a cylinder with an acceleration a in an otherwise still fluid should be given by

$$F = ma + \rho r_0^2 \pi a = (m + m')a \tag{4.15}$$

and therefore the hydrodynamic mass of a circular cylinder will be given by

$$m' = \rho \pi r_0^2 \qquad (4.16)$$

Traditionally, the hydrodynamic mass is written as

$$m' = \rho C_m A \qquad (4.17)$$

in which A is the cross-sectional area of the body ($A = \pi r_0^2$ for the preceding example) and the coefficient C_m is called the hydrodynamic-mass coefficient. C_m for a circular cylinder is (Eq. 4.16):

$$C_m = 1 \qquad (4.18)$$

Appendix II lists the values of the hydrodynamic-mass coefficients for various two- and three-dimensional bodies.

Example 4.2:   Hydrodynamic mass for a circular cylinder near a wall

When the cylinder is placed near a wall (the pipeline problem), the hydrodynamic mass will obviously be influenced by the close proximity of the wall. Yamamoto et al. (1974) has developed a potential flow solution to account for this effect. Their result is reproduced in Fig. 4.4. As is seen, the hydrodynamic-mass coefficient C_m increases with decreasing the gap between the cylinder and the bed. It is further seen that C_m goes to unity, its asymptotic value, as $e/D \rightarrow \infty$. Yamamoto et al. noted that considerations were given for flows accelerating both perpendicular and parallel to the wall; it was found that C_m determined from the theory was the same regardless of the flow direction.

Finally, it may be mentioned that simple algorithms for calculating hydrodynamic mass for cylinders placed near an arbitrarily shaped scoured sea bed were given by Hansen (1990). Hansen's calculations cover also groups of cylinders. A number of examples including multiple riser configurations were given also in Jacobsen and Hansen (1990).

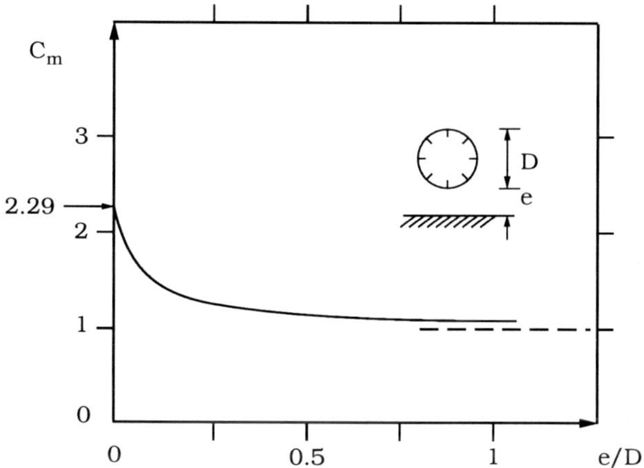

Figure 4.4 Hydrodynamic-mass coefficient for a circular cylinder near a wall. Yamamoto et al. (1974).

4.1.2 The Froude-Krylov force

As seen in the previous section, when a body is moved with an acceleration a in   still water, there will be a force on the body, namely the hydrodynamic-mass force. This force is caused by the acceleration of the fluid in the *immediate surroundings* of the body. When the body is held stationary and the water is moved with an acceleration a, however, there will be two effects. First, the water will be accelerated in the immediate neighbourhood of the body in the same way as in the previous analysis. Therefore, the previously mentioned hydrodynamic mass will be present. The second effect will be that the accelerated motion of the fluid in the *outer-flow region* will generate a pressure gradient according to

$$\frac{\partial p}{\partial x} = -\rho \frac{dU}{dt} \qquad (4.19)$$

where U is the velocity far from the cylinder. This pressure gradient in turn will produce an additional force on the cylinder, which is termed the **Froude-Krylov force**. The force on the body due to this pressure gradient can be calculated by the following integration:

$$F_p = -\int_S p \, d\mathbf{S} \qquad (4.20)$$

where $\mathbf{S}$ is the surface of the body. From the Gauss theorem, Eq. 4.20 can be written as a volume integral

$$F_p = -\int_V \frac{\partial p}{\partial x} dV \qquad (4.21)$$

Using that the pressure gradient is constant and given by Eq. 4.19 this gives

$$F_p = \rho V \, \dot{U} \qquad (4.22)$$

in which $\dot{U}$ is

$$\dot{U} = \frac{dU}{dt} \qquad (4.23)$$

For a *cylinder* with the cross-sectional area A and with unit length, F_p will be

$$F_p = \rho A \, \dot{U} \qquad (4.24)$$

For a *sphere* with diameter D, on the other hand, F_p will be

$$F_p = \rho \left(\frac{\pi D^3}{6} \right) \dot{U} \qquad (4.25)$$

In the case when the body moves in an otherwise still water, there will be no pressure gradient created by the acceleration of the outer flow (Eq. 4.19), therefore the Froude-Krylov force will not exist in this case.

4.1.3 The Morison equation

Now the total in-line force can be formulated for an accelerated water environment where the cylinder is held stationary. The total force, F, is given by Eq. 4.2 with the hydrodynamic-mass force given by Eq. 4.17 and the Froude-Krylov force by Eq. 4.24. Therefore F will be written as

$$F = \frac{1}{2}\rho C_D D U |U| + \rho C_m A \, \dot{U} + \rho A \, \dot{U} \qquad (4.26)$$

Drag   Hydro-   Froude-
force   dynamic Krylov
         mass force force

The preceding equation can be written in the following form

$$F = \frac{1}{2}\rho C_D D U \mid U \mid + \rho(C_m + 1)A\,\dot{U} \qquad (4.27)$$

By defining a new coefficient, C_M, by

$$C_M = C_m + 1 \qquad (4.28)$$

Eq. 4.27 will read as follows

$$F = \frac{1}{2}\rho C_D D U |U| + \rho C_M A\,\dot{U} \qquad (4.29)$$

This equation is known as the Morison equation (Morison, O'Brien, Johnson and Schaaf, 1950).

The new force term, $\rho C_M A\,\dot{U}$, is called the inertia force and the new coefficient C_M is called the inertia coefficient. (In the case of a circular cylinder exposed to an oscillatory flow with small KC numbers such as $O(1)$, $C_M(=C_m+1)$, tends to the value 2, since the flow is unseparated in this case (Fig. 3.15) and therefore the potential-flow value of C_m, namely $C_m = 1$ (Eq. 4.18), can be used).

In the case when the body moves relative to the flow in the in-line direction (this may occur, for example, when the body is flexibly mounted) the Morison equation, from Eq. 4.26, can be written as

$$F = \frac{1}{2}\rho C_D D(U - U_b) \mid U - U_b \mid + \rho C_m A\left(\dot{U} - \dot{U_b}\right) + \rho A\,\dot{U} \qquad (4.30)$$

Drag force	Hydro- dynamic mass force	Froude- Krylov force

where U_b is the velocity of the body in the in-line direction. Clearly, the Froude-Krylov force must be based on $\dot{U}$ rather than $(\dot{U} - \dot{U_b})$, because this force is associated with the absolute motion of the fluid rather than the motion of the fluid relative to the body.

The drag force F_D versus the inertia force F_I

From Eq. 4.29, it is seen that there is a 90° phase difference between the maximum value of F_D and the maximum value of F_I, which is schematically illustrated in Fig. 4.5. This phase difference should be taken into consideration if the maximum value of the in-line force is of interest.

The ratio between the maximum values of the two forces, on the other hand, can be written from Eq. 4.29 as

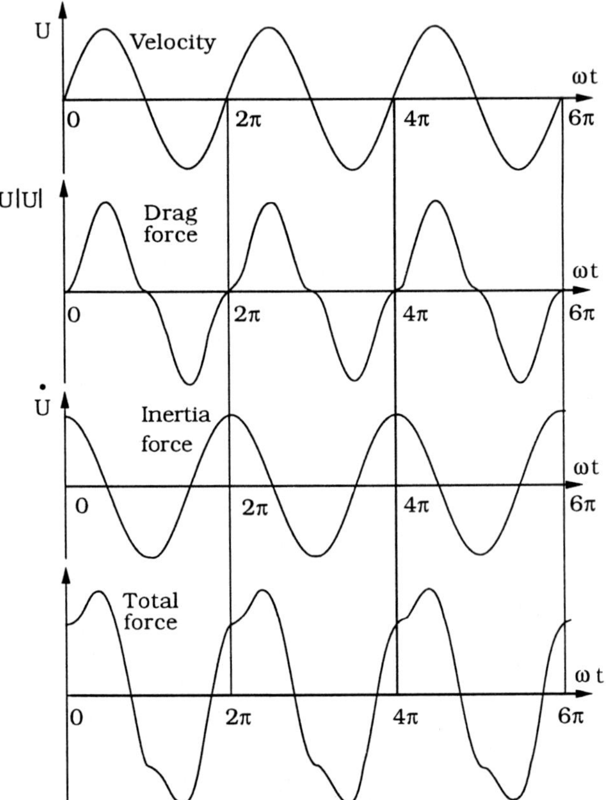

Figure 4.5 Time variation regarding the drag- and the inertia force in oscillatory flows.

$$\frac{F_{I,\max}}{F_{D,\max}} = \frac{C_M \frac{\pi}{4} D^2 \omega U_m}{\frac{1}{2} C_D D U_m^2} = \pi^2 \frac{D}{U_m T} \frac{C_M}{C_D} = \frac{\pi^2}{KC} \frac{C_M}{C_D} \tag{4.31}$$

For small KC numbers, the inertia coefficient C_M can be taken as $C_M = 2$, as mentioned in the preceding section. Therefore, the force ratio in the preceding equation, taking $C_D \cong 1$, becomes

$$\frac{F_{I,\max}}{F_{D,\max}} = \frac{20}{KC} \tag{4.32}$$

This means that, for small KC values, the inertia component of the in-line force is large compared with the drag component, thus in such cases the drag can be neglected. However, as the KC number is increased, the separation begins to occur (Fig. 3.15), and therefore the drag force becomes increasingly important. As a rough guide we may consider the range of the Keulegan-Carpenter number $0 < KC \ll 20 - 30$ as the *inertia-dominated regime*, while $KC > 20 - 30$ as the *drag-dominated regime*.

Finally, it may be mentioned that, in some cases such as in the calculation of damping forces for resonant structural vibrations, the drag force becomes so important that even the small contribution to the total force must be taken into consideration.

4.1.4   In-line force coefficients

Example 4.3:   Asymptotic theory

For very small KC numbers (such as $KC \ll 1$) combined with sufficiently large Re numbers (such as $Re \sim O(1)$ or larger, but not too large for the boundary layer to be in turbulent regime), it is possible for the case of non-separating flow to develop an asymptotic theory for determining the in-line force coefficients (Bearman, Downie, Graham and Obasaju, 1985b). The procedure used in this asymptotic theory is as follows: 1) Calculate the in-line force on the cylinder due to the oscillating flow, using the potential-flow theory; 2) calculate the oscillating, laminar boundary layer over the surface of the cylinder; 3) determine the perturbation to the outer flow caused by the predicted oscillating laminar boundary layer; and finally 4) calculate the in-line force on the cylinder induced by this perturbation. This together with the potential-flow in-line force (in Step 1) will be the total in-line force on the cylinder.

1) Potential-flow solution:

This can be obtained by solving Laplace's equation. Let the resulting solution be $W_0(z)$ where $W_0(z)$ is the complex potential, defined by

$$W_0(z) = \phi + i\psi \qquad (4.33)$$

in which ϕ is the potential function, ψ is the stream function and z is the complex coordinate (Fig. 4.6)

$$z = x + iy = re^{i\theta} \qquad (4.34)$$

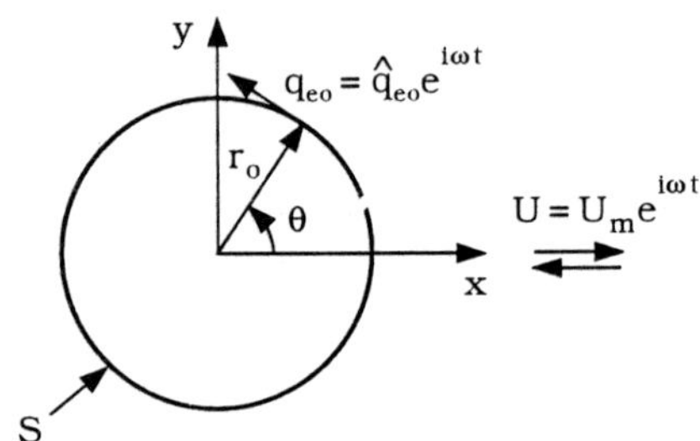

Figure 4.6 Definition sketch for potential-flow solution.

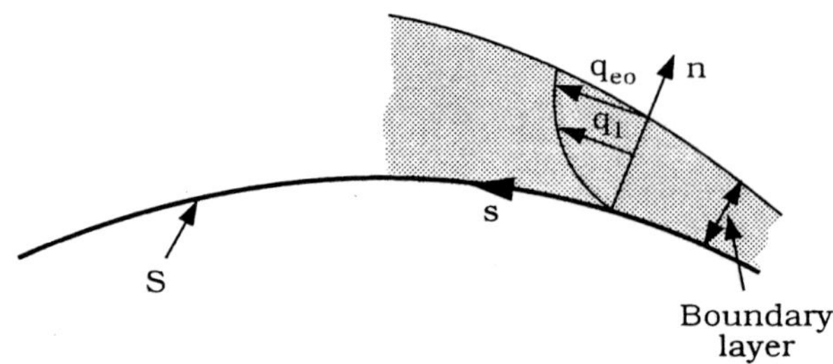

Figure 4.7 Definition sketch for the boundary layer developing on the cylinder surface.

(Milne-Thomson, 1962, Section 6.0). In the case of a circular cylinder, $W_0(z)$ is given by

$$W_0(z) = U(t)\left(z + \frac{r_0^2}{z}\right) \tag{4.35}$$

(Milne-Thomson, 1962, Section 6.22), and the velocity $U(t)$ in the preceding equation for the present case is given by

$$U(t) = U_m e^{i\omega t} \tag{4.36}$$

The in-line force on the cylinder due to this flow can be calculated, using the Blasius formula (Milne-Thomson, 1962, Section 6.41):

$$F_0 = -i\rho \frac{\partial}{\partial t} \oint_S W_0(z)\,dz \tag{4.37}$$

Inserting Eq. 4.35 in the preceding equation, the force due to this potential flow is obtained as

$$F_0 = 2\rho A \, \dot{U} \tag{4.38}$$

in which $A = \pi r_0^2$.

Since $U = U_m e^{i\omega t}$, then F_0 will be

$$F_0 = 2\rho A U_m i\omega e^{i\omega t} \tag{4.39}$$

2) Perturbation due to the boundary layer:

The speed due to the potential flow is calculated by

$$q_0 = \sqrt{u^2 + v^2} = |dW_0/dz| \tag{4.40}$$

Let q_{eo} be the speed on the surface S of the body (Fig. 4.6). From Eqs. 4.35, 4.36 and 4.40, q_{eo} is found as follows

$$q_{eo} = |\, dW_0/dz|_S = \hat{q}_{eo} e^{i\omega t} \tag{4.41}$$

in which $\hat{q}_{eo}$, the amplitude of q_{eo}, is

$$\hat{q}_{eo} = 2U_m \sin\theta \tag{4.42}$$

In response to the velocity q_{eo}, an oscillatory boundary layer will develope on S (Fig. 4.7). In the case when $KC \ll 1$, and $Re \sim O(1)$ or larger (so that the flow can be represented by an outer potential flow and an inner laminar boundary-layer flow), the boundary layer can be approximated to that which occurs on a plane wall. The velocity in such a boundary layer is given as (Batchelor, 1967, p. 354)

$$q_1 = \hat{q}_1 e^{i\omega t} \tag{4.43}$$

in which

$$\hat{q}_1 = \hat{q}_{eo}\left(1 - e^{-(1+i)\alpha n}\right) \tag{4.44}$$

Here, α is

$$\alpha = \left(\frac{\omega}{2\nu}\right)^{1/2} \tag{4.45}$$

and n being the local coordinate (Fig. 4.7) measured normal to the surface S of the body.

This boundary layer will perturb the previously predicted potential-flow force in the following two ways: 1) The wall shear stress caused by the boundary layer will contribute to the force (the friction force); and 2) the growth of the boundary layer will perturb the outer flow, and this will in turn perturb the pressure on the surface of the body, resulting in an additional contribution to the force.

The friction force:

The in-line component of the force due to the wall shear stress on S (the friction force) is

$$F_f = \int_{\theta=0}^{2\pi} \tau_w \sin\theta ds \tag{4.46}$$

in which

$$\tau_w = \mu \frac{\partial q_1}{\partial n}\bigg|_{n=0} = \mu(1+i)\alpha q_{eo} \tag{4.47}$$

and s being the local coordinate (Fig. 4.7) measured along the surface S in the direction of θ. Inserting Eq. 4.47 into Eq. 4.46, and using Eqs. 4.41 and 4.42, F_f is obtained as follows

$$F_f = \frac{1}{2}(1+i)\rho\omega D^2 U_m \left(\frac{\pi}{\beta}\right)^{1/2} e^{i\omega t} \tag{4.48}$$

in which

$$\beta = \frac{1}{\pi}\frac{D^2\omega}{2\nu} = \frac{Re}{KC} \tag{4.48a}$$

The force due to pressure perturbation

The growth of the boundary layer is not uniform over the surface S of the body. If δ^* is the displacement thickness of the boundary layer,

$$\delta^* = \int_0^\infty \left(1 - \frac{q_1}{q_{eo}}\right) dn = \frac{1}{(1+i)\alpha}, \tag{4.49}$$

the product $q_{eo}\delta^*$ will represent the flux deficit at section s (Fig. 4.8). The quantity $\frac{\partial}{\partial s}(q_{eo}\delta^*)ds$ will then represent the difference between the flux deficits at sections s and $s+ds$. This fluid, namely $\frac{\partial}{\partial s}(q_{eo}\delta^*)ds$, is entered into the outer

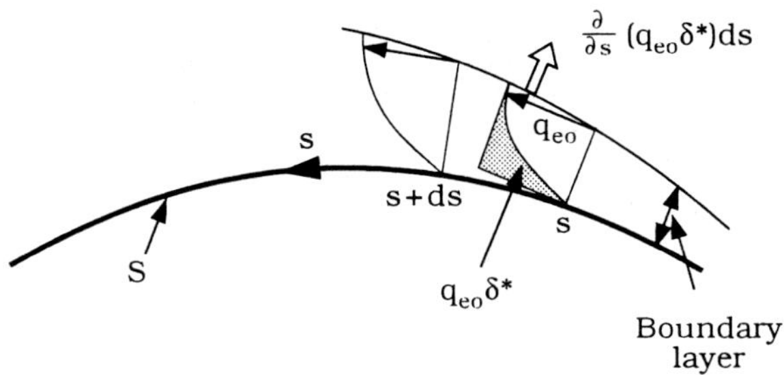

Figure 4.8 Fluid entrainment into the outer potential flow due to growing boundary layer.

potential flow over the length ds (this is the perturbation caused by the boundary layer).

The aforementioned effect can be considered as a source with the strength m determined from the following equation (see Milne-Thomson, 1962, Sections 8.10 and 8.12 for source and its complex potential)

$$2\frac{\partial}{\partial s}\left(q_{eo}\delta^{*}\right)ds = 2\pi m \; ds \tag{4.50}$$

and the corresponding complex potential function can be written as

$$W_1(z) = \oint_S -m\log\left(z - z'(s)\right)ds$$

$$= -\frac{1}{\pi(1+i)\alpha}\oint_S \frac{\partial q_{eo}}{\partial s}\log\left(z - z'(s)\right)ds \tag{4.51}$$

This complex potential will create an additional pressure on the surface S of the body, and the force caused by this additional pressure can be calculated by the Blasius formula

$$F_p = -i\rho\frac{\partial}{\partial t}\oint_S W_1(z)dz - \rho\frac{\partial}{\partial t}\oint Im\{W_1(z)\}dz \tag{4.52}$$

where the second integral represents the contribution from the fact that the stream function of the complex potential, namely $Im\{W_1(z)\}$, is not a constant on S. Using Eq. 4.51, the above integrals were calculated analytically by Bearman et al. (1985b) and the result is

$$F_p = \frac{1}{2}(1+i)\rho\omega D^2 U_m \left(\frac{\pi}{\beta}\right)^{1/2} e^{i\omega t} \tag{4.53}$$

As seen from Eqs. 4.48 and 4.53, the friction force and the pressure force apparently are equal.

3) Total in-line force and in-line force coefficients

The total in-line force is obtained from Eqs. 4.39, 4.48 and 4.53 as

$$F = F_0 + F_f + F_p =$$

$$= 2\rho A U_m i\omega\, e^{i\omega t} +$$

$$+ (1+i)\rho\omega D^2 U_m \left(\frac{\pi}{\beta}\right)^{1/2} e^{i\omega t} \tag{4.54}$$

The same force due to the Morison formulation is

$$F = \frac{1}{2}\rho C_D D U |U| + \rho C_M A \,\dot{U} \tag{4.55}$$

Inserting Eq. 4.36 in Eq. 4.55 and making the approximation that, over a flow cycle, $e^{i\omega t}|e^{i\omega t}| \simeq (8/(3\pi))e^{i\omega t}$, the Morison force can be written as

$$F = \frac{1}{2}\rho C_D D U_m^2 \left(\frac{8}{3\pi} e^{i\omega t}\right) +$$

$$+ \rho C_M A U_m \omega i e^{i\omega t} \tag{4.56}$$

From Eqs. 4.54 and 4.56, the in-line force coefficients are found as follows

$$C_M = 2 + 4(\pi\beta)^{-1/2} \tag{4.57}$$

$$C_D = \frac{3}{2}\pi^3 (KC)^{-1}(\pi\beta)^{-1/2} \tag{4.58}$$

Stokes (1851) was the first to develop an analytical solution for the case of a cylindrical body oscillating sinusoidally in a viscous fluid. His solution is given in the form of a series expansion in powers of $(Re/KC)^{-1/2}$. The results of the asymptotic theory given in the preceding paragraphs are the same as the Stokes' results to $O[(Re/KC)^{-1/2}]$. Subsequently, Wang (1968) extended Stokes' analysis to $O[(Re/KC)^{-3/2}]$, implementing the method of inner and outer expansions.

Fig. 4.9 compares the results of the asymptotic theory with those of experiments by Sarpkaya (1986a) for the value of the $\beta(= Re/KC)$ parameter of 1035. As is seen, the theory shows remarkable agreement with the experiments for very small values of KC where the flow remains attached (cf. Fig. 3.15).

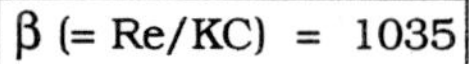

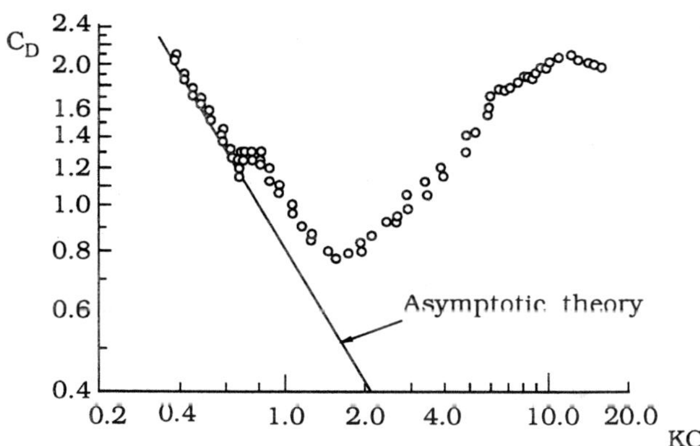

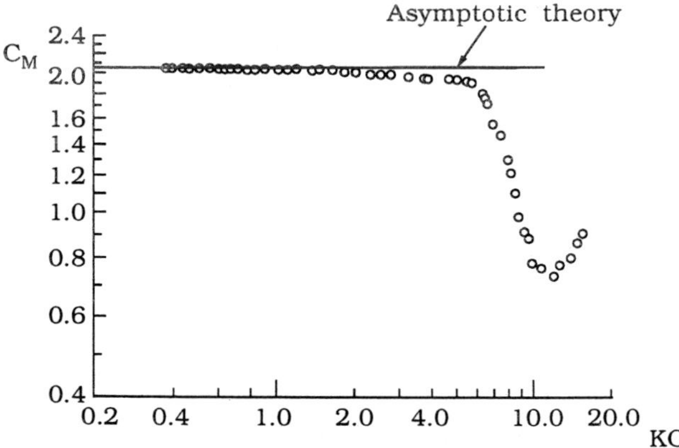

Figure 4.9 Drag and inertia coefficients *vs* Keulegan-Carpenter number. $Re/KC = 1035$. Experiments from Sarpkaya (1986a). Asymptotic theory (Eqs. 4.58 and 4.57).

Measurements of C_D and C_M coefficients

The preceding analysis indicates that the in-line cofficients are dependent on two independent variables, namely the Reynolds number and the Keulegan-Carpenter number. The theory gives the explicit form of this dependence. However, this is for the combination of very small KC numbers and sufficiently large Re numbers only. Although there are several numerical codes developed to calculate flow around and forces on a cylinder in oscillatory flows (Chapter 5), these are still at the development stage and therefore not fully able to document the way in which the force coefficients vary with KC and Re. Hence, the experiments appear to be the most reliable source of information with regard to the force coefficients at the present time.

There are various techniques to determine the coefficients C_D and C_M experimentally. For periodic flows, the most suitable technique may be "the method of least squares". The principle idea of this method is that the C_D and C_M coefficients are determined in such a way that the mean-squared difference between the predicted (by the Morison formula) and the measured force is minimum. A brief description of the method of least squares is given below.

Let $F_m(t)$ be the measured in-line force at any instant t. Likewise, let $F_p(t)$ be the predicted in-line force corresponding to the same instant, namely

$$F_p(t) = \frac{1}{2}\rho C_D D U(t)|U(t)| + \rho C_M A \, \dot{U}(t) \tag{4.59}$$

Let, for convenience:

$$f_d = \frac{1}{2}\rho C_D D \quad \text{and} \quad f_i = \rho C_M A \tag{4.60a, b}$$

Therefore, the predicted force:

$$F_p(t) = f_d U(t)|U(t)| + f_i \, \dot{U}(t) \tag{4.61}$$

Now, let ε^2 be the sum of the difference between the predicted force and the measured force over the total length of the record:

$$\varepsilon^2 = \sum \left[F_p(t) - F_m(t) \right]^2$$

$$= \sum \left[f_d U(t)|U(t)| + f_i \, \dot{U}(t) - F_m(t) \right]^2 \tag{4.62}$$

For ε^2 to be minimum:

$$\frac{\partial \varepsilon^2}{\partial f_d} = 0 \quad , \quad \frac{\partial \varepsilon^2}{\partial f_i} = 0 \tag{4.63}$$

The first equation leads to:

$$f_d\left(\sum U^4(t)\right) + f_i\left(\sum U(t)|U(t)|\,\dot{U}(t)\right) = \sum U(t)|U(t)|F_m(t) \qquad (4.64)$$

and the second equation leads to:

$$f_d\left(\sum U(t)|U(t)|\,\dot{U}(t)\right) + f_i\left(\sum \dot{U}^2(t)\right) = \sum \dot{U}(t)F_m(t) \qquad (4.65)$$

where the summation is taken over the total record length. Eqs. 4.64 and 4.65 form two simultaneous equations with f_d and f_i as unknowns. Solving for f_d and f_i, the in-line force coefficients C_D and C_M can be determined from Eqs. 4.60a and b, respectively.

For a *sinusoidal flow*, it can be shown that the method of least squares gives C_D and C_M as follows:

$$C_D = \frac{8}{3\pi}\,\frac{1}{\rho D U_m^2}\,\int_0^{2\pi} F_m\,\cos(\omega t)\,|\cos(\omega t)|\,d(\omega t) \qquad (4.66)$$

$$C_M = -\frac{2KC}{\pi^3}\,\frac{1}{\rho D U_m^2}\,\int_0^{2\pi} F_m\,\sin(\omega t)d(\omega t) \qquad (4.67)$$

Given the time series of the measured force $F_m(t)$, the C_D and C_M coefficients can therefore be worked out from the preceding equations.

Another technique regarding the experimental determination of C_D and C_M coefficients is the Fourier analysis. This latter technique yields identical C_M values. As for C_D, the C_D values obtained by the Fourier analysis differ only slightly from those obtained by the method of least squares (Sarpkaya and Isaacson, 1981).

Keulegan and Carpenter (1958) were the first to determine the C_D and C_M coefficients for a cylinder exposed to real waves (using the Fourier analysis). Subsequently, Sarpkaya (1976a) made an extensive study of the forces on cylinders exposed to sinusoidally varying oscillatory flows (created in an oscillatory U-shaped tube) with the purpose of determining the force coefficients in a systematic manner as functions of the Keulegan-Carpenter number and the Reynolds number as well as the relative roughness of the cylinder.

Variation of C_D and C_M with KC number

The variation of C_D and C_M with KC has already been illustrated in Fig. 4.9 in conjunction with the asymptotic theory (Example 4.3). The range of KC covered in the figure was rather small. Fig. 4.10 illustrates this variation, covering a much wider range of KC number up to about 60. The Reynolds number for the

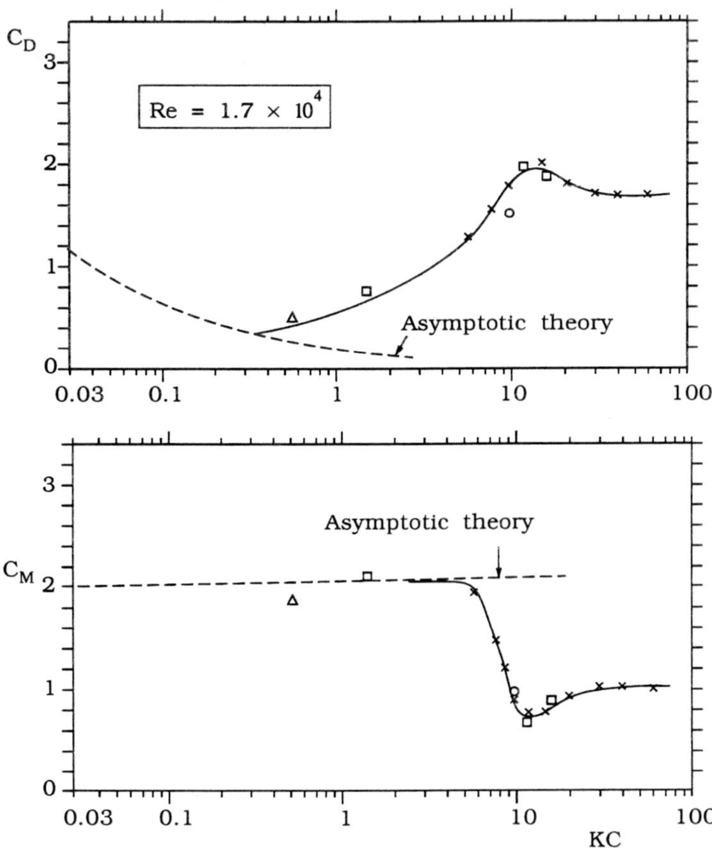

Figure 4.10 Variation of in-line force coefficients with KC number for a given Re number, namely $Re = 1.7 \times 10^4$. Data from ×: Sarpkaya (1976a), o, □: Bearman et al. (1985a), and △: Anatürk (1991). Asymptotic theory (Eqs. 4.57 and 4.58).

data given in the figure is constant ($Re = 1.7 \times 10^4$). The results of the asymptotic theory for the same Re number are also included in the figure.

First consider the *drag coefficient*. As seen from the figure, there are three distinct regimes in the variation of C_D with KC: 1) $KC \lesssim 0.3$, 2) $0.3 \lesssim KC \lesssim 13$ and 3) $KC \gtrsim 13$.

In the first regime, namely $KC < 0.3$, the drag coefficient must be governed by the asymptotic theory summarized in Example 4.3, as the conditions for the application of the asymptotic theory are fully satisfied, namely KC is very small, Re is sufficiently large, and the flow remains attached (Fig. 3.15). Unfortunately, no experimental data exist in the literature for this particular Re number in this range of KC to confirm the validity of the application of the asymptotic theory.

When $KC \cong 0.3$ is reached, separation begins to occur. Therefore, the drag will no longer be governed by the asymptotic theory. Hence, the C_D variation will begin to diverge from the line representing the asymptotic theory in Fig. 4.10. The figure indicates that this regime of C_D variation with KC extends up to $KC \simeq 13$. Apparently, this latter value of KC coincides with that corresponding to the upper boundary of the transverse-vortex-street regime described in Section 3.2. When KC is increased beyond $KC \sim 13$, the transverse vortex street will disappear, and the shed vortices will form a vortex street lying parallel to the direction of the oscillatory motion, in much the same way as in steady current. Therefore the drag coefficient will in this regime ($KC > 13$) not change very extensively with KC.

Regarding the *inertia coefficient*, C_M, from Fig. 4.10, here, too, there are three different regimes, namely: 1) $KC \stackrel{<}{\sim} 6$, 2) $6 \stackrel{<}{\sim} KC \stackrel{<}{\sim} 13$ and 3) $KC \stackrel{>}{\sim} 13$, the boundary between the first two regimes, namely $KC = 6$, being different, however, from that corresponding to the drag coefficient C_D.

As for the first regime, $KC < 6$, the asymptotic theory predicts the C_M coefficient extremely well. However, when KC reaches the value of approximately 6, an abrupt fall occurs in C_M (the so-called inertia crisis). This abrupt fall continues over the range from $KC = 6$ to 13.

$KC \sim 6$ coincides with the lower limit of the vortex-shedding regimes (Section 3.2). The rapid change in C_M at this value of KC number may therefore be attributed to the vortex shedding. The interaction between the vortex shedding and the hydrodynamic process generating the hydrodynamic mass may produce this observed, sudden drop in the C_M coefficient. The reduction in C_M is so large that, subtracting the Froude-Krylov part of C_M, namely unity, from the measured values of C_M, it is found that the inertia cofficient ($C_m = C_M - 1$) will take even negative values for KC values around $KC = 10$, as seen from Fig. 4.10.

As for the third regime in the variation of C_M with KC, namely the range $KC \stackrel{>}{\sim} 13$, the vortex street formed by the shed vortices in this range lies parallel to the direction of the oscillatory motion, as mentioned previously. Therefore the change in C_m (or C_M) with increasing KC in this range will not be very extensive.

Effect of Re number on C_D and C_M

Fig. 4.11 presents the in-line-force-coefficient data, illustrating the effect of Re. The drag coefficient diagram includes also C_D versus Re variation for steady currents (Fig. 2.7) to facilitate comparison. The figure is based on the results

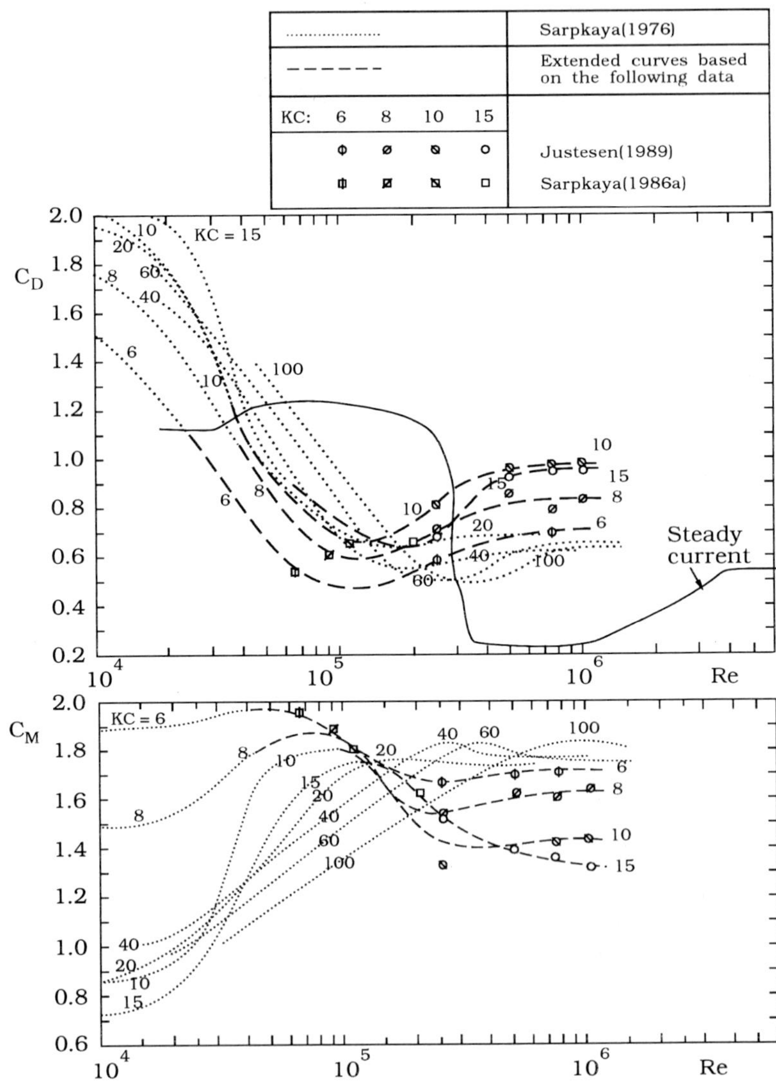

Figure 4.11 In-line force coefficients for a free, smooth cylinder. Steady current C_D variation is reproduced from Fig. 2.7 which is originally taken from Schewe (1983). Oscillatory flow data are from Sarpkaya (1976a), Sarpkaya (1986a) and Justesen (1989).

of the extensive study of Sarpkaya (1976a and 1986a) and the study of Justesen (1989).

It is apparent from the figure that the drag coefficient varies with Re in the same manner as in steady currents. However, the drop in C_D with Re (which is known as the drag crisis in steady currents, see Section 2.2) does not occur as abruptly as in steady currents.

For a given KC number, C_D first experiences a gradual drop with increasing Re number. Similar to the steady currents, this range of Re number may be interpreted as the lower transition regime (see Section 2.2). Subsequently a range of Re number is reached where C_D remains approximately constant. This may be interpreted as the supercritical Re-number regime. Following that, C_D begins to increase with an increase in Re, interpreted as the upper transition Re-number regime. Finally, the C_D coefficient reaches a plateau where it remains approximately constant with increasing Re. This latter regime, on the other hand, may be interpreted as the transcritical Re-number regime.

Regarding the inertia coefficient in Fig. 4.11, the general trend is opposite to that observed for C_D. Where C_D experiences high values, C_M experiences low ones. The increase in C_M may be due to the weak vortex-shedding regime which takes place in the supercritical flow regime and particularly in the upper-transition flow regime.

Example 4.4:   Effect of friction on C_D and C_M

In Chapter 2, based on the experimental data obtained for steady currents, it was demonstrated that, for most of the practical cases, the friction drag is only a small fraction of the total drag (Fig. 2.4).

Regarding the oscillatory flows, unfortunately no data are available in the existing literature, therefore no conclusion can be drawn with regard to the effect of friction on the in-line force. Nevertheless, this effect may be assessed, utilizing Justesen's (1991) theoretical analysis. The results depicted in Fig. 4.12 are from the work of Justesen (1993, private communication), which is an extension of Justesen (1991) where a numerical solution was obtained to a stream function-vorticity formulation of the Navier-Stokes equations for the flow around a circular cylinder at small KC numbers in the subcritical Reynolds number range. Although the results are limited to small Re numbers, they nevertheless illustrate the influence of the friction on the force coefficients.

Regarding the drag coefficients, Fig. 4.12 indicates that the friction is extremely important for small KC numbers. As a matter of fact, the contribution of friction to the total drag is 50% for very small KC numbers ($KC = O(1)$ or less), as predicted by the asymptotic theory (Example 4.3). As KC is increased, however, the diagram indicates that the effect of friction on the drag gradually

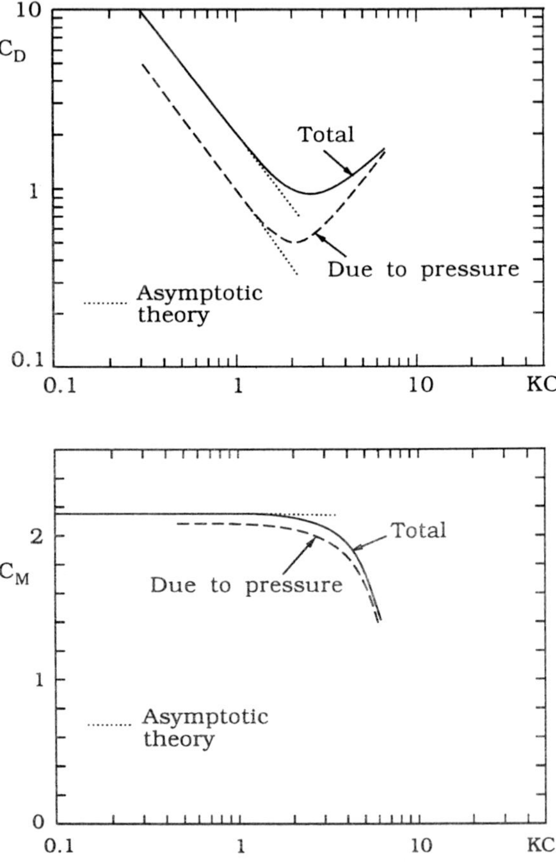

Figure 4.12 Effect of friction on the force coefficients. $\beta(= Re/KC) = 196$. From numerical solution of Navier-Stokes equations in the sub-critical Re number range. Justesen (1993, private communication), which is an extension of Justesen (1991). Asymptotic theory: Eqs. 4.57 and 4.58.

decreases; at $KC = 6$, for example, the friction drag becomes less than 10% of the total drag. Therefore, for large KC numbers, the drag portion of the in-line force may be considered to be due to pressure alone.

Regarding the inertia coefficient, on the other hand, it is seen from Fig. 4.12 that the friction-generated inertia force is only a very small fraction of the

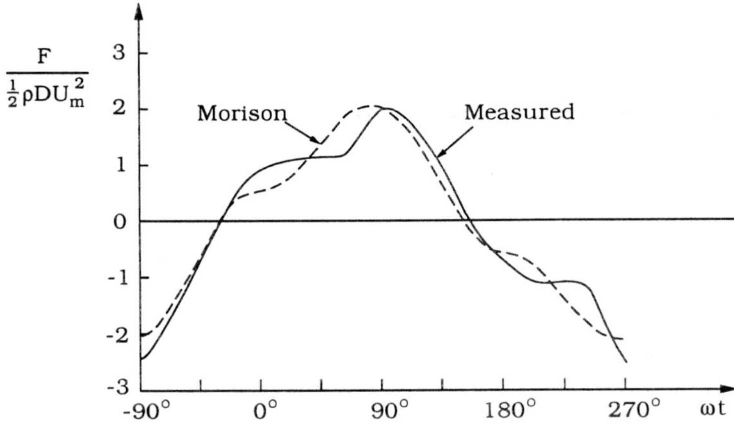

Figure 4.13 Comparison of measured and Morison-predicted in-line force.
$KC = 14$, $Re = 2.8 \times 10^4$. Sarpkaya and Isaacson (1981).

total inertia force (less than 4% at best). Therefore it may be neglected in most
of the practical cases.

4.1.5 Goodness-of-fit of the Morison equation

Fig. 4.13 gives a comparison between the measured and Morison-predicted
in-line forces. Clearly, the Morison representation is not extremely satisfactory
with respect to the measured variation of the in-line force. The question how well
the Morison equation represents the measured in-line force has been the subject
of several investigations (Sarpkaya and Isaacson, 1981).

In order to assess the applicability of the Morison equation, one may introduce a goodness-of-fit parameter, δ, defined by

$$\delta = \frac{\int\limits_0^{T_t} (F_m - F_p)^2 \, dt}{\int\limits_0^{T_t} F_m^2 \, dt} \tag{4.68}$$

in which F_m and F_p are the measured and the predicted (by Morison's equation)
forces, respectively, and T_t is the total duration of data sampling. Fig. 4.14 shows
a typical variation of δ with respect to KC. As is seen, δ increases from zero

for small KC to a maximum at $KC = 12$ where δ attains a value of $\delta \cong 0.12$, and with further increase in KC, δ decreases again. Clearly, the ability of the Morison equation to predict the force depends heavily on the KC number. In the inertia-dominated region, δ is rather small, therefore the Morison representation is rather good, but when the flow is separated, the Morison equation can not provide a complete description of the force variation. To tackle this problem Sarpkaya introduced a four-term Morison equation which may be written as

$$\frac{F}{\frac{1}{2}\rho DU_m^2} = (\pi^2/KC)\, C_M \sin\theta - C_D \cos\theta|\cos\theta|+$$

$$\Lambda^{-1/2}[0.01 + 0.1\exp\{-0.08(KC - 12.5)^2\}]\cos[3\theta-$$

$$\Lambda^{-1/2}(-0.05 - 0.35\exp\{-0.04(KC - 12.5)^2\})]+$$

$$\Lambda^{-1/2}[0.0025 + 0.053\exp\{-0.06(KC - 12.5)^2\}]\cos[5\theta-$$

$$\Lambda^{-1/2}(0.25 + 0.6\exp\{-0.02(KC - 12.5)^2\})] \tag{4.69}$$

in which $\theta = \omega t$ and $\Lambda = (2 - C_M)/(KC\, C_D)$.

The results have shown that, in this way, a significant improvement has been obtained. (Sarpkaya (1981) and Sarpkaya and Wilson (1984)).

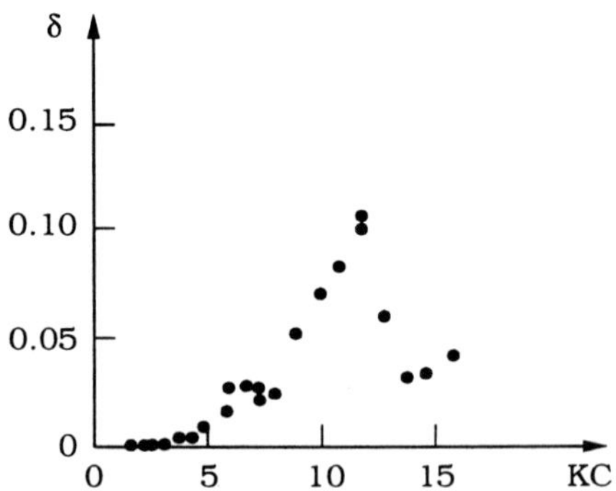

Figure 4.14 Goodness-of-fit parameter δ as function of KC. $Re = 5 \times 10^5$. Smooth cylinder. Justesen (1989).

4.2 Lift force in oscillatory flow

When a cylinder is exposed to an oscillatory flow, it may undergo a lift force (Fig. 4.1). This lift force oscillates at a fundamental frequency different from the frequency of the oscillatory flow. The time variation of the force is directly related to the vortex motions around the cylinder, as has already been discussed in Section 3.2.

Obviously, if the flow around the cylinder is an unseparated flow (very small KC numbers, Figs. 3.15 and 3.16), then no lift will be generated.

Fig. 4.15 illustrates the emergence and subsequently the development of the lift force as the KC number is increased from zero. The figure indicates that, while the lift force first comes into existence when KC becomes 4 (which is due to the asymmetry in the formation of the wake vortices; see Fig. 3.2.e), well-established lift-force regimes are formed only after KC is increased to the value of 6-7, the value of KC number beyond which vortex shedding is present.

When the analysis of the lift force is considered, the most important quantities are the fundamental lift frequency and the magnitude of the lift force.

Regarding the fundamental lift frequency, this has been discussed in details in Sections 3.2 and 3.3, and the normalized fundamental lift frequency $N_L(= f_L/f_w)$, namely the number of oscillations in the lift per flow cycle, has been given in Table 3.1 and in Fig. 3.16.

As regards the magnitude of the lift force, there are two approaches. In one, the maximum value of the lift force is considered, while in the other the root-mean-square (r.m.s.) value of the lift force is adopted to represent the magnitude of the lift force. These may, in terms of the force coefficients, be written in the following forms:

$$F_{L\mathrm{max}} = \frac{1}{2}\rho C_{L_{\mathrm{max}}} D U_m^2 \tag{4.70}$$

and

$$F_{L\mathrm{rms}} = \frac{1}{2}\rho C_{L\mathrm{rms}} D U_m^2 \tag{4.71}$$

in which $F_{L\mathrm{max}}$ and $F_{L\mathrm{rms}}$ are the maximum- and r.m.s.-values of the lift force, respectively, while $C_{L\mathrm{max}}$ and $C_{L\mathrm{rms}}$ are the corresponding force coefficients. If the time variation of the lift force is approximated by a sinusoidal variation, then the two coefficients will be linked by the following relation

$$C_{L\mathrm{max}} = \sqrt{2}C_{L\mathrm{rms}} \tag{4.72}$$

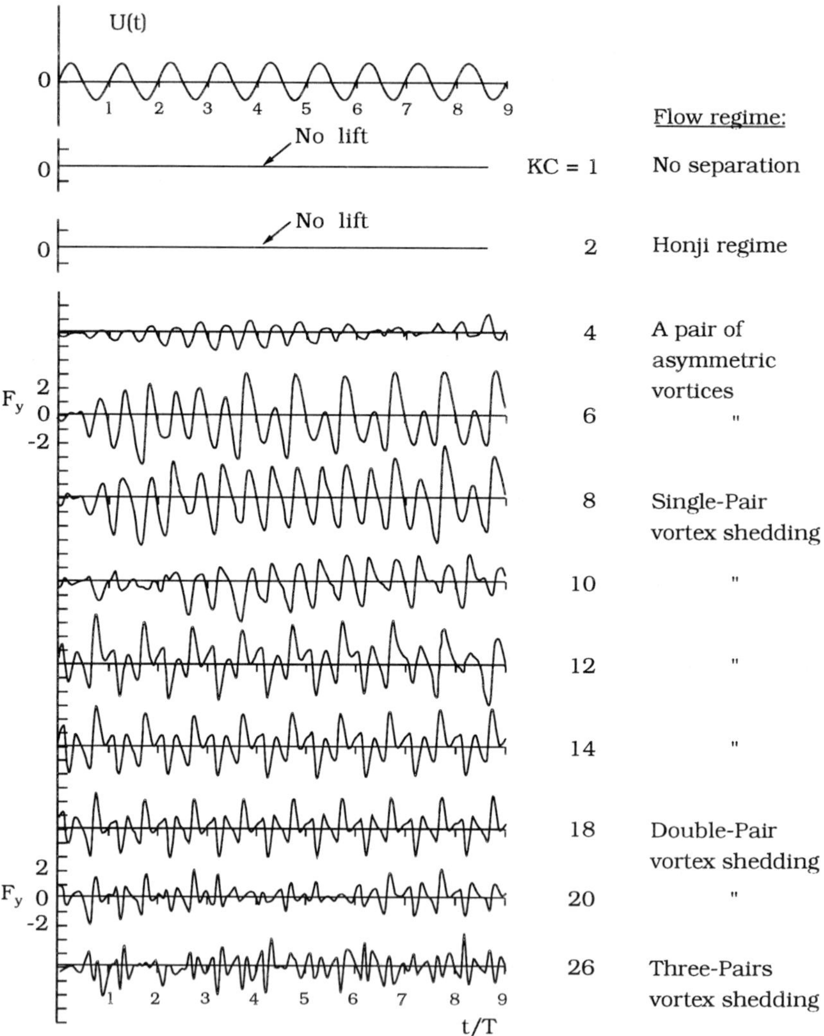

Figure 4.15 Computed lift force traces over nine periods of oscillation at various KC-values for $\beta(= Re/KC) = 196$. Justesen (1991). For the various flow regimes indicated in the figure, see Figs. 3.15 and 3.16.

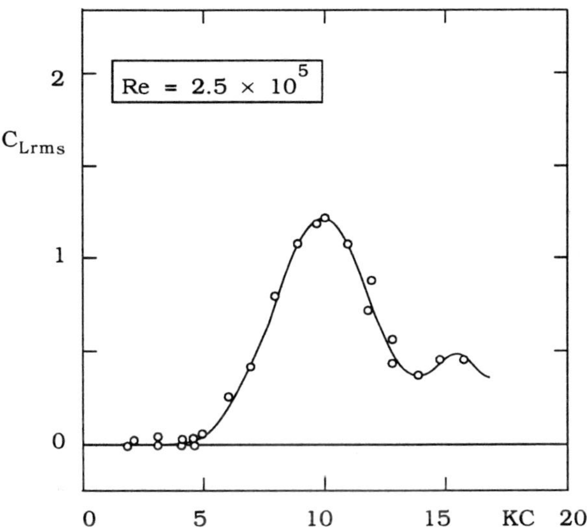

Figure 4.16 Variation of r.m.s. lift-force coefficient as function of KC number. Experimental data from Justesen (1989).

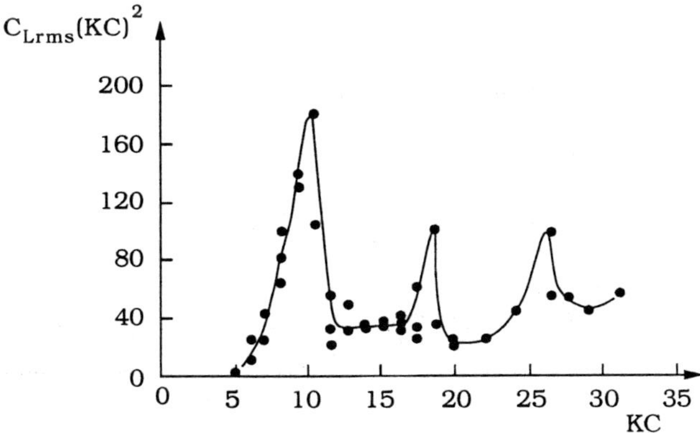

Figure 4.17 Lift force r.m.s. as function of KC for a given value of $\beta(= Re/KC) = 730$. Williamson (1985).

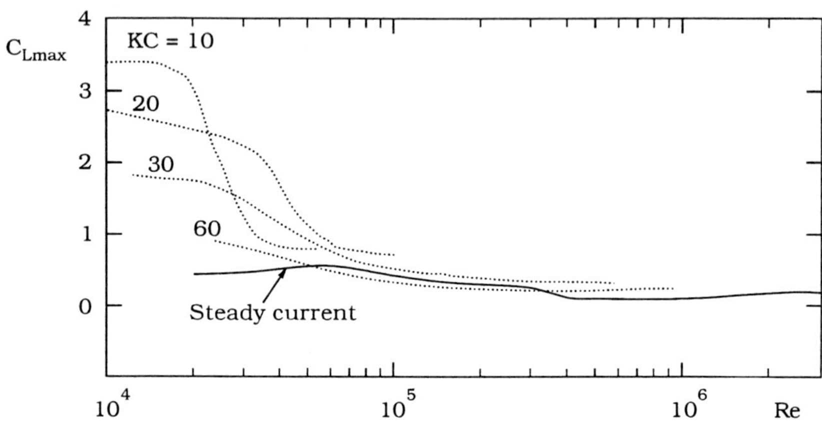

Figure 4.18 Maximum lift coefficient for a free, smooth cylinder. Oscillatory flow data from Sarpkaya (1976a). Steady-current C_L variation is reproduced from Fig. 2.8 where $(\overline{C_L'^2})^{1/2}$ is multiplied by $\sqrt{2}$ to get the maximum lift coefficient, assuming that the lift varies sinusoidally with time.

Both C_{Lmax} and C_{Lrms} are functions of KC and Re. Fig. 4.16 gives C_{Lrms} as a function of KC number for $Re = 2.5 \times 10^5$ (Justesen, 1989). The figure indicates that the lift force experiences two maxima, one at KC around 10 and a slight maximum at KC around 16. This behaviour has been observed previously also by authors such as Maull and Milliner (1978), Williamson (1985), and Sarpkaya (1986b, 1987). In Williamson's (1985) representation, the product $C_{Lrms}(KC)^2$ (rather than C_{Lrms}) has been plotted as a function of KC. This obviously magnifies the aforementioned effect significantly. Williamson's diagram is reproduced here in Fig. 4.17. The figure clearly shows that C_{Lrms} attains maximum values at $KC \cong 11$, 18 and 26. Williamson points out that these peaks probably reflect an increase in the repeatability of the shedding patterns. Each peak corresponds to a certain pattern of shedding; namely, the first peak corresponds to the single-pair regime ($7 < KC < 15$), the second to the double-pair regime ($15 < KC < 24$), and the third to the three-pairs regime ($24 < KC < 32$). Apparently, these peaks coincide with the KC numbers at which large spanwise correlations are measured. The minima in the diagram, on the other hand, correspond to the KC numbers where the spanwise correlation is measured to be relatively low, cf. Fig. 3.28. As discussed in Section 3.5 in relation to Fig. 3.28, the preceding behaviour is linked to the fact that the correlation is measured to be large (and apparently C_{Lrms} experiences maximum values) at certain KC numbers because these KC

numbers lie in the centre of the corresponding KC regimes, while the correlation is measured to be low (and, as a result, C_{Lrms} experiences minimum values) at certain KC numbers because these KC numbers lie at the boundaries between the neighbouring KC regimes.

Finally, Fig. 4.18 presents the lift-force data, illustrating the effect of Re number on the lift force. The figure includes also the steady current data which are reproduced from Fig. 2.8 to facilitate comparison. As is seen, the effect of Re is quite dramatic (see the discussion in Section 2.3).

4.3　Effect of roughness

When the cylinder surface is rough, the roughness will affect various aspects of the flow, such as the hydrodynamic instabilities (vortex shedding and interaction of vortices), the separation angle, the turbulence level, the correlation length, and the vortex strength. In addition to these effects, it increases the cylinder diameter, and the projected area. Therefore it must be anticipated that the effect of roughness upon the force coefficients can have some influence.

Fig. 4.19 shows the influence of roughness on the in-line force coefficients. The data come from the work by Justesen (1989). It must be emphasized that the experimental system in Justesen's work was maintained the same for all the three experiments indicated in the figure, and the experiments were performed under exactly the same flow conditions. It is only the cylinder roughness which was changed. Therefore, the change in the force coefficients is directly related to the change in the roughness.

The figure shows that the drag coefficient increases and the inertia coefficient decreases when the cylinder is changed from a smooth cylinder to a rough one with $k/D = 3 \times 10^{-3}$. Furthermore, it is clear that C_D increases with increasing roughness. Apparently C_M is not influenced much with a further increase in the roughness.

Regarding the increase in C_D with increasing roughness, this may be interpreted in the same way as in steady currents, considering that the Reynolds number of the tests, namely $Re = 5 \times 10^5$, is in the post-critical range (see Figs. 2.11 and 2.14, and also the discussion in Section 2.2).

Regarding the decrease in C_M, on the other hand, a clear explanation is difficult to offer. The non-linear interaction between the vortex shedding and the hydrodynamic process generating the hydrodynamic mass – the mechanism behind the reduction in the hydrodynamic mass in the vortex-shedding-regime KC numbers – must occur more strongly in the case of rough cylinder, since the reduction in C_M is much larger in this case than in the case of smooth cylinder.

Fig. 4.20 illustrates the effect of cylinder roughness on C_D and C_M when KC is kept constant ($KC = 20$, in the presented figure), while Re is changed. It

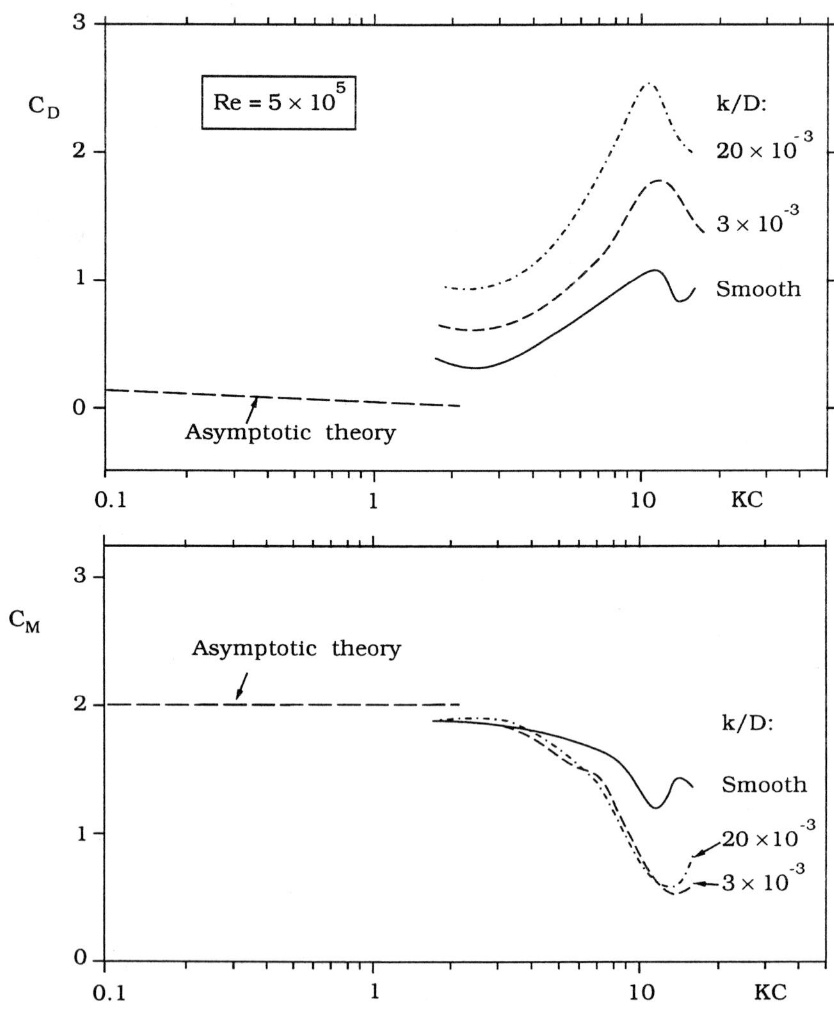

Figure 4.19 Effect of roughness on in-line force coefficients. Experimental data from Justesen (1989). Asymptotic theory: Eqs. 4.57 and 4.58.

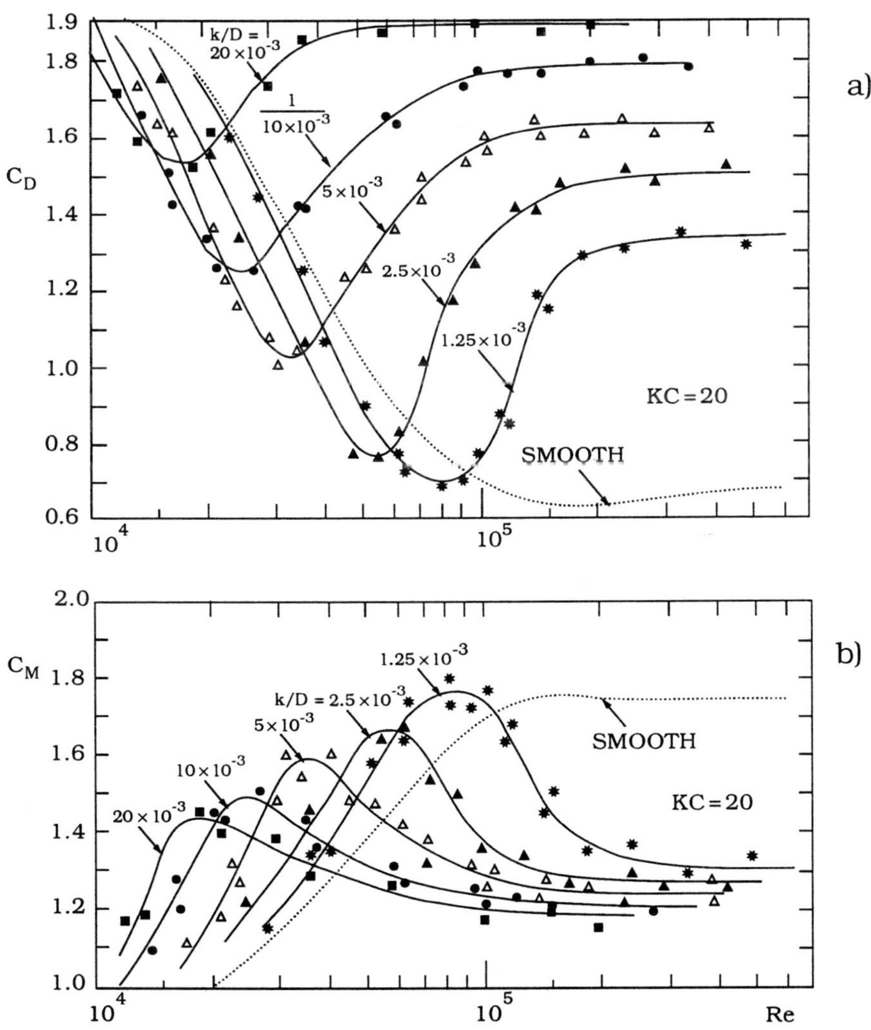

Figure 4.20 Effect of roughness on C_D and C_M versus Re variation. Sarpkaya (1976a).

is interesting to note that the way in which C_D versus Re variation changes with respect to the roughness is quite similar to that observed in the case of steady currents (Fig. 2.11).

As far as the lift force is concerned, Fig. 4.21 illustrates the effect of the change in roughness on the lift coefficient. Note that the depicted data are from the same study as in Fig. 4.19.

Again, the effect is there. It appears that the lift generally increases when the cylinder is changed from a smooth cylinder to a rough one. Similar results were obtained also by Sarpkaya in his work where the parameter $\beta(=Re/KC)$ was kept constant while KC was changed (see Sarpkaya (1976a) and Sarpkaya and Isaacson (1981)).

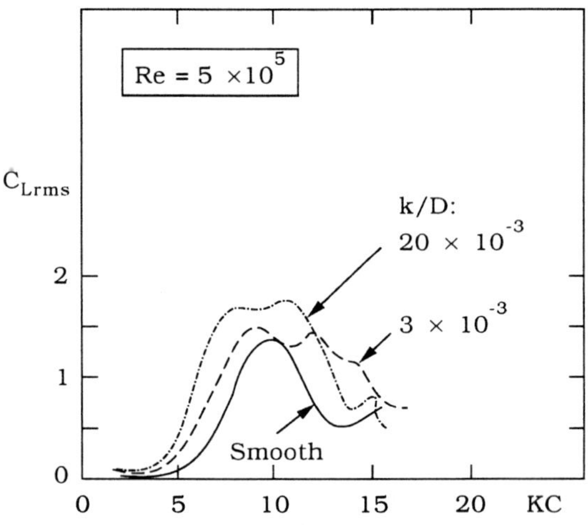

Figure 4.21 Effect of roughness on lift coefficient. Experimental data from Justesen (1989).

Finally, it may be noted that the subject has been investigated very extensively since the mid seventies. This is among other things because of its importance in practice where the roughness is caused by the marine growth. For further implications of the effect of surface roughness on the force coefficients, the reader is referred to the following work: Sarpkaya (1976a, 1977b, 1986b, 1987, 1990), Rodenbusch and Gutierrez (1983), Kashara, Koterayama and Shimazaki (1987), Justesen (1989), Wolfram and Theophanatos (1989), Wolfram, Javidan and Theophanatos (1989) and Chaplin (1993a) among others.

4.4   Effect of coexisting current

If current coexists together with waves, the presence of current may affect the waves. The problem of wave-current interaction is an important issue in its own right. Detailed reviews of the subject are given by Peregrine (1976), Jonsson (1990) and Soulsby, Hamm, Klopman, Myrhaug, Simons and Thomas (1993).

In the following discussion, for the sake of simplicity, we shall consider that the oscillatory flow, which simulates the waves, remains unchanged in the presence of a superimposed current. Let U_c be the velocity of the current. The key parameter of the study will therefore be the ratio of the current velocity to the maximum value of the velocity of the oscillatory flow, namely U_c/U_m. Although there are several alternatives with regard to the definition of the Reynolds number and the Keulegan-Carpenter number in the present case, the definitions adopted in the case of pure oscillatory flow, namely, $Re - U_m D/\nu$ and $KC - U_m T_w/D$ may be maintained.

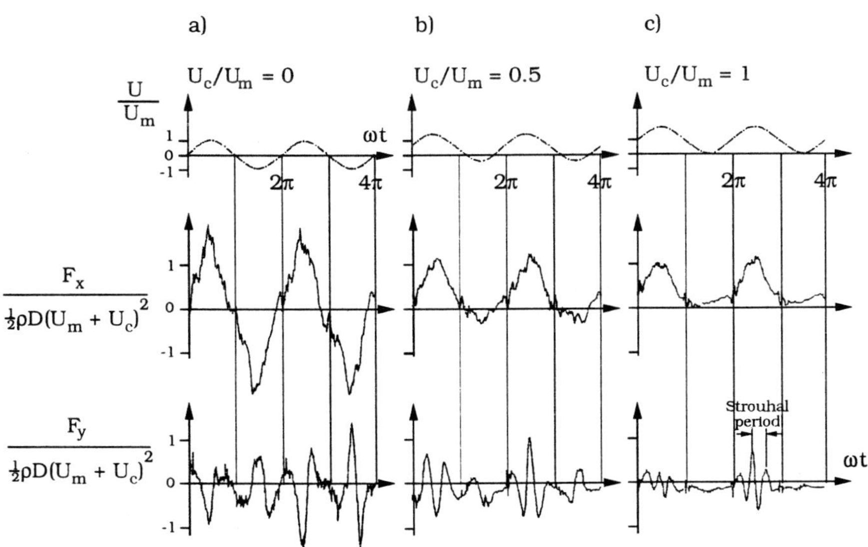

Figure 4.22 Force time series in the case of coexisting current. $KC = 20$.
Sumer et al. (1992).

The effect of coexisting current on forces has been investigated by several authors. These investigations include those by Moe and Verley (1980), Sarpkaya and Storm (1985), Justesen, Hansen, Fredsøe, Bryndum and Jacobsen (1987), Bearman and Obasaju (1989) and Sumer, Jensen and Fredsøe (1992).

The effect of current on forces can be described by reference to Fig. 4.22. The force traces depicted in the figure are taken from the study of Sumer et al. (1992) where the oscillatory flow was generated by the carriage technique, while the current was achieved by recirculating water in the flume.

From the figure the following observations can be made:

1) The in-line force varies with respect to time in the same fashion as the flow velocity.

2) The way in which the lift force varies with time during the course of one flow cycle changes markedly as the parameter U_c/U_m is changed from 0 to 1. For $U_c/U_m = 0.5$, the portion of the flow period where the flow velocity $U < 0$ is just long enough to accomodate shedding from both the upper and the lower sides of the cylinder; this is characterized by one positive and one negative lift force in the lift force trace, Fig. 4.22.

For $U_c/U_m = 1$, however, the figure shows that the shedding disappears (which is characterized by the non-oscillating portions of the lift force traces) when the oscillatory component of the motion is in the direction opposite to the current.

3) During the time periods when the vortex shedding exists, the figure indicates that the Strouhal relation

$$St = \frac{f_v D}{(U_c + U_m)} \tag{4.73}$$

is satisfied provided that the velocity is taken as the sum of the current velocity U_c and the wave velocity U_m. Here f_v is the average vortex-shedding frequency.

Regarding the in-line force coefficients, the Morison equation may be adopted in the present case in the same format as in Eq. 4.29, but with the velocity $U(t)$ defined now in the following way

$$U = U_c + U_m \sin(\omega t) \tag{4.74}$$

Fig. 4.23 presents the C_D and C_M coefficients as functions of the parameter U_c/U_m.

The drag coefficient generally decreases with the ratio U_c/U_m. It approaches, however, the asymptotic value (shown with dashed lines in the diagram) measured for steady current for the same surface roughness and the same Re number, as $U_c/U_m \to \infty$, as expected.

The inertia coefficient, C_M is apparently not very sensitive to U_c/U_m except for the $KC = 5$ case. The discrepancy between the results of Sumer et al.'s (1992) study and those of Sarpkaya and Storm (1985) may be attributed to the differences in the roughness and also in the Re number of the experiments. Also,

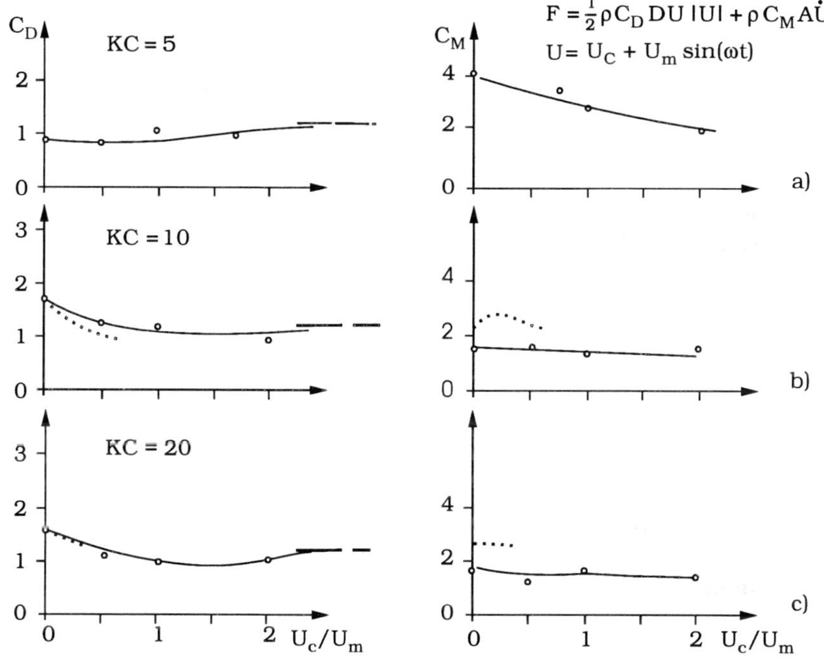

Figure 4.23 Effect of coexisting current on in-line force coefficients. Data from Sumer et al. (1992), $Re = 3 \times 10^4$ and $k/D = 4 \times 10^{-3}$. Dotted curves: Sarpkaya and Storm (1985), $k/D = 10^{-2}$ and $Re = 1.8 \times 10^4$ for $KC = 10$ and 3.6×10^4 for $KC = 20$. Dashed lines: Asymptotic values for steady current for $k/D = 4 \times 10^{-3}$ ($k_s/D = 10 \times 10^{-3}$) and $Re = 3 \times 10^4$ taken from Achenbach and Heinecke (1981) (see Fig. 2.11).

the forces that have been predicted in Sumer et al.'s study are from the pressure measurements at the middle section of the cylinder while, in the study of Sarpkaya and Storm, they were measured by the force transducers over a finite length of the cylinder.

Fig. 4.24 illustrates the influence of current on the lift coefficient. The lift coefficient is defined in the same way as in Eq. 4.70 with U_m replaced now by $U_c + U_m$. The figure indicates that C_{Lmax} decreases markedly when the current is superimposed on the oscillatory flow. Yet, as the ratio U_c/U_m increases, the lift coefficient might be expected to approach its asymptotic value obtained for the

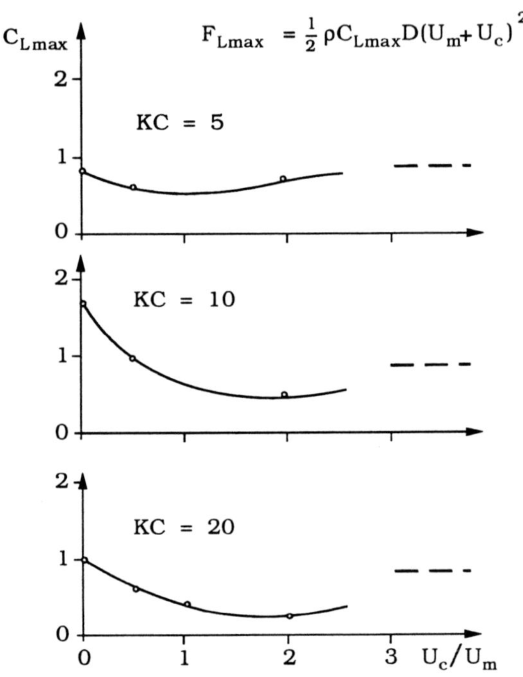

Figure 4.24 Effect of coexisting current on lift coefficient. Data from Sumer et al. (1992). $k/D = 4 \times 10^{-3}$, $Re = 3 \times 10^4$. Dashed lines: Asymptotic values for steady current for $Re = 3 \times 10^4$ taken from Fig. 2.15 where the given r.m.s. value of the lift is multiplied by $\sqrt{2}$ to obtain C_{Lmax}.

current-alone case (indicated in the figure with dashed lines). Although the data for $KC = 5$ and $KC = 10$ indicate that this is indeed the case, the maximum value of the tested range of U_c/U_m is too small to demonstrate this for $KC = 20$.

It may be concluded from the presented results that the superposition of a small current on waves may generally reduce the force coefficients. As the current component of the combined waves-and-current flow becomes increased, however, the force coefficients tend to approach their asymptotic values measured for the case of current alone.

4.5   Effect of angle of attack

It has been seen in Section 2.6 that the so-called *independence or cross-flow principle* (namely the normal component of force, F_N, (see Fig. 2.18) is expressed in terms of the normal component of the flow, U_N, with a force coefficient which is independent of the angle of attack, θ) is generally applicable for steady currents.

The relationship expressing the independence principle, Eq. 2.14, may be extended to oscillatory flows in the form of the Morison equation:

$$F_N = \frac{1}{2}\rho C_D D U_N |U_N| + \rho C_M A \, \dot{U}_N \qquad (4.75)$$

The question, however, is whether the force coefficients C_D and C_M are constants (independent of θ), in line with the steady-current case.

For large KC numbers, the inertia portion of the force is not important. Since the oscillatory flow in this case resembles the steady current, it is therefore expected that the cross-flow principle is valid here, and hence C_D may be independent of θ.

At the other extreme, namely for small KC numbers, on the other hand, the drag portion of the force is insignificant. In this case, the flow behaves like a potential flow, and hence the cross-flow principle must be valid here, too, meaning that the inertia coefficient C_M might be expected to approach the potential-flow value, namely $C_M = 2$, regardless of the value of θ.

Fig. 4.25 illustrates the effect of θ on the force coefficients. Here KC and Re are defined in terms of the normal component of the velocity, U_{Nm}:

$$KC = \frac{U_{Nm} T_w}{D} \text{ and } Re = \frac{U_{Nm} D}{\nu} \qquad (4.76)$$

The data apparently seem to confirm the argument put forward in the preceding paragraphs; i.e., 1) the drag coefficient C_D appears to be independent of θ for large KC numbers (such as $KC \gtrsim 20$), and 2) the inertia coefficient C_M approaches the potential-flow value, $C_M \rightarrow 2$ for small KC numbers (such as $KC \lesssim 8$), regardless of the value of θ.

The differences observed in the range $8 \lesssim KC \lesssim 20$ in Fig. 4.25 may be attributed to the disruption of the transverse-vortex-street regime ($8 < KC < 15$) for the values of angle of attack $\theta = 45°$ and $\theta = 60°$. Even a small deviation from $\theta = 90°$ seems to influence the force coefficients. A deviation from $\theta = 90°$ means that there exists a flow component parallel to the axis of the cylinder. This would eventually disrupt the transverse vortex street, leading to the observed differences in the force coefficients for flow angles different from $90°$.

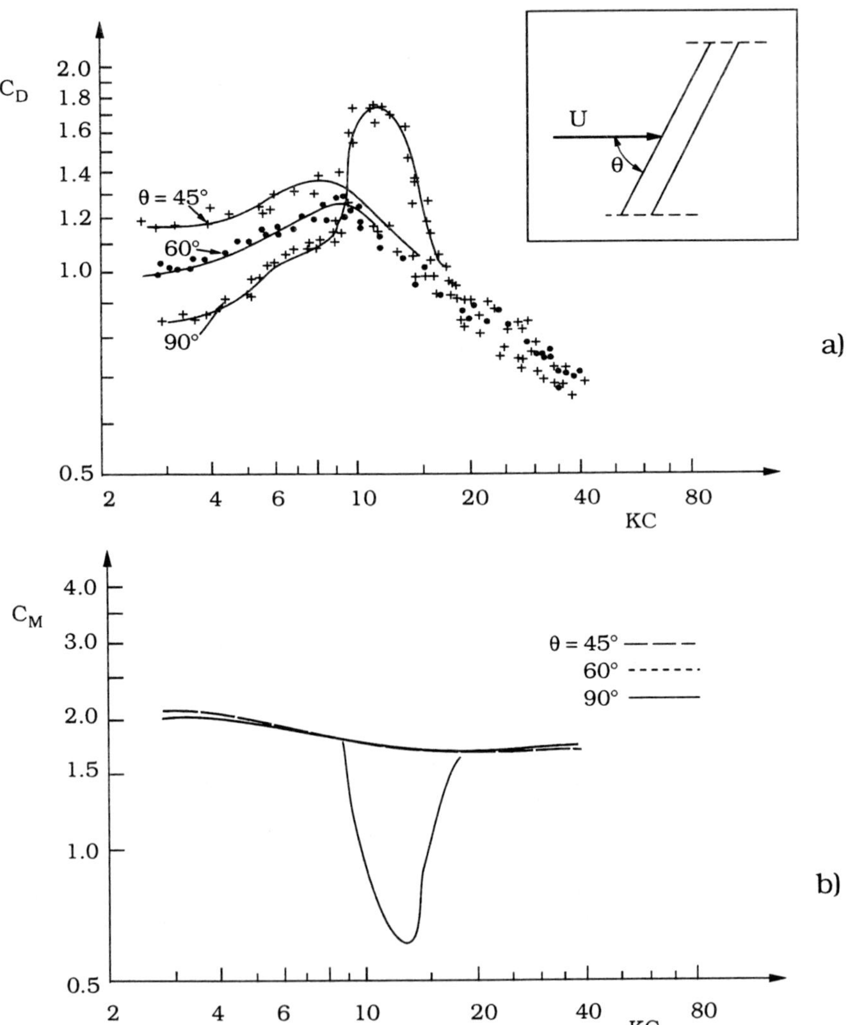

Figure 4.25 Effect of angle of attack on in-line force coefficients. Definitions of C_D, C_M, KC and Re, see Eqs. 4.75 and 4.76. The Reynolds number of the tests is such that $Re/KC = 4000$. (a): Sarpkaya et al. (1982). (b): Sarpkaya et al. (1982) as modified by Garrison (1985).

For further information about the effect of angle of attack, reference may be made to Chakrabarti, Tam and Wolbert (1977), Sarpkaya, Raines and Trytten (1982) and Garrison (1985).

Kozakiewicz et al. (1995) have made a study of the effect of angle of attack on forces acting on a cylinder placed near a plane wall. They tested three values of θ, namely $\theta = 90°$, $60°$ and $45°$, and three values of clearance between the cylinder and the wall, $e/D = 0$, 0.1 and 1.8, e being the clearance for a rather wide range of KC $4 \leq KC \leq 65$. Their results indicate that, for the tested range of θ, the force coefficients C_D, C_M and C_L are practically independent of θ, even in the range $8 \lesssim KC \lesssim 30$. As noted above, the difference observed for this range of KC number for a free cylinder are due to the disruption of the transverse vortex-street regime when θ is changed from $90°$ to $45°$ and $30°$. Now, in the case of a near-wall cylinder, this vortex-flow regime does not exist at all, not even for the case of perpendicular pipe ($\theta = 90°$), owing to the close proximity of the wall to the pipe. Therefore, no change in the force coefficients should be expected. Sumer et al. (1991) give the limiting value of e/D for the disappearance of the transverse-vortex-street regime for $\theta = 90°$ as $e/D \cong 1.7$-1.8, see Section 3.4).

4.6 Effect of orbital motion

Until now forces on a cylinder in a plane oscillatory flow have been studied. Clearly, real waves differ from the case of plane oscillatory flow in several aspects. An important difference between the two cases is that while the water particles in the case of plane oscillatory flow travel over a straight-line trajectory, the trajectory of the orbital motion of water particles in the case of waves is elliptical where the ellipticity of the motion may vary between 0 (the straight-line motion) and 1 (the circular motion). Hence it may be anticipated that the forces on a cylinder subject to a real wave, may be influenced by the presence of the orbital motion.

This section will give a detailed account of the subject. First, the vertical-cylinder case and subsequently the horizontal cylinder case will be studied. The cylinder diameter is assumed to be so small compared to the wave length that effects of diffraction can be neglected (see Chapter 6).

4.6.1 Vertical cylinder

Figs. 4.26 and 4.27 depict two kinds of data related to the in-line force; one for small Re numbers (Fig. 4.26) and the other for large Re numbers (Fig. 4.27), taken from Stansby, Bullock and Short (1983) and Bearman et al. (1985a), respectively. In the figures, the plane oscillatory flow results (from Sarpkaya (1976a) and Justesen (1989), respectively) are also included, to facilitate comparison. The in-line coefficients, C_D and C_M, in the figures are defined in the same way as in Eq. 4.29, U being the horizontal component of the velocity.

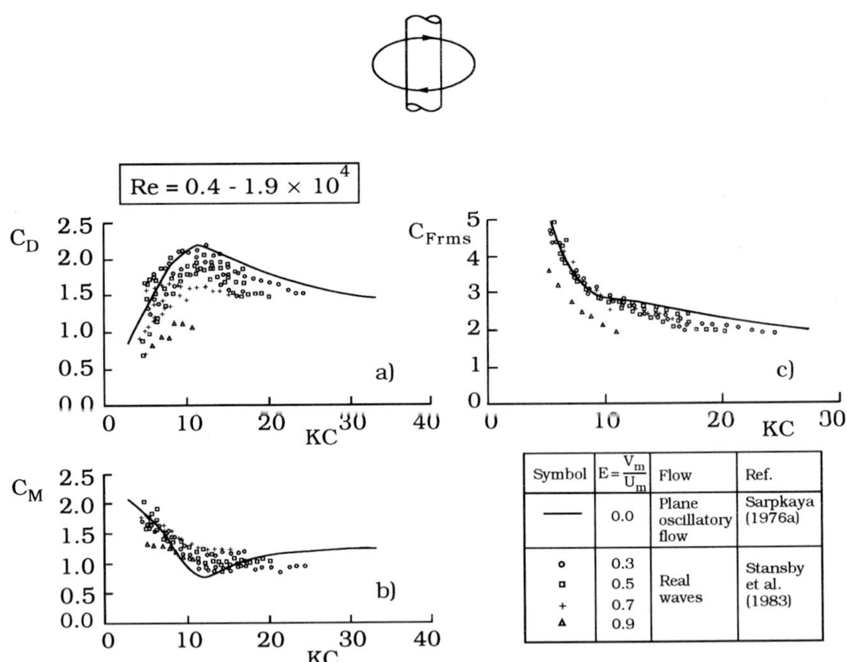

Figure 4.26 Effect of orbital motion on in-line force for vertical cylinders for small Re numbers. (a): Drag coefficient. (b): Inertia coefficient. (c): Force coefficient for the total in-line force. Sarpkaya data in (a) and (b) are for $\beta(=Re/KC) = 784$. The Sarpkaya curve in (c) is worked out from C_D and C_M values given by Sarpkaya for $\beta(=Re/KC) = 784$.

In Fig. 4.26, the quantity E, defined by

$$E = \frac{V_m}{U_m} \quad , \tag{4.77}$$

is the parameter which characterizes the ellipticity of the orbital motion. V_m and U_m are the maximum values of vertical and horizontal components of particle

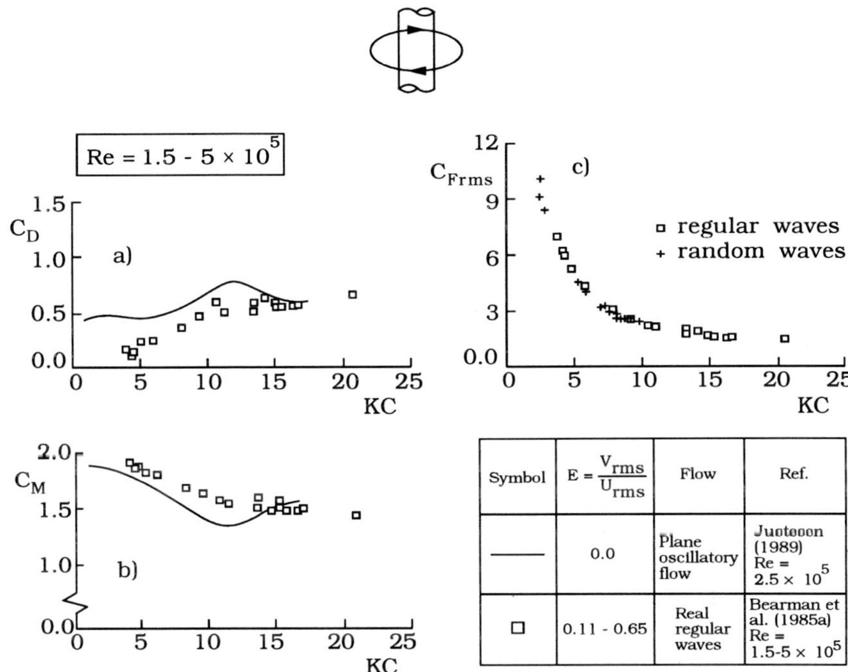

Symbol	$E = \dfrac{V_{rms}}{U_{rms}}$	Flow	Ref.
——	0.0	Plane oscillatory flow	Juotooon (1989) Re = 2.5×10^5
□	0.11 - 0.65	Real regular waves	Bearman et al. (1985a) Re = $1.5\text{-}5 \times 10^5$

Figure 4.27A Effect of orbital motion on in-line force for vertical cylinders for large *Re* numbers. (a): Drag coefficient. (b): Inertia coefficient. (c): Force coefficient for the total in-line force.

velocity, respectively. In Fig. 4.27A, the ellipticity is given in terms of r.m.s. values of the velocity components rather than the maximum values, in conformity with the original notation of the authors (Bearman et al., 1985a). The quantity C_{Frms} in the figures, on the other hand, is the force coefficient corresponding to the total *in-line* force, defined by

$$F_{rms} = \frac{1}{2}\rho C_{Frms} D U_{rms}^2 \tag{4.78}$$

in which F_{rms} is the root-mean-square (r.m.s.) value of the in-line force per unit length of the cylinder, and U_{rms} is the r.m.s. value of the horizontal velocity at the level where the force is measured.

For the *small-Re-number experiments* (Fig. 4.26), as far as C_D and C_M are concerned, it is difficult to find any clear trend with respect to the ellipticity of the motion, the scatter being quite large. However, when the data are plotted in terms of $C_{F\mathrm{rms}}$, they collapse on a narrow band, with the exception of $E = 0.9$. This latter diagram indicates that the total in-line force is hardly influenced by the ellipticity of the orbital motion unless the ellipticity is extremely large, namely $E > 0.7 - 0.8$. For such large E values the data indicate that there will be a reduction in the total in-line force by an amount in the order of magnitude of 20-30%.

As for *the large-Re-number* experiments (Fig. 4.27A), the effect of orbital motion is indistinguishable for the reported range of E, namely $E = 0.11 - 0.65$. Also, it may be noticed that the C_D and C_M variation obtained by Justesen (1989) in plane oscillatory flows ($E = 0$) for a Re number which lies approximately at the centre of Bearman et al.'s Re range is not extremely different from that of Bearman et al.'s real-wave results.

From the preceding discussion it may be concluded that the total in-line force is practically uninfluenced by the orbital motion, unless the ellipticity of the motion is quite large ($E > 0.7 - 0.8$). In the latter case there may be a reduction in the total in-line force by an amount in the order of magnitude of 20-30%, with respect to the value calculated using the plane oscillatory flow data, meaning that the plane-oscillatory-flow calculations remain on the conservative side for these ellipticity values.

Fig. 4.27B presents the data related to *the lift force*. Although Bearman et al. (1985a) report that the dependence on ellipticity E is indistinguishable from their data with E ranging from 0.11 to 0.65, the figure indicates, however, that the lift may be different from that measured in the case of plane oscillatory flow ($E = 0$) as measured in Justesen's (1989) study. A close examination of the figure shows that this deviation occurs in the range of KC from 7 to 13. As seen in Section 3.2, the range of KC number $7 < KC < 13$, known as the single pair vortex-shedding regime for plane oscillatory flows, is the range where the so-called transverse vortex regime prevails. The observed deviation from the plane oscillatory flow in this range of KC number may be attributed to the disruption of the transverse vortex street in the case of real waves with ellipticities different from zero. Outside this range, however, the agreement between the results obtained in the case of plane oscillatory flow and those obtained in the case of real waves appears to be rather good. Presumably this leads to the conclusion that the lift force is practically uninfluenced by the orbital motion with the exception of the KC range $7 < KC < 13$, where the lift force is reduced quite considerably with respect to that experienced in the case of plane oscillatory flow.

The vertical-cylinder problem has been investigated rather extensively in the past, Ramberg and Niedzwecki (1979), Chakrabarti (1980) and Sarpkaya (1984). The wave parameters in Chakrabarti's (1980) study were such that the waves were closer to the shallow-water regime, while Ramberg and Niedzwecki's were close to or in the deep-water regime. Nevertheless, the results of these two studies are in accord with Stansby et al.'s study (presented in Fig. 4.26) in the sense that the in-line force is practically uninfluenced by the orbital motion in Chakrabarti's

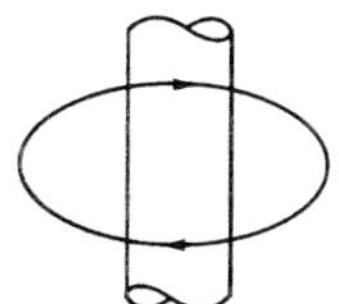

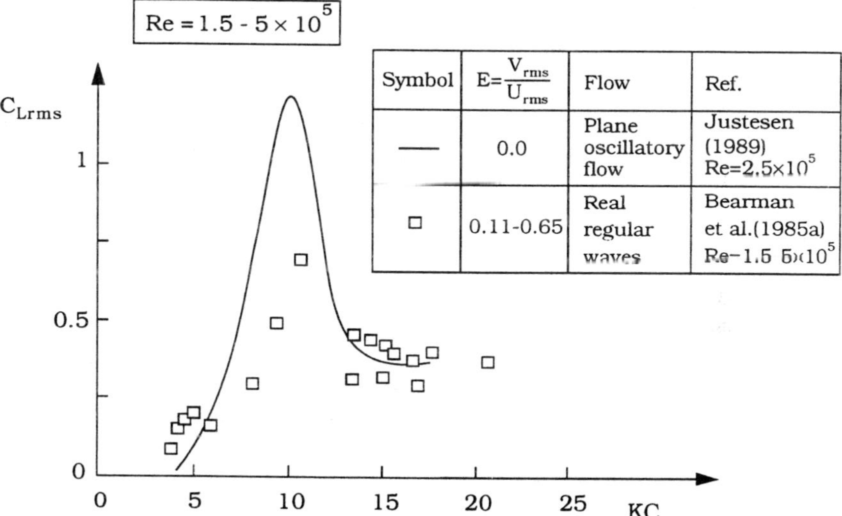

Symbol	$E=\dfrac{V_{rms}}{U_{rms}}$	Flow	Ref.
——	0.0	Plane oscillatory flow	Justesen (1989) $Re=2.5\times10^5$
□	0.11-0.65	Real regular waves	Bearman et al.(1985a) $Re=1.5-5\times10^5$

Figure 4.27B Effect of orbital motion on lift force (transverse force) for vertical cylinders for large Re numbers.

(1980) study (small E values) while it is considerably overestimated by the plane-oscillatory-flow calculations in Ramberg and Niedzwecki's study (large E values such as $E > 0.8 - 0.9$). Sarpkaya (1984), on the other hand, simulated the orbital motion by oscillating the cylinder along its axis in a plane oscillatory flow that takes place in a direction perpendicular to the cylinder axis. Sarpkaya's results show a very distinct trend of the variation of the force coefficients C_D and C_M as function of the ellipticity parameter, E. He reports a decrease in the total force with increasing ellipticity.

Example 4.5: In-line force on a vertical pile in the surface zone

When the Morison equation is used, it will be found that the in-line force on a vertical pile is maximum at the level of the wave crest. However, the analysis of the field data (Dean, Dalrymple and Hudspeth, 1981) show that the force is maximum at an elevation somewhat below the water surface at the wave crest, becoming zero at an elevation somewhat above the wave crest (see Fig. 4.28). This observation was later confirmed by the laboratory study of Tørum (1989).

The reason behind this behaviour is the surface runup in front of the cylinder and the surface rundown at the back, presumably leading to a maximum below the crest elevation. The previously mentioned studies indicated that the location of the force maximum lies approximately $U_m^2/(2g)$ below the crest level, while the location of zero force lies approximately $U_m^2/(2g)$ above the crest level in which U_m is the maximum value of the horizontal velocity at the wave crest.

As regards the in-line force coefficients for the region above the mean water level, Tørum (1989) recommends the following. 1) As for the C_D values, use C_D values as below the mean water level and 2) as for the C_M values, use the C_M variation given in Fig. 4.28.

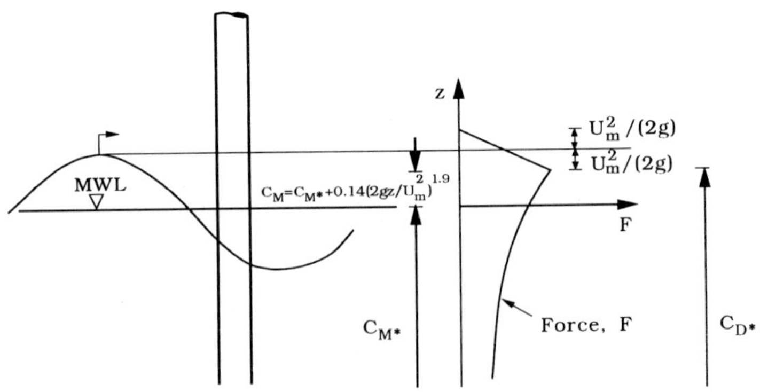

C_{M*} =⎫ Values relevant to the prototype
C_{D*} =⎭ Reynolds number and Keulegan-Carpenter number
U_m = Maximum water-particle velocity at the crest

Figure 4.28 Recommended design C_D and C_M values in surface zone area (Tørum, 1989).

4.6.2 Horizontal cylinder

Fig. 4.29 presents the results of Bearman et al.'s study (1985a) for the case of horizontal cylinder with regard to the in-line force coefficients C_D and C_M for two different Re number intervals in the post-critical Re number range. The range of ellipticity E in these experiments is from 0.15 to 0.75. The figure includes also the results of Justesen's (1989) plane-oscillatory-flow study ($E = 0$) for the corresponding Reynolds numbers. Although the scatter in Bearman et al.'s data is quite large, it is difficult to speak of any definite trend with respect to the ellipticity of the orbital motion from the data. Fig. 4.30 presents the data from the same study (Bearman et al.'s) related to the total force, namely $F_T = (F^2 + F_L^2)^{1/2}$, in terms of the corresponding force coefficient defined by

$$F_{Trms} - \frac{1}{2}\rho C_{Trms} D U_{Trms}^2 \tag{4.79}$$

in which

$$F_{Trms} = \left(F_{rms}^2 + F_{Lrms}^2\right)^{1/2} \tag{4.80}$$

and

$$U_{Trms} = \left(U_{rms}^2 + V_{rms}^2\right)^{1/2} \tag{4.81}$$

where F and F_L are the in-line and lift force components while U and V are the horizontal and vertical components of the particle velocity, respectively. This figure, too, shows that the influence of the orbital motion on the force is not distinguishable. For small Reynolds numbers, however, a systematic reduction in the total in-line force with the ellipticity has been reported by Maull and Norman (1979). Maull and Norman's result is reproduced in Fig. 4.31.

Several investigators simulated the wave-induced, orbital flow around the horizontal cylinder by driving the cylinder in an elliptical orbit in an otherwise still water, Holmes and Chaplin (1978), Chaplin (1981), Grass, Simons and Cavanagh (1985) and Chaplin (1988b). Chaplin's (1988b) results for two different values of the ellipticity are plotted in Fig. 4.32.

While the real-wave data of Fig. 4.30 show practically no evidence about the sensitivity of the results to orbit shape, the data obtained by the mechanical simulation of the orbital flow (Fig. 4.32) indicate a systematic decrease in the total force with increasing wave ellipticity. This has been interpreted by Bearman et al. (1985a) as follows. In the case of mechanical simulation of orbital flow, the motion is exactly periodic and, in the absence of any mass transport, the cylinder inevitably encounters its own wake, and therefore experiences a reduction in the incident velocity relative to the cylinder. They suggest that this effect, a feature

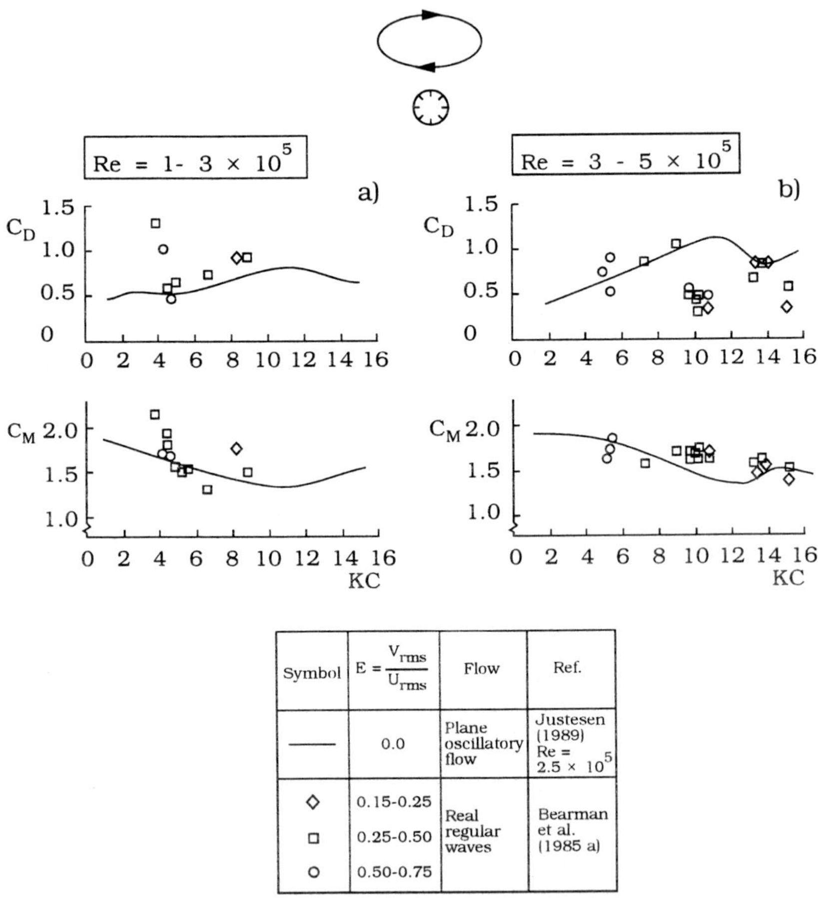

Symbol	$E = \dfrac{V_{rms}}{U_{rms}}$	Flow	Ref.
———	0.0	Plane oscillatory flow	Justesen (1989) Re = 2.5×10^5
◇	0.15-0.25	Real regular waves	Bearman et al. (1985 a)
□	0.25-0.50		
○	0.50-0.75		

Figure 4.29 Effect of orbital motion on in-line force for horizontal cylinders.

of the method of mechanical simulation, may be reduced by small currents or by slight irregularities in the waves.

One other method of mechanical simulation of orbital flow is to oscillate the cylinder placed in a plane oscillatory flow, in a direction perpendicular to the flow. This method was used by Sarpkaya (1984). Similar to the previously mentioned work, Sarpkaya, too, found that the net result is a decrease in the total in-line force with increasing ellipticity.

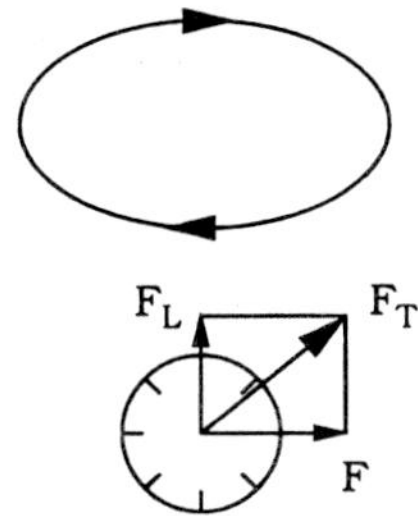

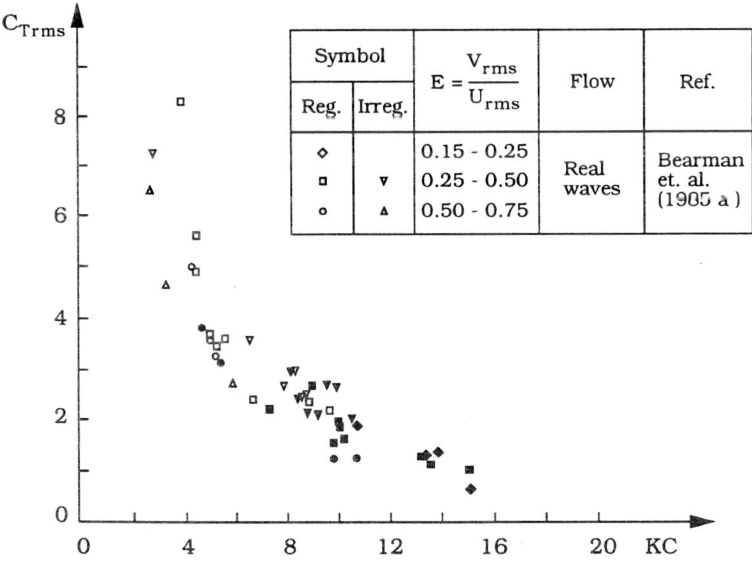

| Symbol | | $E = \dfrac{V_{rms}}{U_{rms}}$ | Flow | Ref. |
Reg.	Irreg.			
◇		0.15 - 0.25		Bearman
□	▽	0.25 - 0.50	Real waves	et. al.
○	▲	0.50 - 0.75		(1985 a)

Figure 4.30 Effect of orbital motion on total (resultant) force for horizontal cylinders. The force coefficient is defined by Eqs. 4.79 and 4.81. $Re = 1 - 3 \times 10^5$ for empty symbols and $3 - 5 \times 10^5$ for solid symbols.

Finally, it may be mentioned that, even when the force coefficients are available (Fig. 4.29), the Morison equation alone provides a very poor approximation to the loading in either horizontal or vertical direction in the case of a horizontal cylinder in orbital flows (Fig. 4.33) for large KC numbers where vortex shedding

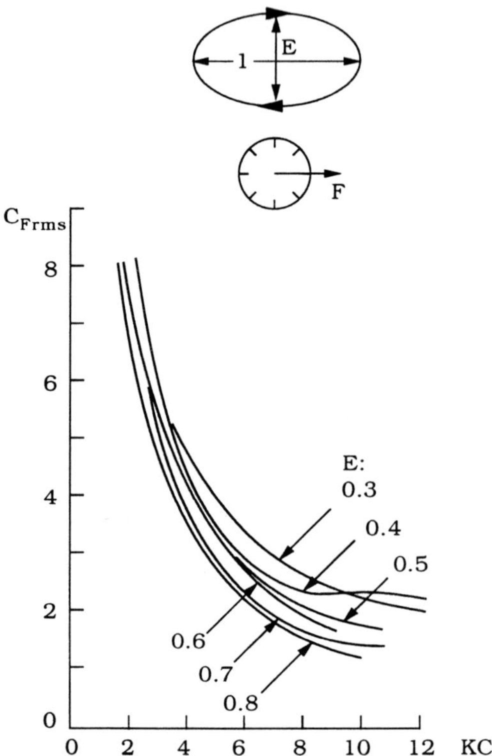

Figure 4.31 Effect of orbital motion on in-line force for horizontal cylinders. The orbital motion is characterized by the ellipticity E defined by $E = V_m/U_m$. $Re = 4 \times 10^3$. Maull and Norman (1979).

occurs. This is because the vortex shedding makes a very important contribution to the loading, and obviously the Morison equation fails to represent this effect. Bearman et al. (1985a) give a detailed discussion of this aspect of the problem.

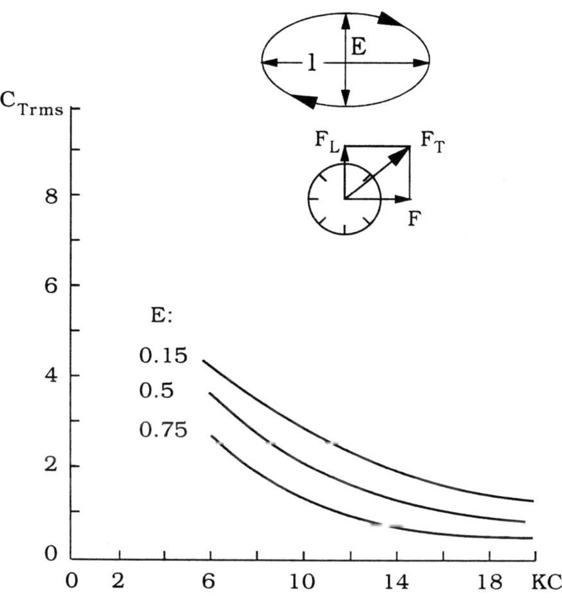

Figure 4.32 Effect of orbital motion on total (resultant) force for horizontal cylinders from experiments where the orbital-motion effect is obtained by mechanical simulation, driving the cylinder in elliptical orbit. E = ellipticity of the orbit. $Re = 1.5 - 2.2 \times 10^5$. Chaplin (1988b).

Example 4.6:   Forces on horizontal cylinders in orbital flows at low KC numbers

In practice, forces on horizontal cylinders in orbital flows in the inertia regime, particularly at rather small KC numbers, may become important. Application areas include, for example, horizontal pontoons of semi-submersibles and tension-leg platforms. In the inertia regime, the drag is insignificant, as discussed in the preceding sections. Therefore the total force is, to a large extent, determined by the inertia force. The inertia force itself may undergo substantial reductions in the case when the cylinder is subject to an orbital flow (or equivalently when

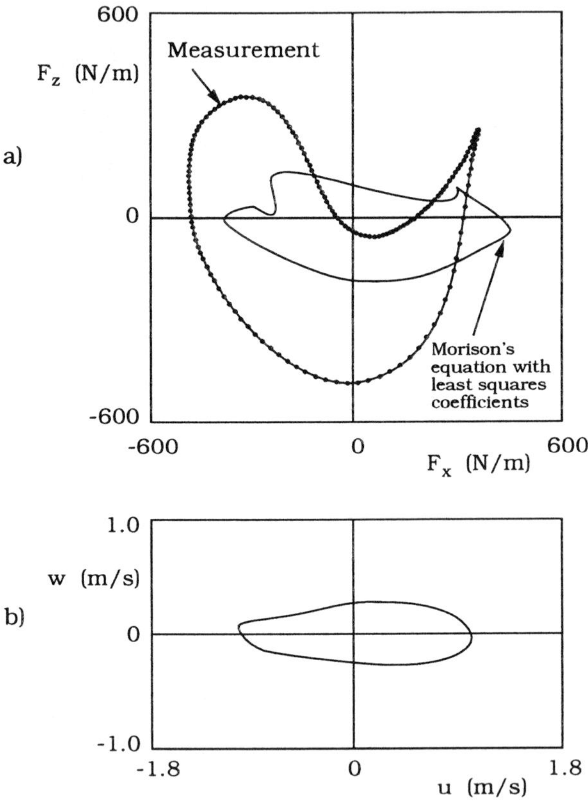

Figure 4.33 (a): Horizontal cylinder: polar representation of the total force vector, averaged over about 30 waves; comparison with the least-squares Morison's equation. (b): Horizontal cylinder: polar representation of the velocity vector for the same run. Bearman et al. (1985a).

it executes an orbital motion in a fluid initially at rest). This occurs at low KC numbers; the inertia coefficient can take values as small as 50% of that experienced in the case of planar oscillatory flow as measured by Chaplin (1984). Fig. 4.34 shows the results of Chaplin's experiments, in which real waves were used, where the diameter of the test cylinder was small compared with the wave length (i.e., outside the diffraction flow regime). The wave-induced flow was an almost

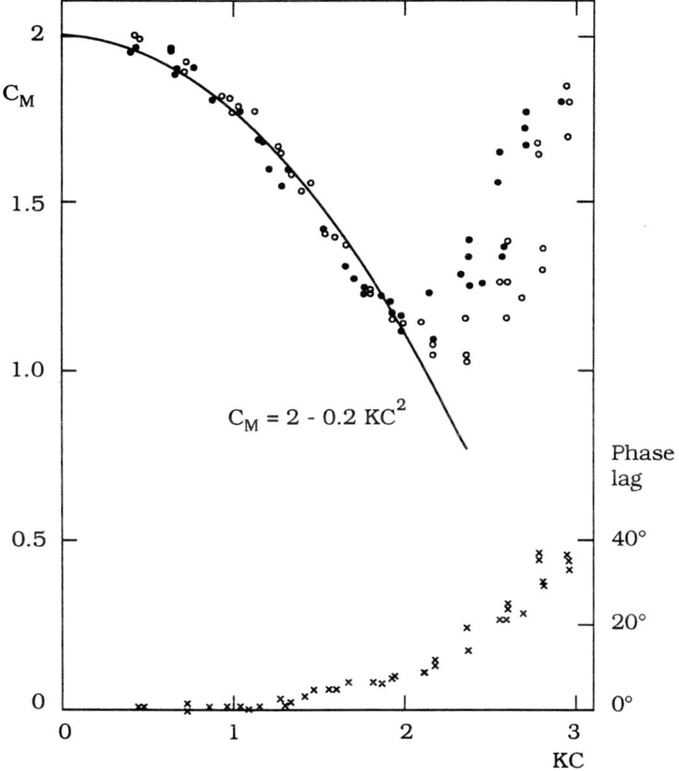

Figure 4.34 Inertia coefficient for a horizontal cylinder subject to an orbital flow: •, horizontal force; ○, vertical force. Phase lag of the force (ocurring at the wave frequency) with respect to the acceleration of the incident flow: ×. Ellipticity, $E = 0.92$. $\beta = 7600$. $L/D = 0.047$ (L being the wave length). Chaplin (1984).

circular orbital flow (the ellipticity, E, was 0.92).

Fig. 4.34 shows that C_M begins to decrease already at KC about 0.5, it reaches a minium at KC about 2, and from this point onwards it increases to attain its potential-flow value, 2, at about $KC = 3$. It may be noticed that the drop in C_M in the present case is completely different from that in the case of planar flow (Figs. 4.9 and 4.10). In the latter case, for a substantial drop in C_M, KC needs to be increased to such values as $KC > 6 - 7$.

The observed behaviour in C_M may be attributed to the steady, recirculating streaming which builds up around the cylinder as the cylinder is exposed to waves. The orbital flow around the cylinder may be viewed as the flow around a cylinder which is executing an orbital motion in a fluid initially at rest. As such, the stirring motion of the cylinder will generate a recirculating flow in the fluid. Clearly the cylinder during its motion will encounter this flow, which is in the same direction as the motion of the cylinder itself, meaning that the inertia force on the cylinder will be reduced. This effect is increased, as KC is increased. However, when KC reaches a critical value where the flow separates (namely, $KC = 2$, in the present example, see Fig. 3.15), the aforementioned recirculating streaming will then be disrupted by the formation of the separation vortices in the wake, leading presumably to an increase in the C_M values. With the complete disappearance of the recirculating streaming (apparently at $KC \cong 3$), the potential-flow value of C_M (i.e., 2) will be restored again (Fig. 4.34).

A simple model to describe the inertia coefficient can be worked out on the basis of the preceding considerations. The simplest case is considered; namely, the cylinder executes a *circular* orbital motion in a fluid initially at rest, satisfying

$$U = U_m \cos(\omega t) \quad \text{and} \quad V = -U_m \sin(\omega t) \tag{4.82}$$

The circulation, defined as $\Gamma = \int_c \mathbf{v} \cdot d\mathbf{s}$, which will be generated by the stirring motion of the cylinder may be written as

$$\Gamma = \int_0^{2\pi} \sqrt{U^2 + V^2} \, (a \, d\theta) \tag{4.83}$$

or, from Eq. 4.82,

$$\Gamma = U_m a \int_0^{2\pi} d\theta = 2\pi a U_m \tag{4.84}$$

in which U_m is the tangential velocity of the orbital motion and $2a$ is the stroke of the motion. Since $U_m = a\omega$, then the circulation will be

$$\Gamma = \frac{2\pi U_m^2}{\omega} \tag{4.85}$$

Now, the cylinder is actually subject to two kinds of flow. One is the incident flow, i.e., the flow relative to the cylinder with the velocity components given in Eq. 4.82. The other is the recirculating flow with the circulation given in Eq. 4.85. The flow is illustrated in Fig. 4.35. First the horizontal force on the cylinder is considered. The flow is decomposed in the manner as sketched in Figs. 4.35b and 4.35c.

The U component of the flow velocity will induce an inertia force in the horizontal direction, equal to $2\rho A \, \dot{U}$ (the factor 2 being the conventional inertia

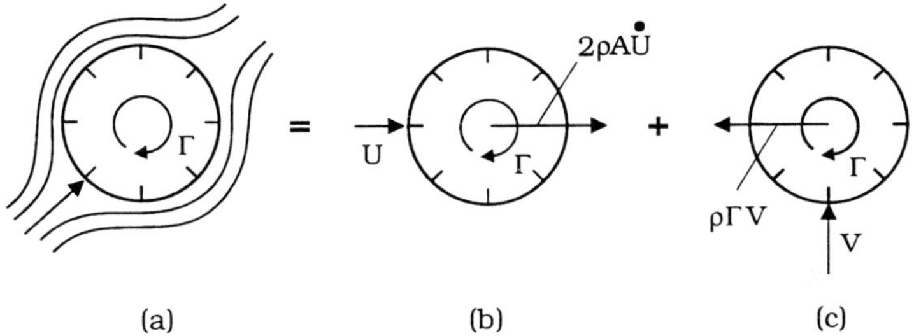

Figure 4.35 Horizontal force acting on a cylinder subject to a circular orbital motion.

coefficient), while the V component of the velocity combined with the circulation Γ will induce a lift force, i.e., a force perpendicular to the incident velocity V, equal to $\rho\Gamma V$, as shown in Fig. 4.35c. (The latter is known as the Magnus effect, see Batchelor 1967, p. 127). Therefore the total horizontal force, neglecting the drag, will be

$$F = 2\rho A \; \dot{U} - \rho\Gamma V \qquad (4.86)$$

or inserting $F = C_M \rho A \; \dot{U}$ and Eqs. 4.82 and 4.85 into the preceding equation, C_M is found

$$C_M = 2 - \frac{2}{\pi^2} KC^2 \qquad (4.87)$$

or

$$C_M = 2 - 0.2 \, KC^2 \qquad (4.88)$$

Likewise, the inertia coefficient associated with the vertical force, namely $F = C_M \rho A \; \dot{V}$, can be worked out; it can be seen easily that this will lead to the same result as that given in the preceding equation, Eq. 4.88.

The above equation is virtually the same equation as that found by Chaplin (1984) empirically from his force data (Fig. 4.34). As seen, the agreement between this equation and the data in the range $0 \leq KC < 2$ (where the flow is unseparated) is very good.

Chaplin's (1984) study covered an almost idealized flow situation where the wave-induced orbital motion was more or less circular and the Reynolds number was small. In a later study, Chaplin (1988a) carried out similar experiments in a large scale facility where the cylinder was rather large with Re in the range

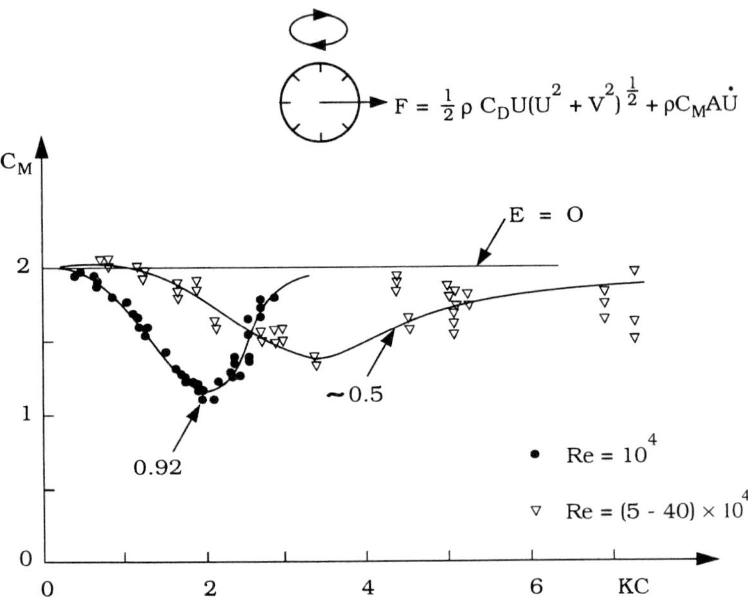

Figure 4.36 The influence of elipticity, E, and the Re number on the inertia coefficient associated with the horizontal force for *low-KC-number flows*. Data: circles (Chaplin, 1984), and triangles (Chaplin, 1988a).

$5 \times 10^4 - 4 \times 10^5$ and the waves were more realistic with ellipticity values even below 0.5. The results of Chaplin's (1988a) study are plotted in Fig. 4.36 together with his earlier data. Two points may be mentioned from the figure: 1) As the ellipticity increases, the reduction in the inertia force increases. 2) As the flow in the cylinder boundary layer becomes turbulent (the large-Re number data) the reduction in C_M spreads over a wider KC range (over a range of KC from 0 to about 3.5 in the case of large-Re-number experiments). This behaviour may be attributed to the fact that the separation is delayed by the turbulence in the boundary layer.

In the case of elliptical orbital flows, the symmetry with respect to x and y axes will break down, therefore the vertical force will be different from the horizontal force. Fig. 4.37 shows the results of Chaplin's (1988a) large-scale facility experiments for the inertia coefficient associated with the vertical force. The data are plotted together with the corresponding data of Chaplin (1984).

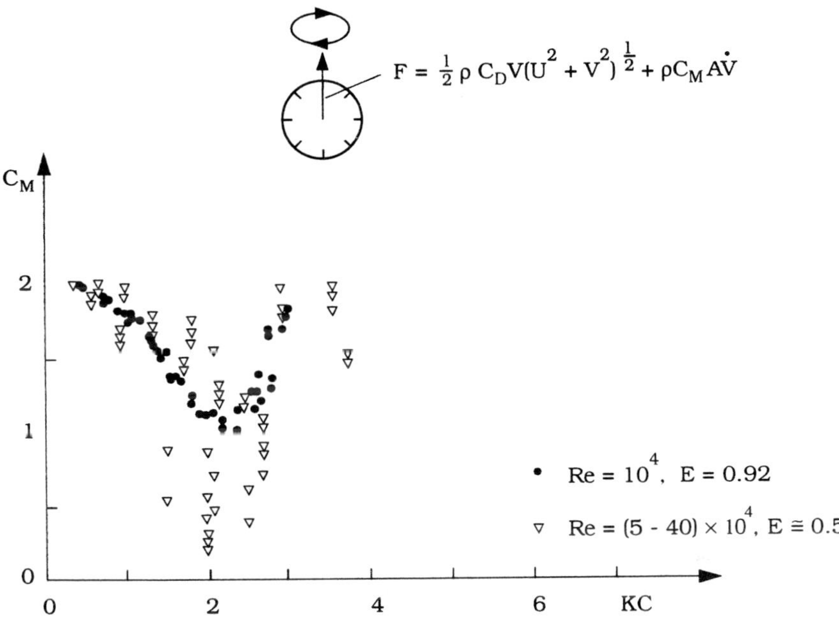

Figure 4.37   The influence of ellipticity, E, and the Re number on the inertia coefficient associated with the vertical force for *low KC-number flows*. Data: circles (Chaplin, 1984), and triangles (Chaplin, 1988a).

The scatter is quite extensive. However, extremely small inertia-coefficient values have been measured. These small values of the force are associated with the high ellipticities. Chaplin (1988a) made an attempt to plot the data in the form of C_{Mx} versus KC_y and C_{My} versus KC_x, to reduce the scatter. This attempt was partially successful.

The issue of low KC-number orbital flows discussed in the present paragraphs has been investigated further by Chaplin (1991 and 1993b), Stansby and Smith (1991) and Stansby (1993). Chaplin (1993b) used a Navier-Stokes code, while Stansby and Smith (1991) and Stansby (1993) used the random vortex method, to obtain the flow field and the forces. For the latter, see Section 5.2.3.

4.7 Forces on a cylinder near a wall

A detailed description of the oscillatory flow around a cylinder placed near a wall is given in Section 3.4. This section focuses on forces on such a cylinder, including the case of a pipeline placed in/over a scour trench.

Force coefficients for a cylinder near a plane wall

Forces on a cylinder near a plane wall and exposed to an oscillating flow has been investigated quite extensively. The first investigation was that of Sarpkaya (1976b), followed by Sarpkaya (1977a) and Sarpkaya and Rajabi (1979). Drag, inertia and lift coefficients on a cylinder placed at various distances from a wall were measured in these studies. Lundgren, Mathiesen and Gravesen (1976) measured the pressure distribution around a wall-mounted cylinder. Jacobsen, et al. (1984), Ali and Narayanan (1986), Justesen et al. (1987) and Sumer et al. (1991) among others have reported measurements regarding the effect of the wall on force coefficients. Forces on cylinders near a plane wall in diffraction regime are examined in Chapter 6 and the effect of irregular waves is described in Chapter 7.

Figs. 4.38 and 4.39 present the force-coefficient data obtained in Sumer et al.'s (1991) study together with Sarpkaya (1977a) and Sarpkaya and Rajabi (1979) data for $Re \cong 10^5$. Also included in the figures is Yamamoto et al.'s (1974) potential-flow solution. The lift coefficients C_{LA} and C_{LT} are defined by

$$\left. \begin{array}{l} F_{yA} = \frac{1}{2}\rho \, C_{LA} \, DU_m^2 \\ F_{yT} = \frac{1}{2}\rho \, C_{LT} \, DU_m^2 \end{array} \right\} \tag{4.89}$$

in which F_{yA} is the maximum value of the lift force away from the wall and F_{yT} that towards the wall. e is the gap between the cylinder and the wall.

Comparison with potential theory

The experimental data on C_M approach the values predicted by the potential theory as $KC \rightarrow 0$. Obviously, this is related to the fact that, for such small KC numbers, no separation will occur, therefore the potential-flow theory predictions of C_M must be approached, as KC goes to zero.

Regarding the asymptotic behaviour of C_L as $KC \rightarrow 0$, for the wall-mounted cylinder ($e/D = 0$), Figs. 3.18c, 3.21d and 3.23d show that the lift is always positive (directed away from the wall), in agreement with the potential-flow theory (Fig. 4.39). See also the discussion in Section 2.7 in relation to Fig. 2.23. Furthermore, the curve representing $e/D = 0$ in the C_{LA} diagram appears to be approaching the potential-flow value, namely $C_{LA} = 4.49$.

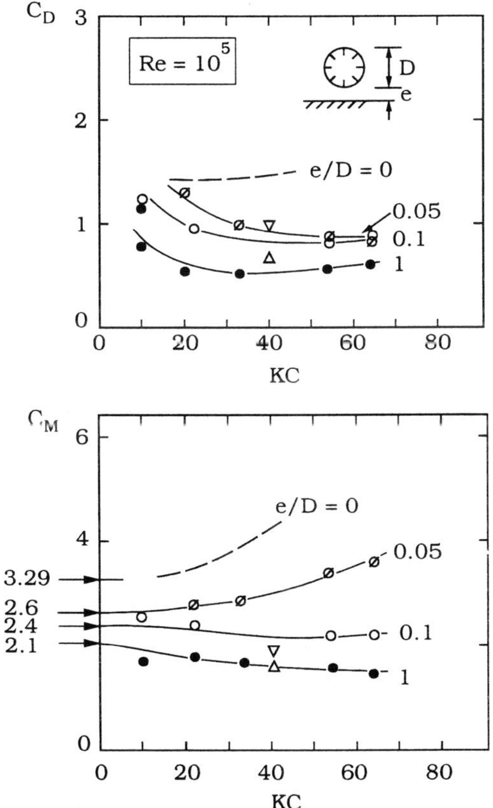

Figure 4.38 Drag and inertia coefficients for a near-wall cylinder. Smooth cylinder. Circles: Sumer et al. (1991) ($Re = 0.8 - 1.1 \times 10^5$); $\triangle$: $e/D = 1$; ∇: $e/D = 0.1$, Sarpkaya (1977a) ($Re = 10^5$); - - - - -: $e/D = 0$, Sarpkaya and Rajabi (1979) ($Re = 1 - 1.1 \times 10^5$). The asymptotic values of C_M for $KC \to 0$ indicated in the figure are the potential-flow solutions due to Yamamoto et al. (1974), reproduced here from Fig. 4.4 where $C_m = C_M - 1$.

However, for a cylinder placed near the wall, even with an extremely small gap ratio such as $e/D = 0.05$, the lift alternates between successive positive and negative peaks (Figs. 3.18b, 3.21c, 3.23c and 3.24c,d). The positive peak in the lift is associated with the movement of the lee-wake vortex over the cylinder during the flow reversals, while the negative peak in the lift is associated with the formation of lee-wake vortex and the high-speed flow in the gap between the cylinder and the wall as discussed in Section 3.4).

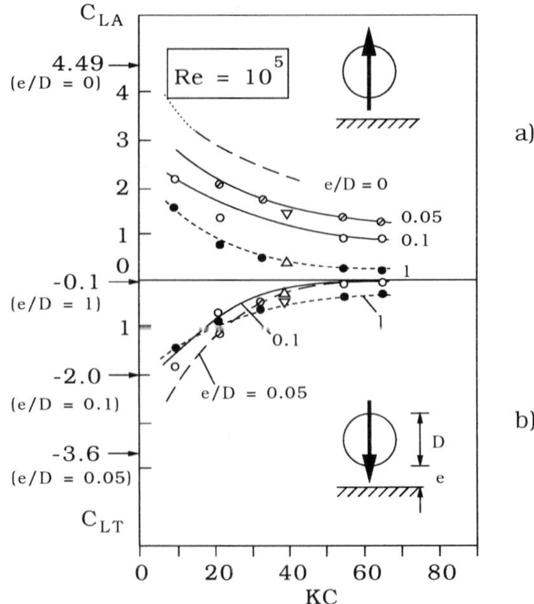

Figure 4.39 Lift-force coefficient for a near-wall cylinder. Symbols are the same as in the previous figure. The asymptotic values of C_L for $KC \to 0$ indicated in the figure are the potential-flow solutions due to Yamamoto et al. (1974).

From the discussion in Section 2.7 in relation to Fig. 2.23, it is apparent that the potential-flow theory in the case of near-wall cylinder does not predict a positive lift but rather a negative lift. The values calculated from the potential flow theory for the gap ratios $e/D = 0.05, 0.1$ and 1 are indicated in Fig. 4.39b. Apparently, as $KC \to 0$, the experimental results seem to be approaching the potential-flow values for $e/D = 0.05$ and 0.1. However, for $e/D = 1$, the experimental C_{LT} values are much lower than the potential-flow value, namely $C_{LT} \cong -0.1$.

Influence of gap ratio

From Figs. 4.38 and 4.39, the data indicate that the force coefficients C_D, C_M and C_{LA} increase as the gap ratio decreases. This is also true for C_{LT} for small $KC(O(10))$. For large KC, however, no clear trend appears. These results generally agree with those of other investigators such as Sarpkaya (1976b, 1977a), Ali and Narayanan (1986) and Justesen et al. (1987).

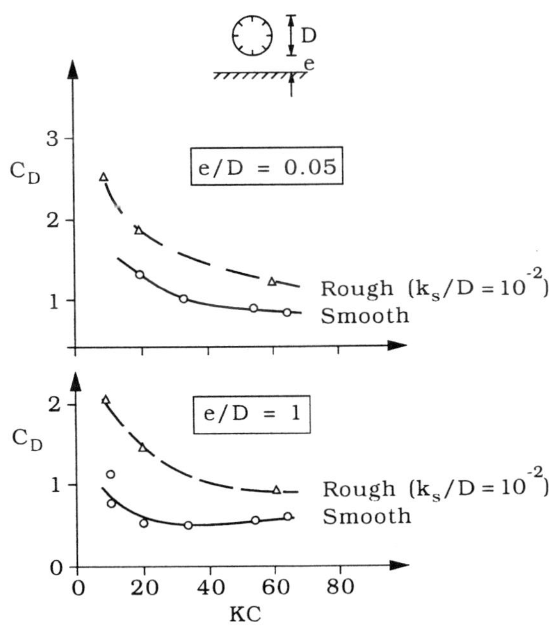

Figure 4.40 Influence of roughness on drag coefficient. $Re = 0.8 \times 10^5 - 1.1 \times 10^5$. Sumer et al. (1991).

Influence of roughness

Figs. 4.40–4.42 compare the force coefficients obtained for the smooth and rough cylinders of Sumer et al.'s (1991) study for the gap ratios $e/D = 1$ and 0.05. Figure 4.40 indicates that C_D increases substantially when the cylinder surface changes from smooth to rough. This is consistent with Sarpkaya's (1976b) wall-free cylinder data corresponding to $Re = 10^5$. Figure 4.41 indicates that C_M does not change significantly with the change of surface roughness for $e/D = 1$.

However, for $e/D = 0.05$, the inertia coefficient increases markedly when the surface of the cylinder changes from smooth to rough. This may be attributed to the retarding effect of the boundary layer at the wall side of the cylinder which may become significant for the inertia coefficient for small gap ratios such as 0.05.

Fig. 4.42 shows that no significant change occurs in the lift coefficients when the surface is changed from smooth to rough. This result appears to be consistent with Sarpkaya's (1976a) wall-free cylinder results and also with Sarpkaya and Rajabi's (1979) wall-mounted cylinder results.

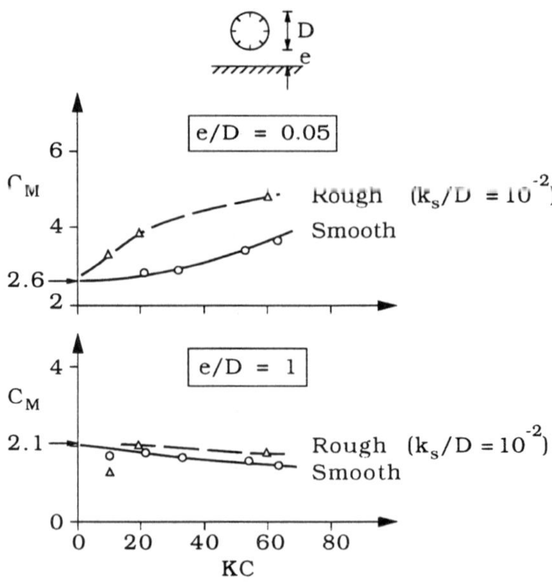

Figure 4.41 Influence of roughness on inertia coefficient. $Re = 0.8 \times 10^5 - 1.1 \times 10^5$. Asymptotic values for $KC \to 0$ are Yamamoto et al.'s (1974) potential flow solutions. Sumer et al. (1991).

Influence of Re

This was studied by Yamamoto and Nath (1976), and Sarpkaya (1977a). Both studies indicate that the way in which the force coefficients change with Re is much the same as in the case of wall-free cylinder (Figs. 4.11 and 4.18).

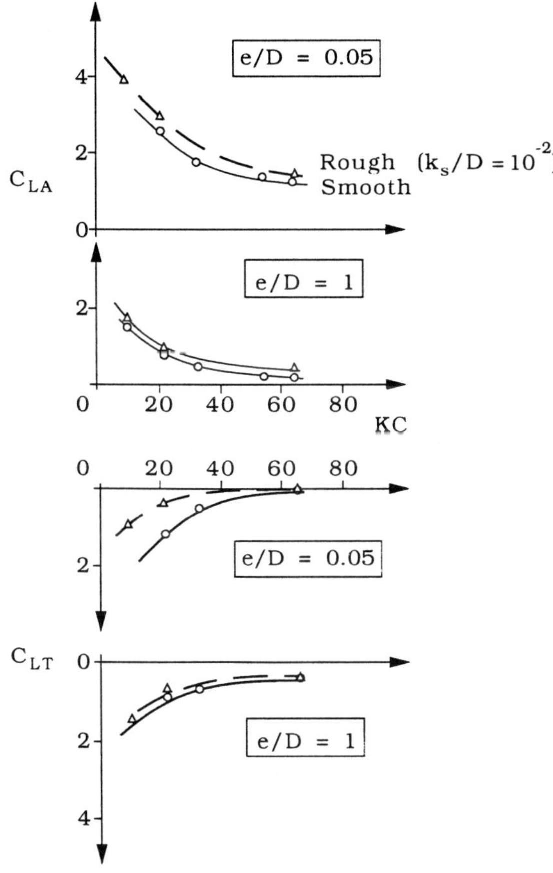

Figure 4.42 Influence of roughness on lift-force coefficients for smooth and rough near-wall cylinders. $Re = 0.8 \times 10^5 - 1.1 \times 10^5$. Sumer et al. (1991).

Wall-mounted cylinder ($e/D = 0$)

Although the force coefficients for a wall-mounted cylinder are given earlier in Figs. 4.38 and 4.39, the covered KC range was somewhat limited.

Fig. 4.43 gives the force coefficients, covering a much broader range of KC number, up to 170 (Bryndum, Jacobsen and Tsahalis, 1992). The figure also

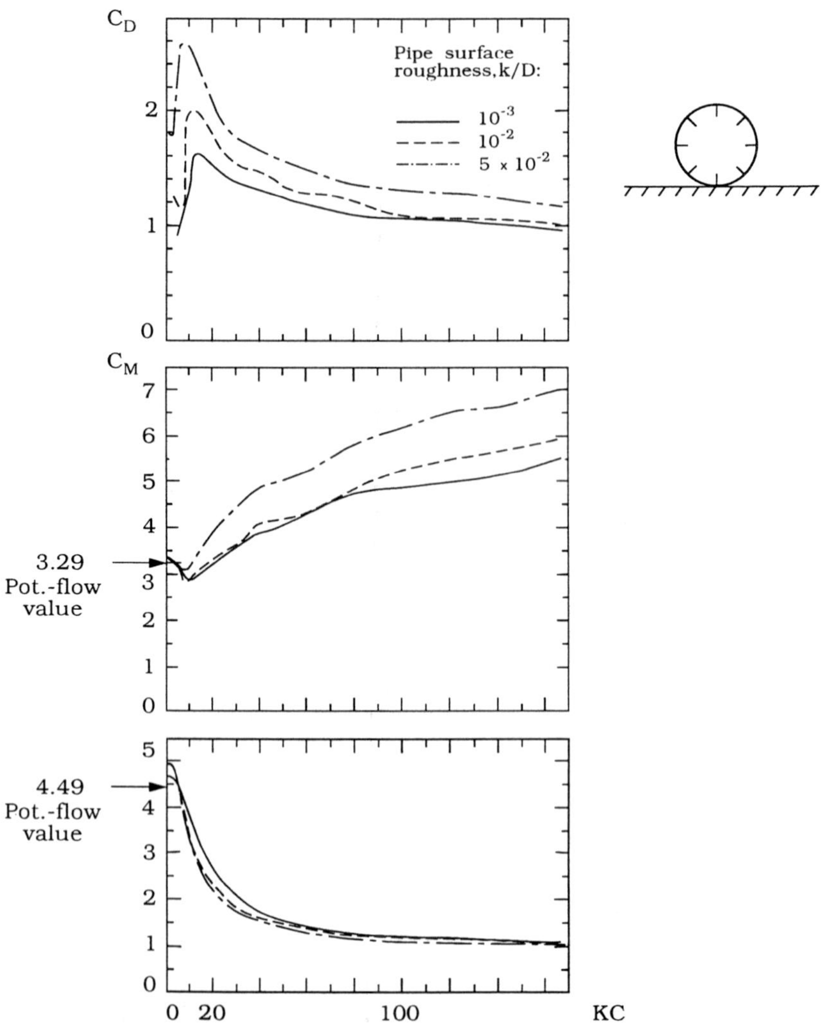

Figure 4.43 Force coefficients for a wall-mounted cylinder.
$Re = (0.5 - 3.6) \times 10^5$. Bryndum et al. (1992).

illustrates the surface-roughness influence. Bryndum et al. examined also other aspects of the problem such as the influence of co-existing current, the Fourier coefficients and phases for the drag and the lift forces and the "extreme" force coefficients, defined by

$$
C_{H,\text{max}} = \left[F_H(t)\right]_{\text{max}} \Big/ \left(\tfrac{1}{2}\rho D U_m^2\right)
$$
$$
C_{V,\text{max}} = \left[F_L(t)\right]_{\text{max}} \Big/ \left(\tfrac{1}{2}\rho D U_m^2\right)
$$

(4.90)

for the horizontal and vertical force components, respectively. An extensive comparison of data was made by Bryndum et al., covering the laboratory tests reported by Sarpkaya and Rajabi (1979), the laboratory tests carried out at the Norwegian Hydrodynamic Laboratories (NHL) and reported in NHL (1985) and the field experiments undertaken off the coast of Hawaii and reported by Grace and Zee (1979).

Force coefficients for pipelines. Partially buried pipes and pipes in trenches

Fig. 4.44 depicts the force coefficients corresponding to the case of a partially buried pipeline, while Fig. 4.45 illustrates the influence of a trench hole (Jacobsen, Bryndum and Bonde, 1989). The force coefficients C_{D0}, C_{M0} and C_{L0} in the figures are those for a pipe resting on a plane bed (Fig. 4.43). As seen, the force coefficients are generally reduced, in some cases quite substantially. The reduction in the force coefficients is due to sheltering effect, as discussed in Section 2.7 in relation to forces on pipelines in trench holes in the case of steady current (Fig. 2.32). The larger the sheltering effect, the larger the reduction in the force coefficients. Jacobsen et al. investigated also the influence of co-existing current on the force coefficients for the partially-buried-pipe case, which indicated the same kind of trend as in Fig. 4.44. In addition to the aforementioned cases, Jacobsen et al. carried out tests on pipelines sliding on the bed.

4.8 Forces resulting from breaking-wave impact

The impact forces on marine structures such as breakwaters, sea walls, piles, etc. generated by breaking waves can attain very large values. Works by Kjeldsen, Tørum and Dean (1986) and Basco and Niedzwecki (1989) show, for instance, that plunging wave forces on a pile can be a factor of 2-3 times larger than the ordinary forces with waves of comparable amplitudes.

Before considering the vertical-pile case, we shall study a simpler case, namely the case of a vertical wall exposed to the action of the impact of a plunging

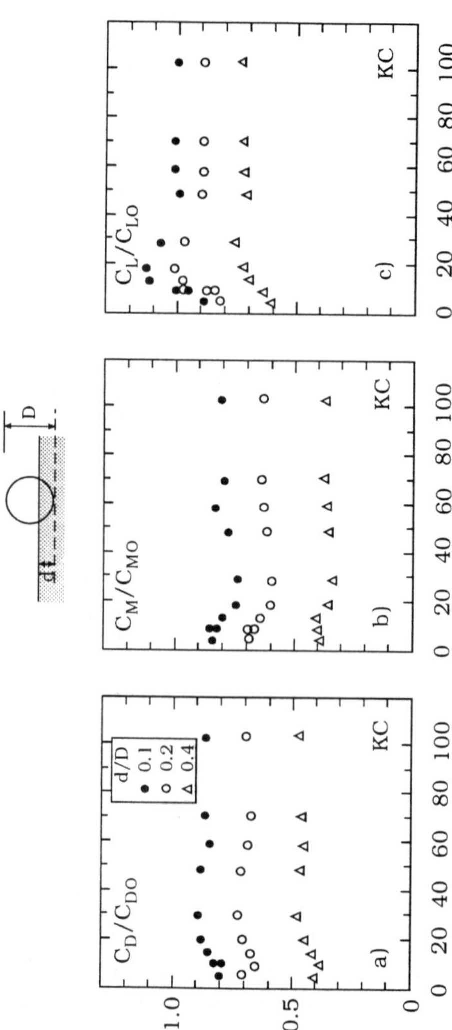

Figure 4.44 Relative force coefficients. Partially buried pipelines. C_{D0}, C_{M0} and C_{L0} are the force coefficients for a pipe resting on a plane bed (see Fig. 4.43). Jacobsen et al. (1989)

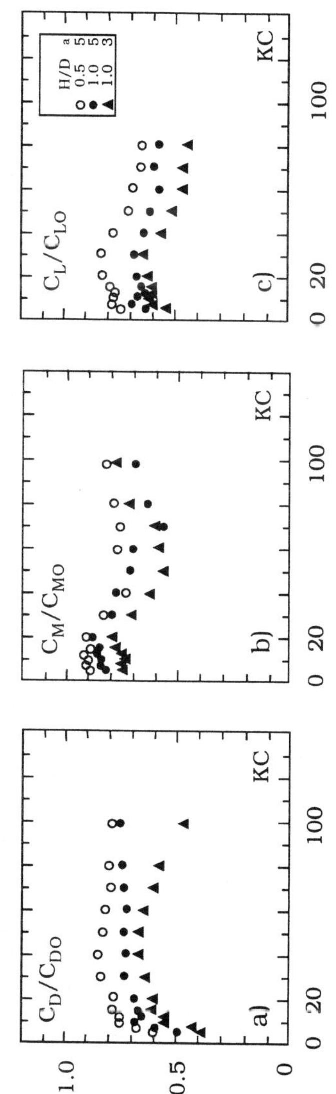

Figure 4.45 Relative force coefficients. Pipeline in open trenches. C_{D0}, C_{M0} and C_{L0} are the force coefficients for a pipe resting on a plane bed (see Fig. 4.43). Jacobsen et al. (1989).

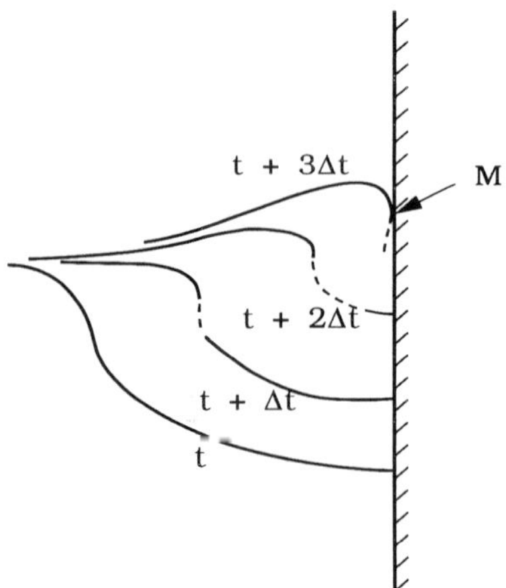

Figure 4.46 Breaking-wave profiles until the instant of impact. Vertical-wall case. Chan and Melville (1988).

breaker, sketched in Fig. 4.46. The figure illustrates the breaking-wave profiles at progressive times with interval Δt where Δt is in the order of magnitude of $0.02T$, T being the wave period, Chan and Melville (1988). As the wave approaches the wall, the breaking wave (the wave profile corresponding to time $t + 3\Delta t$) will impinge on the wall at a certain location, Location M. The impingement of the water on the wall will exert an impulsive pressure on the wall at M, the impact pressure. As the wave progresses, the impact pressure will be experienced on the wall over a larger and larger wall area.

Fig. 4.47 shows the time series of the pressure measured at the point of initial impact. As seen, the pressure increases impulsively, and then it exhibits an oscillatory character as it decreases after the peak. While the impulsive increase is due to the impact, called *hammer shock* (Lundgren, 1969), the oscillatory character of the pressure variation is linked with the air trapped in the water during the course of impact of the water mass (see, for example Lundgren, 1969, Chan and Melville, 1988).

First of all, the impact characteristics are dependent on the particular lo-

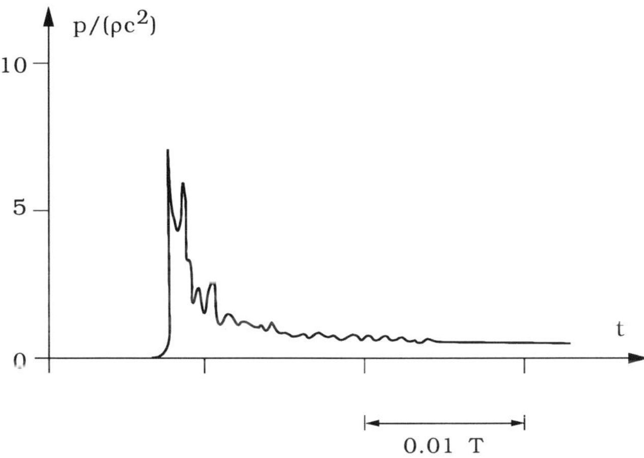

Figure 4.47 Pressure time series at the point of initial impact. Vertical wall.
Chan and Melville (1988).

cation of the wall relative to the location of the wave breaking. Fig. 4.48 summarizes the impact characteristics with the wall location. The most critical location is where the wave plunging develops just before the impact (Fig. 4.48c). Chan and Melville reports that, in this case, the direction of the crest is approximately horizontal. No impact pressures are generated for the locations in Figs. 4.48a and 4.48f. This is simply because wave breaking occurs too late for the case depicted in Fig. 4.48a, and it occurs too early for that in Fig. 4.48f.

Second, pressures at the critical location are the highest. The normalized maximum impact pressures, $p/(\rho c^2)$, typically range from 3 to 10 in which c is the wave celerity, $c = L/T$, with the corresponding rise time being in the range $0.0005T$ to $0.002T$. The obtained peak pressures are comparable to those of the others (see Table 4.1). The broad range of the measured peak pressure, a feature common to all the other studies as well (see Table 4.1), indicates the strong randomness in the process. This is due partly to the randomness in the wave breaking process (and hence due to the randomness in the dynamics of the trapped air) and partly to the randomness in the air-entrapment process. This will result in strong "turbulence" in the measured pressure signal, revealing the observed broad range of pressures.

Third Fig. 4.49 displays the impact pressure distribution over the depth at the location where the largest pressure peaks are experienced. Here, $z = 0$

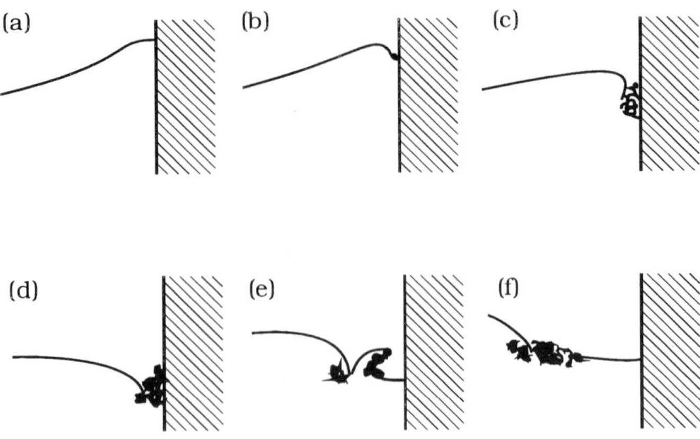

Figure 4.48 Schematics of breaking waves incident on a vertical wall. Chan and Melville (1988).

is the stationary water level. As seen, the maximum pressure occurs at about $z/L = 0.05$.

Fourth, Chan and Melville's results as well as the results of the others (Table 4.1) indicate that the impact pressure scales with ρc^2. This can be inferred from simple impulse-momentum considerations. The impulse-momentum equation for the control volume shown in Fig. 4.50 can, to a first approximation, be written as

$$\rho q c = p a \tag{4.91}$$

in which q is the rate of flow per unit width, $q = cA$, and a and A are the corresponding areas. Hence, a crude estimate of the pressure can be obtained from the preceding equation as $p/(\rho c^2) = A/a$, illustrating that the pressure scales with ρc^2. Clearly, the ratio A/a is much larger than unity, since at the instant of impingement, the impact occurs through the focusing of the incident wave front onto the wall (Chan and Melville, 1988, p.127), revealing the range observed in the experiments (Table 4.1), namely $p/(\rho c^2) = 3 - 10$.

Chan, Cheong and Tan (1995) extended Chan and Melville's study to the case of **vertical cylinders**. Figs. 4.51-4.53 display three sequences of photographs, illustrating the way in which the incoming wave impinges on the cylinder. In Fig. 4.51, the wave impinges on the cylinder before wave breaking occurs, while, in Fig. 4.53 it impinges on the cylinder long after wave breaking occurs. Therefore,

Table 4.1 Comparison of peak impact pressures.

Investigator	Typical range of peak pressures $p_m/(\rho c^2)$	Structure
Kjeldsen & Myrhaug (1979)	1-2	Vertical plate suspended above SWL (deep water).
Kjeldsen (1981)	1-3	Inclined plate suspended above SWL (deep water).
Ochi & Tsai (1984)	1.4	Surface-piercing cylinder (deep water).
Bagnold (1939)	11-40 (highest 90)	Surface-piercing plate on a sloping beach.
Hayashi & Hattori (1958)	3-15	Surface-piercing plate on a sloping beach.
Weggel & Maxwell (1970)	8-20 (highest 40)	Surface-piercing plate on a sloping beach.
Kirkgoz (1982)	8-20	Surface-piercing plate on a sloping beach.
Blackmore & Hewson (1984)	0.5-4	Seawall (prototype structure).
Chan & Melville (1988)	3-10 (highest 21)	Surface-piercing plate (deep water).

in these two cases, no significant impact pressure develops, as demonstrated by the pressure measurements of Chan et al. (1995), whereas, in Fig. 4.52 (the critical case, somewhat similar to that given in Fig. 4.48c), the impingement of breaking wave is such that very high impulsive impact pressures are generated. The pressure measurements of Chan et al. (1995) indicate that the impact pressure is the highest at the instant corresponding to Fig. 4.52b.

Fig. 4.54 gives the measured time series of pressure on the upstream edge of the cylinder. The pressure characteristics are basically similar to those observed for the vertical wall situation.

Chan et al. (1995) observed that the impact pressure decreased gradually with the azimuthal angle, θ, where $\theta = 0$ corresponds to the upstream edge of the cylinder. Also observed is the fact that the occurrence of peak pressures is delayed for locations of larger azimuthal angles, consistent with the motion of the wave crest around the cylinder. The observed extent of the area where the impact pressures $p/(\rho c^2)$ are larger than 3 is $-22.5° < \theta < +22.5°$. One final point as

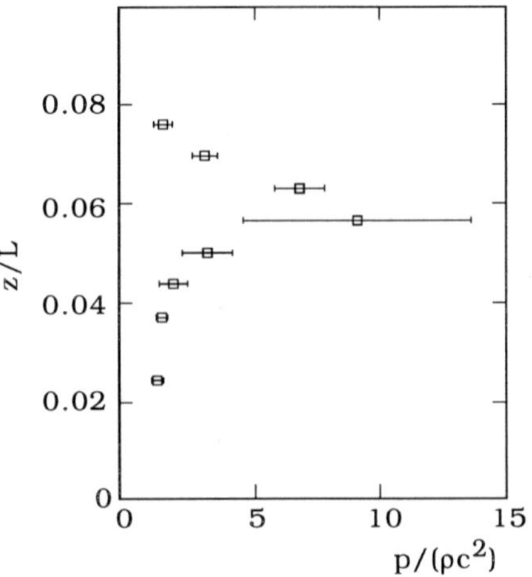

Figure 4.49 Vertical distribution of impact pressure at the location where the largest pressure peaks are experienced. Vertical wall. Chan and Melville (1988).

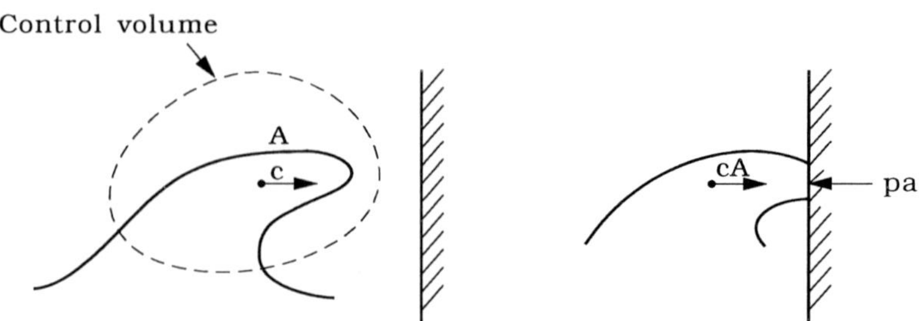

Figure 4.50 Definition sketch for the application of the impulse-momentum principle.

regards the azimuthal variation of the impact pressures is that it is not always $\theta = 0^0$ where maximum impact pressures occur; Chan et al. (1995) observed that the maximum pressures can occur off the symmetry line $\theta = 0^0$, at such θ values as high as 15^0. This is due to the turbulence referred to earlier.

The resulting impact force was estimated in Chan et al.'s (1995) study by

$$f = \int_{\Delta z} \int_{\theta} pr_0 \cos(\theta) d\theta dz \tag{4.92}$$

in which r_0 is the radius of the cylinder, and Δz is the vertical extent of the impact zone (cf. Fig. 4.49). Subsequently, the force coefficients C_s are calculated from:

$$f = \frac{1}{2}\rho C_s \, \Delta z \, Dc^2 \tag{4.93}$$

For example, the C_s value obtained at the instant of peak pressure occurrence at $\theta = 0^0$ is $C_s = 7.0$, while that obtained at the instant of peak pressure occurrence at $\theta = 15^0$ is $C_s = 11.4$.

Fig. 4.54 landscape figure caption in tcst-hj

An estimate of the force coefficient C_s can be made, adopting the method of Kaplan and Silbert (1976). The in-line impact force *per unit height of the cylinder* in the impact zone (Fig. 4.55) just after the impact will be

$$F = \frac{d(m'U)}{dt} \tag{4.94}$$

in which the drag force and the Froude-Krylov force are neglected, since we are interested in the force at the instant of impact ($t, x \to 0$). Here, U is the horizontal component of the velocity and m' is the hydrodynamic mass, corresponding to the hatched area (section a-a) in Fig. 4.55. The right hand-side of the preceding equation can be written as

$$F = m'\frac{dU}{dt} + U\left(\frac{\partial m' dx}{\partial x \, dt}\right) \tag{4.95}$$

Since the velocity U can be considered constant, equal to the wave celerity, c, the equation becomes

$$F = c^2\frac{\partial m'}{\partial x} \tag{4.96}$$

m' is given by Taylor (1930) (see Kaplan and Silbert, 1976)

$$m' = \frac{1}{2}\rho r_0^2\left[\frac{2\pi^3}{3}\frac{(1-\cos\theta)}{(2\pi-\theta)^2} + \frac{\pi}{3}(1-\cos\theta) + (\sin\theta - \theta)\right] \tag{4.97}$$

At the instant of impact ($x \to 0$), it can be shown that

$$\left(\frac{\partial m'}{\partial x}\right)_{x\to 0} = \rho r_0 \pi \tag{4.98}$$

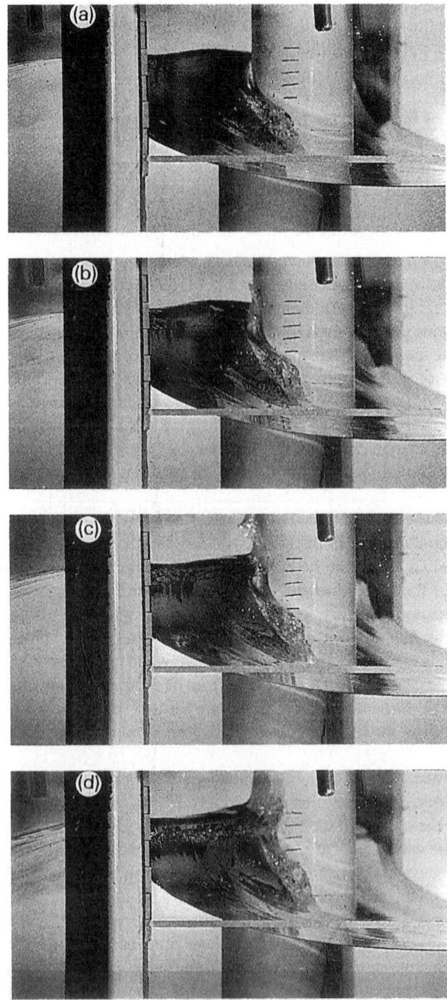

Figure 4.51 Development of wave plunging when cylinder is located at $x/L = 4.764$, $\Delta t =$ (a) 0, (b) 0.01 s ($0.008T$), (c) 0.02 s ($0.016T$), (d) 0.03 s ($0.023T$). x is the distance from the wave pedal. Chan et al. (1995) with permission - see Credits.

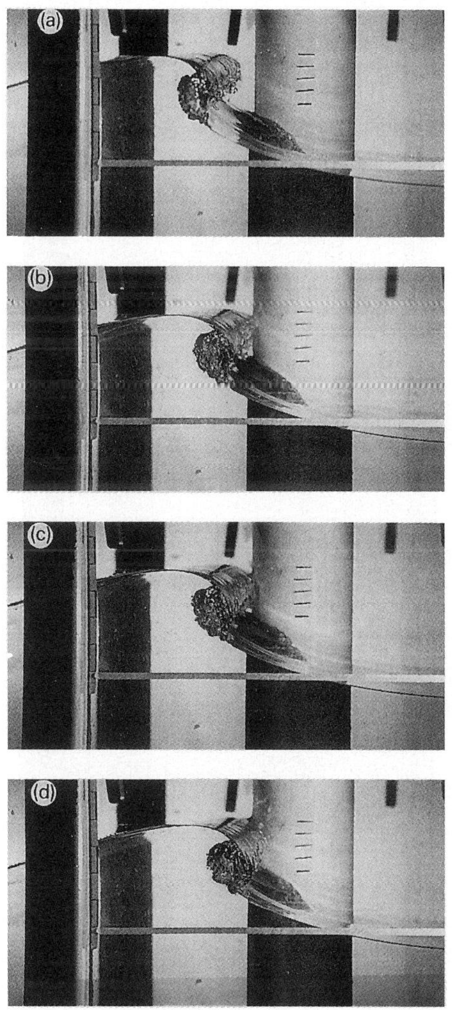

Figure 4.52 Development of wave plunging when cylinder is located at $x/L = 4.885$, $\Delta t =$ (a) 0, (b) 0.01 s ($0.008T$), (c) 0.02 s ($0.016T$), (d) 0.03 s ($0.023T$). Chan et al. (1995) with permission - see Credits.

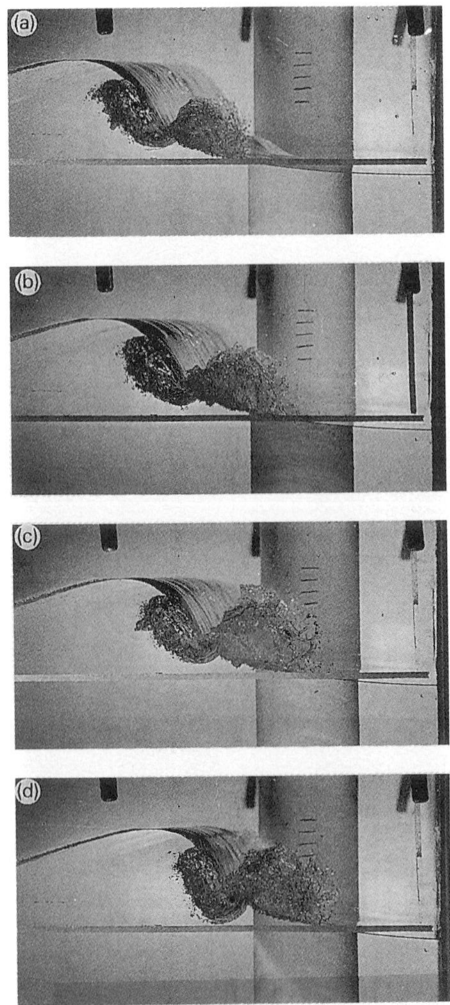

Figure 4.53 Development of wave plunging when cylinder is located at $x/L = 5.047$, $\Delta t =$ (a) 0, (b) 0.01 s $(0.008T)$, (c) 0.02 s $(0.016T)$, (d) 0.03 s $(0.023T)$. Chan et al. (1995) with permission - see Credits.

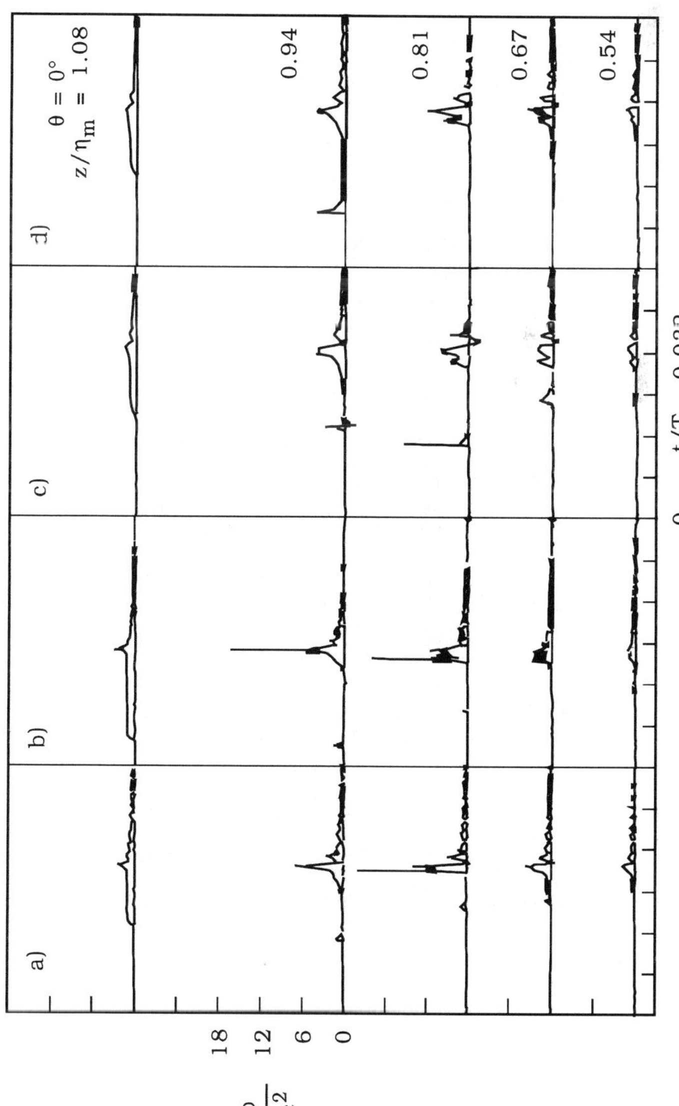

Figure 4.54 Simultaneous pressure time histories recorded from repeated experiments (a, b, c, d) at cylinder location $x/L = 4.885$. η_m is the undisturbed crest elevation. Chan et al. (1995).

Using the force coefficient definition in Eq. 4.93 and recalling that the force F is calculated per unit height of the impact zone, the force coefficient is obtained as $C_s = \pi$, the commonly used value in the empirical models (Goda et al., 1966 (referred to in the paper by Sawaragi and Nochino, 1984) and Wiegel, 1982). As seen, the experimentally obtained values of the force coefficient C_s is a factor of 2-4 larger than the theoretical estimate of C_s. (Similar results were obtained also by Sawaragi and Nochino, 1984). This may be attributed partly to the effect of trapped air.

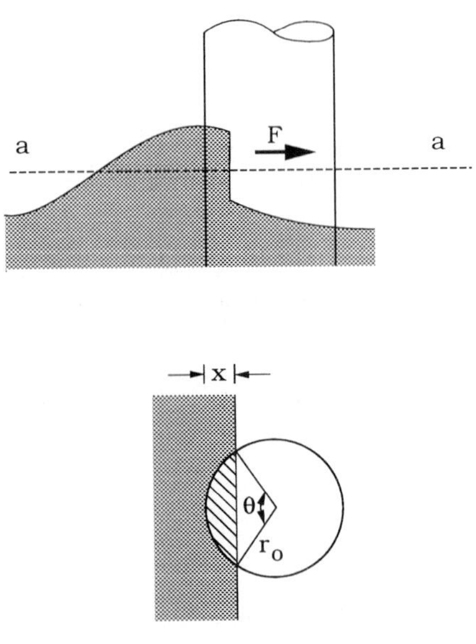

Section a-a

Figure 4.55 Definition sketch.

The previously mentioned studies have been extended by Chan, Cheong and Gin (1991) to the case of a horizontal beam, and by Chan (1993) to the case of a large horizontal cylinder in the splash zone where the structures were placed just above the still water level and exposed to plunging waves. Oumeraci, Klammer and Partenscky (1993) have, for the case of a vertical wall simulating a caisson breakwater, demonstrated that the impact pressure changes, depending on the breaker type. Criteria have been developed for wave breaking and breaker-type classification in this latter study. The breaking-wave impact pressure has been further elaborated by researchers such as Hattori, Arami and Yui (1994), Chan

(1994), Goda (1994) and Oumeraci and Kortenhaus (1994) in conjunction with the vertical-wall breakwaters. Sawaragi and Nochino (1984) studied the case of a vertical cylinder for both the spilling type breaker and the plunging type breaker; the former gave smaller peak pressures in most cases. Tanimoto, Takashi, Kaneko and Shiota (1986) studied the impact forces of breaking waves on an inclined pile. Endresen and Tørum (1992) and Yuksel and Narayanan (1994) studied breaking-wave forces on pipelines on the seabed.

REFERENCES

Achenbach, E. and Heinecke, E. (1981): On vortex shedding from smooth and rough cylinders in the range of Reynolds numbers 6×10^3 to 5×10^6. J. Fluid Mech., 109:239-251.

Ali, N. and Narayanan, R. (1986): Forces on cylinders oscillating near a plane boundary. Proc. 5th Int. Offshore Mechanics & Arctic Engineering (OMAE) Symp., Tokyo, Japan, III:613-619.

Anatürk, A. (1991): An experimental investigation to measure hydrodynamic forces at small amplitudes and high frequencies. Applied Ocean Research, 13(4):200-208.

Bagnold, R.A. (1939): Interim report on wave pressure research. J. Inst. Civil Engrs., 12:201-226.

Basco, D.R. and Niedzwecki, J.M. (1989): Breaking wave force distributions and design criteria for slender piles. OTC 6009, pp. 425-431.

Batchelor, G.K. (1967): An Introduction to Fluid Dynamics. Cambridge University Press.

Bearman, P.W., Chaplin, J.R., Graham, J.M.R., Kostense, J.K., Hall, P.F. and Klopman, G.(1985a): The loading on a cylinder in post-critical flow beneath periodic and random waves. Proc. 4th Int. Conf., In: Behaviour of Offshore Structures, Delft, Elsevier, Ed. J.A. Battjes, Developments in marine technology, 2, pp. 213-225.

Bearman, P.W., Downie, M.J., Graham, J.M.R. and Obasaju, E.D. (1985b): Forces on cylinders in viscous oscillatory flow at low Keulegan-Carpenter numbers. J. Fluid Mech., 154:337-356.

Bearman, P.W. and Obasaju, E.D. (1989): Transverse forces on a circular cylinder oscillating in-line with a steady current. Proc. 8th Int. Conf. on Offshore Mechanics and Arctic Engineering, OMAE, The Hague, March 19-23, 1989, 2:253-258.

Blackmore, P.A. and Hewson, P.J. (1984): Experiments on full scale wave impact pressures, Coastal Engrg., 8:331-346.

Bryndum, M.B., Jacobsen, V. and Tsahalis, D.T. (1992): Hydrodynamic forces on pipelines: Model tests. J. Offshore Mechanics and Arctic Engineering, Trans. ASME, 114:231-241.

Chakrabarti, S.K., Tam, W.A. and Wolbert, A.L. (1977): Wave forces on inclined tubes. Coastal Engineering, 1:149-165.

Chakrabarti, S.K. (1980): In-line forces on a fixed vertical cylinder in waves. J. Waterway, Port, Coastal and Ocean Div., ASCE, 106(WW2):145-155.

Chan, E.S. (1993): Extreme wave action on large horizontal cylinders located above still water level. Proc. 3rd Int. Offshore and Polar Eng. Conf., Singapore, 6-11 June, 1993, III:121-128.

Chan, E.S. (1994): Mechanics of deep water plunging-wave impact o vertical structures. Coastal Engineering, 22(1,2):115-134.

Chan, E.S. and Melville, W.K. (1988): Deep water plunging wave pressures on a vertical plane wall. Proc. R. Soc., London, A417:95-131.

Chan, E.S., Cheong, H.F. and Gin, K.Y.H. (1991): Wave impact loads on horizontal structures in the splash zone. Proc. ISOPE '91, Edinburgh, 3:203-209.

Chan, E.S., Cheong, H.F. and Tan, B.C. (1995): Laboratory study of plunging wave impacts on vertical cylinders. Coastal Engineering, 25:87-107.

Chaplin, J.R. (1981): Boundary layer separation from a cylinder in waves. Proc. International Symposium on Hydrodyn. in Ocean Engrg., Trondheim, 1981, 1:645-666.

Chaplin, J.R. (1984): Non-linear forces on a horizontal cylinder beneath waves. J. Fluid Mech., 147:449-464.

Chaplin, J.R. (1988a): Non-linear forces on horizontal cylinders in the inertia regime in waves at high Reynolds numbers. Proc. Int. Conf. on Behaviour of Offshore Structures (BOSS '88), Trondheim, June 1988, 2:505-518.

Chaplin, J.R. (1988b): Loading on a cylinder in uniform oscillatory flow: Part II - Elliptical orbital flow. Applied Ocean Research, 10(4):199-206.

Chaplin, J.R. (1991): Loading on a horizontal cylinder in irregular waves at large scale. Int. J. of Offshore and Polar Engrg., Dec. 1991, 1(4):247-254.

Chaplin, J.R. (1993a): Planar oscillatory flow forces at high Reynolds numbers. J. Offshore Mech. and Arctic Eng., ASME, 115:31-39.

Chaplin, J.R. (1993b): Orbital flow around a circular cylinder. Part 2. Attached flow at larger amplitudes. J. Fluid Mech., 246:397-418.

Dean, R.G., Dalrymple, R.A. and Hudspeth, R.T. (1981): Force coefficients from wave projects 1 and 11. Data including free-surface effects. Society of Petroleum Engineers Journal. December 1981, pp. 777-786.

Endresen, H.K. and Tørum, A. (1992): Wave forces on a pipeline through the surf zone. Coastal Engineering, 18:267-281.

Garrison, C.J. (1985): Comments on the cross-flow principle and Morison's equation. J. Waterway, Port, Coastal and Ocean Eng., ASCE, 111(6):1075-1079.

Goda, Y., Haranaka, S. and Kitahata, M. (1966): Study on impulsive breaking wave forces on piles. Rep. Port Harbour Res. Inst., 6(5):1-30.

Goda, Y. (1994): Dynamic response of upright breakwaters to impulsive breaking wave forces. Coastal Engineering, 22(1,2):134-158.

Grace, R.A. and Zee, G.T.Y. (1981): Wave forces on rigid pipes using ocean test data. J. Waterway, Port, Coastal and Ocean Division, ASCE, 107(WW2):71-92.

Grass, A.J., Simons, R.R. and Cavanagh, N.J. (1985): Fluid loading on horizontal cylinders in wave type orbital oscillatory flow. Proc. 4th Offshore Mechanics and Arctic Engrg. Symp., Dallas, TX., 1:576-583.

Hansen, E.A. (1990): Added mass and inertia coefficients of groups of cylinders and of a cylinder placed near an arbitrarily shaped seabed. Proc. 9th Offshore Mechanics and Arctic Engrg., Houston, TX, Vol. 1, Part A, pp. 107-113.

Hattori, M., Arami, A. and Yui, T. (1994): Wave impact pressure on vertical walls under breaking waves of various types. Coastal Engineering, 22(1,2):57-78.

Hayashi, T. and Hattori, M. (1958): Pressure of the breaker against a vertical wall. Coastal Engineering in Japan, 1:25-37.

Holmes, P. and Chaplin, J.R. (1978): Wave loads on horizontal cylinders. Proc. 16th International Conf. on Coastal Engrg., Hamburg, 1978, 3:2449-2460.

Jacobsen, V., Bryndum, M.B. and Fredsøe, J. (1984): Determination of flow kinematics close to marine pipelines and their use in stability calculations. Proc. 16th Annual Offshore Technology Conf., Paper OTC 4833, 3:481-492.

Jacobsen, V., Bryndum, M.B. and Bonde, C. (1989): Fluid loads on pipelines: Sheltered or sliding. Proc. 21st Annual Offshore Technology Conf., Paper OTC 6056, 3:133-146.

Jacobsen, V. and Hansen, E.A. (1990): The concepts of added mass and inertia forces and their use in structural dynamics. Proc. 22nd Annual Offshore Technology Conf., Houston, TX, May 7-10, 1990, Paper OTC 6314, 2:419-430.

Jonsson, I.G. (1990): Wave Current Interactions. In: The Sea, eds. B. Le Mehauté and D.M. Hanes, Wiley-Interscience, N.Y., Chapter 9A:65-120.

Justesen, P., Hansen, E.A., Fredsøe, J., Bryndum, M.B. and Jacobsen, V. (1987): Forces on and flow around near-bed pipelines in waves and current. Proc. 6th Int. Offshore Mechanics and Arctic Engrg. Symp., ASME, Houston, TX, March 1-6, 1987, 2:131-138.

Justesen, P. (1989): Hydrodynamic forces on large cylinders in oscillatory flow. J. Waterway, Port, Coastal and Ocean Engineering, ASCE, 115(4):497-514.

Justesen, P. (1991): A numerical study of oscillating flow around a circular cylinder. J. Fluid Mech., 222:157-196.

Kaplan, P. and Silbert, M.N. (1976): Impact forces on platform horizontal members in the splash zone. 8th Annual Offshore Technology Conf., Houston, TX, May 3-6, 1976, OTC 2498, pp. 749-758.

Kasahara, Y., Koterayama, W. and Shimazaki, K. (1987): Wave forces acting on rough circular cylinders at high Reynolds numbers. Proc. 19th Offshore Technology Conf., Houston, TX, OTC 5372, 1:153-160.

Keulegan, G.H. and Carpenter, L.G. (1958): Forces on cylinders and plates in an oscillating fluid. J. Research of the National Bureau of Standards, Vol. 60, No. 5, Research paper 2857, pp. 423-440.

Kirkgoz, M.S. (1982): Shock pressure of breaking waves on vertical walls. J. Waterway, Port, Coastal and Ocean Div., ASCE, 108(WW1):81-95.

Kjeldsen, S.P. and Myrhaug, D. (1979): Breaking waves in deep water and resultant wave forces. Proc. 11th Offshore Tech. Conf., Houston, TX, paper 3646, pp. 2515-2522.

Kjeldsen, S.P. (1981): Shock pressures from deep water breaking waves. Proc. Int. Symp. on Hydrodynamics, Trondheim, Norway, pp. 567-584.

Kjeldsen, S.P., Tørum, A. and Dean, R.G. (1986): Wave forces on vertical piles caused by 2 and 3 dimensional breaking waves. Proc. 20th Int. Conf. Coastal Engineering, Taipei, ASCE, New York, pp. 1929-1942.

Kozakiewicz, A., Fredsøe, J. and Sumer, B.M. (1995): Forces on pipelines in oblique attack. Steady current and waves. Proc. 5th Int. Offshore and Polar Engineering Conf., The Hague, Netherlands, June 11-16, 1995, Vol. II:174-183.

Lundgren, H. (1969): Wave shock forces: An analysis of deformations and forces in the wave and in the foundation. Research and Wave Action. Proc. Symposium, Delft, Vol. 2, Paper 4.

Lundgren, H., Mathiesen, B. and Gravesen, H. (1976): Wave loads on pipelines on the seafloor. Proc. 1st Int. Conf. on the Behaviour of Offshore Structures, BOSS 76, 1:236-247.

Maull, D.J. and Milliner, M.C. (1978): Sinusoidal flow past a circular cylinder. Coastal Engineering, 2:149-168.

Maull, D.J. and Norman, S.G. (1979): A horizontal circular cylinder under waves. Proc. Symp. on Mechanics of Wave-Induced Forces on Cylinders, Bristol, ed. T.L. Shaw, Pitman, pp. 359-378.

Milne-Thomson, L.M. (1962): Theoretical Hydrodynamics. Macmillan.

Moe, G. and Verley, R.L.P. (1980): Hydrodynamic damping of offshore structures in waves and current. 12th Annual Offshore Technology Conf., Paper No. OTC 3798, Houston, TX, May 5-8, 1980, 3:37-44.

Morison, J.R., O'Brien, M.P., Johnson, J.W. and Schaaf, S.A. (1950): The forces exerted by surface waves on piles. J. Petrol. Technol., Petroleum Transactions, AIME, (American Inst. Mining Engrs.), 189:149-154.

NHL (Norwegian Hydrodynamic Laboratories) (1985): Design of Pipelines to Resist Ocean Forces. Final Report on Joint Industry R & D Program, 1985.

Ochi, M.K. and Tsai, C.H. (1984): Prediction of impact pressure induced by breaking waves on vertical cylinders in random seas. Appl. Ocean Res., 6:157-165.

Oumeraci, H., Klammer, P. and Partenscky, H.W. (1993): Classification of breaking wave loads on vertical structures. ASCE, J. Waterway, Port, Coastal and Ocean Engineering, 119(4):381-396.

Oumeraci, H. and Kortenhaus, A. (1994): Analysis of the dynamic response of caisson breakwaters. Coastal Engineering, 22(1,2):159-182.

Peregrine, D.H. (1976): Interaction of water waves and currents. Advances in Applied Mechanics, 16:9-117.

Ramberg, S.E. and Niedzwecki, J.M. (1979): Some uncertainties and errors in wave force computations. Proc. 11th Offshore Technology Conf., Houston, TX, 3:2091-2101.

Rodenbusch, G. and Gutierrez, C.A. (1983): Forces on cylinders in two-dimensional flow. Tech. Report, Vol. 1, BRC 13-83, Bellaire Research Center (Shell Development Co.), Houston, TX.

Sarpkaya, T. (1976a): In-line and transverse forces on smooth and sand-roughened cylinders in oscillatory flow at high Reynolds numbers. Naval Postgraduate School, Monterey, CA, Tech. Rep. NPS-69SL76062.

Sarpkaya, T. (1976b): Forces on cylinders near a plane boundary in a sinusoidally oscillating fluid. Trans. ASME, J. Fluids Engng., 98:499-505.

Sarpkaya, T. (1977a): In-line and transverse forces on cylinders near a wall in oscillatory flow at high Reynolds numbers. Proc. 9th Annual Offshore Technology Conf., Houston, TX, Paper OTC 2898, 3:161-166.

Sarpkaya, T. (1977b): In-line and transverse forces on cylinders in oscillatory flow at high Reynolds numbers. Jour. Ship Research, 21(4):200-216.

Sarpkaya, T. and Rajabi, F. (1979): Hydrodynamic drag on bottom-mounted smooth and rough cylinders in periodic flow. Proc. 11th Annual Offshore Technology Conf., Houston, TX, Paper OTC 3761, 2:219-226.

Sarpkaya, T. (1981): Morison's Equation and the Wave Forces on offshore structures. Naval Civil Engineering Laboratory Report, CR82.008, Port Hueneme, CA.

Sarpkaya, T. and Isaacson, M. (1981): Mechanics of Wave Forces on Offshore Structures. Van Nostrand Reinhold Company.

Sarpkaya, T., Raines, T.S. and Trytten, D.O. (1982): Wave forces on inclined smooth and rough circular cylinders. Proc. 14th Offshore Technology Conf., Houston, TX, OTC 4227, pp. 731-736.

Sarpkaya, T. (1984): Discussion of "Quasi 2-D forces on a vertical cylinder in waves", (paper No. 17671 by P.K. Stansby et al.). J. Waterway, Port, Coastal and Ocean Engineering, 110(1):120-123.

Sarpkaya, T. and Wilson, J.R. (1984): Pressure distribution on smooth and rough cylinders in harmonic flow. Proc. Ocean Structural Dynamics, Corvallis, OR, 1984, pp. 341-355.

Sarpkaya, T. and Storm, M. (1985): In-line force on a cylinder translating in oscillatory flow. Applied Ocean Research, 7(4):188-196.

Sarpkaya, T. (1986a): Force on a circular cylinder in viscous oscillatory flow at low Keulegan-Carpenter numbers. J. Fluid Mech., 165:61-71.

Sarpkaya, T. (1986b): In-line and transverse forces on smooth and rough cylinders in oscillatory flow at high Reynolds numbers. Technical Report No. NPS-69-86-003, Naval Postgraduate School, Monterey, CA.

Sarpkaya, T. (1987): Oscillating flow about smooth and rough cylinders. J. Offshore Mechanics and Arctic Engineering, ASME, 109:307-313.

Sarpkaya, T. (1990): On the effect of roughness on cylinders. Proc. 9th Offshore Mech. and Arctic Engrg. Conf., Feb. 18-22, 1990, Houston, TX, 1(A):47-55.

Sawaragi, T. and Nochino, M. (1984): Impact forces of nearly breaking waves on a vertical circular cylinder. Coastal Engineering in Japan, 27:249-263.

Schewe, G. (1983): On the force fluctuations acting on a circular cylinder in crossflow from subcritical up to transcritical Reynolds numbers. J. Fluid Mech., 133:265-285.

Soulsby, R.L., Hamm, L., Klopman, G., Myrhaug, D., Simons, R.R. and Thomas, G.P. (1993): Wave-current interaction within and outside the bottom boundary layer. Coastal Engineering, 21:41-69.

Stansby, P.K., Bullock, G.N. and Short, I. (1983): Quasi 2-D forces on a vertical cylinder in waves. J. Waterway, Port, Coastal and Ocean Eng., ASCE, 109(1):128-132.

Stansby, P.K. and Smith, P.A. (1991): Viscous forces on a circular cylinder in orbital flow at low Keulegan-Carpenter numbers. J. Fluid Mech., 229:159-171.

Stansby, P.K. (1993): Forces on a circular cylinder in elliptical orbital flows at low Keulegan-Carpenter numbers. Applied Ocean Res., 15:281-292.

Stokes, G.G. (1851): On the effect of the internal friction of fluids on the motion of pendulums. Trans. Cambridge Phil. Soc., Vol.9, Part II, pp. 8-106.

Sumer, B.M., Jensen, B.L. and Fredsøe, J. (1991): Effect of a plane boundary on oscillatory flow around a circular cylinder. J. Fluid Mech., 225:271-300.

Sumer, B.M., Jensen, B.L. and Fredsøe, J. (1992): Pressure measurements around a pipeline exposed to combined waves and current. Proc. 11th Offshore Mechanics and Arctic Engineering Conf., Calgary, Canada, June 7-11, 1992, V-A:113-121.

Tanimoto, K., Takashi, S., Kaneko, T. and Shiota, K. (1986): Impact force of breaking waves on an inclined pile. 5th Int. OMAE Symp., Tokyo, Japan, 1:235-241.

Taylor, J.L. (1930): Some hydrodynamical inertia coefficients. Philosophical Magazine, Series 7, 9:161-183.

Tørum, A. (1989): Wave forces on pile in surface zone. ASCE, J. Waterway, Port, Coastal and Ocean Engineering, 115(4):547-565.

Wang, C.Y. (1968): On high-frequency oscillatory viscous flows. J. Fluid Mech., 32:55-68.

Weggel, J.R. and Maxwell, W.H.C. (1970): Experimental study of breaking wave pressures. Proc. Offshore Tech. Conf., TX, OTC 1244, pp. 175-188.

Wiegel, R.L. (1982): Forces induced by breakers on piles. Proc. 18th Int. Conf. Coastal Engineering, Cape Town, ASCE, New York, pp. 1699-1715.

Williamson, C.H.K. (1985): Sinusoidal flow relative to circular cylinders. J. Fluid Mech., Vol. 155, p. 141-174.

Wolfram, J. and Theophanatos, A. (1989): The loading of heavily roughened cylinders in waves and linear oscillatory flow. Proc. 8th Offshore Mechanics and Arctic Engineering Conf., The Hague, March 19-23, 1989, pp. 183-190.

Wolfram, J., Javidan, P. and Theophanatos, A. (1989): Vortex shedding and lift forces on heavily roughened cylinders of various aspect ratios in planar oscillatory flow. Proc. 8th Offshore Mechanics and Arctic Engineering Conf., The Hague, March 19-23, 1989, pp. 269-278.

Yamamoto, T., Nath, J.H. and Slotta, L.S. (1974): Wave forces on cylinders near plane boundary. J. Waterway, Port, Coastal Ocean Engng. Div., ASCE, 100:345-360.

Yamamoto, T. and Nath, J.H. (1976): High Reynolds number oscillating flow by cylinders. Proc. 15th Int. Conf. on Coastal Engrg., III:2321-2340.

Yuksel, Y. and Narayanan, R. (1994): Breaking wave forces on horizontal cylinders close to the sea bed. Coastal Engineering, 23:115-133.

Chapter 5. Mathematical and numerical treatment of flow around a cylinder

The mathematical/numerical treatment of flow around cylinders has been improved significantly with the increasing capacity of computers. This chapter treats the mathematical/numerical modelling of flow past cylinders; three categories are examined: 1) the methods involving the direct solutions of the Navier-Stokes equations, 2) the vortex methods, and 3) the methods involving the hydrodynamic stability analysis.

5.1 Direct solutions of Navier-Stokes equations

The direct solution of the complete flow equation is until now restricted only to the low Reynolds number case, where the flow is *laminar*. Numerical solution of the N.-S. equation at higher Reynolds number including turbulent features is under way (Spalart and Baldwin (1987) achieved a solution of the oscillatory boundary layer over a plane bed up to $Re \sim 10^5$ using direct simulation).

5.1.1 Governing equations

The motion of fluid around a body is governed by the Navier-Stokes equations

$$\rho\left(\frac{\partial \mathbf{u}}{\partial t} + \mathbf{u} \cdot \nabla \mathbf{u}\right) = -\nabla p + \mu \nabla^2 \mathbf{u} \tag{5.1}$$

and the continuity equation

$$\nabla \cdot \mathbf{u} = 0 \tag{5.2}$$

Here $\mathbf{u}$ is the velocity vector, p the pressure, ∇ the vector gradient, ∇^2 the Laplacian operator, ρ the fluid density and μ the fluid viscosity. Dots represent the scalar multiplication of two vector quantities (Batchelor, 1967).

Past work regarding the solution of the Navier-Stokes equations in relation to flow around cylinders are summarized in Table 5.1.

5.1.2 The Oseen (1910) and Lamb (1911) solution

The pioneering work in conjuction with the viscous-fluid flow around bluff bodies dates back as early as 1851; Stokes (1851) treated the case of a spherical body and determined the flow field around and the drag on the spherical body. He achieved this under the assumption that the motion is extremely slow (the creeping motion) so that $Re \ll 1$. In this case, the inertia forces will be small compared with the viscous forces, therefore Eq. 5.1 can be approximated to

$$0 = -\nabla p + \mu \nabla^2 \mathbf{u} \tag{5.3}$$

Stokes obtained a solution to this linear equation and computed the drag, F_D, on the *spherical body* as

$$C_D = \frac{F_D}{\frac{1}{2}\rho\left(\frac{\pi D^2}{4}\right)U^2} = \frac{24}{Re} \quad ; \; Re \ll 1 \tag{5.4}$$

in which C_D is the drag coefficient, U the velocity of the body and D the diameter of the body.

The basic ideas behind Stokes analysis is outlined in Example 5.1.

Table 5.1 A partial list of the past work regarding the solution of the two-dimensional Navier-Stokes equations for flow around a cylinder in steady current.

Author	Re	Cylinder	Remarks
Oseen (1910) and Lamb (1911)	$Re \ll 1$	Circular	–
Thom (1933)	10 and 20	"	–
Kawaguti (1953)	40	"	–
Apelt (1961)	40 and 44	"	–
Fromm & Harlow (1963)	$15 \leq Re \leq 6000$	Rectangular	For $Re \tilde{<} 40$ flow remained steady after the introduction of perturbation to excite vortex shedding
Keller & Takami (1966)	2, 4, 10 and 15	Circular	–
Son & Hanratty (1969)	40, 200 and 500	"	No perturbation to excite vortex shedding; only steady-state solutions
Dennis & Chang (1970)	$5 \leq Re \leq 100$	"	"
Jordan & Fromm (1972)	100, 400 and 1000	"	Vortex shedding is excited by a perturbation

Table 5.1 continued......

Author	Re	Cylinder	Remarks
Davis & Moore (1982)	$100 \leq Re \leq 2800$	Square	No perturbation; vortex-shedding is excited by round-off errors
"	250 and 1000	Square, rectangular	1) " 2) Effect of angle of attack, effect of shear, effect of aspect ratio
Braza, Chassaing & Minh (1986)	100, 200 and 1000	Circular	Vortex shedding is excited by a perturbation
Lecointe & Piquet (1989)	$140 \leq Re \leq 2000$	"	"
Braza, Chassaing & Minh (1990)	$2000 \leq Re \leq 10000$	"	"
Franke, Rodi & Schönung (1990)	$40 \leq Re \leq 5000$ $70 \leq Re \leq 300$	Circular Square	No perturbation; vortex shedding is excited by round-off errors
Wang & Dalton (1991a)	$300 \leq Re \leq 1000$	Circular	Vortex shedding is excited by a perturbation. Calculations are extended so as to cover the decelerated-flow
Braza, Nogues & Persillon (1992)	20000 and 30000	"	–

Example 5.1:   Drag on a sphere at small Reynolds number

The sphere is held stationary and the fluid moves with a velocity U in the negative direction of the x-axis (Fig. 5.1). The spherical coordinate system is chosen. Only two coordinates, namely, r and θ, will be involved due to the axisymmetric character of the problem

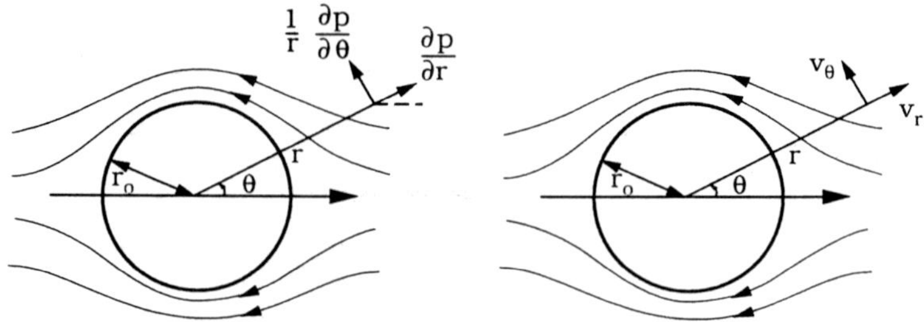

Figure 5.1 Definition sketch. Flow around a sphere.

By taking the divergence of both sides of Eq. 5.3

$$0 = -\nabla^2 p + \mu \nabla^2 (\nabla \cdot \mathbf{u}) \tag{5.5}$$

and using Eq. 5.2, the pressure is found to satisfy the Laplace equation:

$$\nabla^2 p = 0 \tag{5.6}$$

A general solution to the Laplace equation (Eq. 5.6) can be given as an infinite series of spherical harmonics. However, in the present problem, it turns out that the previously mentioned infinite series solution is unnecessary, and that the solution corresponds to a doublet flow

$$p = -\frac{\alpha}{r^2} \cos \theta \tag{5.7}$$

which is known to be a spherical harmonic (Milne-Thomson, 1962, Section 16.1). Here α is a constant.

Now the outer boundary conditions demand that the flow approaches to a uniform stream

$$\psi \to -\frac{1}{2}Ur^2 \sin^2\theta \quad \text{as} \quad r \to \infty \tag{5.8}$$

(Milne-Thomson, 1962, Section 15.22). Hence a general expression for the stream function can be sought in the following form

$$\psi = -f(r)\sin^2\theta \tag{5.9}$$

in which f is an unknown function.

Now consider the x-component of the equation of motion, Eq. 5.3,

$$\frac{\partial p}{\partial x} = \mu \nabla^2 u \tag{5.10}$$

and insert the following identities into the above equation

$$\frac{\partial p}{\partial x} = \frac{\partial p}{\partial r}\cos\theta - \frac{1}{r}\frac{\partial p}{\partial\theta}\sin\theta \quad \text{(see Fig.5.1)} \tag{5.11}$$

$$u = v_r\cos\theta - v_\theta\sin\theta \quad \text{(see Fig.5.1)} \tag{5.12}$$

$$\nabla^2 = \frac{1}{r^2}\frac{\partial}{\partial r}\left(r^2\frac{\partial}{\partial r}\right) + \frac{1}{r^2\sin\theta}\frac{\partial}{\partial\theta}\left(\frac{\partial}{\partial\theta}\sin\theta\right) \tag{5.13}$$

(the Laplace operator in spherical polar coordinates)

$$v_r = -\frac{1}{r\sin\theta}\frac{1}{r}\frac{\partial\psi}{\partial\theta} \quad \text{and} \quad v_\theta = \frac{1}{r\sin\theta}\frac{\partial\psi}{\partial r} \tag{5.14}$$

(in spherical polar coordinates)

in which p is given by Eq. 5.7 and ψ is given by Eq. 5.9. This yields

$$\frac{2\alpha}{r}\cos^2\theta - \frac{\alpha}{r}\sin^2\theta =$$

$$= \mu\left[-\frac{4f}{r^2} + 2f''\right]\cos^2\theta - \mu\left[\frac{2f'}{r} - \frac{4f}{r^2} - rf'''\right]\sin^2\theta \tag{5.15}$$

By setting the factors in front of $\sin^2\theta$ and $\cos^2\theta$ equal to zero, the following two ordinary differential equations are obtained:

$$r^2 f'' - 2f = \frac{\alpha}{\mu}r \quad \text{and} \quad r^3 f''' - 2rf' + 4f = -\frac{\alpha r}{\mu} \tag{5.16}$$

which both have the solution

$$f = \beta_1\, r^2 - \frac{\alpha r}{2\mu} + \frac{\beta_2}{r} \tag{5.17}$$

in which β_1 and β_2 are arbitrary constants.

From the outer boundary condition, namely $f \to \frac{1}{2}Ur^2$, Eqs. 5.8 and 5.9,

$$\beta_1 = \frac{1}{2}U \tag{5.18}$$

On the surface of the sphere, on the other hand, $v_r = v_\theta = 0$ which, from Eq. 5.14 reads

$$f(r_0) = 0 \quad \text{and} \quad f'(r_0) = 0 \tag{5.19}$$

in which the constans α and β_2 are found as follows

$$\alpha = \frac{3}{2}\mu U r_0 \quad \text{and} \quad \beta_2 = \frac{1}{4}U r_0^3 \tag{5.20}$$

The velocity components are therefore

$$v_r = U \cos\theta \left[1 - \frac{3}{2}\frac{r_0}{r} + \frac{1}{2}\left(\frac{r_0}{r}\right)^3 \right]$$

$$v_\theta = U \sin\theta \left[1 - \frac{3}{4}\frac{r_0}{r} - \frac{1}{4}\left(\frac{r_0}{r}\right)^3 \right] \tag{5.21}$$

The force on the sphere due to pressure will be (using Eqs. 5.7 and 5.20)

$$F_p = -2\pi r_0^2 \int_0^\pi p \sin\theta \cos\theta \, d\theta = 2\pi \mu r_0 U \tag{5.22}$$

The force on the sphere due to friction, on the other hand, will be

$$F_f = -2\pi r_0^2 \int_0^\pi \tau_{r\theta} \sin^2\theta \, d\theta = 4\pi \mu r_0 U \tag{5.23}$$

in which $\tau_{r\theta}$ is calculated from $\tau_{r\theta} = \mu \partial v_\theta / \partial r$, yielding

$$\tau_{r\theta} = \mu \frac{\partial v_\theta}{\partial r} = \frac{3}{2}\mu \frac{U}{r_0} \sin\theta \tag{5.24}$$

Hence, the total force from Eqs. 5.23 and 5.24 will be

$$F = F_p + F_f = 6\pi \mu r_0 U \tag{5.25}$$

which, in terms of drag coefficient, can be written as in Eq. 5.4.

As a final remark, the solution (Eq. 5.21) is self-consistent at positions near the sphere in the sense that the inertia forces are small compared with the viscous forces, justifying the creeping motion assumption leading to Eq. 5.3. However, the inertia forces corresponding to the solution (5.21) become comparable with viscous forces at distances from the sphere of order r_0/R (Batchelor, 1967, p. 232). (The solution is clearly not valid at such large distances). This is called Oseen's paradox. We shall return to this problem in the next example.

Example 5.2:   Drag on a circular cylinder at small Reynolds number

A solution to Eqs. 5.2 and 5.3 may be sought for a circular cylinder in the same way as for a sphere.

The pressure is given by the following equation (in place of Eq. 5.7)

$$p = -\frac{\alpha}{r}\cos\theta \tag{5.26}$$

in which $(r,\,\theta)$ are the polar coordinates (Fig. 4.3). The analogue of Eq. 5.9 is

$$\psi = -f(r)\sin\theta \tag{5.27}$$

The differential equations satisfied by the function f (the analogues of Eqs. 5.16) are

$$r^2 f'' + r f' - f = \frac{\alpha}{\mu} r \quad \text{and} \quad r^3 f''' + r^2 f'' - 2r f' + 2f = -\frac{\alpha}{\mu} r \tag{5.28}$$

which both have the solution

$$f = \frac{1}{2}\frac{\alpha}{\mu} r \ln r - \beta_1 r - \frac{\beta_2}{r} \tag{5.29}$$

in which β_1 and β_2 are arbitrary constants.

On the surface of the cylinder, $v_r = v_\theta = 0$, i.e.,

$$v_r = -\frac{1}{r}\frac{\partial\psi}{\partial\theta} = 0 \quad \text{and} \quad v_\theta = \frac{\partial\psi}{\partial r} = 0 \tag{5.30}$$

or, from Eq. 5.27

$$f(r_0) = 0 \quad \text{and} \quad f'(r_0) = 0 \tag{5.31}$$

From the latter two equations, the constants β_1 and β_2 are found as follows

$$\beta_1 = \frac{1}{4}\frac{\alpha}{\mu} + \frac{1}{2}\frac{\alpha}{\mu}\ln r_0 \quad \text{and} \quad \beta_2 = -\frac{\alpha r_0^2}{4\mu} \tag{5.32}$$

Hence, the velocity components v_r and v_θ are

$$v_r = -\frac{1}{r} f \cos\theta \quad \text{and} \quad v_\theta = -f'\sin\theta \tag{5.33}$$

in which

$$f = \frac{1}{2}\frac{\alpha}{\mu}\left[r\ln r - \left(\frac{1}{2} + \ln r_0\right)r + \frac{1}{2}r_0^2\frac{1}{r}\right] \tag{5.34}$$

The force on the cylinder due to pressure will be (using Eq. 5.26)

$$F_p = -\int_0^{2\pi} p(r_0\,d\theta)\cos\theta = \pi\alpha \tag{5.35}$$

and the force due to friction will be

$$F_f = -\int_0^{2\pi} \tau_{r\theta}(r_0\,d\theta)\sin\theta = \pi\alpha \tag{5.36}$$

in which $\tau_{r\theta}$ is calculated from $\tau_{r\theta} = \mu\partial v_\theta/\partial r$, giving

$$\tau_{r\theta} = \mu\frac{\partial v_\theta}{\partial r} = -\frac{\alpha}{r_0}\sin\theta \tag{5.37}$$

From Eqs. 5.35 and 5.36, the total force on the cylinder will be

$$F = F_p + F_f = 2\pi\alpha \tag{5.38}$$

The remaining arbitrary constant α has to be determined from the outer boundary condition. However, no choice of α will make $\mathbf{u}$ go to the constant value corresponding to the undisturbed flow, as $r \to \infty$, since f diverges as $\ln r$ when r is large (Eq. 5.34). It can be shown that the inertia force becomes comparable with the viscous force at large distances from the cylinder, and the solution (5.34) is thus not a self-consistent approximation to the flow field at large values of r (Oseen's paradox). Clearly some approximation to the equation of motion at large r is needed, and Eq. 5.34 must match with the solution of this approximate equation at large distances from the cylinder.

It can be shown that this approximation to the equation of motion is

$$-\rho\mathbf{U}\cdot\nabla\mathbf{u} = -\nabla p + \mu\nabla^2\mathbf{u} \tag{5.39}$$

This, together with the equation of continuity, Eq. 5.2, are known as the Oseen equations (Oseen, 1910). The calculations due to Lamb (1911) show that Eq. 5.39 has a solution which, near the cylinder, approximates to the solution (Eq. 5.34) provided that the constant in Eq. 5.34 is chosen as (Batchelor, 1967, p. 246)

$$\alpha = \frac{2U\mu}{\ln(7.4/Re)} \tag{5.40}$$

Thus the drag coefficient, from Eqs. 5.38 and 5.40 will be

$$C_D = \frac{8\pi}{Re \ln(7.4/Re)} \quad ; \quad Re \ll 1 \tag{5.41}$$

This relation is in good agreement with experiments for values of Re up to about 0.5 (Fig. 2.7).

5.1.3 Numerical solutions

The N.–S. equations and the continuity equation, Eqs. 5.1 and 5.2, for a two-dimensional flow in a Carterian co-ordinate system are

$$\frac{\partial u}{\partial t} + u\frac{\partial u}{\partial x} + v\frac{\partial u}{\partial y} + \frac{\partial(p/\rho)}{\partial x} = \nu\left(\frac{\partial^2 u}{\partial x^2} + \frac{\partial^2 u}{\partial y^2}\right), \tag{5.42}$$

$$\frac{\partial v}{\partial t} + u\frac{\partial v}{\partial x} + v\frac{\partial v}{\partial y} + \frac{\partial(p/\rho)}{\partial y} = \nu\left(\frac{\partial^2 v}{\partial x^2} + \frac{\partial^2 v}{\partial y^2}\right), \tag{5.43}$$

$$\frac{\partial u}{\partial x} + \frac{\partial v}{\partial y} = 0, \tag{5.44}$$

in which u and v are the components of velocity along the x and y directions, respectively.

It is more convenient to write the N.–S. equations in terms of the stream function, ψ, and the vorticity function, ω, defined by

$$u = \frac{\partial\psi}{\partial y} \tag{5.45}$$

$$v = -\frac{\partial\psi}{\partial x} \tag{5.46}$$

$$\omega = \frac{\partial v}{\partial x} - \frac{\partial u}{\partial y} \tag{5.47}$$

The continuity equation (Eq. 5.44) is satisfied automatically by Eqs. 5.45 and 5.46. Regarding the N.–S. equation (Eqs. 5.42 and 5.44), eliminating the pressure from these equations and making use of Eqs. 5.45 - 5.47, the following equation is obtained

$$\frac{\partial\omega}{\partial t} + u\frac{\partial\omega}{\partial x} + v\frac{\partial\omega}{\partial y} = \nu\left(\frac{\partial^2\omega}{\partial x^2} + \frac{\partial^2\omega}{\partial y^2}\right) \tag{5.48}$$

This equation is known as the **vorticity-transport equation**.

Inserting Eqs. 5.45 and 5.46 into Eq. 5.47, on the other hand, the so-called **Poisson equation** is obtained

$$\frac{\partial^2 \psi}{\partial x^2} + \frac{\partial^2 \psi}{\partial y^2} = -\omega \ , \tag{5.49}$$

Eqs. 5.45–5.49, or their polar co-ordinate counterparts, constitute the basic equations used in a numerical solution of N.–S. equations. These equations are to be solved with the boundary conditions on the cylinder surface and at a boundary far away from the cylinder (the outer boundary). The requirements are: On the cylinder surface the no-slip and impermeability conditions must be satisfied while, at the outer boundary, the velocity components must be identical to those of the undisturbed flow.

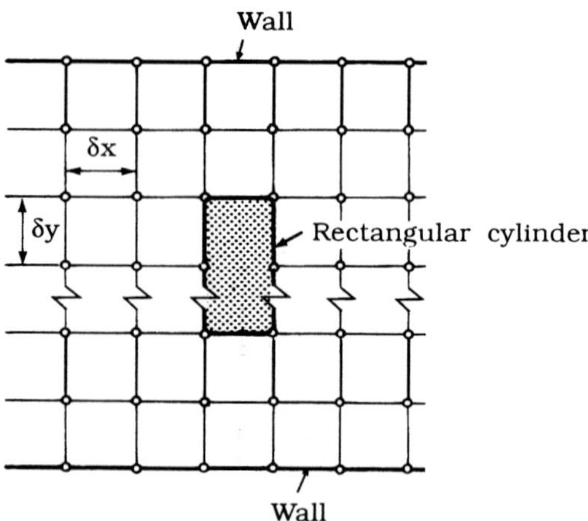

Figure 5.2 Portion of computational region showing finite-difference mesh and its relation to solid boundaries.

The basic principles of such a numerical study may be described by the following example, which is taken from the work by Fromm and Harlow (1963), (see Table 5.1). A rectangular cylinder with a large aspect ratio is impulsively accelerated to a *constant velocity* in a channel of finite width. A finite-difference mesh of cells of sides δx and δy, dividing the spatial region of interest in the manner shown in Fig. 5.2, is introduced.

In this way, the continuous flow field can be described by a finite number of quantities. The basic steps involved in advancing the solution from time t to time $t + \delta t$ are as follows:

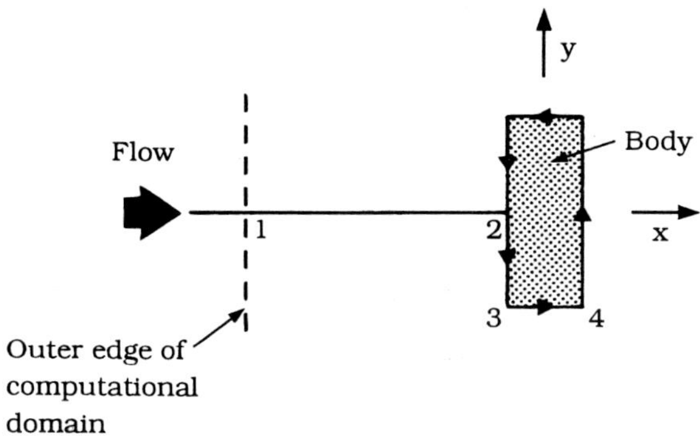

Figure 5.3 The pressure distribution is determined by numerically integrating the momentum equations (Eqs. 5.50 and 5.51) over 1234...

1. At the beginning, all required quantites are available in the computer memory.

2. For each "vorticity" point, a new value of ω is found by use of a finite-
difference approximation of Eq. 5.48.

3. For each "stream-function" point, a new value of ψ is found from a finite-
difference approximation of Eq. 5.49. (The method of solution involves a
succession of iterations).

4. Implementing Eqs. 5.45 and 5.46, the new components of velocity are found,
where care is taken in the entire procedure that the results are consistent
with the finite-difference form of Eq. 5.47.

5. Given the velocity and the vorticity field, the pressure is then calculated,
using the following equations:

On $y = constant$ lines:

$$\frac{p_B}{\rho} + \frac{q_B^2}{2} = \frac{p_A}{\rho} + \frac{q_A^2}{2}$$

$$-\int_A^B \frac{\partial u}{\partial t}\,dx + \int_A^B v\omega\,dx - \int_A^B \nu \frac{\partial \omega}{\partial y} \tag{5.50}$$

in which A and B are two points on the y-constant line.

On $x = constant$ lines:

$$\frac{p_D}{\rho} + \frac{q_D^2}{2} = \frac{p_C}{\rho} + \frac{q_C^2}{2}$$

$$- \int_C^D \frac{\partial v}{\partial t} dy - \int_C^D u\omega dy + \int_C^D \nu \frac{\partial \omega}{\partial x} \qquad (5.51)$$

This equation is a version of the energy equation in a viscous fluid (they can easily be obtained from Eqs. 5.42, 5.43, 5.45, 5.46, 5.47 and 5.49).

a)

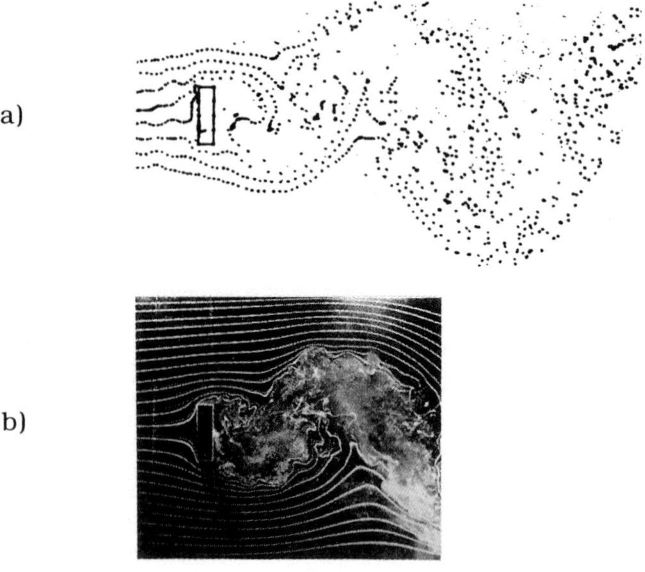

b)

Figure 5.4 Snap shot of flow around a rectangular cylinder. a) Numerical results by solution of the 2D N.–S. equations $Re = 6000$. The cylinder-height-to-channel-width ratio $(= D/H) = 1/6$. b) Experiment. Fromm and Harlow (1963) with permission - see Credits.

To get the pressure on the body surface, Eq. 5.50 is first applied on line 12 (Fig. 5.3), then Eq. 5.51 on line 23, then Eq. 5.50 on line 34 and so on. To get the wall shear stress on the body surface

$$\tau = \mu \frac{\partial u}{\partial y} \text{ (on horizontal lines)} \tag{5.52}$$

and

$$\tau = \mu \frac{\partial v}{\partial x} \text{ (on vertical lines)} \tag{5.53}$$

must be applied. Integrating the pressure and wall shear stress distributions around the cylinder surface gives the instantaneous resultant force on the cylinder.

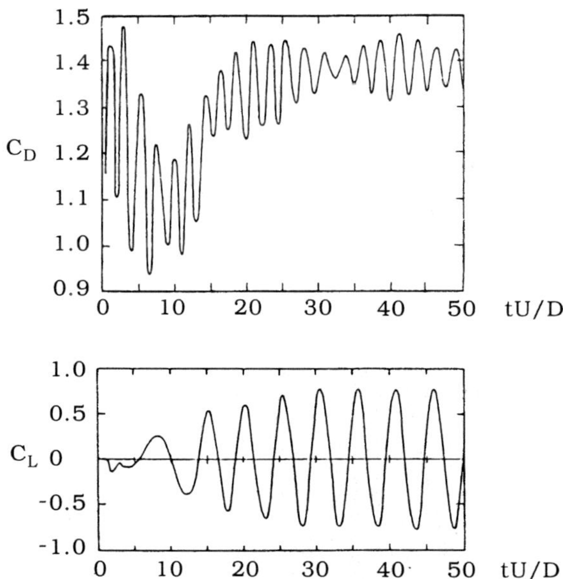

Figure 5.5 Time series of drag and lift coefficients for a circular cylinder obtained numerically from the solution of the 2D N.–S. equations in steady current. $Re = 200$. Braza et al. (1986).

Although the underlying principles of a numerical solution of the N.–S. equations for flow around a cylinder may appear to be quite straightforward, there are numerous details involved in the solution procedure to ensure that the solution is both stable and sufficiently accurate: these details are related to various aspects of the problem such as the boundary conditions; the choice of δx, δy and, δt; the stability of the finite-difference equations; the introduction of a perturbation to initiate the vortex shedding within a short time interval; and so on. Also, the finite-difference scheme used in the solution of the equations may have a direct influence on the end results (Borthwick, 1986).

Fig. 5.4a gives a snapshot of the flow obtained in the study of Fromm and Harlow (1963) while Fig. 5.4b gives that from an actual experiment. As seen, the numerical results reveal the main features of the flow quite well.

Fig. 5.5 illustrates the time series of the drag and lift coefficients for a circular cylinder obtained numerically by solving the N.–S. equations for $Re = 200$ (Braza et al., 1986). The forces reach a steady state with periodic oscillations after a transient time interval. The vortex shedding is excited in Braza et al.'s study by a physical perturbation imposed numerically.

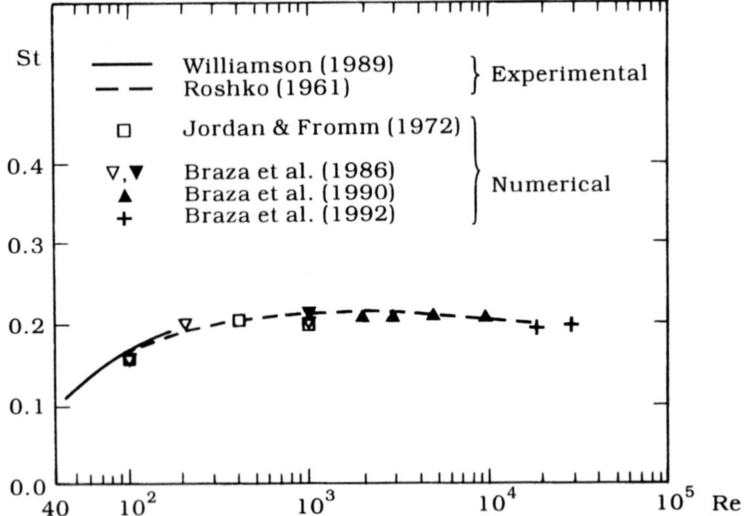

Figure 5.6 Strouhal number for a circular cylinder in steady current. Numerical results are from the solutions of the 2D N.–S. equations.

Fig. 5.6 compares the numerically obtained results regarding the Strouhal number with the experiments in the case of circular cylinder. Likewise, Fig. 5.7 compares the mean drag coefficient obtained numerically with the experiments. The numerical data in the figures are all from the solutions of the 2D N.–S. equations. The agreement between the numerical results and the experiments is quite good.

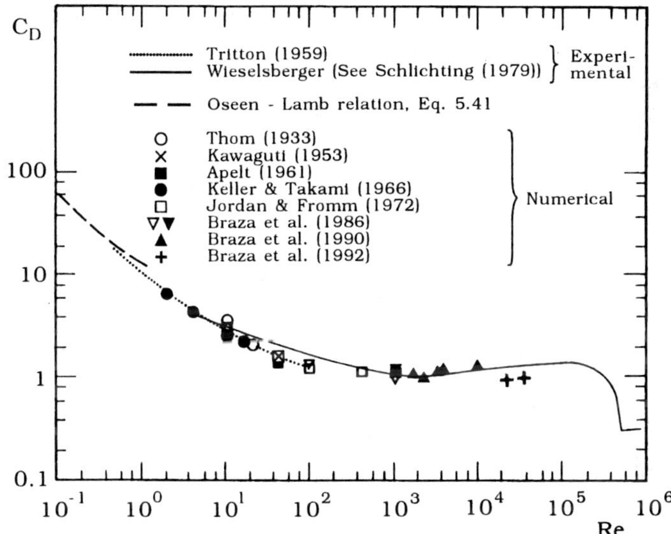

Figure 5.7 Mean drag coefficient for a circular cylinder in steady current. Numerical results are from the solutions of the 2D N.–S. equations.

Turbulent flow

Until now, the numerical solution of the *two-dimensional* N.–S. equations has been discussed. It is known, however, that the flow around the cylinder is two-dimensional only when $Re < 200$. For larger Re numbers, the vortex shedding occurs in cells and therefore the flow is three-dimensional (Fig. 1.26 and Sections 1.1 and 1.2.2). Hence, for such Re numbers, the 2D N.–S. solution is only an approximation. Although the 2D N.–S. solutions give fairly good agreement with the measurements with regard to the gross-flow parameters (Figs. 5.4, 5.6 and 5.7), this is not so, however, for the lift force for instance; see Fig. 5.8. As seen from Fig. 5.8, the lift force is grossly overpredicted. This may be due partly to the 2D computations: in the real flow, the presence of cells implies that the lift does not take place concurrently along the whole length of the cylinder, thus reducing the average lift. (Note that the two values plotted in Fig. 5.8 were obtained, using two different grid sizes in Braza et al.'s (1990) study).

For Re numbers smaller than 3×10^5 (but larger than 300), the flow is turbulent in the wake (Fig. 1.1). When Re is increased further, turbulence begins to spread into the boundary layer (Fig. 1.1 g-i). So, in this situation, the instan-

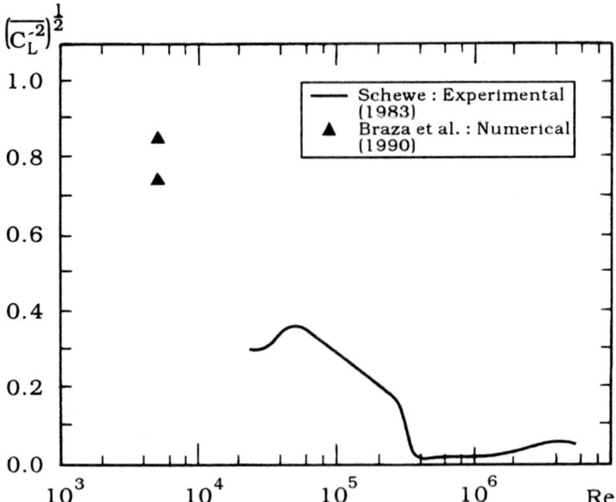

Figure 5.8 R.m.s. value of oscillating lift in steady current. Numerical results are from the solutions of the *two-dimensional* N.–S. equations.

taneous flow is three-dimensional not only in the wake but also in boundary layer itself. It is possible to carry out 3D computations where the 3D N.–S. equations are solved numerically. This method, called the direct numerical simulation of N.–S. equations, is presently feasible only for relatively small Re numbers; for large Re numbers, the scales of the dissipative part of turbulent motion are so small that this kind of small scale motion can not be resolved in a numerical calculation (the number of grid points required to resolve this motion increases approximately with Re^3) (Rodi, 1992). We shall return to the issue of 3D computations later in Section 5.2.

It is clear from the preceding discussion that, for relatively large Re numbers (where the flow in the cylinder boundary layer is turbulent), the direct numerical simulation of the N.–S. equations is not feasible. Similar arguments can be reasoned also for the case of rough-surface cylinders. So, in such situations, it may be desirable to solve the flow equations in such a way that the turbulence effects are modelled by use of a turbulence model such as an eddy-viscosity model or a Reynolds-stress-equation model or a large-eddy simulation model. An account of such a model (Justesen, 1990) is given in the next section. A review of the turbulence models as applied to flow past bluff bodies in *steady current* has been given by Rodi (1992).

5.1.4 Application to oscillatory flow

Stokes (1851) was the first to develop an analytical solution for the 2D N.–S. equations for the case of a cylindrical body oscillating sinusoidally in a viscous fluid, as has already been pointed out in conjunction with the asymptotic theory described in Example 4.3. (Recall that the results of the asymptotic theory in Example 4.3 are the same as the Stokes' theory to $O[(Re/KC)^{-1/2}]$). Wang (1968) later extended Stokes' analysis to $O[(Re/KC)^{-3/2}]$.

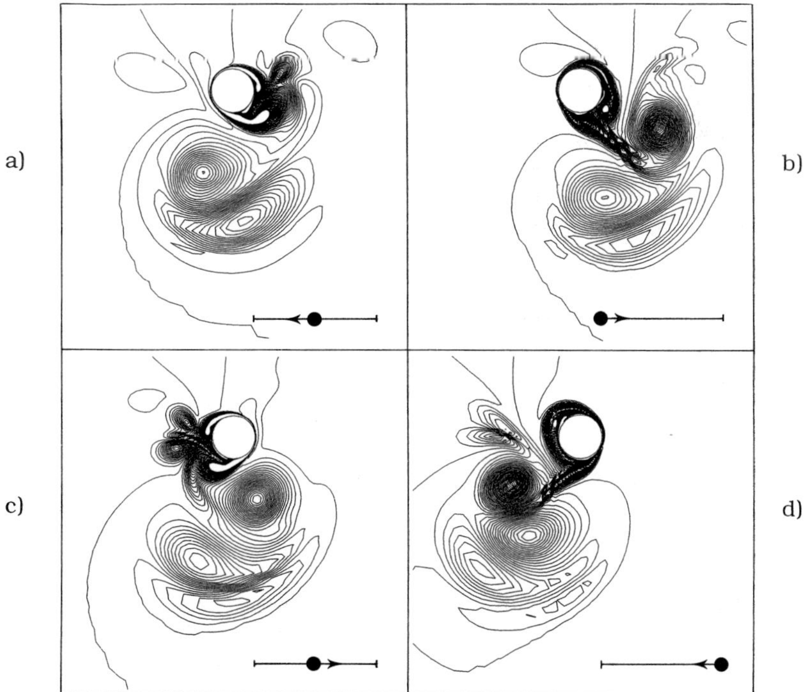

Figure 5.9 Computed vorticity contours due to N.–S. solution for $KC = 8$ and $\beta = 196$. Four instances are shown: (a-d) $\frac{1}{2}\pi$; π; $\frac{3}{2}\pi$ and 2π respectively. Justesen (1991) with permission - see Credits.

Regarding the numerical treatment of the problem, the equations which are to be integrated numerically are the same as those given in the previous section, namely, the vorticity transport equation (Eq. 5.48) and the Poisson equation (Eq. 5.49). The outer flow velocity is now a periodic function of time:

$$U = U_m \sin(\omega t) \tag{5.54}$$

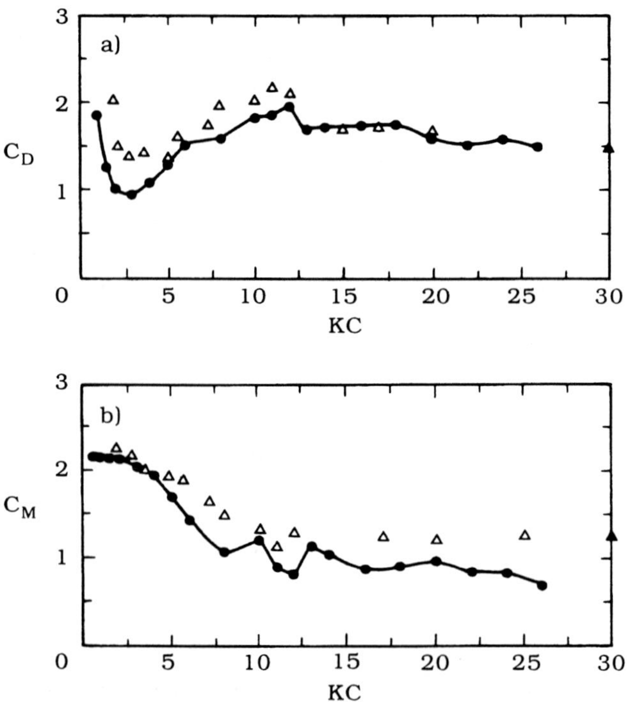

Figure 5.10 Circles: Computed in-line force coefficients due to N.–S. solution of Justesen (1991). $\beta = 196$; Triangles: Experiments by Obasaju et al. (1988). (a) Drag coefficient; (b) inertia coefficient.

Baba and Miyata (1987) were the first to attempt at solving the N.–S. equations for a sinusoidal flow. They presented two calculations; one for the combination of $KC = 5$ and $Re = 1000$, and the other for $KC = 7$ and $Re = 700$. In both calculations, the wake was symmetric in contrast to observations (Fig.

3.16). Murashige, Hinatsu and Kinoshita (1989) have made similar calculations for three KC numbers, $KC = 5, 7$ and 10, for Re numbers around 10^4. In the latter work, the flow was perturbed, to trigger asymmetry for relatively small KC numbers and eventually to excite vortex shedding for larger KC numbers. Apparently, these authors were able to obtain the transverse-vortex street regime (Figs. 3.6a and 3.7) for $KC = 10$. Later Wang and Dalton (1991b) made similar calculations for KC ranging from 1 to 12 and Re ranging from 100 to 3000. The latter authors reported their results also in Zhang, Dalton and Wang (1991).

Justesen (1991) has made an extensive study of oscillating flow around a circular cylinder, solving the N.–S. equations numerically for a wide range of KC, namely $0 < KC < 26$, and for three values of $\beta(= Re/KC)$ in the range 196-1035. Fig. 5.9 shows the computed vorticity contours for $KC = 8$ and $\beta = 196$. The presence of the transverse-vortex street is quite evident. Justesen also computed the conventional force coefficients (for all three β values). Figs. 5.10 and 5.11 compare Justesen's numerical results for $\beta = 196$ with the results from experiments.

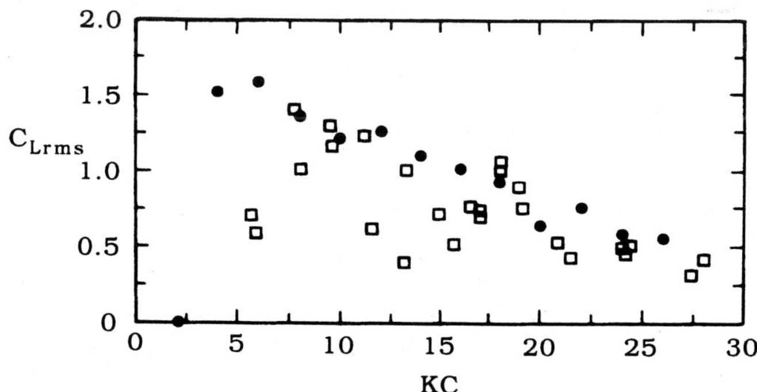

Figure 5.11 Lift. $\beta = 196$. Circles: Justesen's N.–S. solution (1991). Squares: Experiments by Maull and Milliner (1978) for $\beta = 200$. Lift force in Maull and Millner's experiments was measured by strain gauges and represents the force on the total length of the cylinder, L, the ratio L/D being approximately 18.

In Justesen's calculations, the Reynolds number was kept rather small ($\beta = 196$) such that the effect of transition and turbulence remain as small as possible. For that reason, the computations were stopped at $KC = 26$. Although it may be argued that even $KC = 26$ may be too high for the turbulent effects to be negligible (see Fig. 3.15), the agreement between the numerical results and the experimental data is rather good with regard to the in-line coefficients (Fig. 5.10). This may be due to the fact that the flow is turbulent only in parts of the oscillation cycle or in the wake such that the boundary layer separation is predominantly laminar. Regarding the lift coefficients (Fig. 5.11), there is some discrepancy. Apparently, the numerical predictions of lift agree extremely well with the experimental data at KC numbers $KC \cong 10$, 18 and 26 where large spanwise correlations are measured (Fig. 3.28). However, for KC values where the spanwise correlation is small the flow is strongly three-dimensional and therefore it is expected that a 2-D model is not able to handle such cases. This is quite evident from Fig. 5.11 (c.f. Fig. 3.28). Finally, Fig. 5.12 presents Justesen's results regarding C_D and C_M coefficients obtained for the highest β-value, namely $\beta = 1035$.

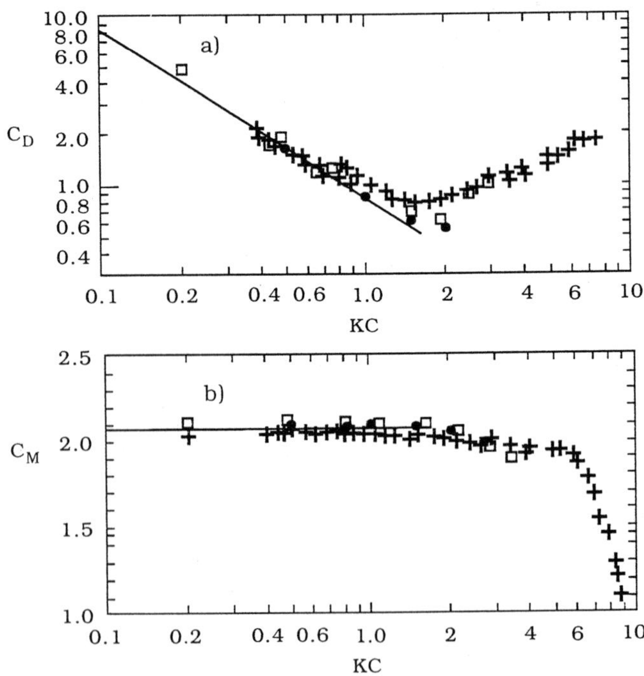

Figure 5.12 (a) Drag coefficient; (b) Inertia coefficient. Circles: Computed in-line force coefficients due to N.–S. solution of Justesen (1991). $\beta = 1035$; Squares: Discrete vortex method by Stansby and Smith (1989); Crosses: Experiments by Sarpkaya (1986); —, asymptotic theory (Eqs. 4.57 and 4.58).

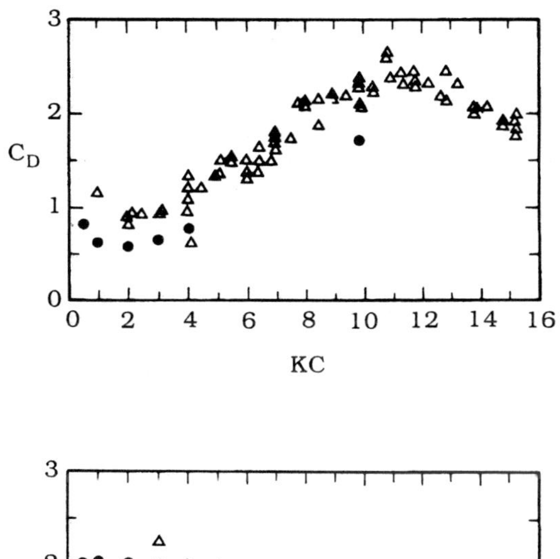

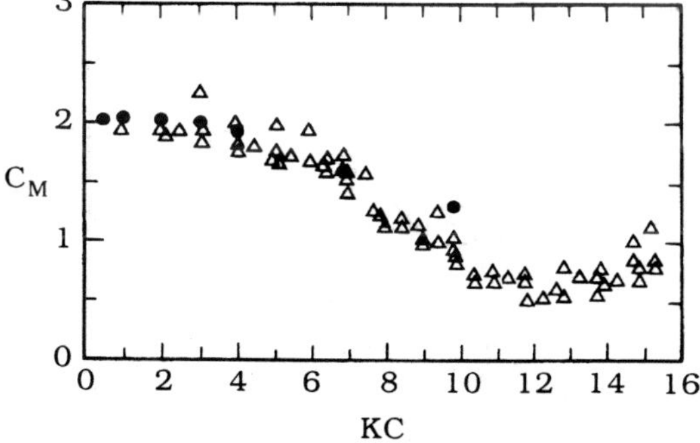

Figure 5.13 Circles: Computed in-line force coefficients due to turbulent N.–
S. solution of Justesen (1990). $k_s/D = 4.8 \times 10^{-2}$. Triangles:
Experiments by Justesen (1989).

Justesen (1990) treated also the case of *turbulent flow* where the turbulence effects were modelled by use of a one-equation turbulence model for a rough cylinder. The equations are essentially the same as in the case of laminar flow, namely the vorticity transport equation and the Poisson equation. The only difference is that, in the present case, the vorticity transport equation includes also the so-called turbulent viscosity, ν_T. ν_T is modelled by a one-equation model. This presumably adds one more equation (namely, the equation for turbulent energy) to the set of equations which is to be solved. Justesen (1990) carried out his calculations for KC numbers up to $KC = 10$ for a cylinder roughness of $k_s/D = 48 \times 10^{-3}$. Fig. 5.13 compares his numerical results with the results of experiments reported in Justesen (1989).

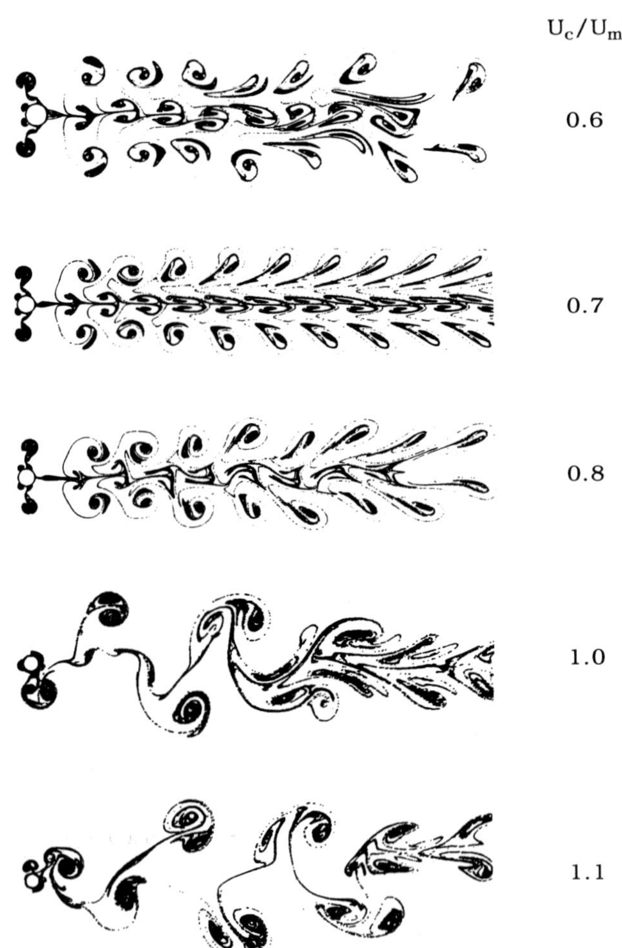

Figure 5.14 Streaklines obtained from N.–S. solutions. For combined oscillatory-flow and current environment $KC = 4$. $\beta = 200$. Sarpkaya et al. (1992) with permission - see Credits.

The N.-S. solutions have been obtained also for the case of *combined oscillatory flow and current* (Sarpkaya, Putzig, Gordon, Wang and Dalton, 1992). The calculations were carried out for $KC = 1$–6 with $\beta = 200 (Re = 800$–$1200)$ for various values of $U_c/U_m = 0$–1.2 in which U_c is the current velocity. The results have revealed the existence of a wake feature in the interval $U_c/U_m = 0.6$–0.8 for $KC = 4$ (Fig. 5.14) different from those in steady currents and in oscillatory flows. Furthermore, Sarpkaya et al. obtained reasonable agreement with the experiments regarding the in-line coefficients for $KC = 4$–6.

Recently Badr et al. (1995) have reported the results of a numerical solution to the N.-S. equations for $Re = 10^3$, $KC = 2$ and 4, and for $Re = 10^4$ and $KC = 2$. As mentioned in Chapter 3, their results revealed the presence of steady streaming patterns (shown in Fig. 3.38) even in the case of separated flows.

5.2   Discrete vortex methods

In practice, large difficulties are encountered for solving the N.-S. equations using the finite-difference or finite-element methods. One of the major difficulties is that the number of grid points (therefore, the amount of computation) required to obtain a solution increases with increasing Reynolds number, and may become prohibitive at large Reynolds numbes, as mentioned earlier. It is therefore of interest to develop a grid-free (or almost grid-free) numerical method. A simple method offering an alternative to the finite-difference method is the discrete vortex method.

The equations to be solved are exactly the same as in the preceding section, namely the vorticity-transport equation (Eq. 5.48) and the Poisson equation (Eq. 5.49):

$$\frac{\partial \omega}{\partial t} + u \frac{\partial \omega}{\partial x} + v \frac{\partial \omega}{\partial y} = \nu \left(\frac{\partial^2 \omega}{\partial x^2} + \frac{\partial^2 \omega}{\partial y^2} \right) \tag{5.55}$$

$$\frac{\partial^2 \psi}{\partial x^2} + \frac{\partial^2 \psi}{\partial y^2} = -\omega \tag{5.56}$$

In principle, the only difference between the vortex methods and the finite-difference methods is that the solution to the vorticity-transport equation (Eq. 5.55) in case of vortex methods is obtained through a numerical simulation of convective diffusion of discrete vortices generated on the cylinder boundary (the numerical simulation of vorticity transport).

In the following, attention will be concentrated first on the simulation of vorticity transport. This will follow by the description of the underlying principles of the vortex method as applied to flow around a cylinder. The section ends with illustration of several examples selected from the literature, covering both the steady current and oscillatory-flow situations.

5.2.1 Numerical simulation of vorticity transport

There is an analogy between the convective diffusion of any passive quantity such as concentration (or temperature in the case of heat transfer) and the transport of vorticity. Both processes are governed by the same differential equation. This is seen in Table 5.2 where other elements of the analogy are also indicated. In the table, C is the concentration of the passive quantity and K is the diffusion coefficient (C and K in the case of heat transfer are the temperature and the thermal conductivity, respectively, Crank (1975)).

Table 5.2 Analogy between the convective diffusion of passive quantity and that of vorticity.

Convective diffusion of a passive quantity such as mass or heat	Transport of vorticity
C: Concentration (or temperature)	ω: Vorticity
Convective diffusion equation: (5.57) $\frac{\partial C}{\partial t} + u\frac{\partial C}{\partial x} + v\frac{\partial C}{\partial y} = K\left(\frac{\partial^2 C}{\partial x^2} + \frac{\partial^2 C}{\partial y^2}\right)$	Vorticity transport equation: (5.58) $\frac{\partial \omega}{\partial t} + u\frac{\partial \omega}{\partial x} + v\frac{\partial \omega}{\partial y} = \nu\left(\frac{\partial^2 \omega}{\partial x^2} + \frac{\partial^2 \omega}{\partial y^2}\right)$
Standard deviation of particle position: $\sqrt{\overline{r^2}} = \sqrt{2K\,\delta t}$ (5.59)	Standard deviation of vortex position: $\sqrt{\overline{r^2}} = \sqrt{2\nu\,\delta t}$ (5.60)
Lagrangian simulation with particles: $C = N/A$ (5.61)	Lagrangian simulation with vortices: $\omega = \Gamma/A$ (5.62)

Numerical simulation of convective diffusion

An alternative to solving the convective diffusion equation (Eq. 5.57 in Table 5.2) is the Lagrangian simulation of convective diffusion process by a random-walk model. This method has been developed over the last decades (Bugliarello (1971), Sullivan (1971) among others) and is now a powerful numerical tool used in the problems related to diffusion of mass in flow environments.

The aforementioned simulation may be described by the following simple example. Consider the diffusion of mass from a continuous point source (Fig.

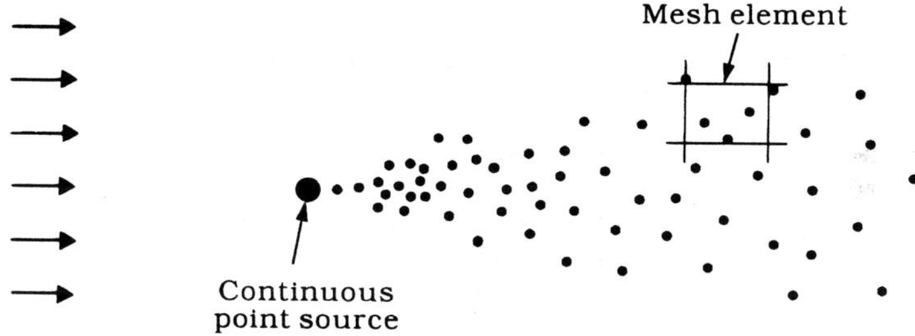

Figure 5.15 Diffusion of passive quantity from a point source.

5.15). The diffusing mass in the example can be considered as a cloud of large number of "particles". Each particle actually follows two basic steps, namely

1) a convective step determined by the velocity of the field corresponding to the position of the particle, and

2) a random diffusive step (Fig. 5.16). The magnitude and the direction of the random diffusive step is selected from a Gaussian process with a standard deviation set equal to $\sqrt{2K\,\delta t}$ in which δt is the small time interval during which the particle takes its step (Eq. 5.59 in Table 5.2). In the simulation, many such particles released from the source point are followed as they travel through the statistical field variables.

The concentration, C, can then be calculated, in principle, from the number of particles found in a mesh element by $C = N/A$ (Eq. 5.61 in Table 5.2) in which N is the number of particles in the mesh element and A is the area of the same mesh element (Fig. 5.15). It can be shown that the concentration obtained in this way (for large number of particles) is equivalent to that found from the solution of the convective diffusion equation (Eq. 5.57 in Table 5.2).

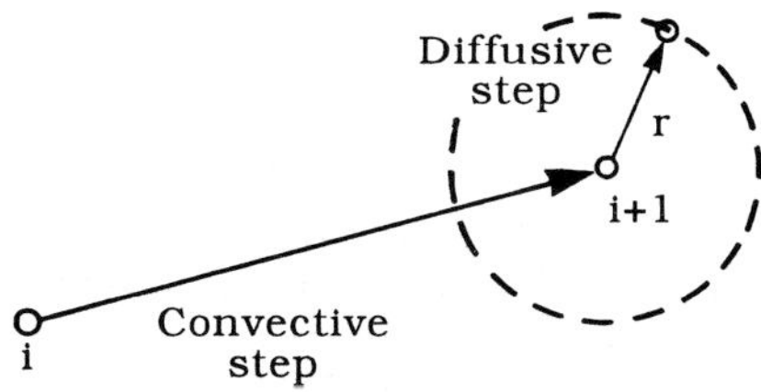

Figure 5.16 Random walk of a particle.

Numerical simulation of convective diffusion of vorticity

In the case of vorticity transport, the diffusing "mass" of vorticity may be considered as a cloud of large number of vortex "particles", analogous to the diffusion of particles described in the previous section. The vortex "particle" may be termed the vortex blob or the discrete vortex. Obviously, these discrete vortices must be generated on the boundaries, each vortex being assigned with a certain strength and a direction of rotation. In the case of a cylinder the vortex generation takes place on the surface of the cylinder, Fig. 5.17.

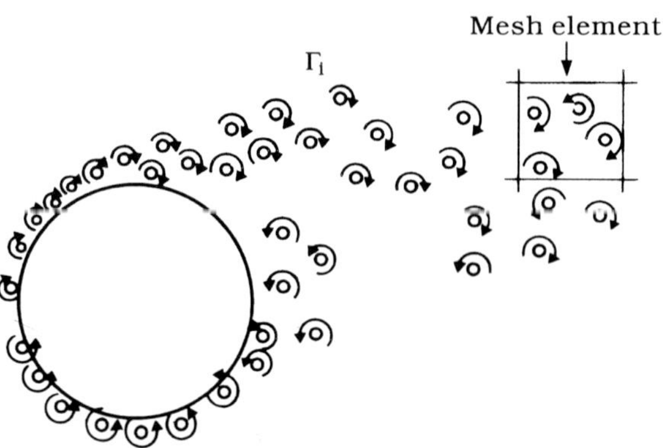

Figure 5.17 Discrete vortices released from the cylinder surface.

As in the case of diffusing passive particles described in the preceding section, the discrete vortices introduced into the flow from the boundaries follow two basic steps: a convective step and a diffusive step (Fig. 5.16). The convective step is determined by the velocity of the field corresponding to the position of the discrete vortex, while the diffusive step is selected from a Gaussian process with a standard deviation equal to $\sqrt{2\nu\,\delta t}$. (Recall the analogy between the diffusion coefficient K and the kinematic viscosity ν, Table 5.2. The diffusion here corresponds to *molecular (Brownian)* diffusion; in the case of turbulent flow, ν has to be replaced by $\nu + \nu_T$, where ν_T is the turbulence viscosity, to simulate the *turbulent* diffusion, see Section 5.2.3). Many such vortices are followed, and the vorticity, ω, can, in principle, be calculated by $\omega = \Gamma/A$ (Eq. 5.62 in Table 5.2) in which Γ is the sum of the strengths of the vortices found in a mesh element (Fig. 5.17) and A is the area of the mesh element itself.

Finally it should be noted that the aforementioned scheme was shown to converge to the solution of the N.–S. equations (Chorin, Hughes, McCracken and Marsden, 1978)

5.2.2 Procedure used in the implementation of discrete vortex method

As has already been mentioned, the principal idea behind the discrete vortex method is to achieve the solution of the vorticity-transport equation (Eq. 5.55) through the numerical simulation of vorticity transport. For this, the following procedure is used.

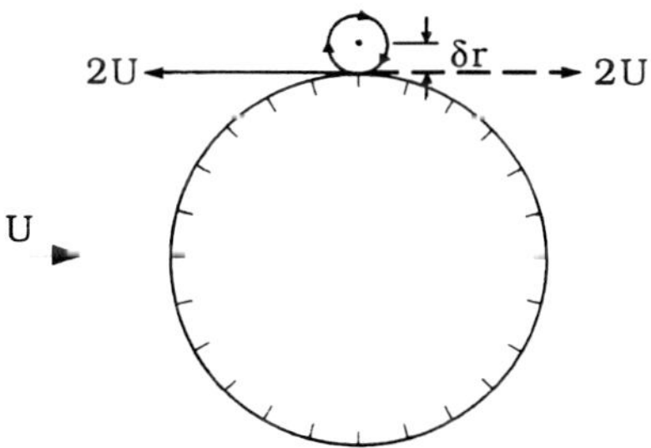

Figure 5.18 The vortex-induced velocity at the surface of the cylinder to cancel the existing velocity so that the no-slip boundary condition can be fulfilled on the surface at that particular location.

1. First use the potential flow solution and work out the velocity on the cylinder surface. (To avoid numerical difficulties, a gradual (timewise) increase in the velocity to the value U may be contemplated in the computations).

2. Introduce discrete vortices just above the cylinder surface. For this, determine the strengths of these vortices such that the no-slip condition is satisfied on the surface. For example, for the vortex which will be introduced at the top edge of the cylinder at the initial instant (Fig. 5.18), the strength of the vortex should be

$$\Gamma = 2\pi \; \delta r \; (2U) \tag{5.63}$$

and the direction of rotation should be clock-wise, so that the velocity just at that point on the cylinder surface would be zero:

$$+2U \quad + \quad \left(-\frac{\Gamma}{2\pi\delta r}\right) = +2U - 2U = 0 \tag{5.64}$$

from the   from the
potential flow solution introduced vortex

(Introducing these vortices can be refined. It can be either taken at the first mesh point or distributed over several mesh points, using the boundary-layer theory.)

3. Move the vortices according to the random-walk model described in the previous section.

4. Distribute the strengths of vortices on the mesh according to a specified scheme. For example, according to a weighting scheme which is widely used in simulation studies, a vortex located at Point P in the mesh element illustrated in Fig. 5.19 generates vorticity at Point i:

$$\omega_i = \frac{A_i\Gamma}{A^2} \quad , \quad i = 1, 2, 3, 4 \tag{5.65}$$

in which A is the total area of the mesh and A_i are the areas indicated in Fig. 5.19

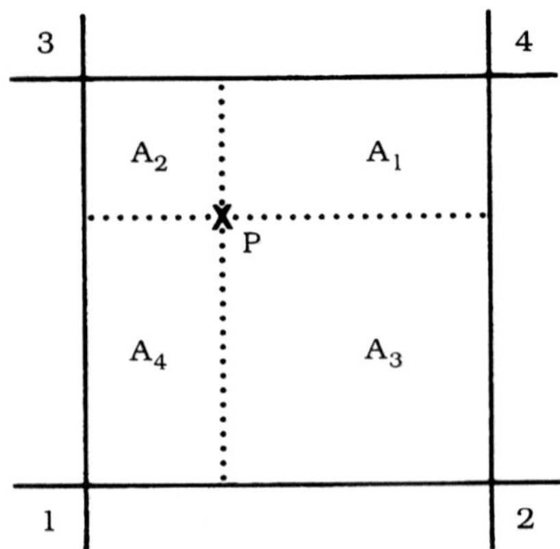

Figure 5.19 Vorticity values at the mesh points 1, 2, 3 and 4, caused by the vortex at Point P, are calculated according to the scheme in Eq. 5.65.

5. Given the vorticity values at the mesh points, solve the Poisson equation (Eq. 5.56) numerically and obtain the new velocity components at the mesh points by

$$u = \frac{\partial \psi}{\partial y} \quad \text{and} \quad v = -\frac{\partial \psi}{\partial x} \tag{5.66}$$

6. Restore the no-slip boundary condition at the surface of the cylinder by introducing a new set of discrete vortices at the cylinder boundary (Fig. 5.20) and repeat the steps 3 to 6.

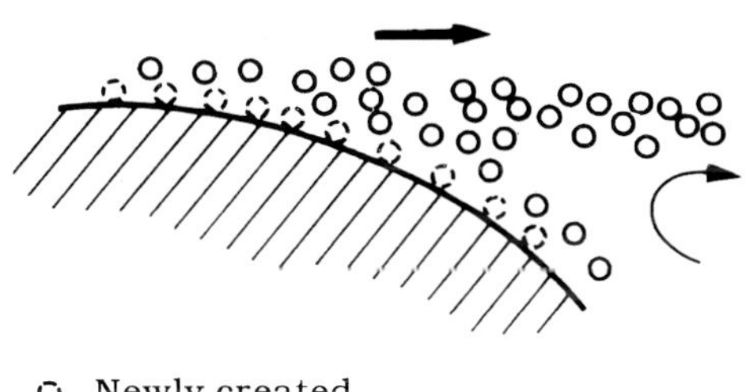

Figure 5.20 Vortex creation to satisfy the no-slip condition.

7. At any time t, given (a) the position of the $i\,th$ vortex in terms of polar coordinates r_i, θ_i (Fig. 5.21), (b) the velocity components that the $i\,th$ vortex experiences in the x and y directions, namely u_i, v_i, and (c) the vortex strength, Γ_i, corresponding to the $i\,th$ vortex, the force components F_x and F_y on the cylinder may be calculated using the following expressions

$$F_x = \sum_{i=1}^{N} \Gamma_i \left[\frac{u_i \sin(2\theta_i) - v_i \cos(2\theta_i)}{r_i^2} \right] + 2F_{xs} \tag{5.67}$$

$$F_y = -\sum_{i=1}^{N} \Gamma_i \left[\frac{v_i \sin(2\theta_i) + u_i \cos(2\theta_i)}{r_i^2} \right] + 2F_{ys} \tag{5.68}$$

in which N is the total number of vortices, and F_{xs} and F_{ys} are the force due to surface shear stress (skin friction) in x and y directions, respectively. The skin friction force is obtained from the surface shear stress which is actually proportional to the surface vorticity. The quantities appearing in Eqs. 5.67 and 5.68 are obtained by having (a) the position of discrete vortices, (r_i, θ_i), (b) the velocity of discrete vortices, (u_i, v_i) and (c) the vortex strength, Γ_i. The method is due to Quartapelle and Napolitano (1983). It reduces to the aforementioned convenient form for the case of a *circular cylinder* (Stansby and Slaouti, 1993).

These are the typical steps taken in the implementation of the discrete vortex method. There are, however, numerous details which need to be taken into

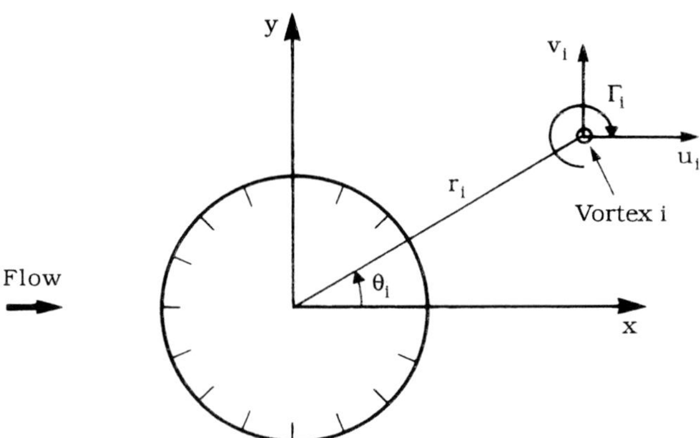

Figure 5.21 Definition sketch for the force calculation in the discrete vortex method.

account, such as the choice of the time step δt in the random-walk simulation, the number of vortices introduced per time step, the extent of the mesh (or meshes), the mesh size and so on. (Smith and Stansby, 1988).

— — — — — — —

The vortex method was originally proposed by Rosenhead (1931) and further developed in recent time by Chorin (1973 and 1978). In the version that Chorin presented, the velocity is calculated by directly summing the influence of all the other vortices. This may be computationally prohibitive, since there are very many vortices ($O(10^4)$) in the flow. To avoid this, the so-called **vortex-in-cell (or cloud-in-cell)** method has been devised (Christiansen (1973) and Baker (1979)). In this method, the contribution of each vortex to the vorticity at the mesh points is calculated (in the manner as described in Step 4 above) and then the velocity is obtained by solving the Poisson equation (Step 5 above). Therefore, the disadvantage of the method requiring a large number of vortices is compensated by this kind of efficient vortex handling.

The vortex methods where vorticity is created only at separation point have also been developed. In this case, the method requires knowledge of separation locations and therefore these methods may be suitable for bodies with sharp edges.

The advantages of vortex methods over the other methods to solve the N.–S. equations may be summarized as follows: 1) First of all, the inviscid theory could be employed (Step 3 above); 2) the numerical diffusion problems associated with the vorticity gradient terms in Eulerian schemes are to a large degree avoided; 3) there are no zone assumptions which could, for instance, require matching of an outer flow to an inner flow; and finally 4) the method is relatively stable and well suited to vectorisation on supercomputers (Stansby and Isaacson, 1987).

A detailed review of the vortex methods has been given by Leonard (1980) and Sarpkaya (1989).

Small Re-number simulation by the discrete vortex method

The vortex shedding is two-dimensional in the range $40 < Re < 200$ (Section 1.1). Therefore this range of Re number would offer the possibility of true application of the method, since no three-dimensionality is present. Stansby and Slaouti (1993) did computations of the flow around a circular cylinder for Re numbers ranging from 60 to 180, using the discrete-vortex method. They were able to reproduce the Reynolds number dependence of the Strouhal number as that obtained by the careful experiments of Williamson (1989) (see Fig. 1.9 for the latter experimental data). Comparison is reproduced here in Table 5.3.

Table 5.3 Strouhal numbers for $Re = 60$–180 computed by Stansby and Slaouti (1993) by the discrete vortex method. Experimental data from Williamson (1989).

Re	60	100	140	180
Computed Strouhal number	0.139	0.166	0.180	0.192
Experimental Strouhal number	0.135	0.164	0.180	0.191

The force coefficients including the skin-friction drag and the skin-friction lift obtained by Stansby and Slaouti (1993) are shown in Fig. 5.22 for the tested lowest and highest Re numbers, $Re = 60$ and 180. The mean drag coefficient values are in very good agreement with those obtained by the N.–S. solutions presented in Fig. 5.7. Also, it may be mentioned that Stansby and Slaouti made a detailed comparison between their results and the results obtained from the finite-element and the spectral methods and found an agreement within 2–4%.

Regarding the lift coefficient, no experimental data are available for such small Re numbers. Comparison of the results with those found from the previously mentioned methods show, however, that the agreement is within 10–12% (Stansby and Slaouti (1993)).

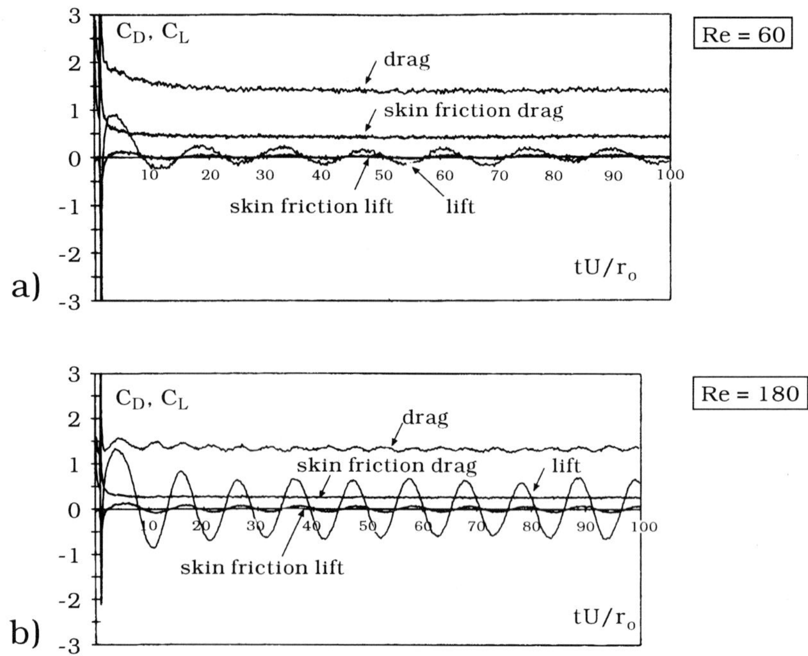

Figure 5.22 Force variation with time computed from the vortex method. r_o is the cylinder radius. Stansby and Slaouti (1993).

5.2.3 Application areas

Steady current

When $Re > 300$, the flow becomes three-dimensional (Section 1.1). In such situations, the implementation of the vortex method in the way as described in the preceding paragraphs may not be entirely correct. To account for the effects of three-dimensionality of the flow, the concept of circulation reduction has been introduced in the calculations (Sarpkaya and Shoaff, 1979). Discrete vortex models show that the concentrated vortices in the wake contain about 80% of the shed vorticity, while experiments show that this figure is around 60% (Sarpkaya and Shoaff, 1979). A model of circulation reduction basically seeks to dissipate vorticity so that the 20% more reduction in circulation can be realized in the calculations. Apparently, this concept worked well and gave good agreement with the experiments (see also Sarpkaya, 1989).

In offshore-engineering practice the Reynolds number is rather high and the surface roughness may be rather large, therefore the flow is normally postcritical. Special vortex methods have been developed to handle such situations, Smith and Stansby (1989) and Yde og Hansen (1991). In Smith and Stansby's work, the turbulent flow is simulated in a thin boundary region around the cylinder by superimposing random walks on the convection of point vortices in this region. In the calculation of random walks, the molecular viscosity, ν, is replaced by an effective viscosity, ν_e, which is equal to $\nu_e = \nu + \nu_T$ in which ν_T is the turbulence viscosity. In the model, ν_T is determined from the vorticity distribution through an algebraic turbulence model.

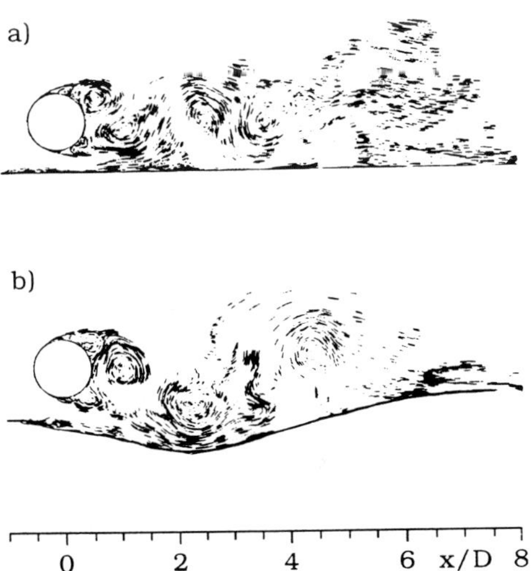

Figure 5.23 Vorticity field obtained through the cloud-in-cell vortex method; $e/D = 0.4$ in which e is the gap between the pipe and the bed. Sumer et al. (1988).

In the work of Yde and Hansen (1991), on the other hand, a turbulent boundary-layer model (based on Fredsøe's wave boundary layer model (1984), which assumes a logarithmic velocity distribution in the boundary layer) has been included. The key point in Yde and Hansen's method is that the discrete vortices are introduced at the "centroid of the vorticity" in the boundary layer. To pinpoint where these points lie across the boundary layer thickness, the boundary-layer calculation needs to be performed at each time step, to get the boundary-layer thickness. The model is capable of giving the Reynolds number dependence and the roughness dependence in the transcritical flow regimes through the assumed logarithmic velocity distribution in the boundary layer.

The vortex methods have been implemented quite extensively in various areas of fluid engineering, covering from offshore to aerospace-engineering applications, such as flow around multiple cylinders (Skomedal, Vada and Sortland (1989), Yde and Hansen (1991)), oscillatory flow around cylinders (see next Section), flow around arbitrary shaped and sharp edged bodies (Scolan and Faltinsen, 1994), flow around a pipeline over a scoured bed (Sumer, Jensen, Mao and Fredsøe, 1988), to name but a few. Fig. 5.23 illustrates the vorticity field around a pipeline over a plane bed (Fig. 5.23a) and a scoured bed (Fig. 5.23b) obtained by cloud-in-cell vortex method. In this latter simulation, the vortices are released steadily into the flow from the boundaries, namely the pipe surface and the bed. The strength of these vortices are calculated in such a way that the zero normal velocity and zero slip conditions are satisfied together on the pipe surface and also that the zero normal velocity condition is satisfied on the bed.

Oscillatory flows and waves

The vortex methods have been implemented widely for prediction of flows around cylinders subject to waves. Stansby and Dixon (1983) extended Chorin's (1973) method so as to cover the case of oscillatory flows. Later, similar works were carried out by Stansby and Smith (1989), Skomedal et al. (1989) and Graham and Djahansouzi (1989).

Fig. 5.12 shows a comparison between the results of Stansby and Smith (1989) and those of other methods (namely, Justensen's (1991) N.–S. solution and the results of the asymptotic theory described in Example 4.3) and the experiments. The agreement between the discrete vortex method, the N.-S. solution and the asymptotic theory appears to be rather good. The vortex-method results agree quite well with the experiments except the KC range between 1 and 2.5. This may be linked to the 3D Honji vortices and transitional flow regimes (b and c in Fig. 3.15) experienced in $1 < KC < 2.5$ for $\beta = 1035$.

In the previously mentioned studies, the Reynolds number was kept rather small to satisfy the laminar-flow conditions. As noted in the preceding section, special vortex methods have been developed to cope with the situations where the postcritical flow regimes prevail with the boundary layer being partially or completely turbulent; Hansen, Yde and Jacobsen (1991) used the algorithm presented in Yde and Hansen (1991) to investigate the flow around single and multiple cylinders subject to unsteady and oscillatory flows. Two, four and eight cylinders were investigated with $Re = 10^5 - 5 \times 10^6$ and $k_s/D = 0 - 30 \times 10^{-3}$ and with various angles of attack. Valuable information was obtained with regard to, among others, the influence of spacing between the cylinders on loading. Fig. 5.24 illustrates how an impulsively-started flow develops around two cylinders in tandem arrangement.

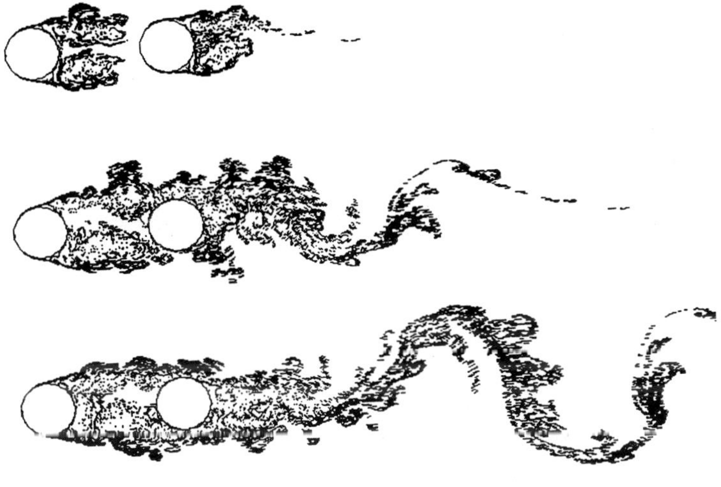

Figure 5.24 Simulated impulsively-started flow around two cylinders in tandem. Hansen et al. (1991).

The vortex methods have been used for the case of orbital flow as well, Stansby and Smith (1991) and Stansby (1993). The latter authors conducted the discrete-vortex simulations for low KC numbers and low β numbers (see Example 4.5 for a full discussion of the forces on cylinder in orbital flows at low KC numbers). Fig. 5.25 shows the *steady* streamlines, averaged over a number of cycles for various values of ellipticity, E, and the KC number, taken from Stansby (1993). While, for zero ellipticity (i.e., the planar oscillatory flow), the streamlines clearly illustrate the steady streaming pattern studied earlier in Section 3.6 (Fig. 3.38), this pattern is disrupted with increasing E, and eventually degenerates into a steady, recirculating streaming in the case of circular orbital motion (for $E = 1$). Fig. 5.26, on the other hand, shows the vorticity picture with the background streamlines as obtained in Stansby and Smith's study (1991). Both Stansby's and Stansby and Smith's works show a substantial reduction in the inertia force, in full accord with the previously mentioned observations (Example 4.5). Stansby (1993) gives also numerically obtained drag coefficients in addition to the inertia coefficient data.

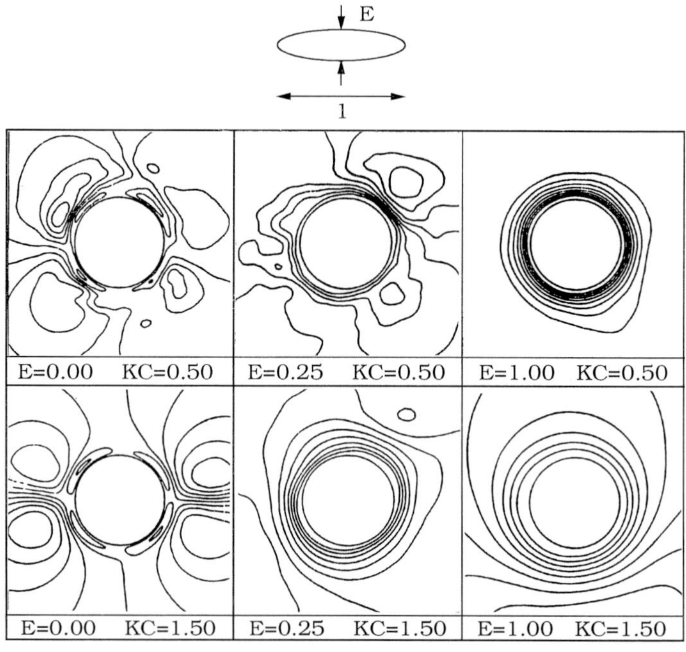

Figure 5.25 Steady streamlines for orbital flow, averaged over cycles 16–20 for $E > 0$ and over cycles 10–14 for $E = 0$. Stansby (1993).

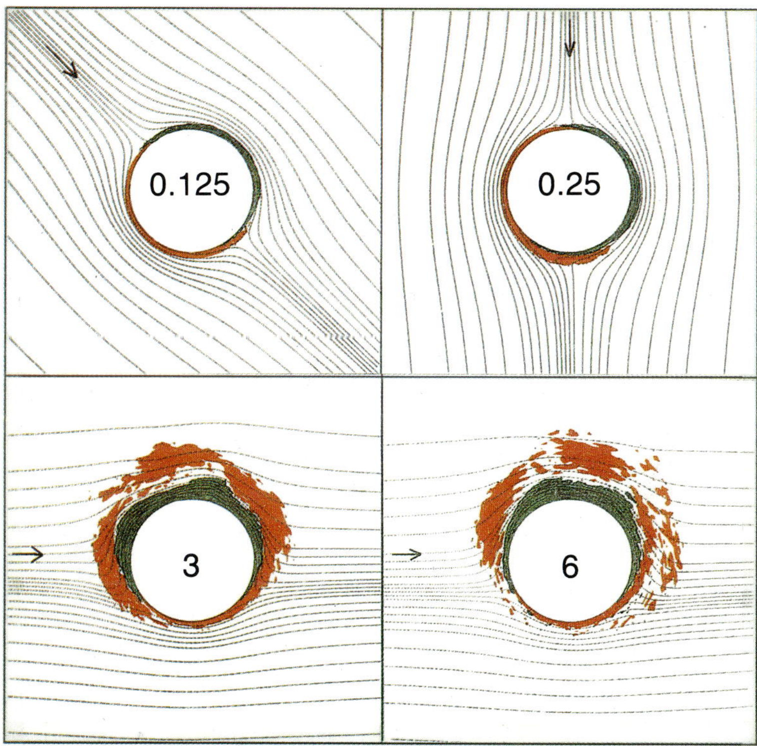

Figure 5.26 Streamline and vorticity contours for uniform, circular, onset flow with $KC = 1.5$, at various t/T, shown by the number in the cylinder. T is the wave period. The arrow on the streamline shows the incident flow direction. The green area shows vorticity of clockwise rotation, the red area vorticity of anticlockwise rotation. Stansby and Smith (1991) with permission - see Credits.

5.3 Hydrodynamic stability approach

The formation of vortex shedding behind a cylinder may be viewed as an instability of the flow in the wake. The instability emerges because the presence of the wake behind the cylinder introduces two shear layers as sketched in Fig. 5.27. Shear layers are known to be unstable, and the familiar hydrodynamic stability analysis can be employed to predict the frequency and the spacing of the vortex shedding. Such an analysis has been carried out by Triantafyllou et al. (1986 and 1987) for a circular cylinder. The following paragraphs will summarize this work.

Assuming a two-dimensional and parallel flow with the velocity components given by

$$u = U(y) + u'   \tag{5.69}$$

$$v = 0 + v'   \tag{5.70}$$

and the pressure

$$p = P + p'   \tag{5.71}$$

and writing the infinitesimal disturbances introduced in the velocity components, namely u' and v', in terms of a stream function ψ' as

$$u' = \frac{\partial \psi'}{\partial y}   \tag{5.72}$$

$$v' = -\frac{\partial \psi'}{\partial x}   \tag{5.73}$$

and furthermore neglecting the quadratic terms, the N.–S. equations and the continuity equation (Eqs. 5.42–5.44) lead to the so-called Orr-Sommerfeld equation (Schlichting, 1979, p. 460):

$$(kU - \omega)(\phi'' - k^2\phi) - kU''\phi =$$
$$= -i\nu(\phi'''' - 2k^2\phi'' + k^4\phi)   \tag{5.74}$$

Here ϕ is defined as the amplitude in the stream function of the disturbance flow

$$\psi'(x,\ y,\ t) = \phi(y)e^{i(kx - \omega t)}   \tag{5.75}$$

in which k is the wave number, ω is the angular frequency of the introduced disturbance, and i is the imaginary unit ($= \sqrt{-1}$). Eq. 5.74 is the basic equation for the stability analysis. When the mean flow $U(y)$ is specified, the solution of the equation (i.e., the eigen solutions) give ω and k:

$$\omega = \omega_r + i\,\omega_i \tag{5.76}$$

$$k = k_r + i\,k_i \tag{5.77}$$

If ω_i is positive, it will represent the growth rate of the introduced disturbance *in time* (cf. Eq. 5.75), otherwise it will represent the decay rate. Likewise, k_i expresses the growth rate *in space* of the disturbance when it is negative and the decay rate otherwise.

Triantafyllou, Triantafyllou and Chryssostomidis (1986, 1987) considered the inviscid version of the Orr-Sommerfeld equation, known as the Rayleigh equation:

$$(kU - \omega)\,(\phi'' - k^2\phi) - kU''\phi = 0 \tag{5.78}$$

with the velocity profile $U(y)$ given by

$$\frac{U(y)}{U_0} = 1 - A + A\tanh\left\{a\left[\left(\frac{y}{D}\right)^2 - b\right]\right\} \tag{5.79}$$

in which U_0 is the mean flow velocity as $y \to \infty$, and A, a and b are curve-fitting parameters determined from the actual, measured mean velocity profiles (see Fig. 5.27 for definition sketch).

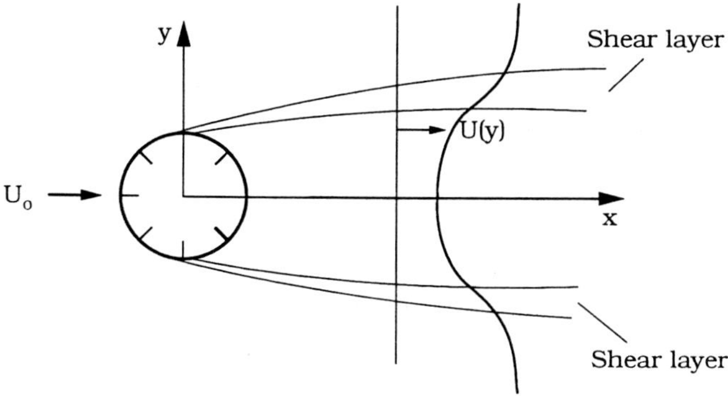

Figure 5.27 Velocity profile considered in the hydrodynamic stability analysis.

The so-called parallel flow assumption has been made in Triantafyllou et al.'s study. Namely, the mean flow is assumed to vary gradually with the distance x, so that locally the instability properties of the wake can be adequately represented by the instability properties of a parallel flow (namely, a constant velocity profile extending over an infinite x distance) having the same mean velocity profile as the local wake section considered. Hence, whether the flow is unstable has been determined as function of the distance x.

Triantafyllou et al. did the calculations for three families of $U(y)$ profiles. The first two, one for $Re = 30$ and the other for $Re = 56$, were taken from Kovásznay's (1949) measurements. The third one, taken from Cantwell (1976), corresponded to a turbulent wake with a Re number equal to 140.000 ("pseudo-laminar" flow calculations). Although Triantafyllou et al. considered the inviscid Orr-Sommerfeld equation, it is clear that the Reynolds number dependence is intrinsic in the analysis through the considered velocity distributions.

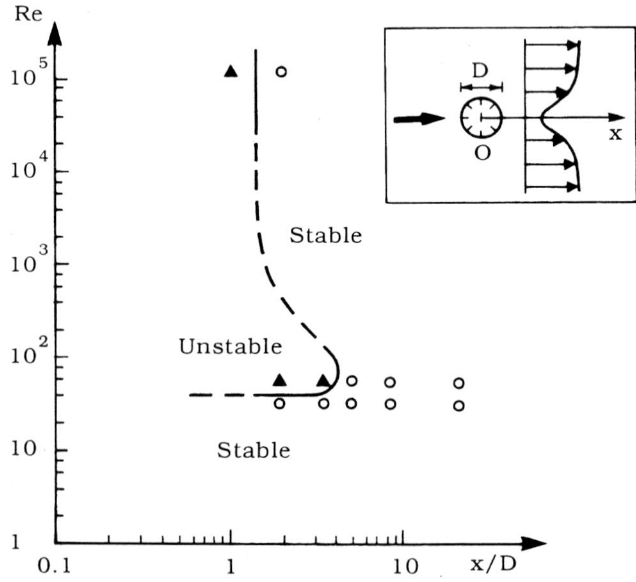

Figure 5.28 Stability diagram for flow past a cylinder by Triantafyllou et al. (1987). Triangles: Absolute instability. Circles: Convective instability.

Fig. 5.28 displays the results of Triantafyllou et al.'s analysis. In the figure, the "unstable" region is the region of absolute instability while the "stable" region is that where there is only convective instability (i.e., a wave that grows as it travels; when the disturbance is convected away, however, the oscillations will eventually die out). Fig. 5.28 shows the following. 1) The flow is unstable (i.e., vortex shedding occurs) if $Re \stackrel{\sim}{>} 40$. This is because when the Reynolds number becomes so large (larger than about 40), the dissipative (or damping) action of viscosity then becomes relatively weak. This leads to the change in the mode of flow in the form of vortex shedding. Regarding the critical value of Re, namely $Re = 40$, this value is in good agreement with experiments (see Section 1.1). 2) Furthermore, it is seen that the streamwise extent of the region of instability decreases with increasing Re. Triantafyllou et al. related this to the so-called formation region, which determines the frequency of vortex formation. Apparently, the results regarding the size of the region of instability are consistent with the corresponding dimensions reported for the formation region (Triantafyllou et al., 1986 and 1987).

At the x-sections where there is instability, the corresponding values of ω_r and k_r would give the frequency and the spacing of the vortex shedding, respectively:

$$f = \frac{\omega_r}{2\pi} \quad , \quad \lambda = \frac{2\pi}{k_r} \tag{5.80}$$

The results obtained by Triantafyllou et al. (1987) regarding the above quantities are summarized in Table 5.4. As seen, the Strouhal frequencies obtained by means of the stability analysis agree remarkably well with the experimental data given in Fig. 1.9.

Table 5.4 Frequency and spacing of vortex shedding obtained through the stability analysis of Triantafyllou et al. (1987).

Re	x/D	$\omega_r D/U_o$	$k_r D$	$St = \frac{fD}{U_o}$	$\frac{\lambda}{D}$
56	2.0	0.83	1.1	0.13	5.7
	3.5	0.83	1.45	0.13	4.3
	5.0	0.83	1.2	0.13	5.2
	8.0	0.83	1.05	0.13	6.0
	20.0	0.83	0.90	0.13	7.0
1.4×10^5	1.0	1.3	2.2	0.21	2.9
	2.0	1.3	1.9	0.21	3.3

It may be noted that Triantafyllou et al. (1987) developed a model of the wake, based on the results of their instability analysis, which is able to obtain good estimates of the steady and unsteady forces on the cylinder.

Finally, it may be mentioned that a similar analysis, but only for a laminar wake and with a different velocity profile expression, was undertaken by Nakaya (1976) with some limited results, indicating that the wake flow may become unstable for Re number above a value of about 40-50.

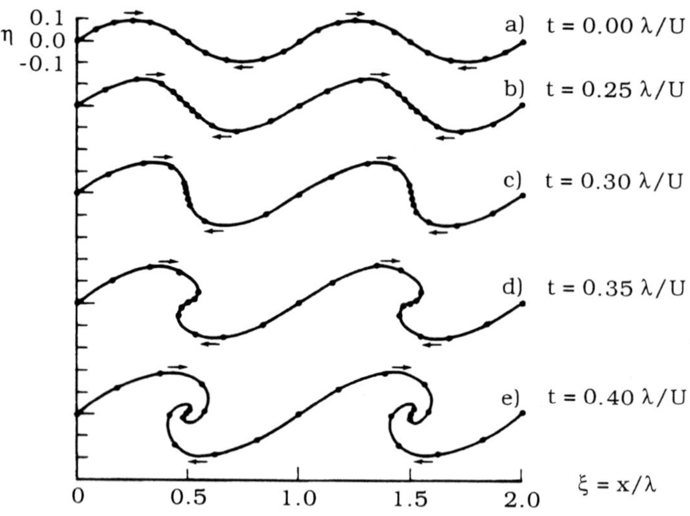

Figure 5.29 Instability of shear layer. Rosenhead (1931).

In the context of hydrodynamic stability, it would be interesting to recall some of the previously mentioned information given in Section 5.1 in relation to the direct solution of N.–S. equations. The knowledge on hydrodynamic stability regarding the flow around a cylinder may be obtained directly from the solution of N.–S. equations. In fact, Fromm and Harlow's (1963) calculations did indicate that, for $Re < 40$, the flow around a rectangular cylinder remained stable (i.e., no shedding developed) after the introduction of a small perturbation in the form of an artificial increase in the value of the vorticity just in front of the cylinder. For $Re > 40$, however, their calculations showed that the flow became unstable to such small perturbations; they reported that within a fairly short time after the introduction of the perturbation, the shedding process began to occur. Apparently, to achieve the flow instability, introduction of small artificial perturbation in one form or another is a common practice used in the numerical solution of the N.–S. equations, unless the round-off errors in the calculations excite the vortex shedding process (Table 5.1).

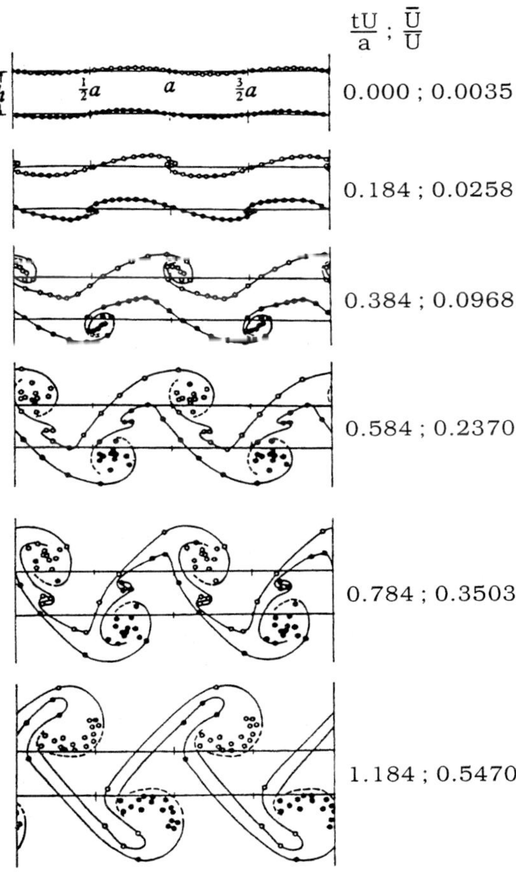

Figure 5.30 Vortex street formation with $h/a = 0.281$, $A = -0.0250a$, $\gamma = (\tanh \pi h/a)$, $n = 21$, and $\Delta t = 0.004a/U$. Abernathy and Kronauer (1962). $\overline{U}$: the mean horizontal velocity of translation of the vortex system.

Instability of two parallel concentrated shear layers

Another approach to study the instability of the wake flow is to assume that the wake flow may be simulated by two parallel shear layers, where the shear is concentrated into a single step in flow velocity (rather than the more smooth distribution as given by Eq. 5.79).

Regarding the instability of shear layers in general, the work in this area dates back as early as 1879; earlier studies of Rayleigh (1879) showed that parallel shear flows are unstable. Rosenhead (1931) studied the instability of a shear layer with an infinitesimal thickness using the vortex method. Rosenhead's study showed that 1) the shear layer is unstable to small disturbances, 2) the initially sinusoidal disturbance grows asymmetric, and 3) the vorticity in the shear layer eventually concentrates in vortices (Fig. 5.29). The frequency associated with the aforementioned shear-layer instability could not be predicted through the method of Rosenhead since the effect of diffusion was not taken into consideration; this frequency is known to depend on the momentum thickness of the shear layer (Ho and Huerre, 1984).

The method of Rosenhead (1931) was later adopted by Abernathy and Kronauer (1962) to study the instability of two parallel shear layers, simulating the wake flow behind a bluff body. This study was successfull in demonstrating that the vorticity in the shear layers concentrates into vortices and further that the vortices are eventually arranged in a staggered configuration, reminiscent of Kármán street (Fig. 5.30). Similar to Rosenhead's study, the frequency or the spacing associated with the instability could not be obtained by the applied method.

Abernathy and Kronauer studied in detail the instability of the two shear layers for various values of the parameter h/a in which a is the wave length of the initial disturbance and h the distance between the shear layers. They found that the pattern of vortex street formation did not change with h/a. They observed, however, that $h/a = 0.28$ is the smallest shear-layer spacing for which only two clouds form per wave length. This value coincides with the value obtained by Kármán (1911 and 1912) as the stability condition for two infinite rows of point vortices in a staggered configuration where h is the spacing of the two arrays of vortices and a the distance between the vortices on the same array.

Example 5.3: Kármán's stability analysis

Single infinite row:

For reasons of simplicity, first consider an infinite row of vortices located at the points 0, $\pm a$, $\pm 2a$, ..., each with strength κ (Fig. 5.31).

The complex potential of 2n + 1 vortices nearest the origin (including the one at the origin) is

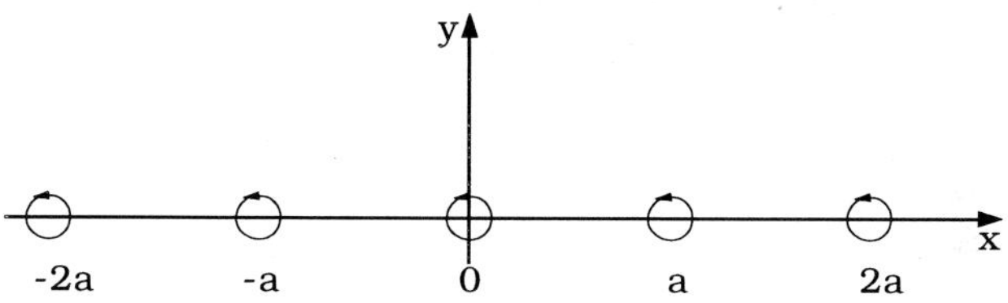

Figure 5.31 A single row of vortices.

$$w_n = i\kappa \ln z + i\kappa \ln(z - a) + \ldots + i\kappa \ln(z - na)$$
$$+ i\kappa \ln(z + a) + \ldots + i\kappa \ln(z + na) \tag{5.81}$$

in which, for example, the term $i\kappa \ln(z - a)$ represents the contribution to w_n of the vortex located at $z = a + i0 = a$ (Milne-Thomson, 1962, Section 13.71). Combining the terms

$$w_n = i\kappa \ln \left\{ \frac{\pi z}{a} \left(1 - \frac{z^2}{a^2}\right) \left(1 - \frac{z^2}{2^2 a^2}\right) \ldots \left(1 - \frac{z^2}{n^2 a^2}\right) \right\}$$
$$+ i\kappa \ln \left\{ \frac{a}{\pi} a^2 (2^2 a^2) \ldots (n^2 a^2) \right\} \tag{5.82}$$

and omitting the second term (because it will not contribute to the velocity, since it is constant):

$$w_n = i\kappa \ln \left\{ \frac{\pi z}{a} \left(1 - \frac{z^2}{a^2}\right) \left(1 - \frac{z^2}{2^2 a^2}\right) \ldots \left(1 - \frac{z^2}{n^2 a^2}\right) \right\} \tag{5.83}$$

From the identity

$$\sin\left(\frac{\pi z}{a}\right) = \frac{\pi z}{a} \left(1 - \frac{z^2}{a^2}\right) \left(1 - \frac{z^2}{2^2 a^2}\right) \ldots \left(1 - \frac{z^2}{n^2 a^2}\right) \ldots \tag{5.84}$$

(Abramowitz and Stegun, 1965, Formula 4.3.89), the complex potential in Eq. 5.83, when $n \to \infty$, will be

$$w - i\kappa \ln\left(\sin\left(\frac{\pi z}{a}\right)\right) \tag{5.85}$$

The complex velocity at the vortex $z = 0$ induced by the *remaining* vortices of the infinite row is

$$u - iv = \left(-\frac{dw}{dz}\right)_{z=0}$$

$$= -\frac{d}{dz}\left\{i\kappa \ln\left(\sin\left(\frac{\pi z}{a}\right)\right) - i\kappa \ln z\right\}_{z=0}$$

$$= -i\kappa\left(\frac{\pi}{a}\cot\frac{\pi z}{a} - \frac{1}{z}\right)_{z=0} = 0 \tag{5.86}$$

Hence, the vortex at $z = 0$ is at rest, and therefore all the vortices are at rest, meaning that the row induces no velocity in itself.

Two infinite rows in a staggered configuration. Kármán vortex street

In order to consider the two shear layers in the downstream wake, we now consider two infinite rows of vortices in a staggered configuration *at time $t = 0$* (Fig. 5.32). The vortices in the rows have equal strengths, namely κ, but opposite rotation. Also, note that the ones in the upper row are at points $ma + \frac{1}{2}ih$ ($m = 0,\ \pm1,\ \pm2,...$) and those in the lower row at the points $(n + \frac{1}{2})a - \frac{1}{2}ih$ ($n = 0,\ \pm1,\ \pm2,....$).

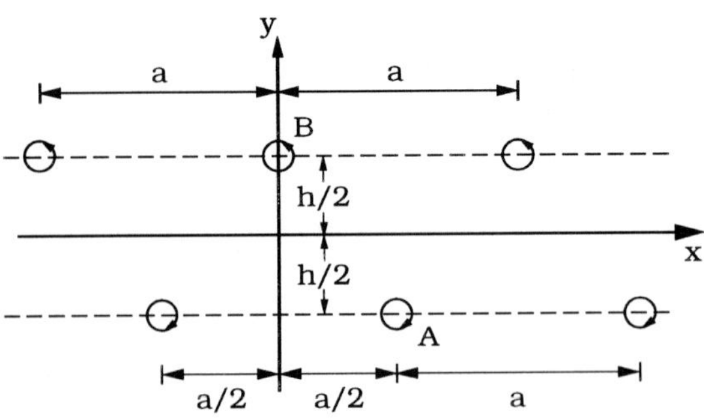

Figure 5.32 Two infinite row of vortices.

The complex potential for this arrangement of vortices at time $t = 0$ is therefore

$$w = i\kappa \ln\left[\sin\left(\frac{\pi}{a}\left(z - \frac{ih}{2}\right)\right)\right] +$$

$$+ i(-\kappa)\ln\left[\sin\left(\frac{\pi}{a}\left(z - \frac{a}{2} + \frac{ih}{2}\right)\right)\right] \tag{5.87}$$

in which the first term is the contribution of the upper row, while the second term is that of the lower row (see Eq. 5.85).

The velocity of the system may be calculated as follows. The velocity of the vortex at $z = \frac{1}{2}a - \frac{1}{2}ih$ (Vortex A):

$$u - iv = \left[-\frac{dw}{dz}\right]_{z=\frac{1}{2}a-\frac{1}{2}ih}$$

$$= -\frac{d}{dz}\left\{i\kappa\left[\sin\left(\frac{\pi}{a}\left(z - \frac{ih}{2}\right)\right)\right]\right\}_{z=\frac{1}{2}a-\frac{ih}{2}} \tag{5.88}$$

(on taking only the term in w associated with the upper row, as the lower row does not induce any velocity in itself, as discussed in the preceding paragraphs). Hence

$$u - iv = \left[-\frac{dw}{dz}\right]_{z=\frac{1}{2}a-\frac{1}{2}ih} = -i\frac{\kappa\pi}{a}\cot\left(\frac{\pi}{2} - \frac{i\pi h}{a}\right) \tag{5.89}$$

Using,

$$\tanh x = -i\tan(ix) \tag{5.90}$$

(Abramowitz and Stegun, 1965, formula 4.5.9), Eq. 5.89

$$u - iv = \left[-\frac{dw}{dz}\right]_{z=\frac{1}{2}a-\frac{1}{2}ih} = \frac{\kappa\pi}{a}\tanh\left(\frac{\pi h}{a}\right) \tag{5.91}$$

This indicates that the vortex moves in the x-direction with this velocity, and so do all the vortices of the lower row, meaning that the lower row advances with velocity

$$V = \frac{\kappa\pi}{a}\tanh\left(\frac{\pi h}{a}\right) \tag{5.92}$$

and, likewise, the upper row advances with the same velocity.

The stability analysis. The procedure of Kármán's stability analysis is basically as follows: 1) displace the vortices slightly according to a periodic disturbance and 2) determine whether the displacement of vortices ever grow (instability) or otherwise (stability). The governing equation used for the analysis is simply the equation of motion for any one of the vortices:

$$\frac{d\bar{z}}{dt} = u - iv \tag{5.93}$$

in which $\bar{z} = x - iy$, the conjugate complex of z, the location of that particular vortex, and $u - iv$ is the complex velocity induced by all the other vortices at that point.

Now, *first*, move the vortices slightly with the following displacements

$$z_m - \gamma \cos(m\phi) \tag{5.94}$$

$$z'_n = \gamma' \cos\left(\left(n + \frac{1}{2}\right)\phi\right) \tag{5.95}$$

in which z_m and z'_n are the displacements for the upper and lower vortices, respectively, γ and γ' are small complex numbers, and ϕ is $0 < \phi < 2\pi$.

Second, work out the velocity of, for example, the vortex at $z = 0 + \frac{1}{2}ih$ at time $t = 0$ (namely, Vortex B). The contributions to this velocity from the vortices corresponding to $\pm m$ in the upper row, will be

$$u - iv = -\frac{dw}{dz} = -\frac{d}{dz}\left\{ i\kappa \ln\left[z - \left(am + \frac{ih}{2} + z_m\right)\right] + \right.$$
$$\left. + i\kappa \ln\left[z - \left(-am + \frac{ih}{2} + z_{-m}\right)\right]\right\}_{z=0+\frac{ih}{2}+z_0}$$

$$= -i\kappa\left[\frac{1}{z_0 - z_m - ma} + \frac{1}{z_0 - z_{-m} + ma}\right]$$

or expanding by the binomial theorem and
retaining the first powers of z_0, z_m, z_{-m}

$$= -i\kappa\frac{z_m + z_{-m} - 2z_0}{m^2 a^2} \tag{5.96}$$

and those from the vortices coresponding to $-n - 1$ and n in the lower row

$$u - iv = -\frac{dw}{dz} = i\kappa\left[\frac{1}{z_0 - z'_n - (n + 1/2)a + ih} + \frac{1}{z_0 - z'_{-n-1} + (n + 1/2)a + ih}\right]$$

or, by the binomial expansion and retaining the first powers of z_0, z'_{-n-1}, z'_n

$$= i\kappa\left[-\frac{z_0 - z'_{-n-1}}{[(n + 1/2)a + ih]^2} - \frac{z_0 - z'_n}{[(n + 1/2)a - ih]^2}\right.$$

$$\left. - \frac{1}{(n + 1/2)a - ih} + \frac{1}{(n + 1/2)a + ih}\right] \tag{5.97}$$

From Eqs. 5.95 and 5.96, and using Eqs. 5.93 and 5.94, the total velocity of the vortex is found as

$$u - iv = \sum_{m=1}^{\infty} \frac{2\kappa i}{a^2} \frac{\gamma(1 - \cos(m\phi))}{m^2} -$$

$$- \sum_{n=0}^{\infty} \frac{2\kappa i \left[\gamma - \gamma' \cos\left((n+1/2)\phi\right)\right]\left[(n+1/2)^2 - k^2\right]}{a^2 \left[(n+1/2)^2 + k^2\right]^2}$$

$$+ \sum_{n=0}^{\infty} \frac{\kappa}{a^2} \frac{2ka}{(n+1/2)^2 + k^2} \tag{5.98}$$

in which

$$k = \frac{h}{a} \tag{5.99}$$

Third, apply the equation of motion (5.93) for the considered vortex (Vortex B) for which $d\bar{z}/dt$ is

$$\frac{d\bar{z}}{dt} = V + \frac{d\bar{z}_0}{dt}$$

$$= V + \frac{d}{dt}\left(\bar{\gamma}\cos(0.\phi)\right) = V + \frac{d\bar{\gamma}}{dt} \tag{5.100}$$

and, from Eqs. 5.93, 5.98 and 5.100, one gets

$$\frac{2\kappa i}{a^2} \sum_{m=1}^{\infty} \frac{\gamma(1 - \cos(m\phi))}{m^2} - \sum_{n=0}^{\infty} \frac{2\kappa i\left[\gamma - \gamma' \cos(n+1/2)\theta\right]\left[(n+1/2)^2 - k^2\right]}{a^2\left[(n+1/2)^2 + k^2\right]^2}$$

$$+ \sum_{n=0}^{\infty} \frac{\kappa}{a^2} \frac{2ka}{(n+1/2)^2 + k^2} = V + \frac{d\bar{\gamma}}{dt} \tag{5.101}$$

Using the identity

$$\sum_{n=0}^{\infty} \frac{1}{(n+1/2)^2 + k^2} = \frac{\pi}{2k}\tanh(\pi k) \tag{5.102}$$

(Gradshteyn and Ryzhik, 1965, formula 1.421.2), and recalling Eq. 5.92, the equation of motion (5.101) will be

$$\frac{d\bar{\gamma}}{dt} = \frac{2i\kappa}{a^2}(A\gamma + C\gamma') \tag{5.103}$$

in which A and C are

$$A = \sum_{m=1}^{\infty} \frac{1 - \cos(m\phi)}{m^2} - \sum_{n=0}^{\infty} \frac{(n + \frac{1}{2})^2 - k^2}{\left[(n + \frac{1}{2})^2 + k^2\right]^2} \tag{5.104}$$

$$C = \sum_{n=0}^{\infty} \frac{\left[(n + \frac{1}{2})^2 - k^2\right] \cos\left[(n + \frac{1}{2})\phi\right]}{\left[(n + \frac{1}{2})^2 + k^2\right]^2} \tag{5.105}$$

For a vortex in the lower row, replacing κ with $-\kappa$ and interchanging γ and γ', the counterpart of Eq. 5.103 is obtained as

$$\frac{d\bar{\gamma}'}{dt} = \frac{-2i\kappa}{a^2}(A\gamma' + C\gamma) \tag{5.106}$$

The *fourth step* in the analysis is to solve Eqs. 5.103 and 5.106 to get γ and γ', the two unknowns of the problem. For this, differentiate Eq. 5.103 with respect to t:

$$\frac{d^2\bar{\gamma}}{dt^2} = \frac{2i\kappa}{a^2}\left(A\frac{d\gamma}{dt} + C\frac{d\gamma'}{dt}\right) \tag{5.107}$$

The conjugate of the above equation is then

$$\frac{d^2\gamma}{dt} = -\frac{2i\kappa}{a^2}\left(A\frac{d\bar{\gamma}}{dt} + C\frac{d\bar{\gamma}'}{dt}\right) \tag{5.108}$$

and using Eqs. 5.103 and 5.106, the following differential equation is obtained for γ:

$$\frac{d^2\gamma}{dt} - \frac{4\kappa^2}{a^4}\left(A^2 - C^2\right)\gamma = 0 \tag{5.109}$$

A similar equation may be obtained also for γ'. Now, a trial solution for 5.109 is

$$\gamma = G\exp\left(\frac{2\kappa}{a^2}\lambda t\right) \tag{5.110}$$

which yields

$$\lambda^2 - \left(A^2 - C^2\right) = 0 \tag{5.111}$$

The discriminant of this second degree equation is

$$\Delta = 4\left(A^2 - C^2\right) \tag{5.112}$$

if $\Delta > 0$, λ will be real, therefore the motion will be unstable.

Now, consider the case when $\phi = \pi$, which gives the maximum disturbance (Eqs. 5.94 and 5.95). In this case, from Eq. 5.105, C becomes nil, therefore from Eq. 5.112

$$\Delta = 4A^2 \tag{5.113}$$

which is always positive, meaning that the motion is always unstable, unless $A = 0$. The latter condition, from Eq. 5.104, reads

$$A = \sum_{m=1}^{\infty} \frac{1 - \cos(\pi m)}{m^2} - \sum_{n=0}^{\infty} \frac{\left(n + \frac{1}{2}\right)^2 - k^2}{\left[\left(n + \frac{1}{2}\right)^2 + k^2\right]^2} = 0 \tag{5.114}$$

The first series in the preceding equation is

$$\sum_{m=1}^{\infty} \frac{1 - \cos(\pi m)}{m^2} = \frac{2}{1^2} + \frac{2}{3^2} + \frac{2}{5^2} + \dots = 2\frac{\pi^2}{8} = \frac{\pi^2}{4} \tag{5.115}$$

(Gradshteyn and Ryzhik, 1965, formula 0.234.2), and the second series, by differentiation of Eq. 5.102 with respect to k,

$$-\sum_{n=0}^{\infty} \frac{\left(n + \frac{1}{2}\right)^2 - k^2}{\left[\left(n + \frac{1}{2}\right)^2 + k^2\right]^2} - \frac{\pi^2}{2\cosh^2(k\pi)} \tag{5.116}$$

and therefore Eq. 5.114 will be

$$\frac{\pi^2}{4} - \frac{\pi^2}{2\cosh^2(k\pi)} = 0 \tag{5.117}$$

yielding

$$k\pi = 0.8814, \quad \text{or} \quad h = 0.281a \tag{5.118}$$

As a conslusion, the motion (or the arrangement of vortices in Fig. 5.32) is always unstable unless the ratio h/a has precisely this value, namely 0.281.

For a more detailed discussion of this topic, reference may be made to Lamb (1945, Article 156). Lamb further shows that, for all values of ϕ from 0 to 2π, the arrangement is stable for $h/a = 0.281$. Also, as another stability problem, Lamb discusses the case of *symmetrical double row*, and shows that this arrangement is always unstable.

Instability of shear layer separating from cylinder

Experiments show that an instability develops in the shear layer separating from the cylinder, where the shear layer rolls into small vortices, when Re becomes higher than about 2000 (Bloor (1964), Gerrard (1978), Wei and Smith (1986),

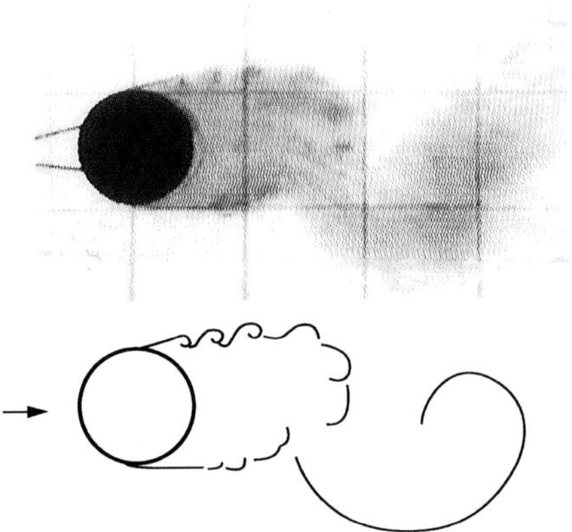

Figure 5.33 Instability of shear layer separating from the cylinder, where the shear layer rolls into small vortices.

Kourta, Boisson, Chassaing and Minh (1987) and Unal and Rockwell (1988)). Fig. 5.33 illustrates the small-scale vortices formed as a result of this instability.

The instability waves corresponding to these small-scale vortices are often called transition waves. The frequency of these waves, f_t, is considerably higher than the frequency of vortex shedding f_v.

Braza, Chassaing and Minh (1990) has studied the aforementioned instability by the numerical simulation of the flow in the range $Re = 2 \times 10^3 - 10^4$ by solving the two-dimensional N.–S. equations. Although the transition mechanism leading to the transition-waves instability is analogous to that generating the instability of a free shear layer (Ho and Huerre, 1984), there may be an interaction between the transition-waves instability and the instability leading to vortex shedding. Braza et al., among other issues, examined this interaction. Fig. 5.34 illustrates the velocity field together with the schematic representation of vortices corresponding to the presented velocity field for $Re = 3000$ obtained in Braza et al.'s study. Fig. 5.35 compares the numerically obtained data on the ratio of f_t/f_v with experiments. From the figure, it is seen that while f_t/f_v is about 5 for $Re = 2 \times 10^3$, it becomes about 18 when $Re \cong 3 \times 10^4$.

3-D instability

Steady current:

Another instability in relation to the flow around cylinders is the onset of three-dimensionality for the Reynolds numbers larger than about 200, see Section 1.1. This phenomenon has been investigated numerically by Karniadakis and

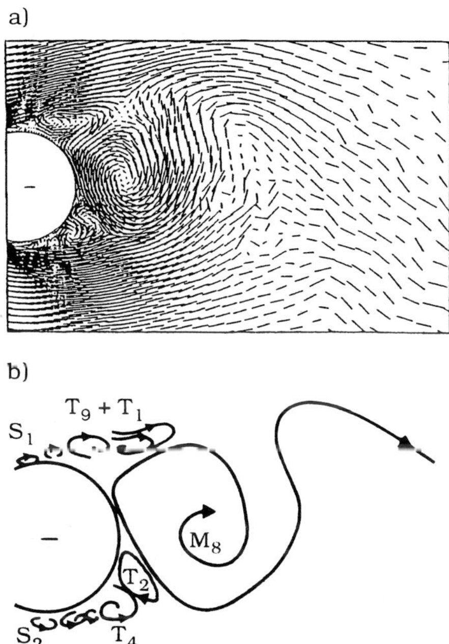

Figure 5.34 (a) Velocity field. (b) Schematic representation of main (M) and secondary (S, T) vortices in the near wake. $Re = 2000$. Braza et al. (1990).

Triantafyllou (1992) by direct simulation of the N.–S. equation in the range of Re, $175 \leq Re \leq 500$. Karniadakis and Triantafyllou's calculations showed that while, for $Re = 175$, the flow remained stable, the instability set in (i.e., the three-dimensionality occurred) when the Reynolds number is increased to $Re = 225$, being consistent with the observations.

Figs. 5.36 and 5.37 show time series of the streamwise and spanwise components of the velocity for the previously mentioned Re numbers. The spanwise component of the instantaneous velocity, w, may be used as a measure of the three-dimensionality. From the time series of w presented in Figs. 5.36 and 5.37, it is seen that, while a noise, initially introduced into the flow, dies out for the case of $Re = 175$, it apparently grows and eventually settles for a constant amplitude in the case of $Re = 225$.

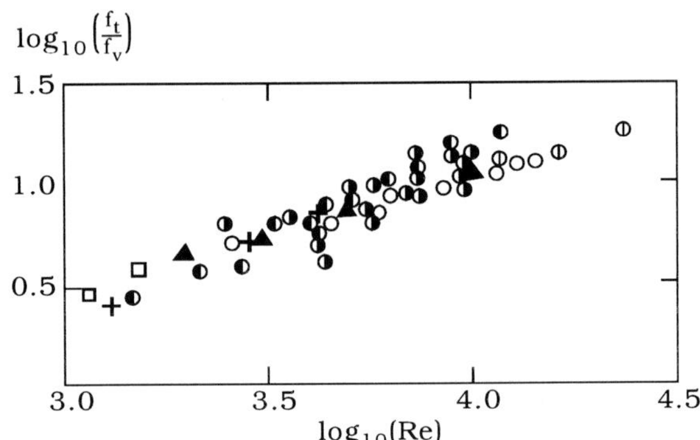

Figure 5.35 Ratio of the transition wave frequency over Strouhal frequency versus Reynolds number. ①, +:Bloor (1964); ▫, Gerrard (1978); ○, Kourta et al. (1987);◑,◐, Wei and Smith (1986);▲ Braza et al.'s (1990) direct numerical simulation. Adapted from Braza et al. (1990).

Further to their direct simulation at $Re = 175$ and 225, Karniadakis and Triantafyllou (1992) have studied the transition to turbulence by conducting the 3-D simulations also for Re numbers $Re = 300$, 333 and 500.

Another three-dimensional stability analysis has been carried out by Noack and Eckelmann (1994). using low-dimensional Galerkin method. Their key results are as follows: 1) The flow is stable with respect to all perturbations for $Re < 54$. 2) While the 2-D perturbations (of the vortex street) rapidly decay, 3-D perturbations with long spanwise wave lengths neither grow nor decay for $54 < Re < 170$. 3) The periodic solution becomes unstable at $Re = 170$ by a perturbation with the spanwise wave length of 1.8 diameters, leading to a three-dimensional periodic flow.

Oscillatory flows:

As seen in Section 3.1, the oscillatory viscous flow becomes unstable to spanwise-periodic vortices above a critical KC number (the Honji instability). This kind of instability was investigated analytically by Hall (1984). Subsequently, Zhang and Dalton (1995) modelled the phenomenon numerically; they obtained a definite 3-D behaviour as regards the variation of vorticity and also they obtained that the sectional lift coefficient has a strong spanwise variation.

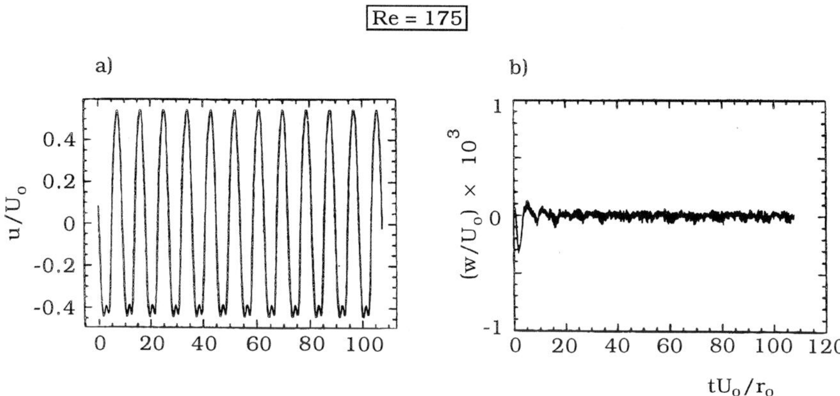

Figure 5.36 Time history of the velocity components at $x/D = 1$; $y/D = 0.075$; $z = 0$ and $\beta = 2.0$. r_0 is the cylinder radius. (a) Streamwise and (b) spanwise components. Karniadakis and Triantafyllou (1992).

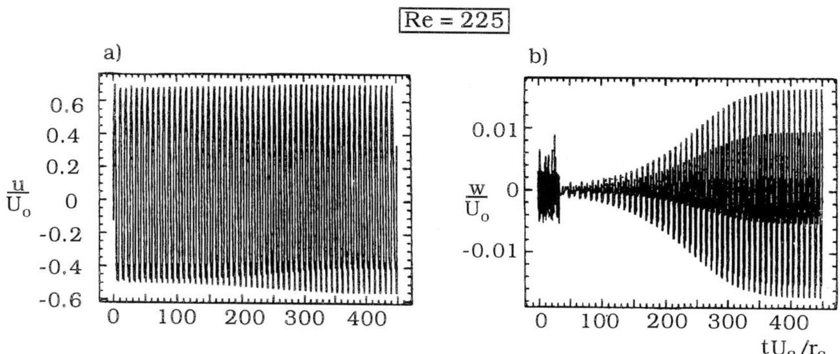

Figure 5.37 Time history of the velocity components at $x/D = 1$; $y/D = 0.075$; $z = 0$; and $\beta = 2.0$. r_0 is the cylinder radius. (a) Streamwise and (b) spanwise components. Karniadakis and Triantafyllou (1992).

REFERENCES

Abernathy, F.H. and Kronauer, R.E. (1962): The formation of vortex street. J. Fluid Mech., 13:1-20.

Abramowitz, M. and Stegun, I.A. (eds.) (1965): Handbook of Mathematical Functions. Dover Publications, Inc., New York.

Apelt, C.J. (1961): The steady flow of a viscous fluid past a circular cylinder at Reynolds numbers 40 and 44. R. & M. No. 3175, A.R.C. Tech. Rep., Ministry of Aviation Aero. Res. Council Rep. & Memo., 1961, 28 p.

Baba, N. and Miyata, H. (1987): Higher-order accurate difference solutions of vortex generation from a circular cylinder in an oscillatory flow. J. Computational Physics, 69:362-396.

Badr, H.M., Dennis, S.C.R., Kocabiyik, S. and Nguyen, P. (1995): Viscous oscillatory flow about a circular cylinder at small to moderate Strouhal number. J. Fluid Mech., 303:215-232.

Baker, G.R. (1979): The "cloud in cell" technique applied to the roll up of vortex sheets. J. Computational Physics, 31:76-95.

Batchelor, G. K. (1967): An Introduction to Fluid Dynamics. Cambridge U. Press.

Bloor, M.S. (1964): The transition to turbulence in the wake of a circular cylinder. J. Fluid Mech., 19:290-304.

Borthwick, A.G.L. (1986): Comparison between two finite-difference schemes for computing the flow around a cylinder. Int. J. for Num. Meth. in Fluids, 6:275-290.

Braza, M., Chassaing, P. and Minh, H.H. (1986): Numerical study and physical analysis of the pressure and velocity fields in the near wake of a circular cylinder. J. Fluid Mech., 165:79-130.

Braza, M., Chassaing, P. and Minh, H.H. (1990): Prediction of large-scale transition features in the wake of a circular cylinder. Phys. Fluids, A2(8):1461-1471.

Braza, M., Nogues, P. and Persillon, H. (1992): Prediction of self-induced vibrations in incompressible turbulent flows around cylinders. Proc. 2nd ISOPE Conf., San Francisco, USA, June 14-19, 1992, 3:284-292.

Bugliarello, G. (1971): Some examples of stochastic modelling for mass and momentum transfer. In: Stochastic Hydraulics (Ed. Chao-Lin Chiu), Proc. 1st Int. Symp. on Stoch. Hyd., Univ. of Pittsburgh, Penn., USA, May 31–June 2, 1971, pp. 39-55.

Cantwell, B.J. (1976): A flying hot wire study of the turbulent near wake of a circular cylinder at a Reynolds number of 140.000. Ph.D.-Thesis, California Institute of Technology, Pasadena, CA.

Chorin, A.J. (1973): Numerical study of slightly viscous flow. J. Fluid Mech., Vol. 57, part 4, pp. 785-796.

Chorin, A.J. (1978): Vortex sheet approximation of boundary layers. J. Computational Physics, 27:128 142.

Chorin, A.J., Hughes, T.J.R., McCracken, M.F. and Marsden, J.E. (1978): Product formulas and numerical algorithms. Communications on Pure and Applied Mathematics, 31:205-256.

Christiansen, J.P. (1973): Numerical simulation of hydrodynamics by the method of point vortices. J. Computational Physics, 13:363-379.

Crank, J. (1975): The mathematics of diffusion. Clarendon Press, Oxford, U.K.

Davis, R.W. and Moore, E.F. (1982): A numerical study of vortex shedding from rectangles. J. Fluid Mech., 116:475-506.

Dennis, S.C.R. and Chang, G.-Z. (1970): Numerical solutions for steady flow past a circular cylinder at Reynolds numbers up to 100. J. Fluid Mech., Vol. 42, part 3, pp. 471-489.

Franke, R., Rodi, W. and Schönung, B. (1990): Numerical calculation of laminar vortex-shedding flow past cylinders. J. Wind Engineering and Industrial Aerodynamics, 35:237-257.

Fredsøe, J. (1984): Turbulent boundary layer in wave-current motion. J. Hydraulic Engineering, ASCE, 110(8):1103-1120.

Fromm, J.E. and Harlow, F.H. (1963): Numerical solution of the problem of vortex street development. Phys. of Fluids, July 1963, 6(7):975-982.

Gerrard, J.H. (1978): The wakes of cylindrical bluff bodies at low Reynolds number. Phil. Transactions of the Royal Soc. London, Series A, 288(A1354):351-382.

Gradshteyn, I.S. and Ryzhik, I.M. (1965): Table of integrals, series and products. Academic Press, N.Y. and London.

Graham, J.M.R. and Djahansouzi, B. (1989): Hydrodynamic damping of structural elements. Proc. 8th Int. Conf. OMAE. The Hague, The Netherlands, 2:289-293.

Hall, P. (1984): On the stability of the unsteady boundary layer on a cylinder oscillating transversely in a viscous fluid. J. Fluid Mech., 146:347-367.

Hansen, E.A., Yde, L. and Jacobsen, V. (1991): Simulated turbulent flow and forces around groups of cylinders. Proc. 23rd Annual OTC, Houston, TX, May 6-9, 1991, Paper No. 6577, pp. 143-153.

Ho, C.-H. and Huerre, P. (1984): Perturbed free shear layers. Ann. Rev. Fluid Mech., 16:365-424.

Jordan, S.K. and Fromm, J.E. (1972): Oscillatory drag, lift and torque on a circular cylinder in a uniform flow. Phys. of Fluids, 15(3):371-376.

Justesen, P. (1990): Numerical modelling of oscillatory flow around a circular cylinder. 4th Int. Symp. on Refined Flow Modelling and Turbulence Measurements, Wuhan, China, Sept. 1990, pp. 6-13.

Justesen, P. (1991): A numerical study of oscillating flow around a circular cylinder. J. Fluid Mech., 222:157-196.

Kármán, Th. von (1911): Über den Mechanismus des Widerstandes, den ein bewegter Körper in einer Flüssigkeit erfährt. Nachrichten, Gesellschaft der Wissenschaften, Göttingen, Math.-Phys. Klasse, pp. 509-517.

Kármán, Th. von (1912): Über den Mechanismus des Widerstandes, den ein bewegter Körper in einer Flüssigkeit erfährt. Nachrichten, Gesellschaft der Wissenschaften, Göttingen, Math.-Phys. Klasse, pp. 547-556.

Karniadakis, G.E. and Triantafyllou, G.S. (1992): Three-dimensional dynamics and transition to turbulence in the wake of bluff objects. J. Fluid Mech., 238:1-30.

Kawaguti, M. (1953): Numerical solution of the Navier-Stokes equations for the flow around a circular cylinder at Reynolds number 40. Jour. Phys. Soc. of Japan, 8(6):747-757.

Keller, H.B. and Takami, H. (1966): Numerical studies of steady viscous flow about cylinders. In: Numerical Solutions of Nonlinear Differential Equations. (Ed. D. Greenspan), Proc. of Adv. Symp. Math. Res. Center, U.S. Army at Univ. of Wisconsin, Madison, May 9-11, 1966, John Wiley & Sons, Inc.

Kourta, A., Boisson, H.C., Chassaing, P. and Minh, H.H. (1987): Nonlinear interaction and the transition to turbulence in the wake of a circular cylinder. J. Fluid Mech., 181:141-161.

Kovásznay, L.S.G. (1949): Hot-wire investigation of the wake behind cylinders at low Reynolds numbers. Proc. Royal Soc., A, London, 198:174-190.

Lamb, H. (1911): On the uniform motion of a sphere through a viscous fluid. Philosophical Magazine, Vol. 21, 6th Series, pp. 112-121.

Lamb, H. (1945): Hydrodynamics. Dover Publications, New York.

Lecointe, Y. and Piquet, J. (1989): Flow structure in the wake of an oscillating cylinder. Trans. of ASME, J. of Fluids Engineering, 111:139-148.

Leonard, A. (1980): Review: Vortex methods for flow simulation. J. Computational Physics, 37:289-335.

Maull, D.J. and Milliner, M.C. (1978): Sinusoidal flow past a circular cylinder. Coastal Engineering, 2:149-168.

Milne-Thomson, L.M. (1962): Theoretical Hydrodynamics. 4. ed., Macmillan.

Murashige, S., Hinatsu, M. and Kinoshita, T. (1989): Direct calculations of the Navier-Stokes equations for forces acting on a cylinder in oscillatory flow. Proc. 8th Int. Conf. OMAE, The Hague, The Netherlands, 2:411-418.

Nakaya, C. (1976): Instability of the near wake behind a circular cylinder. J. Phys. Soc. of Japan, Letters, 41(3):1087-1088.

Noack, B.R. and Eckelmann, H. (1994): A global stability analysis of the steady and periodic cylinder wake. J. Fluid Mech., 270:297-330.

Obasaju, E.D., Bearman, P.W. and Graham, J.M.R. (1988): A study of forces, circulation and vortex patterns around a circular cylinder in oscillating flow. J. Fluid Mech., 196:467-494.

Oseen, C.W. (1910): Über die Stokes'sche Formel und über eine verwandte Aufgabe in der Hydrodynamik. Arkiv för Mat., Astron. och Fys., 6(29):1910.

Quartapelle, L. and Napolitano, M. (1983): Force and moment in incompressible flows. AIAA Journal, 21(6):911-913.

Rayleigh (Lord Rayleigh) (1879): On the instability of jets. Proc. London Mathematical Soc., X:4-13.

Rodi, W. (1992): On the simulation of turbulent flow past bluff bodies. J. of Wind Engineering, No. 52, August, pp. 1-16.

Rosenhead, L. (1931): The formation of vortices from a surface of discontinuity. Proc. Roy. Soc. of London, Series A, 134:170-192.

Roshko, A. (1961): Experiments on the flow past a circular cylinder at very high Reynolds number. J. Fluid Mech., 10:345-356.

Sarpkaya, T. (1986): Force on a circular cylinder in viscous oscillatory flow at low Keulegan-Carpenter numbers. J. Fluid Mech., 165:61-71.

Sarpkaya, T. (1989): Computational methods with vortices - - The Freeman Scholar Lecture. J. Fluids Engineering, Trans. ASME, 111:5-52.

Sarpkaya, T. and Shoaff, R.L. (1979): A discrete-vortex analysis of flow about stationary and transversely oscillating circular cylinders. Naval Postgraduate School Tech. Report No: NPS-69SL79011, Monterey, CA.

Sarpkaya, T., Putzig, C., Gordon, D., Wang, X. and Dalton, C. (1992): Vortex trajectories around a circular cylinder in oscillatory plus mean flow., J. Offshore Mech. and Arctic Engineering, Trans. ASME, 114:291-298.

Schewe, G. (1983): On the force fluctuations acting on a circular cylinder in cross-flow from subcritical up to transcritical Reynolds numbers. J. Fluid Mech., 133:265-285.

Schlichting, H. (1979): Boundary-Layer Theory. 7. ed., McGraw-Hill Book Company.

Scolan, Y.-M. and Faltinsen, O.M. (1994): Numerical studies of separated flow from bodies with sharp corners by the vortex in cell method. J. Fluids and Structures, 8:201-230.

Skomedal, N.G., Vada, T. and Sortland, B. (1989): Viscous forces on one and two circular cylinders in planar oscillatory flow. Appl. Ocean Res., 11(3):114-134.

Smith, P.A. and Stansby, P.K. (1988): Impulsively started flow around a circular cylinder by the vortex method. J. Fluid Mech., 194:45-77.

Smith, P.A. and Stansby, P.K. (1989): Postcritical flow around a circular cylinder by the vortex method. J. Fluids and Structures, 3:275-291.

Son, J.S. and Hanratty, T.J. (1969): Numerical solution for the flow around a cylinder at Reynolds numbers of 40, 200 and 500. J. Fluid Mech., Vol. 35, part 2, pp. 369-386.

Spalart, P.R. and Baldwin, B.S. (1987): Direct simulation of a turbulent oscillating boundary layer. NASA Tech. Memo. 89460, Ames Res. Center, Moffett Field, CA.

Stansby, P.K. (1993): Forces on a circular cylinder in elliptical orbital flows at low Keulegan-Carpenter numbers. Appl. Ocean Res., 15:281-292.

Stansby, P.K. and Dixon, A.G. (1983): Simulation of flows around cylinders by a Lagrangian vortex scheme. Appl. Ocean Res., 5(3):167-178.

Stansby, P.K. and Isaacson, M. (1987): Recent developments in offshore hydrodynamics: workshop report. Appl. Ocean Res., 9(3):118-127.

Stansby, P.K. and Smith, P.A. (1989): Flow around a cylinder by the random vortex method. In Proc. 8th Int. Conf. OMAE. The Hague, The Netherlands, 2:419-426.

Stansby, P.K. and Smith, P.A. (1991): Viscous forces on a circular cylinder in orbital flow at low Keulegan-Carpenter numbers. J. Fluid Mech., 229:159-171.

Stansby, P.K. and Slaouti, A. (1993): Simulation of vortex shedding including blockage by the random-vortex and other methods. Int. Journal for Numerical Methods in Fluids, 17:1003-1013.

Stokes, G.G. (1851): On the effect of the internal friction of fluids on the motion of pendulums. Trans. Cambridge Phil. Soc., Vol. 9, Part II, pp. 8-106.

Sullivan, P.J. (1971): Longitudinal dispersion within a two-dimensional shear flow. J. Fluid Mech., Vol. 49:551-576.

Sumer, B.M., Jensen, H.R., Mao, Y. and Fredsøe, J. (1988): Effect of lee-wake on scour below pipelines in current. J. Waterway, Port, Coastal and Ocean Engineering, ASCE, 114(5):599-614.

Thom, A. (1933): The flow past circular cylinders at low speeds. Proc. Roy. Soc., A, 141:651-669.

Triantafyllou, G.S., Triantafyllou, M.S. and Chryssostomidis, C. (1986): On the formation of vortex streets behind stationary cylinders. J. Fluid Mech., 170:461-477.

Triantafyllou, G.S., Triantafyllou, M.S. and Chryssostomidis, C. (1987): Stability analysis to predict vortex street characteristics and forces on circular cylinders., J. OMAE, Trans. ASME, 109:148-154.

Tritton, D.J. (1959): Experiments on the flow past a circular cylinder at low Reynolds numbers. J. Fluid Mech., 6:547-567.

Unal, M.F. and Rockwell, D. (1988): On vortex formation from a cylinder. Part 1. The initial instability. J. Fluid Mech., 190:491-512.

Wang, C.Y. (1968): On high-frequency oscillatory viscous flows. J. Fluid Mech., 32:55-68.

Wang, X. and Dalton, C. (1991a): Numerical solutions for impulsively started and decelerated viscous flow past a circular cylinder. Int. Journal for Numerical Methods in Fluids, 12:383-400.

Wang, X. and Dalton, C. (1991b): Oscillating flow past a rigid circular cylinder: A finite-difference calculation. J. of Fluids Engineering, 113:377-383.

Wei, T. and Smith, C.R. (1986): Secondary vortices in the wake of circular cylinders. J. Fluid Mech., 169:513-533.

Williamson, C.H.K. (1989): Oblique and parallel modes of vortex shedding in the wake of a circular cylinder at low Reynolds number. J. Fluid Mech., 206:579-627.

Yde, L. and Hansen, E.A. (1991): Simulated high Reynolds number flow and forces on cylinder groups. Proc. 10th Int. Conf. OMAE, Stavanger, Norway, June 1991, 1-A:71-80.

Zhang, J., Dalton, C. and Wang, X. (1991): A numerical comparison of Morison equation coefficients for oscillatory flows: sinusoidal and non-sinusoidal. Proc. 10th Int. Conf. OMAE, Stavanger, Norway, June 1991, 1-A:29-37.

Zhang, J. and Dalton, C. (1995): The onset of a three-dimensional wake in two-dimensional oscillatory flow past a circular cylinder. Presented at the 6th Asian Conf. on Fluid Mech., Singapore, 1995.

Chapter 6. Diffraction effect. Forces on large bodies

In the previous chapters, attention has been concentrated on forces on small cylinders where the cylinder diameter, D, is assumed to be much smaller than the wave length L. In this case, the presence of the cylinder does not influence the wave. In the case when D becomes relatively large, however, the body will disturb the incident waves. Consider, for example, a large vertical, circular cylinder placed on the bottom (Fig. 6.1). As the incident wave impinges on the cylinder, a reflected wave moves outward. On the sheltered side of the cylinder there will be a "shadow" zone where the wave fronts are bent around the cylinder, the so-called diffracted waves (Fig. 6.1). As seen, the cylinder disturbs the incident waves by the generation of the reflected and the diffracted waves. This process is generally termed **diffraction**. The reflected and diffracted waves, combined, are usually called the **scattered** waves.

By the process of diffraction the pressure around the body will change and therefore the forces on the body will be influenced.

It is generally accepted that the diffraction effect becomes important when the ratio D/L becomes larger than 0.2 (Isaacson, 1979).

Normally, in the diffraction flow regime, the flow around a circular cylindrical body is unseparated. This can be shown easily by the following approximate analysis. Consider the sinusoidal wave theory. The amplitude of the horizontal component of water-particle motion at the sea surface, according to the sinusoidal wave theory, is (Eq. III.14, Appendix III):

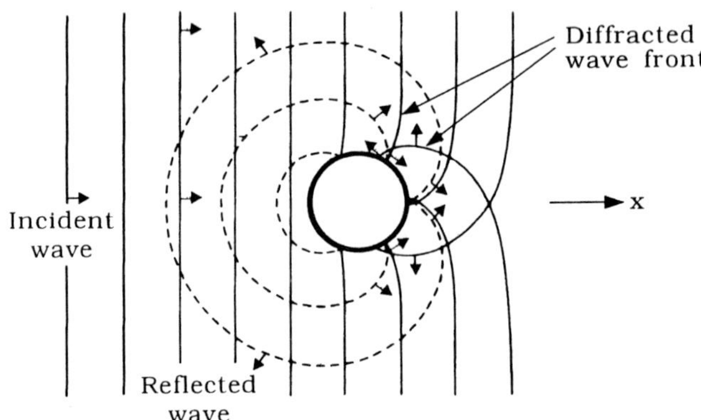

Figure 6.1 Sketch of the incident, diffracted and reflected wave fronts for a
vertically placed cylinder.

$$a = \frac{H}{2}\frac{1}{\tanh(kh)} \tag{6.1}$$

in which H is the wave height, h is the water depth and k is the wave number, i.e.

$$k = \frac{2\pi}{L} \tag{6.2}$$

(Fig. 6.2). The Keulegan-Carpenter number for a vertical circular cylinder will
then be

$$\begin{aligned} KC &= \frac{2\pi a}{D} \\ &= \frac{\pi(H/L)}{(D/L)\tanh(kh)} \end{aligned} \tag{6.3}$$

Obviously the largest KC number is obtained when the maximum wave
steepness is reached, namely when $H/L = (H/L)_{\mathrm{max}}$. The latter may be given
approximately as (Isaacson, 1979)

$$\left(\frac{H}{L}\right)_{\mathrm{max}} = 0.14\tanh(kh) \tag{6.4}$$

Therefore, the largest KC number that the body would experience may, from Eqs.
6.3 and 6.4, be written as

$$KC = \frac{0.44}{D/L} \tag{6.5}$$

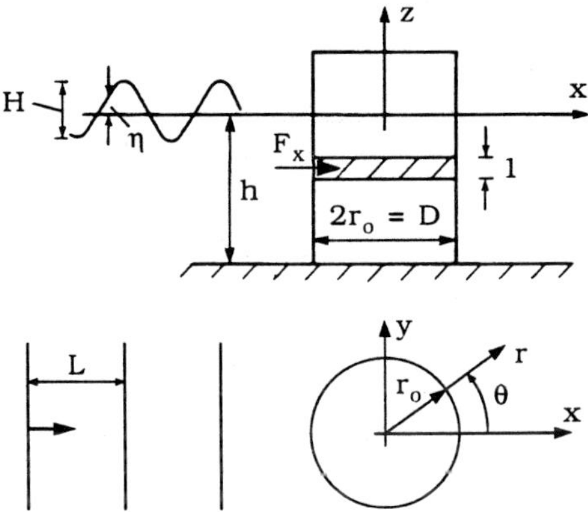

Figure 6.2 Definition sketch for a vertical circular cylinder.

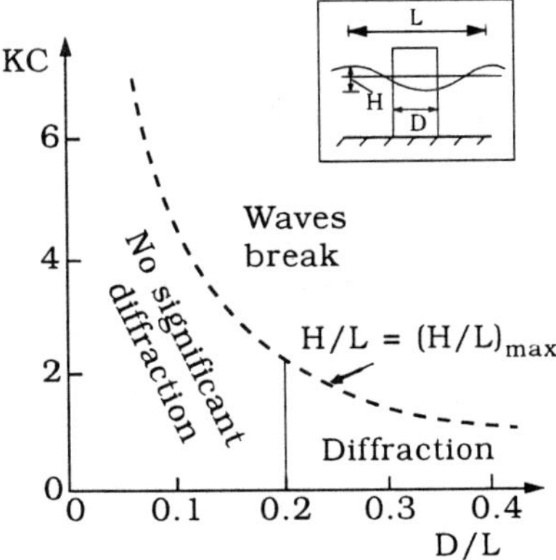

Figure 6.3 Different flow regimes in the $(KC, D/L)$ plane. Adapted from Isaacson (1979).

For the KC numbers larger than this limiting value, the waves will break. Eq. 6.5 is plotted as a dashed line in Fig. 6.3. The vertical line $D/L = 0.2$ in the figure, on the other hand, represents the boundary beyond which the diffraction effect becomes significant. Now, Fig. 6.3 indicates that the KC numbers experienced in the diffraction flow regime are extremely small, namely $KC < 2$. The Reynolds number, on the other hand, must be expected to be extremely large (large compared with $O(10^3)$ in any event). From Fig. 3.15, it is seen that, for $KC < 2$ and $Re \gg O(10^3)$, the flow will be unseparated in most of the cases. When KC number approaches to 2, however, there will be a separation. Yet, the separation under these conditions (small KC numbers) will not be very extensive (Fig. 3.15).

The preceding analysis suggests that the problem regarding the flow around and forces on a large body in the diffraction regime may be analyzed by potential theory in most of the situations, since the flow is unseparated. However, in some cases such as in the calculation of damping forces for resonant vibrations of structures, the viscous effects must be taken into consideration. Obviously, under such conditions, potential-flow theory is no longer applicable.

The discussion given in the preceding paragraphs refers to only *circular* cylinders. When the body has sharp corners the separation will be inevitable. In this case the viscous effects may not be negligible.

6.1   Vertical circular cylinder

This section will describe the diffraction effect, applying potential theory developed by MacCamy and Fuchs (1954). The problem of diffraction of plane waves from a circular cylinder of infinite length has been solved analytically for sound waves (see Morse, 1986, p. 346). MacCamy and Fuchs (1954) applied the known theory with some modifications for water waves incident on a circular pile in the case of finite water depth. The theory is a linear theory and the results are exact to the first order. The theory was initially developed by Havelock (1940) for the special case of infinite water depth.

The analysis given in the following paragraphs is based on the work of MacCamy and Fuchs (1954).

6.1.1   Analytical solution for potential flow around a vertical circular cylinder

Fig. 6.2 shows the definition sketch. The incident wave is coming in from left to right. As it impinges on the cylinder, a reflected wave moves outward from the cylinder, and a diffracted wave forms on the sheltered area (Fig. 6.1). Let ϕ be the total potential function, defined by $u_i = \partial\phi/\partial x_i$. The function ϕ can be found from the following equations:

The continuity equation (the Laplace equation):

$$\nabla^2 \phi = \frac{\partial^2 \phi}{\partial x^2} + \frac{\partial^2 \phi}{\partial y^2} + \frac{\partial^2 \phi}{\partial z^2} = 0 \tag{6.6}$$

No vertical velocity at the bed:

$$\frac{\partial \phi}{\partial z} = 0 \quad \text{at} \quad z = -h \tag{6.7}$$

Bernoulli equation at the surface, where the pressure is constant (linearized):

$$\frac{\partial^2 \phi}{\partial t^2} + g\frac{\partial \phi}{\partial z} = 0 \quad \text{at} \quad z = 0 \tag{6.8}$$

The velocity component normal to the surface of the body (the r-direction) is zero:

$$\frac{\partial \phi}{\partial r} = 0 \text{ at the body surface} \tag{6.9}$$

From the linear feature of potential flow, the total potential function, ϕ, can be written as the sum of two potential functions

$$\phi = \phi_i + \phi_s \tag{6.10}$$

in which ϕ_i is the potential function of the undisturbed incident wave and ϕ_s is that of the scattered (reflected plus diffracted) wave

Potential function for the undisturbed incident wave, ϕ_i

The potential function ϕ_i, is given by the linear theory:

$$\phi_i = -i\frac{gH}{2\omega}\frac{\cosh\big(k(z+h)\big)}{\cosh(kh)}e^{i(kx-\omega t)} \tag{6.11}$$

It can be seen easily that the real part of ϕ_i is the same as the potential function given in Eq. III.16 in Appendix III. It is known that this solution satisfies

$$\nabla^2 \phi_i = 0 \tag{6.12}$$

$$\frac{\partial \phi_i}{\partial z} = 0 \quad \text{at} \quad z = -h \tag{6.13}$$

and

$$\frac{\partial^2 \phi_i}{\partial t^2} + g\frac{\partial \phi_i}{\partial z} = 0 \quad \text{at} \quad z = 0 \tag{6.14}$$

The quantity ω in Eq. 6.11 is the angular frequency and related to k by the dispersion relation (Appendix III, Eq. III.8):

$$\omega^2 = gk \tanh(kh) \tag{6.15}$$

i in Eq. 6.11 is the imaginary unit $i = \sqrt{-1}$. Also, for later use, the expression for the surface elevation (Appendix III, Eq. III.5):

$$\eta = -\frac{1}{g}\left(\frac{\partial \phi}{\partial t}\right)_{z=0} = \frac{H}{2}\cos(\omega t - kx) \tag{6.16}$$

and the velocity components (Appendix III, Eqs. III.10 and III.12):

$$u = \frac{\partial \phi_i}{\partial x} = \frac{\pi H}{T}\frac{\cosh\big(k(z+h)\big)}{\sinh(kh)}\cos(\omega t - kx) \tag{6.17}$$

$$w = \frac{\partial \phi_i}{\partial z} = -\frac{\pi H}{T}\frac{\sinh\big(k(z+h)\big)}{\sinh(kh)}\sin(\omega t - kx) \tag{6.18}$$

in which T is the wave period.

Now, introducing the polar coordinates (Fig. 6.2), ϕ_i can be expressed as

$$\phi_i = -i\frac{gH}{2\omega}\frac{\cosh\big(k(z+h)\big)}{\cosh(kh)}e^{-i\omega t}e^{ikr\cos\theta} \tag{6.19}$$

in which the last term from Abramowitz and Stegun (1965, Eqs. 9.1.44 and 9.1.45) can be written as

$$e^{ikr\cos\theta} = \cos(kr\cos\theta) + i\sin(kr\cos\theta)$$

$$= J_0(kr) + 2\sum_{p=1}^{\infty}(-1)^p J_{2p}(kr)\cos(2p\theta)$$

$$+ i\left\{2\sum_{p=0}^{\infty}(-1)^p J_{2p+1}(kr)\cos\big[(2p+1)\theta\big]\right\}$$

$$= J_0(kr) + \sum_{p=1}^{\infty}2i^p J_p(kr)\cos(p\theta) \tag{6.20}$$

in which $J_p(kr)$ is the Bessel function of the first kind, order p. The Bessel functions are given in tabulated forms in mathematical handbooks (e.g. Abramowitz and Stegun, 1965, Chapter 9) and also in various mathematical softwares as built-in functions (e.g. Mathsoft, 1993, Chapter 12). Fig. 6.4 gives three examples of the Bessel functions, namely J_0, J_1 and J_{10}.

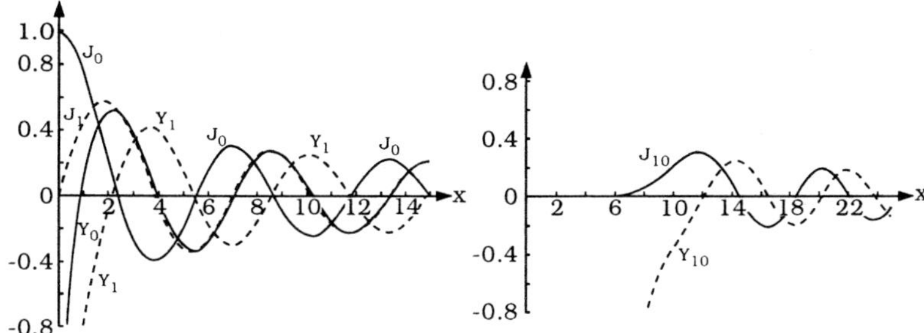

Figure 6.4 Examples of Bessel functions. $J_0(x)$, $Y_0(x)$, $J_1(x)$, $Y_1(x)$, $J_{10}(x)$ and $Y_{10}(x)$.

Inserting Eq. 6.20 in Eq. 6.19, the final form of the incident-wave potential is

$$\phi_i = -i\frac{gH}{2\omega}\frac{\cosh\big(k(z+h)\big)}{\cosh(kh)} \times$$

$$\left[J_0(kr) + \sum_{p=1}^{\infty} 2i^p J_p(kr)\cos(p\theta)\right]e^{-i\omega t} \tag{6.21}$$

Potential function for the scattered wave, ϕ_s

It is assumed that ϕ_s has a form similar to Eq. 6.21. The particular combination appropriate to a wave symmetric with respect to θ (i.e., $\phi_s(-\theta) = \phi_s(\theta)$) is

$$\phi_s = \frac{\cosh\big(k(z+h)\big)}{\cosh(kh)}\sum_{p=0}^{\infty} A_p \cos(p\theta)\Big[J_p(kr) + iY_p(kr)\Big]e^{-i\omega t} \tag{6.22}$$

in which $Y_p(kr)$ is the Bessel function of the second kind, order p (Abramowitz and Stegun, 1965. See also the examples given in Fig. 6.4). In Eq. 6.22, $A_p(p = 0, 1, ...)$ are constants which are to be determined from the boundary conditions. Eq. 6.22 satisfies the Laplace equation

$$\nabla^2\phi_s = \frac{\partial^2\phi_s}{\partial r^2} + \frac{1}{r}\frac{\partial\phi_s}{\partial\theta} + \frac{1}{r^2}\frac{\partial^2\phi_s}{\partial\theta^2} + \frac{\partial^2\phi_s}{\partial z^2} = 0 \tag{6.23}$$

and the boundary conditions

$$\frac{\partial \phi_s}{\partial z} = 0 \quad \text{at} \quad z = -h \tag{6.24}$$

and

$$\frac{\partial^2 \phi_s}{\partial t^2} + g\frac{\partial \phi_s}{\partial z} = 0 \quad \text{at} \quad z = 0 \ , \tag{6.25}$$

Also, Eq. 6.22 has, for large values of r, the form of a periodic wave moving outward in the r-direction with wave number k, and vanishing at $r = \infty$. This can be seen easily from the asymptotic form of the particular combination of the Bessel functions in Eq. 6.22. This combination of J_p and Y_p, known as the Hankel function of the first kind,

$$H_p^{(1)}(kr) = J_p(kr) + iY_p(kr) \tag{6.26}$$

has, for large values of r, the asymptotic form (Abramowitz and Stegun, 1965, Eq. 9.2.3)

$$H_p^{(1)}(kr) \sim \sqrt{\frac{2}{kr}}\, e^{i\left(kr - \frac{2p-1}{4}\pi\right)} \tag{6.27}$$

which reveals that the potential function ϕ_s vanishes at $r = \infty$.

The total potential function, ϕ

The total potential function ϕ is, from Eqs. 6.10, 6.21 and 6.22,

$$\phi = -i\frac{gH}{2\omega}\frac{\cosh\big(k(z+h)\big)}{\cosh(kh)} \times$$

$$\times \left[J_0(kr) + \sum_{p=1}^{\infty} 2i^p J_p(kr)\cos(p\theta) \right] e^{-i\omega t} +$$

$$+ \frac{\cosh\big(k(z+h)\big)}{\cosh(kh)} \times$$

$$\times \sum_{p=0}^{\infty} A_p \cos(p\theta)\left[J_p(kr) + iY_r(kr) \right] e^{-i\omega t} \tag{6.28}$$

This function satisfies the Laplace equation (Eq. 6.6) and the boundary conditions, Eqs. 6.7 and 6.8. The only remaining boundary condition is the zero-normal-velocity condition at the surface of the body, namely Eq. 6.9. Applying this condition, the values of the constants $A_p(p = 0, 1,...)$ are determined. The final form of the potential function is

$$\phi = -i\frac{gH}{2\omega}\frac{\cosh\big(k(z+h)\big)}{\cosh(kh)}\times$$

$$\times \sum_{p=0}^{\infty}\varepsilon_p i^p\left[J_p(kr) - \frac{J_p'(kr_0)}{H_p^{(1)'}(kr_0)}H_p^{(1)}(kr)\right]\cos(p\theta)e^{-i\omega t} \qquad (6.29)$$

in which the derivative terms are

$$J_p'(kr_0) - \left.\frac{dJ_p(\alpha)}{d\alpha}\right|_{\alpha=kr_0} \qquad (6.30)$$

and

$$H_p^{(1)'}(kr_0) = \left.\frac{dH_p^{(1)}(\alpha)}{d\alpha}\right|_{\alpha=kr_0} \qquad (6.31)$$

in which α is a dummy variable. In Eq. 6.29, ε_p is defined as

$$\varepsilon_p = \begin{cases} 1 & p = 0 \\ 2 & p \geq 1 \end{cases} \qquad (6.32)$$

The Bernoulli equation (in linearized form) is used to get the pressure:

$$p = -\rho\frac{\partial\phi}{\partial t} \qquad (6.33)$$

From Eqs. 6.29 and 6.33, the pressure on the cylinder surface is obtained as

$$p = i\frac{\rho gH}{\pi kr_0}\frac{\cosh\big(k(z+h)\big)}{\cosh(kh)}\sum_{p=0}^{\infty}\frac{\varepsilon_p i^p}{H_p^{(1)'}(kr_0)}\cos(p\theta)e^{-i\omega t} \qquad (6.34)$$

To reach this equation, the following identity is used (Spiegel, 1968, Formula 24.135)

$$J_p(\alpha)Y_p'(\alpha) - J_p'(\alpha)Y_p(\alpha) = \frac{2}{\pi\alpha} \qquad (6.35)$$

The free-surface elevation η can be calculated from

$$\eta = -\frac{1}{g}\left(\frac{\partial\phi}{\partial t}\right)_{z=0}$$

and presumably the runup profiles around the cylinder can be worked out accordingly (see Sarpkaya and Isaacson (1981, p. 394) and Isaacson (1979)).

6.1.2 Total force on unit-height of cylinder

Having obtained the wave and flow field around a vertical cylinder, the resulting forces can easily be obtained.

The in-line force acting on a unit height of the cylinder (Fig. 6.2) is

$$F_x = -\int_0^{2\pi} p(r_0\,d\theta)\cos\theta \tag{6.36}$$

Inserting Eq. 6.34 into Eq. 6.36 and carrying out the integration and taking the *real* part only, the force is found as follows:

$$F_x = \frac{2\rho g H}{k}\frac{\cosh\big(k(z+h)\big)}{\cosh(kh)}A(kr_0)\cos(\omega t - \delta) \tag{6.37}$$

in which

$$\delta(kr_0) = -\tan^{-1}\big[Y_1'(kr_0)/J_1'(kr_0)\big] \tag{6.38}$$

$$A(kr_0) = \big[J_1'^2(kr_0) + Y_1'^2(kr_0)\big]^{-1/2} \tag{6.39}$$

Here the derivatives $J_1'(kr_0)$ and $Y_1'(kr_0)$ are calculated in the same fashion as in Eqs. 6.30 and 6.31.

The functions $A(kr_0)$ and $\delta(kr_0)$ can be worked out, using the Bessel-function tables in Abramowitz and Stegun (1965). Figs. 6.5a and 6.5b give the functions $A(kr_0)$ and $\delta(kr_0)$. The function $\delta(kr_0)$ represents the phase difference between the incident wave and the force, and it will be discussed later in the section.

Inertia coefficient

The far-field velocity corresponding to the incident wave is given by Eq. 6.17. From this equation, the maximum acceleration (the absolute value) is obtained as

$$\left|\frac{du}{dt}\right|_m = \frac{\pi H \omega}{T}\frac{\cosh\big(k(z+h)\big)}{\sinh(kh)} \tag{6.40}$$

Now, inserting Eq. 6.40 into Eq. 6.37, F_x may be expressed as

$$F_x = \rho\left[\frac{4A(kr_0)}{\pi(kr_0)^2}\right](\pi r_0^2)\left[\left|\frac{\partial u}{\partial t}\right|_m \cos(\omega t - \delta)\right] \tag{6.41}$$

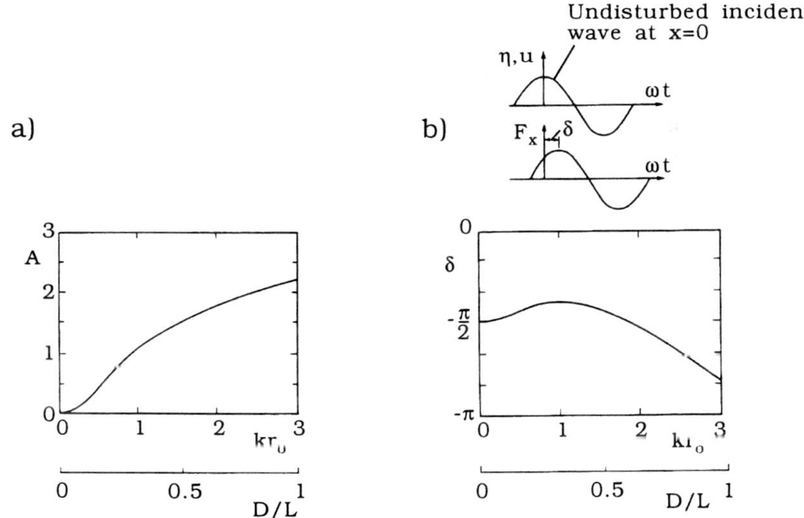

Figure 6.5 (a): The function $A(kr_0)$ in the force expression. (b): The phase
function $\delta(kr_0)$ in the force expression.

This equation has the same form as the Morison equation (Eq. 4.29) with the
drag omitted, namely

$$F_x = \rho C_M(\pi r_0^2)\, \dot{u} \tag{6.42}$$

(However, in Eq. 6.41, the force follows the incident wave crest (passing through
$x = 0$) with a phase delay equal to δ (see Fig. 6.5b)).

Hence the inertia coefficient in the case of diffraction flow regime can, from
Eq. 6.41, be expressed as in the following

$$C_M = \frac{4A(kr_0)}{\pi(kr_0)^2} \tag{6.43}$$

in which $A(kr_0)$ is given by Eq. 6.39. Therefore, the force F_x:

$$F_x = \rho C_M(\pi r_0^2)\left[\left|\frac{\partial u}{\partial t}\right|_m \cos(\omega t - \delta)\right] \tag{6.44}$$

or alternatively,

$$F_x = \frac{\pi}{8}\rho g H k D^2 \frac{\cosh\big(k(z+h)\big)}{\cosh(kh)} C_M \cos(\omega t - \delta) \tag{6.45}$$

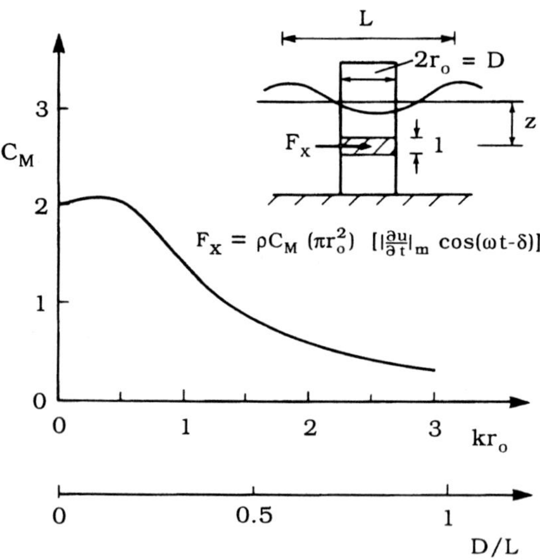

Figure 6.6 The influence of diffraction on the inertia coefficient in the Morison equation.

The inertia coefficient C_M is plotted in Fig. 6.6 as function of kr_0.

First of all, the figure indicates that the diffraction solution approaches the value of 2, the plane potential-flow solution given in Eq. 4.18 (namely $C_M = C_m + 1 = 2$), as $kr_0 \to 0$.

Secondly, C_M begins to be influenced by the diffraction effect after D/L reaches the value of approximately 0.2, in conformity with the previously mentioned limiting value in the beginning of this chapter.

Thirdly, the inertia coefficient decreases with increasing D/L ratio. The physical reason behind this is that the acceleration of flow is maximum over one part of the body while it is not so over the rest of the body. This would obviously give rise to a reduction in the inertia force. As the ratio D/L increases, this effect becomes more and more pronounced, therefore the inertia force will be decreased, as D/L increases.

Fig. 6.7 gives an overview as regards the C_M coefficient. For small cylinders where $D/L \to 0$, the major parameters are the Keulegan-Carpenter number, KC, the Reynolds number, Re, the roughness parameter, k_s/D and the ellipticity of the orbital motion, E; the variation of C_M in this case as function of KC, Re, k_s/D and E are obtained mainly by experiments (see Chapter 4). For large cylinders, on

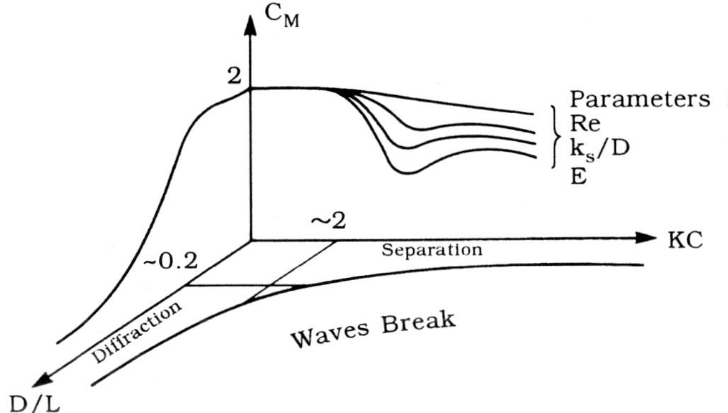

Figure 6.7 Sketch showing C_M as function of major parameters.

the other hand, where $KC \to 0$, the major parameter is D/L; the variation of C_M with D/L is obtained, using the linear diffraction theory. As Fig. 6.7 suggests, C_M actually forms a family of surfaces in the three-dimensional space (C_M, D/L, KC) over the area on the plane (D/L, KC) where the waves do not break.

Phase difference, δ

The phase difference δ between the maximum undisturbed wave-induced flow velocity at $x = 0$ and the maximum force is depicted in Fig. 6.5b which shows that δ goes to $-\pi/2$, as $kr_0 \to 0$, meaning that the force leads over the velocity with a phase difference of $\pi/2$. This is exactly the same result as that obtained for small cylinders ($D/L \ll 1$) for the inertia component of the in-line force (cf. Fig. 4.5). As the diffraction effect begins to influence the force, δ begins to diverge from the value of $-\pi/2$ (Fig. 6.8).

The Froude-Krylov force

The Froude-Krylov force, by definition (Section 4.1.2), can be calculated from the incident-wave potential ϕ_i given in Eq. 6.11:

$$p = -\rho \frac{\partial \phi_i}{\partial t} \tag{6.46}$$

Hence, the pressure on the surface of the body is:

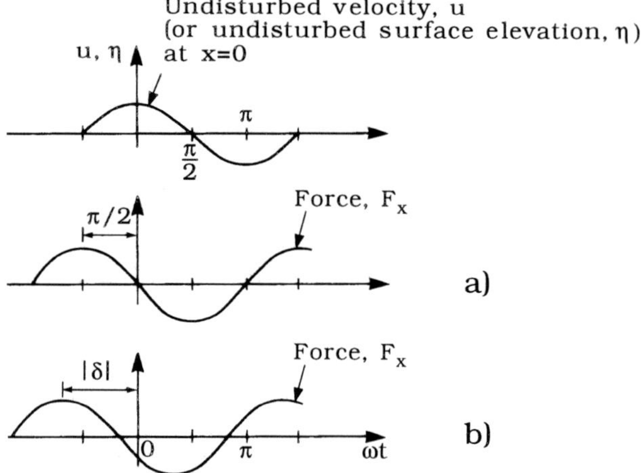

Figure 6.8 Phase difference between the velocity (or the surface elevation) and the force. a) No diffraction ($kr_0 \rightarrow 0$). b) Diffraction where F_x leads over u (or η) with δ different from $\pi/2$ according to Fig. 6.5b.

$$p = \rho g \frac{H}{2} \frac{\cosh\big(k(z+h)\big)}{\cosh(kh)} e^{i(kr_0 \cos\theta - \omega t)} \tag{6.47}$$

Therefore the Froude-Krylov force per unit height of the cylinder will be

$$F_K = -\int_0^{2\pi} p(r_0\, d\theta)\cos\theta$$

$$= -\rho g H \frac{\cosh\big(k(z+h)\big)}{\cosh(kh)} r_0 \pi J_1(kr_0)\sin(\omega t) \tag{6.48}$$

in which the identity

$$J_1(kr_0) = \frac{1}{\pi i}\int_0^\pi e^{ikr_0\cos\theta}\cos\theta\, d\theta \tag{6.49}$$

has been used (Abramowitz and Stegun, 1965, formula 9.1.21). It may be noted that Eq. 6.48 reduces to $F_K = \rho(\pi r_0^2)(du/dt)$, the familiar relation for the

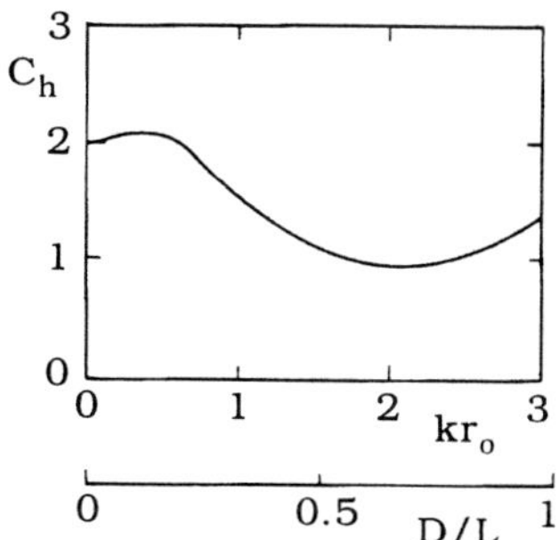

Figure 6.9 Ratio of maximum value of total force to that of Froude-Krylov force for a vertical circular cylinder.

Froude-Krylov force (Eq. 4.24), as $kr_0 \rightarrow 0$, i.e., for small cylinders, where $J_1(kr_0)/(kr_0) \rightarrow 1/2$.

Now, the ratio of the maximum value of the total force to that of the Froude-Krylov force can be calculated from Eqs. 6.37 and 6.48:

$$C_h = \frac{\text{Max}|F_x|}{\text{Max}|F_K|}$$

$$= \frac{2A(kr_0)}{\pi kr_0 J_1(kr_0)} \tag{6.50}$$

The definition of C_h suggests that the total force on the body may be regarded as the product of the Froude-Krylov force on the structure and a coefficient, C_h. This coefficient is plotted against kr_0 in Fig. 6.9. For small cylinders (as $kr_0 \rightarrow 0$), $C_h \rightarrow 2$. This is because the hydrodynamic-mass force and the Froude-Krylov force in this case are equal, as seen in Section 4.1.2.

6.1.3 Total force over the depth and the overturning moment

The total force on the cylinder can be calculated by integrating the force F_x from Eq. 6.37 over the total water depth:

$$F_{x,\text{tot}} = \int_{z=-h}^{0} F_x \, dz$$

$$= \frac{2\rho g H}{k^2} A(kr_0) \tanh(kh) \cos(\omega t - \delta) \tag{6.51}$$

or, in terms of the inertia coefficient (Eq. 6.43):

$$F_{x,\text{tot}} = \frac{\pi}{8} \rho g H D^2 C_M \tanh(kh) \cos(\omega t - \delta) \tag{6.52}$$

Likewise, the Froude-Krylov force acting on the total height of the cylinder, from Eq. 6.48,

$$F_{K,\text{tot}} = \int_{z=-h}^{0} F_K \, dz$$

$$= -\frac{\rho g H}{k^2} \pi k r_0 J_1(kr_0) \tanh(kh) \sin(\omega t) \tag{6.53}$$

and the ratio of the maximum value of the total force to that of the Froude-Krylov force

$$C_h = \frac{\text{Max} |F_{x,\text{tot}}|}{\text{Max} |F_{K,\text{tot}}|} = \frac{2A(kr_0)}{\pi k r_0 J_1(kr_0)} \tag{6.54}$$

The overturning moment, on the other hand, may be evaluated by

$$M = \int_{z=-h}^{0} (z+h)(F_x \, dz)$$

$$= \frac{\pi}{8} \rho g H \frac{D^2}{k} \left[\frac{kh \sinh(kh) + 1 - \cosh(kh)}{\cosh(kh)} \right] \times$$

$$\times C_M \cos(\omega t - \delta) \tag{6.55}$$

The results of the linear diffraction theory described in the preceding paragraphes have been tested against the experiments by several investigators; see, for example, Mogridge and Jamieson (1976). The agreement is generally good.

Second-order effect

A great many number of second-order theories have been developed over the years. An extensive review of these theories have been given by Chakrabarti (1985, 1987). It appears that these theories lack systematic verification against experimental data on the second-order forces. The forces predicted by the second-order theories seem to be slightly larger than the predictions by the linear theory. Chakrabarti (1987), Sarpkaya and Isaacson (1981) and Kriebel (1990, 1992) can be consulted for further information about the second-order effects.

2D and 3D irregular seas

Computer models have been developed for diffraction around vertical circular cylinders, when the incoming waves are irregular 2D-waves (Rao and Raman, 1988) and 2D- or 3D-waves (Skourup, 1994). The MacCamy and Fuchs theory is applied for each wavelet in the spectrum, and superposition is used to obtain the results.

6.2 Horizontal circular cylinder near or on the seabottom. Pipelines

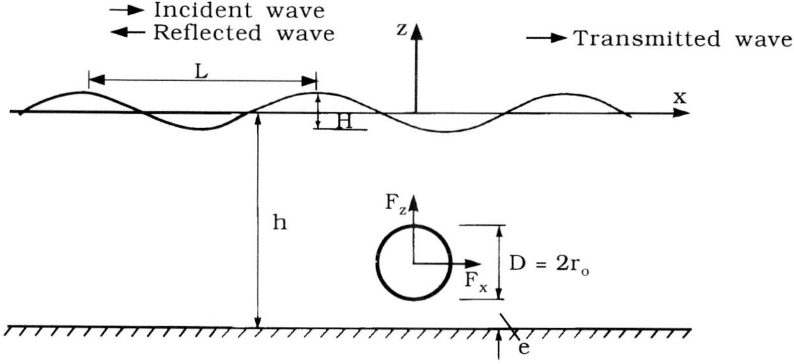

Figure 6.10 Definition sketch.

The diffraction effect may be important also for near-bottom (or on-bottom) horizontal bodies such as pipelines, tunnels, rectangular blocks, etc.. Chakrabarti (1987) gives an extensive review of the subject; the bottom-seated horizontal cylinders (both the half- and full-cylinder situations), the bottom-seated hemispheres and the half- and fully-submerged cylinders are among the cases which are included in Chakrabarti's review.

In this section attention will be concentrated on the horizontal, near-bottom (or on-bottom) cylinder case only.

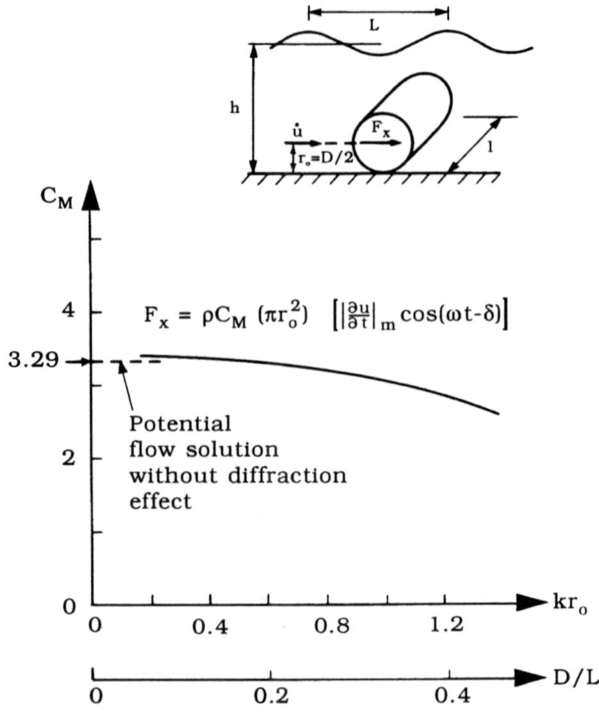

Figure 6.11 Inertia coefficient for a bottom-seated cylinder in the inertia-dominated range. $h/r_0 = 6$. Diffraction effect. Solution is due to Subbiah et al. (1993). The potential-flow solution without diffraction effect is due to Yamamoto et al. (1974) (see Fig. 4.4 for the latter).

This case has drawn considerable attention recently due to its practical application to pipelines (Efthymiou and Narayanan (1980), Jothi Shankar, Raman and Sundar (1985), Cheong, Shankar and Subbiah (1989), Subbiah, Jothi Shankar and Cheong (1993), and Chioukh and Narayanan (1994)). Although pipelines are normally considered to fall into the drag-dominated regime, there is a growing trend in which larger and larger pipelines are installed, both for the disposal of industrial and municipal waste water into the sea and for the transportation of gas and crude oil from offshore platforms. Therefore the inertia-dominated regime where the diffraction effect may be important is not totally out of context.

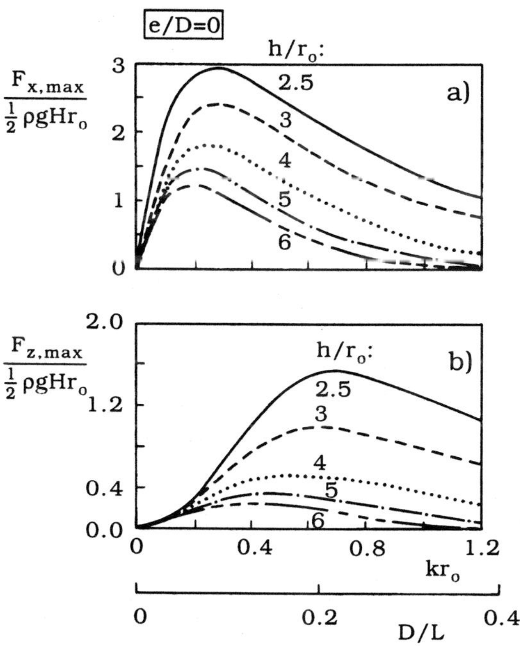

Figure 6.12 Forces per unit length of a *bottom-seated cylinder* in the diffraction regime. Subbiah et al. (1993).

Fig. 6.10 gives the definition sketch. The incident wave is coming from left to right, perpendicular to the pipe. The pipe disturbs the flow by the generation of a reflected wave and a transmitted wave indicated in Fig. 6.10. This process is generally called the diffraction effect in literature, in parallel to the case of vertical cylinder, although the waves are not diffracted in the present case in the sense as described in conjunction with the case of vertical cylinder. In the present case, the force on the cylinder will have also a vertical component due to the asymmetry in the flow.

Fig. 6.11 illustrates how the diffraction effect influences the inertia coefficient in the case of a bottom-seated cylinder (cf. Fig. 6.6). The water-depth-to-

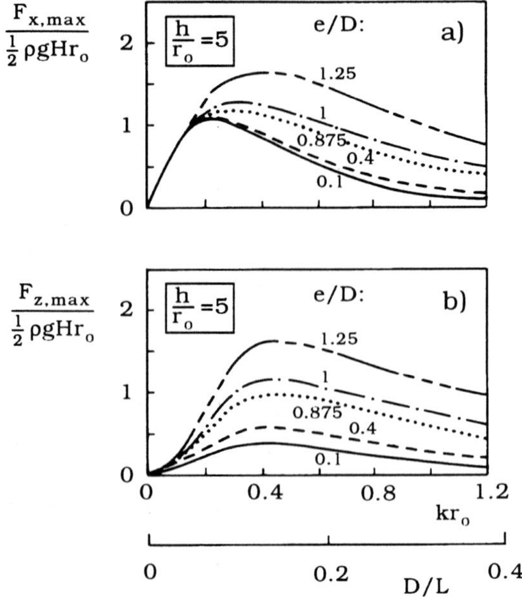

Figure 6.13 Forces per unit length of a *near-bottom cylinder* in the diffraction regime. Subbiah et al. (1993).

cylinder-radius ratio in this particular example is $h/r_0 = 6$. The solution presented in the diagram is due to Subbiah et al. (1993), who used the linearized potential flow theory. The definition of C_M in Fig. 6.11 is the same as in Eq. 6.44. The acceleration $\partial u/\partial t$ in the present case is calculated for the undisturbed flow at the level of the center of the cylinder. (Note that, in Subbiah et al.'s original paper, the solution is presented in a form different from that in Fig. 6.11. Subbiah et al.'s solution has been recast in terms of Eq. 6.44. The acceleration, $\partial u/\partial t$, has been evaluated, using the linear wave theory).

First, the solution is seen to approach the value $C_M = C_M + 1 = 2.29 + 1 = 3.29$, the plane potential-flow solution obtained by Yamamoto et al. (1974) (Fig. 4.4). Second, the diffraction effect begins to make its influence felt when D/L becomes larger than about 0.2, which agrees quite well with the vertical cylinder case, discussed previously.

Fig. 6.12 presents Subbiah et al.'s results regarding the maximum forces in the case of bottom-seated cylinder. As seen, the force decreases as the depth-to-radius ratio, h/r_0, increases. This is due to the decrease in the wave-induced

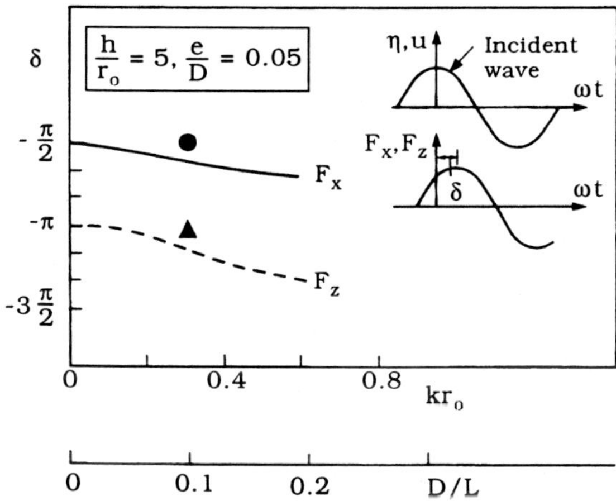

Figure 6.14 Phase of the maximum forces with respect to the incident wave crest (Subbiah et al., 1993). Data points (from Wright and Yamamoto, 1979): ●: F_x, △: F_z where $h/r_0 = 10.6$, $KC = 0.32$ and $e/D = 0.042$.

velocity with h/r_0.

Fig. 6.13 gives the same kind of information for a near-bottom cylinder ($h/r_0 = 5$), while Fig. 6.14 depicts the phase of the maximum forces in this case when $e/D = 0.05$.

The maximum forces increase with increasing e/D. This is explained in the same way as in the previous paragraph. Namely, for a given value of h/r_0, the velocity that the cylinder experiences increases with increasing e/D, meaning that the cylinder with a larger e/D value should experience larger forces. As regards the phase of the maximum force, the phase angle decreases with increasing kr_0 (Fig. 6.14), similar to the case of vertical cylinder with the exception that δ has a slight increase for small values of kr_0.

The latter figure includes also two data points from an experiment carried out by Wright and Yamamoto (1979), which seem to be in reasonable agreement with the theory. Fig. 6.15 depicts the force time series (Fig. 6.15a) corresponding to these data points together with a second set of force time series obtained in the same study but with a much larger KC number, namely $KC = 11$ (Fig. 6.15b). Of particular interest is the change in the behaviour of the time variation in the lift force when the flow regime is changed from the "potential-flow" regime (Fig. 6.15a, $KC = 0.3$) to the separated-flow regime (Fig. 6.15b, $KC = 11$). The positive and negative peaks in the F_z time series in the case of separated flow regime have been

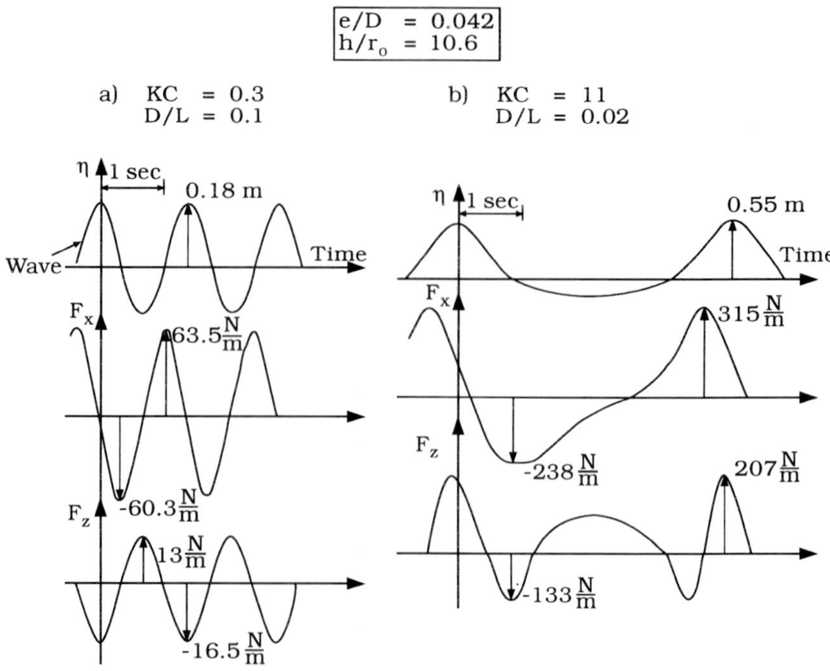

Figure 6.15 Experimentally-obtained force traces for a near-bottom horizontal cylinder. Wright and Yamamoto (1979).

explained in terms of the motion of the lee-wake vortex and the gap flow in Section
3.4 (cf. Figs. 3.21c and 6.15b). Whereas, in the case of potential-flow regime (Fig.
6.15a), the figure indicates that the negative lift is associated with the passage of
the wave crest while the positive lift is associated with that of the wave trough.
Wright and Yamamoto's (1979) study shows a similar picture for a bottom-seated
cylinder, too.

Finally, for the case where waves on a deep water meet a large submerged
circular cylinder, references can be given to Grue and Palm (1985) and Chakrabarti
(1987).

REFERENCES

Abramowitz, M. and Stegun, I.A. (eds.) (1965): Handbook of Mathematical Functions. Dover Publications, Inc., New York.

Chakrabarti, S.K. (1985): Recent advances in high-frequency wave forces on fixed structures. J. Energy Resources Technology, Sept. 1985, 107:315-328.

Chakrabarti, S.K. (1987): Hydrodynamics of Offshore Structures. Computational Mechanics Publications, Springer Verlag.

Cheong, H.F., Jothi Shankar, N. and Subbiah, K. (1989): Inertia dominated forces on submarine pipelines near seabed. J. Hydraulic Res., 27(1):5-22.

Chioukh, N. and Narayanan, R. (1994): Inertia dominated forces on oblique horizontal cylinders in waves near a plane boundary. Coastal Engineering, 22:185-199.

Efthymiou, M. and Narayanan, R. (1980): Wave forces on unburied pipelines. J. Hydraulic Res., 18(3):197-211.

Garrison, C.J. (1984): Nonlinear wave loads on large structures. Proc. 3rd Int. Offshore Mech. and Arctic Engrg. Symposium, ASME, N.Y. Febr. 1984, pp. 128-135.

Grue, J. and Palm, E. (1984): Reflection of surface waves by submerged cylinders. Appl. Ocean Res., 6(1):54-60.

Havelock, T.H. (1940): The pressure of water waves upon a fixed obstacle. Proc. the Royal Soc. of London, Series A. Mathematical and Physical Sciences, 175(A963):409-421.

Isaacson, M. (1979): Wave-induced forces in the diffraction regime. In: Mechanics of Wave-Induced Forces on Cylinders, (Ed. T.L. Shaw). Pitman Advanced Publishing Program, pp. 68-89.

Jothi Shankar, N., Raman, H. and Sundar, V. (1985): Wave forces on large offshore pipelines. Ocean Engineering, 12:99-115.

Kriebel, D.L. (1990): Nonlinear wave interaction with a vertical circular cylinder. Part I: Diffraction Theory. Ocean Engrg., 17(4):345-377.

Kriebel, D.L. (1992): Nonlinear wave interaction with a vertical circular cylinder. Part II: Wave Run-Up. Ocean Engrg., 19(1):75-99.

MacCamy, R.C. and Fuchs, R.A. (1954): Wave forces on piles: A diffraction theory. U.S. Army Corps of Engineers, Beach Erosion Board, Tech. Memo No. 69, 17 p.

Mathsoft (1993): Mathcad Plus 5.0. User's Guide. Mathsoft Inc., Cambridge, MA, 1993.

Mogridge, G.R. and Jamieson, W.W. (1976): Wave loads on large circular cylinders: A design method. Hydraulics Laboratory, Division of Mechanical Engineering, National Research Council Canada, NRC No. 15827, Dec. 1976, 34 p.

Morse, P.M. (1986): Vibration and sound. Published by the American Institute of Physics for the Acoustical Society of America. 3rd printing paperback edition, 1986, 468 p.

Rahman, M. (1984): Wave diffraction by large offshore structures: An exact second-order theory. Appl. Ocean Research, 6(2):90-100.

Rao, P.S.V. and Raman, H. (1988): Wave elevation on large circular cylinders excited by wind-generated random waves. J. Offshore Mech. and Arctic Engrg., 110:48-54.

Sarpkaya, T. (1976): In-line and transverse forces on smooth and sand-roughened cylinders in oscillatory flow at high Reynolds numbers. Naval Postgraduate School, Monterey, C.A., Tech. Rep. NPS-69SL76062.

Sarpkaya, T. and Isaacson, M. (1981): Mechanics of Wave Forces on Offshore Structures. Van Nostrand Reinhold Company.

Skourup, J. (1994): Diffraction of 2-D and 3-D irregular seas around a vertical circular cylinder. Proc. of Offshore Mechanics and Arctic Engineering Conf. (1994 OMAE), ASME, Vol. I, Offshore Technology, 293-300.

Spiegel, M.R. (1968): Mathematical handbook of formulas and tables. New York, McGraw-Hill, 1968, 271 p.

Subbiah, K., Jothi Shankar, N. and Cheong, H.F. (1993): Wave forces on a large horizontal cylinder near a plane boundary. Ocean Engineering, 20(1):77-95.

Wright, J.C. and Yamamoto, T. (1979): Wave forces on cylinders near plane boundaries. J. Waterways, Harbours and Coastal Eng. Division, ASCE, 105(WW1):1-13.

Yamamoto, T., Nath, J.H. and Slotta, L.S. (1974): Wave forces on cylinders near plane boundary. J. Waterways, Harbours and Coastal Eng. Division, ASCE, 100(WW4):345.359.

Chapter 7. Forces on a cylinder in irregular waves

Waves experienced at any point in the sea are not regular. Obviously, this will influence the forces on structures. To what extent the forces are influenced by the irregularity of waves will be the focus of this chapter.

Irregular waves can be treated by use of statistical theories. In the following paragraphs, first, the statistical treatment of irregular waves will be described, and next the forces on cylinders exposed to irregular waves will be outlined.

7.1 Statistical treatment of irregular waves

The wave climate at a given location in the sea may be characterized by a series of short-term (say, 6 hrs) wave climates. These short-term wave climates are termed *sea states*.

Each sea state may be regarded as a *stationary* random process, i.e. the statistical properties of any quantity (such as the surface elevation, for example) are time invariant for the particular sea state considered.

This section will mainly deal with the short-term sea state statistics of surface elevation, wave height and wave period. A brief account of the long-term statistics will be given at the end of this section.

7.1.1 Statistical properties of surface elevation

The surface elevation in the real sea is a random variable (Fig. 7.1). For a sea state lasting a certain period, say 6 hrs, this quantity may be regarded as a stationary random function of time. Two important statistical properties of the surface elevation are its probability density function and its spectrum function.

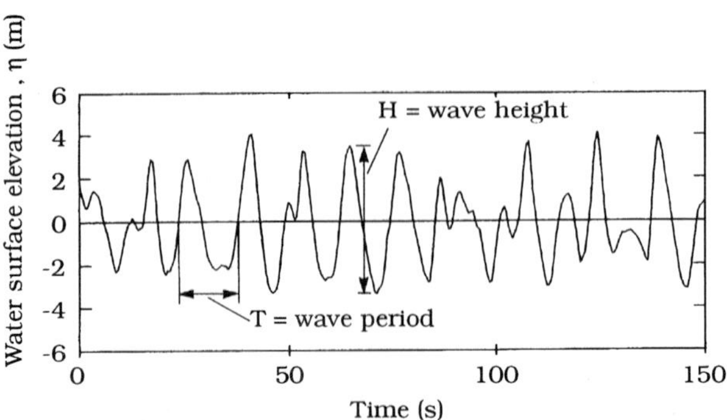

Figure 7.1 An actual wave record from the North Sea.

Probability density function of surface elevation

The probability density function (p.d.f.) of surface elevation, $p(\eta)$, is defined by the following two equations:

$$p(\eta)d\eta = Pr[\eta < \eta' < \eta + d\eta] \tag{7.1}$$

and

$$\int_{-\infty}^{\infty} p(\eta)d\eta = 1 \tag{7.2}$$

in which $Pr[\eta < \eta' < \eta + d\eta]$ is the probability of occurrence of surface elevation between η and $\eta + d\eta$ (see Fig. 7.2).

Observations show that $p(\eta)$ may be characterized by the Gaussian distribution (Fig. 7.3):

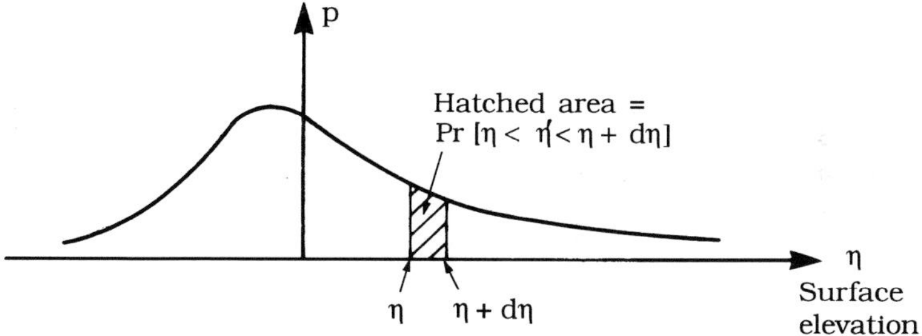

Figure 7.2 Probability density function of surface elevation.

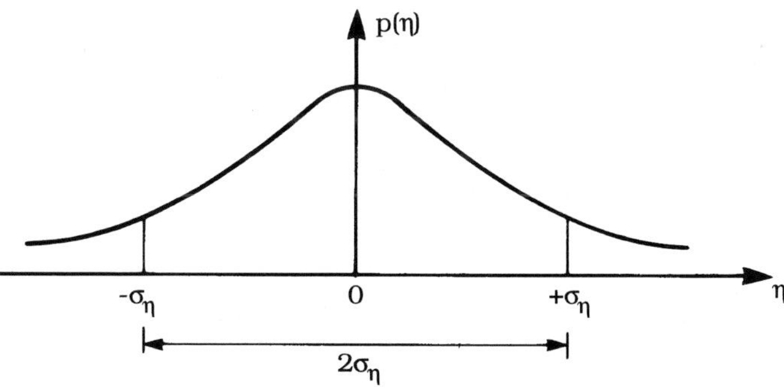

Figure 7.3 Gaussian distribution

$$p(\eta) = \frac{1}{\sqrt{2\pi}\sigma_\eta} \exp\left(-\frac{\eta^2}{2\sigma_\eta^2}\right) \tag{7.3}$$

in which σ_η is the standard deviation of η:

$$\sigma_\eta = \left(\overline{\eta^2}\right)^{1/2} \tag{7.4}$$

where the overbar denotes time averaging. Note that, from the *ergodicity hypothesis*, ensemble averaging and time averaging for a stationary random process are identical:

$$\sigma_\eta^2 = \overline{\eta^2} = \frac{1}{T}\int_0^T \eta^2\,dt \qquad (7.5)$$

in which T is the observation length, and is supposed to be large $\left(O(5\ \text{hrs})\right)$.

The significance of σ_η is that this quantity is a measure of the spreading of η around the zero mean. The probability of occurrence of η outside the range, for example, $-2\sigma < \eta < +2\sigma$ is only 4.6% while that outside the range $-3\sigma < \eta < +3\sigma$ is even much smaller, namely 0.3 %.

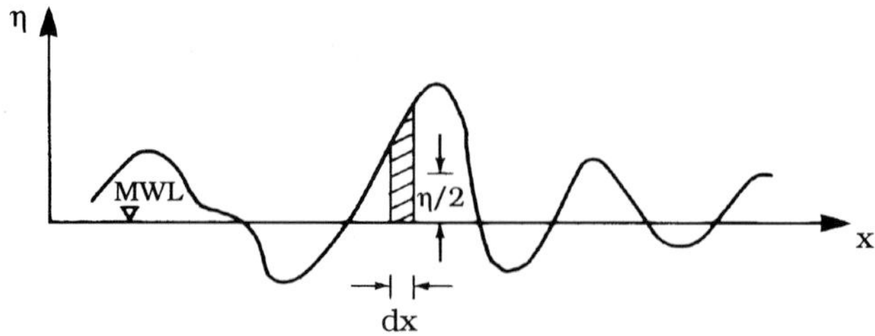

Figure 7.4 Potential energy of water column η is $(\rho g\eta dx)\frac{\eta}{2}$.

From the hydrodynamic stand point, σ_η^2 represents the mean energy per unit area

$$\overline{E} = \frac{1}{2}\rho g\sigma_\eta^2 \qquad (7.6)$$

as will be shown below. The mean wave energy, $\overline{E}$, may be written as (Fig. 7.4):

$$\overline{E} = \lim_{\Lambda\to\infty}\frac{1}{\Lambda}\int_0^\Lambda (\rho g\eta dx)\frac{\eta}{2} = \frac{1}{2}\rho g\left(\lim_{\Lambda\to\infty}\frac{1}{\Lambda}\int_0^\Lambda \eta^2\,dx\right) \qquad (7.7)$$

Since η is a stationary random process, the field of η in space (in the x-direction) must be homogenous, therefore

$$\lim_{\Lambda\to\infty}\frac{1}{\Lambda}\int_0^\Lambda \eta^2\,dx$$

must be constant and identical to

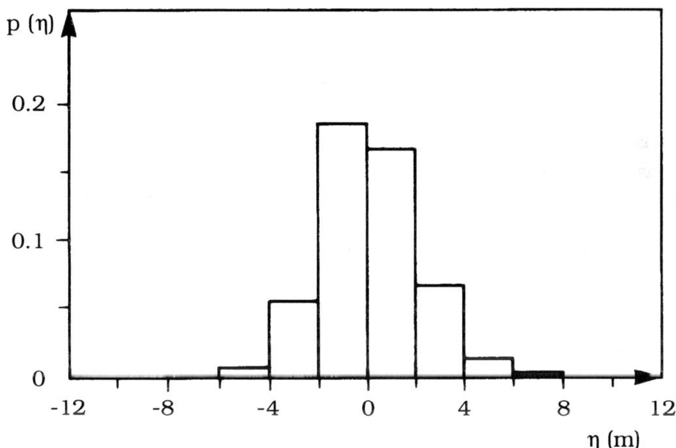

Figure 7.5 Probability density function of surface elevation for the total length of the record referred to in Fig. 7.1. Sample size = 16400 (samples taken at 0.498 s sampling interval).

$$\lim_{\Lambda \to \infty} \frac{1}{\Lambda} \int_0^{\Lambda} \eta^2 dx \equiv \lim_{T \to \infty} \frac{1}{T} \int_0^{T} \eta^2 dt = \sigma_\eta^2 \qquad (7.8)$$

Hence, from Eq. 7.7,

$$\overline{E} = \frac{1}{2} \rho g \sigma_\eta^2$$

If the wave is a sinusoidal wave, then σ_η^2 will be

$$\sigma_\eta^2 = \frac{1}{T_w} \int_0^{T_w} a^2 \sin^2(\omega t) dt = \frac{1}{2} a^2, \qquad (7.9)$$

in which a is the amplitude of η. Thus the mean energy per unit area will be

$$\overline{E} = \frac{1}{2} \rho g \left(\frac{1}{2} a^2 \right) \qquad (7.10)$$

or in terms of wave height

$$\overline{E} = \frac{1}{16} \rho g H^2 \qquad (7.11)$$

which is the expression known from the potential wave theory (Appendix III, Eq. III. 18).

Returning to the p.d.f. of η, it may be noted that, in the case of waves with very large wave heights, crest amplitudes are higher than trough amplitudes, therefore the p.d.f. of η will, in this case, be skewed, as it is seen in Fig. 7.5.

Spectrum function of surface elevation

The spectrum function may be described by reference to the irregular-wave record given in Fig. 7.6. This is actually a simulated wave which is obtained by simple superposition of five regular waves shown in Fig. 7.7. In Fig. 7.7 it is known that $\frac{1}{2}a^2$ represents the energy of regular waves (Eq. 7.10).

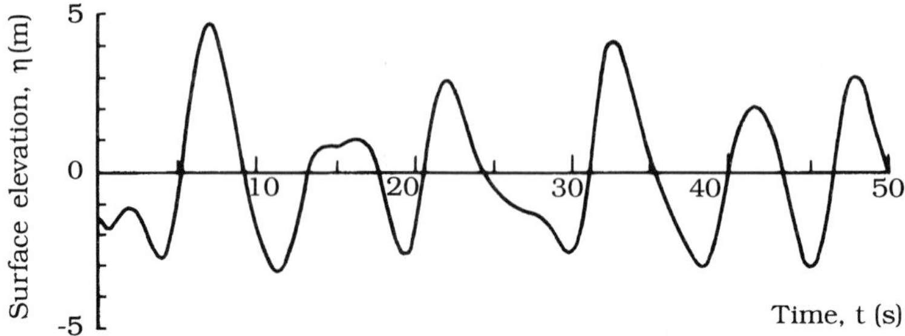

Figure 7.6 Irregular wave record obtained by the superposition of sinusoidal waves shown in the next figure. $\sigma_\eta^2 = 4.0$ m^2 (taken from Goda, 1985).

Now, consider the diagram depicted in Fig. 7.8. This diagram is constructed in such a way that each rectangular area represents one regular wave component given in Fig. 7.7, the area of the rectangle being equal to the corresponding wave energy, namely $\frac{1}{2}a^2$, and the frequency f corresponding to the frequency of the regular wave.

It can be shown that the total area, namely $\sum(\frac{1}{2}a^2)$, is equal to the mean energy of the superposed irregular wave, namely σ_η^2:

$$\sigma_\eta^2 = \sum\left(\frac{1}{2}a^2\right) \tag{7.12}$$

The mathematical details will not be given herein. However, the preceding equation implies that the diagram in Fig. 7.8 would illustrate how the contribution of each regular wave component to the mean total wave energy is distributed with

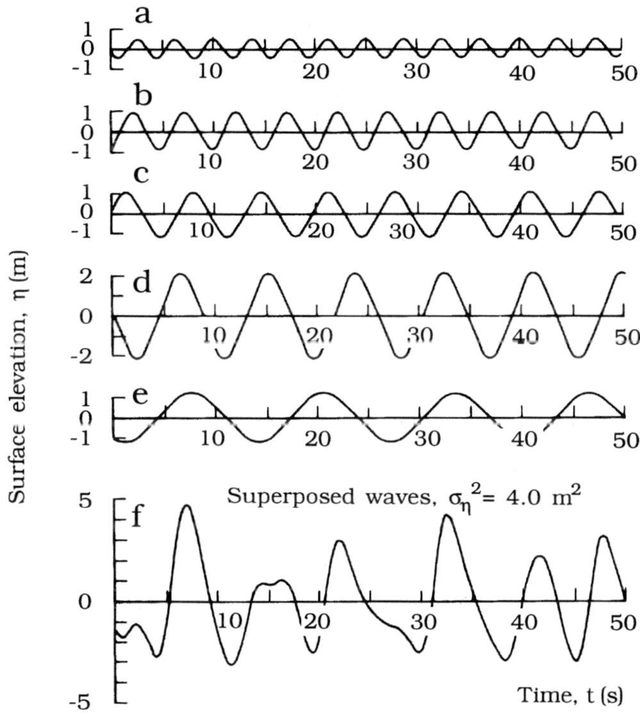

Figure 7.7 Superposition of these five sinusoidal waves gives the irregular wave displayed in the previous figure (taken from Goda, 1985).
(a): f = 0.266 Hz, a = 0.455 m, $\frac{1}{2}a^2 = 0.10$ m^2
(b): f = 0.198 Hz, a = 0.915 m, $\frac{1}{2}a^2 = 0.42$ m^2
(c): f = 0.151 Hz, a = 1.090 m, $\frac{1}{2}a^2 = 0.59$ m^2
(d): f = 0.115 Hz, a = 2.090 m, $\frac{1}{2}a^2 = 2.18$ m^2
(e): f = 0.077 Hz, a = 1.190 m, $\frac{1}{2}a^2 = 0.71$ m^2.

respect to the wave frequency, f. For example, the largest contribution to the total wave energy in Fig. 7.8 comes from the waves with frequency 0.115 Hz.

In reality, there are an infinite number of regular wave components in a given irregular wave record. Therefore, the diagram in Fig. 7.8 will appear as a continuous curve, as illustrated in Fig. 7.9. The quantity $S_\eta(f)$ in the figure is called the spectrum function of surface elevation η, or simply the wave spectrum. As implied by the figure, the area below the wave spectrum curve is equal to the mean wave energy σ_η^2 (Fig. 7.9):

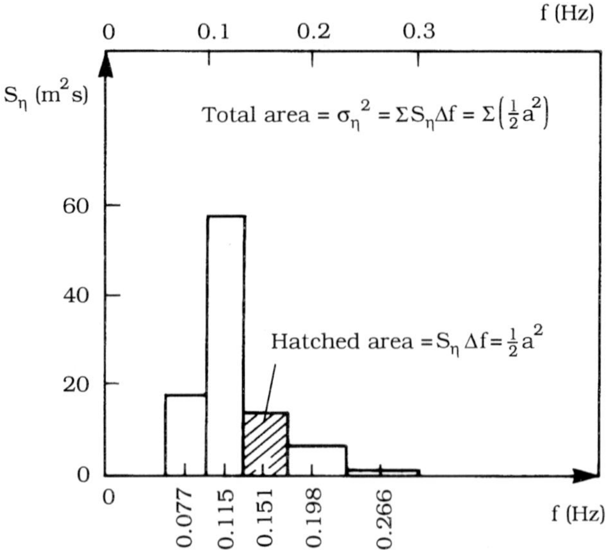

Figure 7.8 Wave spectrum corresponding to the superposed, irregular wave record in Fig. 7.6. Each rectangular area in the above diagram corresponds to one regular wave component shown in Fig. 7.7.

$$\sigma_\eta^2 = \int_0^\infty S_\eta(f)\,df \tag{7.13}$$

The hatched area in the figure therefore represents the contribution to the total wave energy of the waves with frequencies between f and $f + df$.

Formally, $S_\eta(f)$ is defined such that its Fourier transform is equal to the so-called *autocovariance function* of the surface elevation, $R(p)$:

$$R(p) = \int_{-\infty}^\infty e^{i2\pi fp} S_\eta(f)\,df \tag{7.14}$$

in which $R(p)$ is given by

$$R(p) = \lim_{\tau \to \infty} \frac{1}{\tau} \int_{-\tau/2}^{\tau/2} \eta(t)\eta(t+p)\,dt \tag{7.15}$$

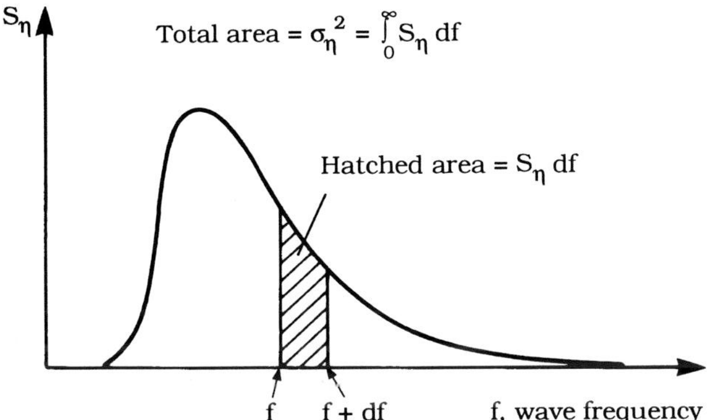

Figure 7.9 Wave spectrum for real waves.

The inverse transformation regarding Eq. 7.14 can be written as

$$S_\eta(f) = \int_{-\infty}^{\infty} R(p)e^{-i2\pi fp}dp \qquad (7.16)$$

It can be seen easily that, when $p = 0$, Eq. 7.14 reduces to Eq. 7.13, since, from Eq. 7.15

$$R(p = 0) = \lim_{\tau \to \infty} \frac{1}{\tau} \int_{-\tau/2}^{\tau/2} \eta(t)\eta(t)dt = \sigma_\eta^2 \qquad (7.17)$$

Eq. 7.16 along with Eq. 7.15 form the basis for the calculation of the wave spectrum, $S_\eta(f)$, from a given time series of $\eta(t)$. There are actually two commonly used methods to calculate the energy spectrum of a wave record, namely the auto-correlation method (see Southworth, 1960) and the Fast Fourier Transform (FFT) technique (see for instance Press et al., 1989). Fig. 7.10 depicts the wave spectrum of the wave record indicated in Fig. 7.1. (Note that the wave record seen in Fig. 7.1 constitutes only a small portion of the actual wave record used in the calculation of the spectrum depicted in Fig. 7.10.)

Sometimes the *spectral moments* are useful in the spectral analysis of waves. The definition of the n *th* moment of the spectrum is given by

$$m_n = \int_{0}^{\infty} f^n S_\eta(f)df \qquad (7.18)$$

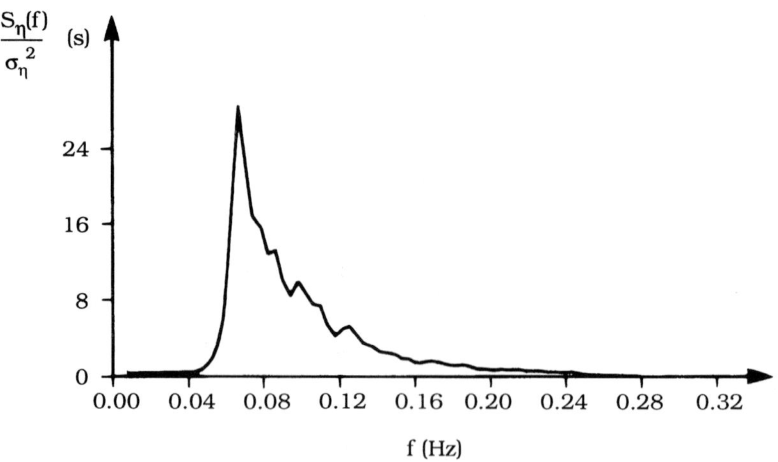

Figure 7.10 Spectrum of surface elevation for the wave record referred to in Fig. 7.1. The total number of waves is 845, corresponding to an observation length of about 2.27 hrs. $\sigma_\eta = 1.94$ m. $f_0 = 0.067$ s^{-1}. $\varepsilon = 0.593$.

The zeroth moment is seen to be identical to the variance of η:

$$m_0 = \int\limits_0^\infty f^0 S_\eta(f)df = \int\limits_0^\infty S_\eta(f)df = \sigma_\eta^2 \tag{7.19}$$

Two important properties of a spectrum function may be its peak frequency f_0 and its width parameter ε. The peak frequency is quite straightforward. The width parameter, on the other hand, is defined by

$$\varepsilon^2 = 1 - \frac{m_2^2}{m_0 m_4} \tag{7.20}$$

Obviously, for regular waves, $\varepsilon = 0$, since $S_\eta(f) = \sigma_\eta^2 \, \delta(f - f_0)$ where δ is the Dirac delta function and therefore

$$\frac{m_2^2}{m_0 m_4} = \frac{\left(\int\limits_0^\infty f^2 S_\eta(f)\,df\right)^2}{\left(\int\limits_0^\infty S_\eta(f)\,df\right)\left(\int\limits_0^\infty f^4 S_\eta(f)\,df\right)} =$$

$$\frac{\left(\int\limits_0^\infty f^2 \delta(f - f_0)\,df\right)^2}{\left(\int\limits_0^\infty \delta(f - f_0)\,df\right)\left(\int\limits_0^\infty f^4 \delta(f - f_0)\,df\right)} = \frac{f_0^4}{f_0^4} = 1 \qquad (7.21)$$

The larger the width parameter, the broader the spectrum. The spectrum is considered a broad-band spectrum if ε is above 0.6. The width parameter for the example given in Fig. 7.10 is $\varepsilon = 0.593$.

Another width parameter which has been used by some authors (see Longoria, Beaman and Miksad, 1991) is defined by

$$q = \sqrt{1 - \frac{m_1 m_1}{m_0 m_2}} \qquad (7.22)$$

For broad-band spectra q approaches 1, while for narrow-band spectra q is close to 0.

Several authors in the past have proposed various models for the wave spectra (see for example Chakrabarti (1987) for a detailed account of these models). Two of the widely used models are the Pierson-Moskowitz spectrum and the JONSWAP spectrum.

The **Pierson and Moskowitz** spectrum is given as (Pierson and Moskowitz, 1964)

$$S_\eta(f) = \frac{\alpha g^2}{(2\pi)^4 f^5} \exp\left(-\frac{B}{f^4}\right) \qquad (7.23)$$

in which $\alpha = 0.0081$, g is the acceleration due to gravity. The quantity B is given by

$$B = 0.74\left(\frac{g}{2\pi U_w}\right)^4 \qquad (7.24)$$

in which U_w is the wind speed. It can be seen easily that the peak frequency lies at

$$f_0 = \left(\frac{4}{5} \times 0.74\right)^{1/4} \frac{1}{2\pi}\frac{g}{U_w} = 0.14\frac{g}{U_w} \qquad (7.25)$$

From Eqs. 7.23 - 7.25, the Pierson-Moskowitz spectrum may be written alternatively in the following normalized form:

$$\frac{S_\eta(f)}{S_\eta(f_0)} = \left(\frac{f}{f_0}\right)^{-5} \exp\left\{1.25\left[1 - \left(\frac{f}{f_0}\right)^{-4}\right]\right\} \tag{7.26}$$

The **JONSWAP** spectrum model has been developed by Hasselmann et al. (1973) during the course of the Joint North Sea Wave Project. This spectrum is an extension of the Pierson-Moskowitz spectrum to give room for much sharper peaks. The JONSWAP spectrum is given by

$$S_\eta(f) = \frac{\alpha g^2}{(2\pi)^4 f^5} \exp\left(-\frac{5}{4}\left(\frac{f}{f_0}\right)^{-4}\right)\gamma^a \tag{7.27}$$

Here f_0 is the peak frequency which is given by

$$f_0 = x_0^{-0.33}\frac{g}{U_w} \tag{7.28}$$

in which x_0 is the fetch parameter:

$$x_0 = \frac{gx}{U_w^2} \tag{7.29}$$

where x is the fetch over which the wind blows (x in length units). Also in Eq. 7.27, α is given by

$$\alpha = 0.076\, x_0^{-0.22} \tag{7.30}$$

α is taken 0.0081 (the same as in the Pierson-Moskowitz spectrum) when the fetch x is unknown. The power a in Eq. 7.27, on the other hand, is given by

$$a = \exp\left[-\frac{(f - f_0)^2}{2\sigma^2 f_0^2}\right] \tag{7.31}$$

in which

$$\sigma = \begin{cases} 0.07 & \text{for} \quad f \le f_0 \\ 0.09 & \text{for} \quad f > f_0 \end{cases} \tag{7.32}$$

The quantity γ in Eq. 7.27 is actually the ratio of the maximum spectral density to that of the corresponding Pierson-Moskowitz spectrum. The mean value of this quantity has been found to be 3.3, varying from 1 to 7. However, the value 3.3 has been recommended for all practical purposes.

The JONSWAP spectrum has been found to be the best approximation to measured spectra in the North Sea. Fig. 7.11 depicts a measured spectrum in the North Sea together with the Pierson-Moskowitz and JONSWAP spectra. As seen, the JONSWAP spectrum represents the measured spectrum better.

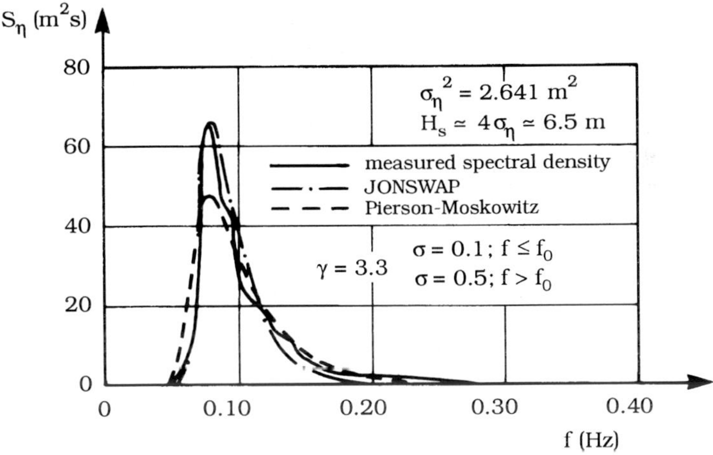

Figure 7.11   Typical wave spectrum measured during a storm in the North Sea. The values shown are special values for the Danish North Sea area. DIF (1984).

Effect of superimposed current on wave spectrum

It is known that when a wave encounters a current, the wave characteristics change. If the current is in the direction of wave propagation, the wave amplitude decreases and the wave length increases, while if it is in the opposite direction, the inverse is true (Longuet-Higgins and Stewart, 1961). In an irregular wave field, the wave characteristics experience similar changes resulting in the modification of frequency and wave-number spectra (Huang, Chen, Tung and Smith (1972) and Tung and Huang (1973)). Huang et al. showed that, under the action of a steady current, the wave spectrum is modified to

$$S_\eta^*(f) = \frac{4}{\alpha_*(1 + \alpha_*)^2} S_\eta(f) \tag{7.33}$$

in which α_* is defined as

$$\alpha_*^2 = 1 + \frac{8\pi U_f}{g} \tag{7.34}$$

and $S_\eta(f)$ is the wave spectrum in the absence of the current. Fig. 7.12 depicts the spectra for three different values of U, illustrating the effect of the steady current on the wave spectrum.

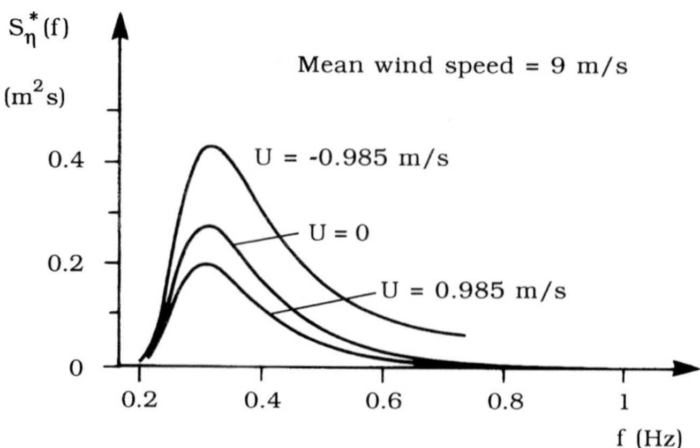

Figure 7.12 Effect of superimposed steady current on wave spectrum. Plus sign for U velocity indicates that the current is in the direction of wave propagation and the minus sign is the opposite. Tung and Huang (1973).

Directional spectrum function

So far, we have seen the spectra related to long crested, two-dimensional irregular waves. The wind-generated waves in the real sea, however, are generally three-dimensional. Fig. 7.13 illustrates how two-dimensional regular waves with different frequencies and directions may generate a three-dimensional wave pattern.

It may be deduced from the preceding figure that a three-dimensional wave in the real sea may be considered to be composed of an infinite number of regular waves with different frequencies and different directions. The wave spectrum in this case will be a function of not only the wave frequency f, but also the wave direction θ, (Fig. 7.13). The mean wave energy in this generalized case will be

$$\sigma_\eta^2 = \int\limits_0^\infty \int\limits_{-\pi}^{\pi} S_\eta(f, \theta) d\theta df \tag{7.35}$$

This is an extension of the previously given relation for one-directional spectrum function $S_\eta(f)$ in Eq. 7.13 to the three-dimensional spectrum. Fig. 7.14 schematically illustrates this three-dimensional spectrum function. The quantity $S_\eta(f, \theta) df d\theta$ represents the contribution to the total wave energy in a three-directional sea from waves with frequency and direction in the small rectangle $df\, d\theta$, centered at (f, θ).

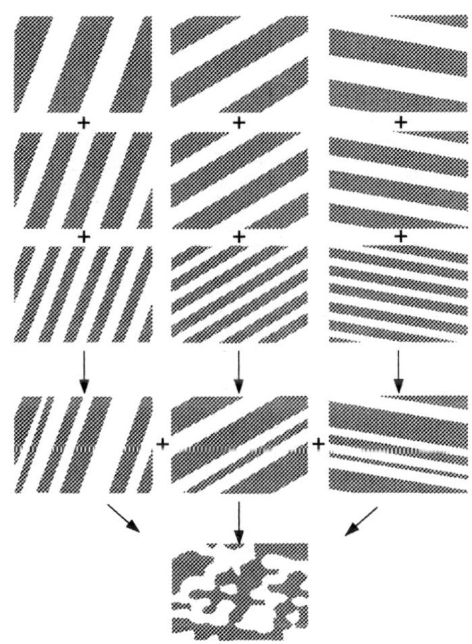

Figure 7.13 Superposition of 9 regular waves with different frequencies and directions (by courtesy of C. Aage of the Inst. of Ocean Engrg., Techn. Univ. of Denmark).

The three-dimensional spectrum may be described by the product of $S_\eta(f)$ and a spreading function $D(f,\theta)$:

$$S_\eta(f,\ \theta) = S_\eta(f)D(f,\theta) \tag{7.36}$$

$D(f,\ \theta)$ is called the **directional spectrum.**

From Eqs. 7.13 and 7.35, the relation between $S_\eta(f)$ and $S_\eta(f,\ \theta)$ is found as

$$S_\eta(f) = \int_{-\pi}^{\pi} S_\eta(f,\theta)\ d\theta \tag{7.37}$$

Furthermore, from Eqs. 7.36 and 7.37, it is seen that the spreading function $D(f,\theta)$ should satisfy the following equation:

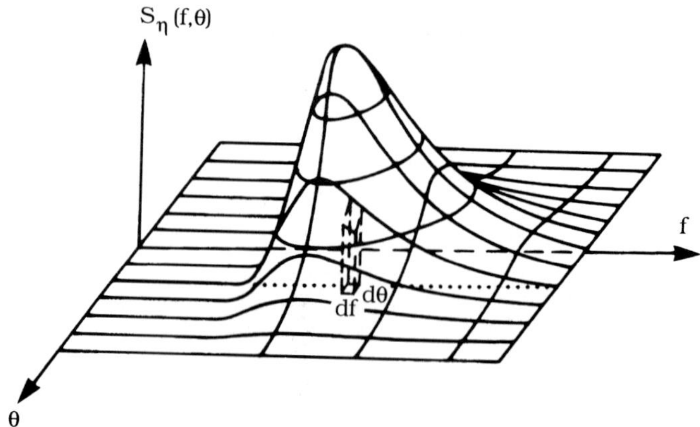

Figure 7.14 Schematic representation of a three-dimensional wave spectrum.

$$\int_{-\pi}^{\pi} D(f,\theta)d\theta = 1 \qquad (7.38)$$

Regarding the explicit form of the function $D(f,\ \theta)$, various expressions have been proposed (see, for example, the reviews given in Sarpkaya and Isaacson (1981) and Sand (1979)). The so-called cosine-power distribution, for example, is one of the expressions used for $D(f,\theta)$:

$$D(f,\theta) = \frac{2^{2s-1}}{\pi} \frac{\Gamma^2(s+1)}{\Gamma(2s+1)} \cos^{2s}\left[\frac{1}{2}(\theta - \theta_m)\right] \qquad (7.39)$$

in which Γ is the Gamma function (Abramowitz and Stegun, 1965, p. 253), θ_m is the mean value of θ and s is a parameter characterizing the degree of spread of θ around the mean. Fig. 7.15 depicts $D(f,\theta)$ as function of θ with various values of the width parameter s.

7.1.2 Statistical properties of wave height

Like the surface elevation η, also the wave height H (the height between the successive crests and troughs, Fig. 7.1) is a random variable. Longuet-Higgins (1952), for a narrow-band Gaussian wave (Eq. 7.3), has shown that the p.d.f. of wave height is given by the Rayleigh distribution:

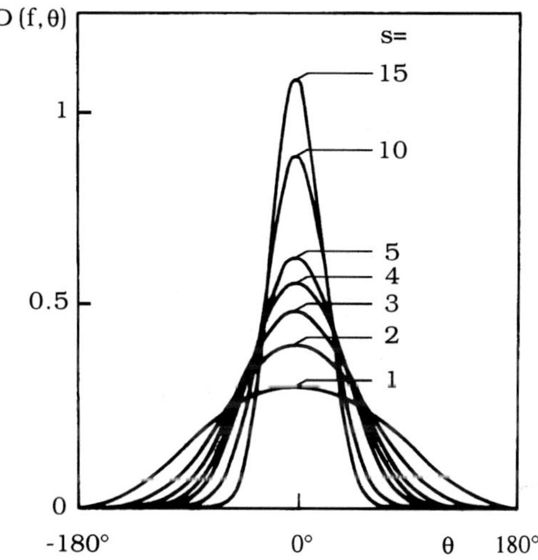

Figure 7.15 Spreading function D, (taken from Sand, 1979).

$$p(H) = \frac{2H}{H_{\rm rms}^2} \exp\left(-\frac{H^2}{H_{\rm rms}^2}\right) \tag{7.40}$$

in which $H_{\rm rms}$ is the root-mean-square of the wave heights:

$$H_{\rm rms} = \overline{H^2} \tag{7.41}$$

As will be seen in the following, the r.m.s.-value of H is related to σ_η, the standard deviation of surface elevation (Eqs. 7.5 and 7.19). In a narrow-band irregular wave, each wave may be approximated by a single sinusoidal wave. Therefore

$$\sigma_\eta^2 = m_0 = \lim_{T \to \infty} \frac{1}{T} \int_0^T \eta^2 dt =$$

$$= \lim_{N \to \infty} \frac{1}{N} \sum_{i=1}^{N} \frac{1}{2\pi} \int_0^{2\pi} \left(\frac{H_i}{2}\right)^2 \sin^2(\omega_i t) d(\omega_i t)$$

$$= \frac{1}{8} \lim_{N \to \infty} \frac{1}{N} \sum_{i=1}^{N} H_i^2 = \frac{1}{8} H_{\rm rms}^2 \tag{7.42}$$

Hence

$$H_{\rm rms} = 2\sqrt{2}\,\sigma_\eta = 2\sqrt{2}\sqrt{m_0} \tag{7.43}$$

Fig. 7.16 shows the p.d.f. of the wave data referred to earlier in Fig. 7.1 along with the Rayleigh distribution.

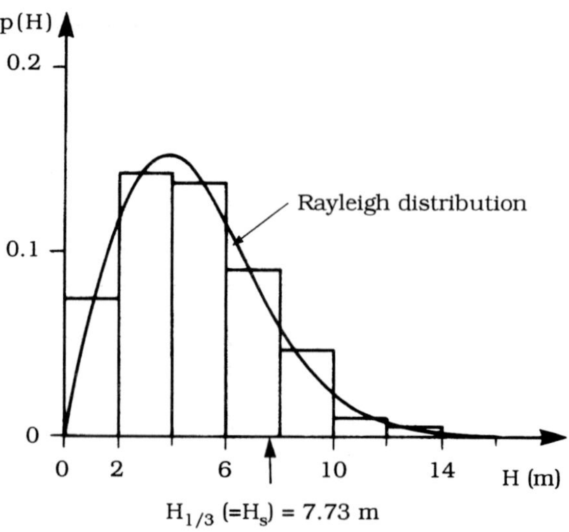

Figure 7.16 Probability density function of wave height for the wave record referred to in Fig. 7.1. Sample size = 845 waves. $\overline{H}$ = 4.77 m. $H_{\rm rms.}$ = 5.43 m. H_s = 7.73 m. H_m = 14.9 m (actual) and 14.2 m (calculated from Eq. 7.49).

One of the most widely used statistical property of H is the **significant wave height**, denoted by H_s or $H_{1/3}$. This quantity is defined as the average of the highest one-third of all waves.

Let H_0 be defined such that

$$Pr\left[H > H_0\right] = \frac{1}{3} \tag{7.44}$$

Hence, the significant wave height $H_{1/3}$ will be the average of the H-values which satisfy Eq. 7.44. This average can be calculated by

$$H_{1/3} = \frac{\int\limits_{H_0}^{\infty} H\, p(H)\,dH}{\int\limits_{H_0}^{\infty} p(H)\,dH} \qquad (7.45)$$

If H is Rayleigh distributed, then H_0 and subsequently the significant wave height $H_{1/3}$ may be calculated easily from Eqs. 7.40, 7.44 and 7.45. The results are

$$H_0 = 1.05\, H_{\mathrm{rms}} \qquad (7.46)$$

and

$$H_{1/3} = H_s = 1.42\, H_{\mathrm{rms}} \qquad (7.47)$$

The significant wave height, in terms of standard deviation of surface elevation, can then be expressed from Eq. 7.43 as

$$H_{1/3} = 4\sigma_\eta = 4\sqrt{m_0} \qquad (7.48)$$

The predicted values of $H_{1/3}(= H_s)$ and σ_η for the example depicted in Fig. 7.10 give $H_{1/3}/\sigma_\eta = 7.73$ m/1.94 m $\cong 4$, revealing the preceding relationship.

The maximum wave height in a wave record was expressed by Longuet-Higgins (1952) in terms of significant wave height $H_{1/3}$ and the number of waves N occurring during the record:

$$\frac{H_m}{H_{1/3}} = \sqrt{\frac{1}{2}\ln N} \qquad (7.49)$$

For example, for $N = 2000$ waves, $H_m \cong 2\, H_{1/3}$.

Finally, it may be noted that Kriebel and Dawson (1993) has developed a theoretical model to account for 1) the *non-linear* increase in the highest wave crests, and 2) the selective reduction of some fraction of these high crests due to *wave breaking*. The model has been verified, using several sets of laboratory data for severe breaking seas which have approximate JONSWAP wave spectra.

7.1.3   Statistical properties of wave period

The wave period T is defined as the time interval between the successive zero-upcrossings of surface elevation (Fig. 7.1). The probability density function of T is given by Longuet-Higgins (1975) on the hypothesis that the sea surface is Gaussian and that the wave spectra is sufficiently narrow:

$$p(\tau) = \frac{1}{2}\left(1 + \tau^2\right)^{-3/2} \tag{7.50}$$

in which τ is the normalized wave period defined by

$$\tau = \frac{T - \overline{T}}{\nu \overline{T}} \tag{7.51}$$

$\overline{T}$ is the mean period defined by

$$\overline{T} = \frac{m_0}{m_1} \tag{7.52}$$

and ν is

$$\nu = \left(\frac{m_2}{m_0}\right)^{1/2} \overline{T} \tag{7.53}$$

which can be interpreted as a parameter describing the width of the wave spectrums, similar to ε introduced earlier (Eq. 7.20). Here m_0, m_1, and m_2 are the spectral moments defined in Eq. 7.18. The distribution given by Eq. 7.50 is a bell-shaped curve (which is not Gaussian).

Table 7.1 Various wave-period definitions in irregular waves.

Mean zero-upcrossing period	T_z	Mean period between successive zero upcrossings.
Mean crest period	T_c	Mean period between successive crests.
Mean period	$\overline{T}$	$\overline{T} = m_0/m_1$, or $\overline{T} = \sqrt{m_0/m_2}$ m_0, m_1 and m_2: the zeroth, first and second spectral moments, respectively.
Peak period	T_0	$T_0 = 1/f_0$ f_0: the peak frequency.
Significant wave period	T_s	Mean period of the highest one-third of all waves.

Regarding the simplest statistical property of the wave period, namely the mean wave period, there are several definitions. These are summarized in Table 7.1. For the example given in Fig. 7.10, the various mean periods found from the relationships given in Table 7.1 are indicated in the caption of Fig. 7.10. The presented values imply the following relationships:

$$T_z = T_c = 0.65\,T_0, \quad \overline{T} = 0.74\,T_0, \quad T_s = 0.92\,T_0 \tag{7.54}$$

If the so-called Bretschneider spectrum is used, the numerical factors in the preceding relationships will be 0.71; 0.77; and 0.95, respectively (Sarpkaya and Isaacson, 1981, p. 515).

Finally, the joint probability density function of wave height and wave periods is given by Longuet-Higgines (1975) as (again, under the hypothesis that the sea surface is Gaussian and that the wave spectrum is sufficiently narrow):

$$p(\xi,\,\tau) = \frac{\xi^2}{\sqrt{2\pi}}\,\exp\left[-\frac{1}{2}\xi^2\left(1+\tau^2\right)\right] \tag{7.55}$$

in which

$$\xi = \frac{a}{\sqrt{m_0}} = \frac{\sqrt{2}(2a)}{\sqrt{22}\sqrt{m_0}} = \frac{\sqrt{2}H}{H_{\rm rms}} \tag{7.56}$$

Fig. 7.17 gives this joint probability as a contour plot.

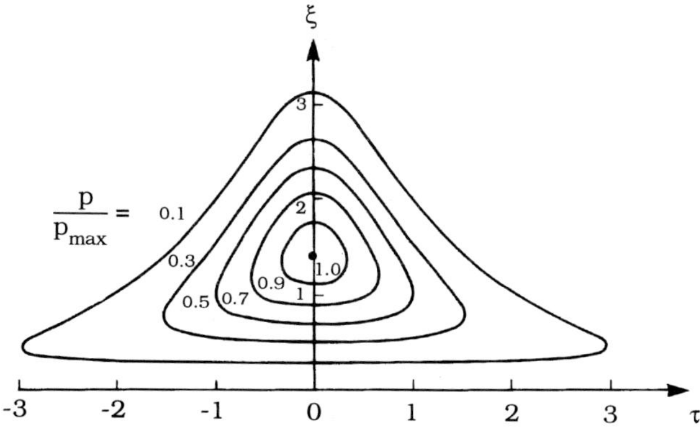

Figure 7.17 Contours of the function $p(\xi,\,\eta) = (2\pi)^{-1/2}\xi^2\,\exp[-\xi^2(1+\tau^2)/2]$ giving the joint probability density of the normalized wave amplitude and wave period. Longuet-Higgins (1975).

7.1.4 Long-term wave statistics

As stated earlier, the wave climate at a given location may be characterized by a series of short-term sea states.

It has been seen in the preceding paragraphes that each sea state may be characterized by a representative wave height (the significant wave height, for example), a representative wave period (the mean period, or the peak period, for example), and additionally perhaps by the mean direction of wave propagation, the wave spectrum and the directional spreading function. Such data may be obtained by direct measurements at a location over some period of time (a few years). This kind of data may also be obtained by wave hindcasting where the evolution of the wave spectrum is traced, by solving the equation of conservation of wave energy numerically, taking into consideration the energy input from the wind, the energy transfer due to wave/wave interactions and the energy dissipation due to wave breaking and bottom friction (see, for example, Abdalla and Özhan, 1993).

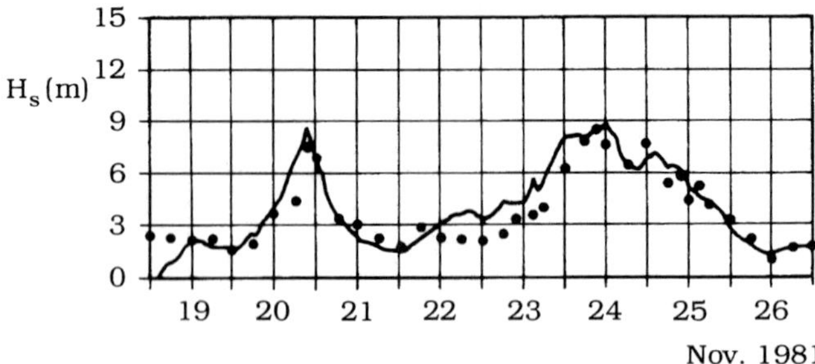

Figure 7.18 Time series of sea states, characterized by the significant wave height, for the location Gorm Field in the Danish sector of the North Sea. Dots: measured. Solid line: computed, using a wave hindcasting model. Abbott (1991).

Fig. 7.18 illustrates the variation of successive sea states, characterized by the significant wave height, over a period of 8 days. The figure indicates quite clearly that the variation of sea states over the long term (tens of years perhaps, in contrast to a few hours) must be taken into consideration to make reliable estimates of the properties of the design wave such as the extreme significant wave height corresponding to a specified return period (50 years, for example) and its associated period. There is a vast amount of work in literature related to the estimate of long-term extreme values. The following references may be consulted for the details of the various methods regarding the long-term wave statistics: Ochi (1981), Isaacson and MacKenzie (1981), Hansen (1981), Muir and El-Shaarawi (1986), Chakrabarti (1987).

As an example, the so-called Weibull distribution is given below. The probability of exceedence of a significant wave height H is according to this distribution,

$$P(H) = \exp\left[-\left(\frac{H-\gamma}{\beta}\right)^m\right] \tag{7.57}$$

in which m is the shape parameter and may be put equal to a value in the range 0.75 to 2.0. The other parameters β and γ are determined from measurements, using the least-square method. This distribution, when plotted on a diagram with scales $x = \ln(H-\gamma)$ and $y = \ln\left[-\ln[P(H)]\right]$, appears as a straight line.

The return period T_R, on the other hand, is given in terms of P as follows:

$$\frac{T_R}{r} = \frac{1}{P} \tag{7.58}$$

where r is the average duration between successive data points, and the encounter probability E of the corresponding wave event occurring during a specified duration L (such as the design life of the structure) is given approximately by (see Isaacson, 1988):

$$E = 1 - \exp(-L/T_R) \tag{7.59}$$

The annual probability of exceedence e is obtained by setting $L = 1$:

$$e = 1 - \exp(-1/T_R) \tag{7.60}$$

in which T_R is in years.

7.2 Forces on cylinders in irregular waves

7.2.1 Force coefficients

In the past, the majority of the work dealing with the effect of irregular waves on forces has considered the actual wave environments (Wiegel et al. (1957), Borgman (1965, 1967, 1972), Jothi Shankar et al. (1987)). This obviously brings into the picture some additional effects such as wave non-linearity and wave asymmetry. One way of eliminating these additional effects is to experiment with a random oscillatory flow, either generated in a water tunnel or simulated by the motion of a carriage in an otherwise still water.

Longoria et al. (1991) present the results of such tests made in a water tunnel with random oscillatory motion.

The drag and *inertia coefficients* were determined for *each cycle* of data, by the least-squares fit of the force time series (cf. Section 4.1.4 under measurements of C_D and C_M coefficients). The results of a typical test are plotted in Fig. 7.19 along with the results obtained in a sinusoidally oscillating flow under corresponding conditions in the same oscillatory water tunnel.

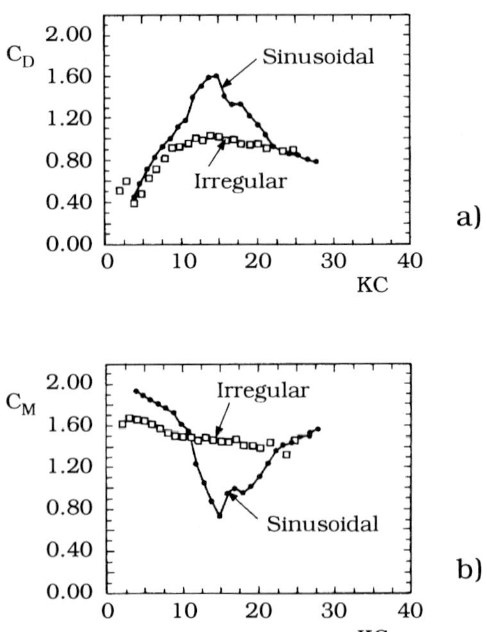

Figure 7.19 Inertia and drag coefficients measured under sinusoidal and irregular (random) oscillatory flow conditions. For sinusoidal flow experiments, $\beta(= Re/KC) = 2323$. For irregular, oscillatory flow conditions the β_r-value is $\beta_r(= Re_r/KC_r) = 2348$, KC_r-value is $KC_r = 8.6$, and q-value is $q = 0.27$. Longoria et al. (1991).

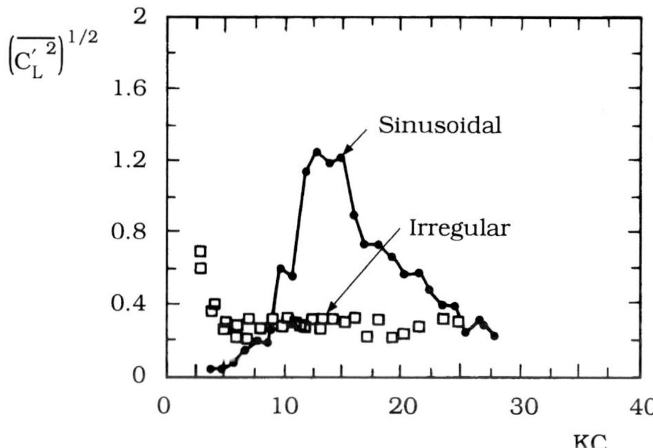

Figure 7.20 Root-mean-square lift force coefficient measured under random flow conditions compared with that obtained in the case of corresponding sinusoidal flow. For flow conditions, see the caption of Fig. 7.19. Longoria et al. (1991).

The nomenclature used in the figure and in the figure caption is as follows. The quantity β_r is the β-value for random flow, defined by

$$\beta_r = \frac{Re_r}{KC_r} = \frac{D^2}{\nu T_z} \tag{7.61}$$

in which T_Z is the mean zero-upcrossing period of the random motion (Table 7.1). Regarding KC_r and Re_r, the irregular-wave results presented in the figure belong to a single random oscillatory-flow test; therefore it is possible to define a representative KC number and a representative Re number, which are statistical analogs of KC and Re:

$$KC_r = \frac{\left(\sqrt{2}\sigma_U\right)T_z}{D}, \quad Re_r = \frac{\left(\sqrt{2}\sigma_U\right)D}{\nu} \tag{7.62}$$

in which σ_U is the measured root-mean-square value of the fluid velocity U for the total length of the random-flow-test velocity record considered. The quantity $\sqrt{2}\sigma_U$ may be regarded as a representative value for the velocity amplitude for random motion, recalling that $\sqrt{2}\sigma_U$ is identical to U_m for sinusoidal flows. Finally, the quantity q is defined by Eq. 7.22. m_0, m_1 and m_2 in the present context are the zeroth, the first and the second spectral moments of velocity $U(t)$ defined in the same fashion as in Eq. 7.18; the quantity q characterizes the width of the velocity spectrum. Regarding the input spectrum in Longoria et al.'s experiments, they used a normalized form of a Pierson-Moskowitz wave-height spectrum (Eq. 7.23) as the control spectrum shape for the flow acceleration spectrum in the water tunnel. It may be noted that the previously mentioned values of KC_r, β_r (or Re_r) and q may be considered to define a design sea state.

Regarding the *lift force coefficient*, the corresponding results of Longoria et al.'s tests are plotted in Fig. 7.20. As seen from the figures (Figs. 7.19 and 7.20), the force coefficients differ significantly under sinusoidal and irregular wave conditions. This is attributed to the tremendous changes in the vortex-flow regimes in the case of random oscillatory flow, as will be shown in the following example.

Example 7.1:   Vortex-flow regimes in random oscillatory flow

Sumer and Kozakiewicz (1995) made a visualization study of flow around a cylinder in a random oscillatory flow. Three kinds of tests were conducted: 1) the regular, sinusoidal oscillatory flow tests, 2) the random oscillatory flow tests with a narrow-band spectrum and, 3) that with a broad band spectrum. A JONSWAP type spectrum was used in the study.

Sumer and Kozakiewicz grouped the regular, sinusoidal oscillatory-flow vortex regimes (observed by Williamson (1985) and described in Chapter 3, Sections 3.1 and 3.2) into three fundamental classes: 1) the *vortex-pair* regime which occurs in the range $O(1) < KC < 7$, 2) the *transverse-vortex-street* regime which occurs in the range $7 < KC < 15$, and 3) the *vortex-street* regime which occurs for $KC > 15$. Note that for $KC < O(1)$, the flow is unseparated (Chapter 3, Section 3.1 and 3.3). These vortex-flow regimes are depicted in Fig. 7.21 for easy reference.

Fig. 7.22 illustrates how much the previously-mentioned vortex-flow regimes are disrupted when the flow changes from regular, sinusoidal oscillatory flow to random oscillatory flow, the degree of irregularity of the random oscillatory flow increasing with increasing ε. In Fig. 7.22, the ordinate, p, is the frequency of occurrence of the fundamental vortex regimes. From Fig. 7.22, the following conclusions can be drawn.

First, for $KC_r = 3$ (Fig. 7.22a), the regular oscillatory flow vortex regime, namely the vortex-pair regime, appears not to be influenced much by the irregularity of the oscillatory flow. Although the transverse-vortex-street regime occurs occasionally, the frequency of occurrence is rather small, around 4%.

Second, for $KC_r = 10$ (Fig. 7.22b), the frequency of occurrence, p, of the regular oscillatory flow vortex regime (i.e., the transverse-vortex-street regime) is reduced drastically with increasing ε. While p is 100% for $\varepsilon = 0$, it reduces to 37% for $\varepsilon = 0.25$, and to only 10% for $\varepsilon = 0.56$.

Third, likewise, for $KC_r = 20$, the regular oscillatory flow vortex regime (i.e., the vortex-street regime) undergoes similar changes. While the frequency of

Regime of vortex motion	Pattern	KC range
Vortex pair		O(1)<KC≤ 7
Transverse vortex street	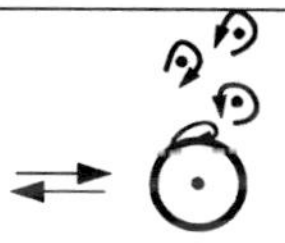	7<KC≤ 15
Vortex street		KC>15

Figure 7.21 Classification of vortex-flow regimes in *regular sinusoidal oscillatory flow.*

occurrence, p, of this regime is 100% for $\varepsilon = 0$, it drops to 52% for $\varepsilon = 0.25$, and to 24% for $\varepsilon = 0.56$.

Fourth, it is evident that, for $KC_r = 10$ and 20, the regular vortex regimes undergo quite a substantial amount of disruption under irregular oscillatory flow conditions. This is explained by Sumer and Kozakiewicz as follows.

A regular vortex-flow regime for $KC > 7$ (for example, the transverse vortex street) is actually a product of regular, repeatable interaction between vortices of two successive half periods in sinusoidal flows (Section 3.2, Figs. 3.6 and 3.7). In the case of irregular oscillatory flows, however, this interaction is partially or at times nearly completely prevented due to the randomly changing successive half periods of the motion. This would presumably result in the disruption of the regular vortex regimes.

The reason why the regular vortex regime in the case of $KC_r = 3$ (i.e., the vortex-pair regime) is disrupted only very little under irregular oscillatory flow conditions is simply because the half periods in this regime are too short for this kind of disruption to occur.

For the case of broad-band spectrum ($\varepsilon = 0.56$) - the case which may be regarded as representative for irregular waves experienced in the real sea - Fig.

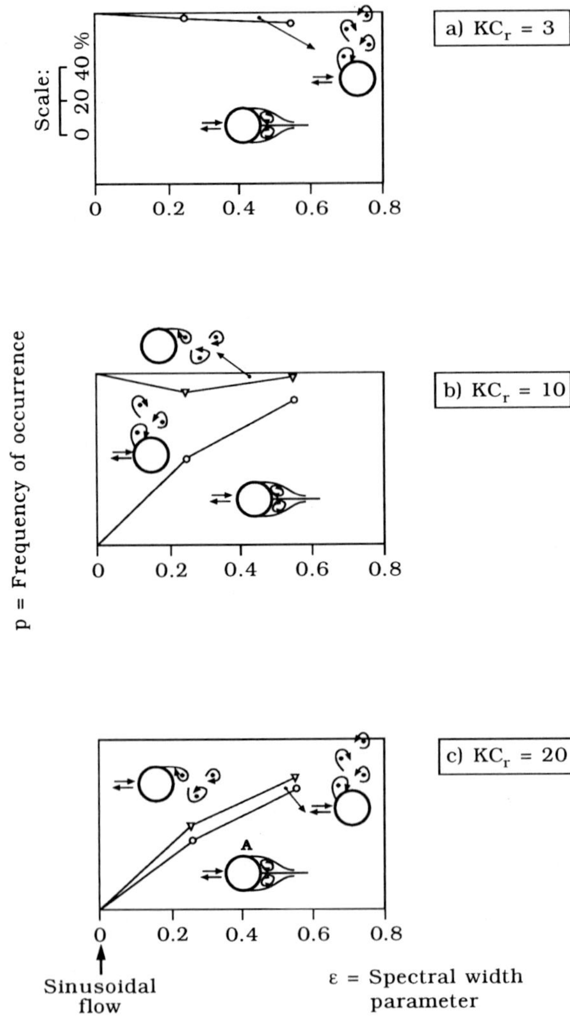

Figure 7.22 Frequency of occurence of vortex-flow regimes, p, as function of spectral width parameter, ε. The tests for $\varepsilon = 0.56$ may be regarded as representative for irregular waves experienced in the real sea with JONSWAP wave spectrum.

7.22 shows that the vortex-pair regime dominates, regardless of the KC number. The frequency of occurrence of this regime is 96% in the case of $KC_r = 3$ (Fig. 7.22a), 88% in the case of $KC_r = 10$ (Fig. 7.22b), and 72% in the case of $KC_r = 20$ (Fig. 7.22c). Sumer and Kozakiewicz emphasizes that in many half periods, the flow resembles the impulsively started cylinder flow where a symmetrical pair of vortices is formed in the wake of the cylinder at the initial phase of its motion.

Implications with regard to force coefficients

Longoria et al.'s (1991) results (Fig. 7.19) show that, in contrast to the regular oscillatory-flow case, the in-line force coefficients, in the case of random oscillatory flow, are maintained roughly constant over the measured range of temporal KC number $O(1) < KC < 30$ with the exception that C_D experiences some variation with KC over $O(1) < KC < 10$. (This variation in C_D may not be very significant, since the contribution of the drag force to the total in-line force in this range of KC is rather small. In some cases however, such as in the calculation of damping forces for resonant vibrations of structures, the drag contribution may become important). Likewise, the lift-force coefficient (Fig. 7.20) is maintained roughly constant over the KC range, $O(5) < KC < 30$. The force coefficients do not change over the measured KC range simply because the same flow regime, namely the vortex-pair regime, predominantly prevails over this range of KC numbers (Fig. 7.22b and 7.22c at $\varepsilon = 0.56$), as described in the preceding paragraphs.

7.2.2 Force spectra

The spectral analysis of the force components may be important, when the distribution of various frequency content of forces is considered. The information on force spectra may be needed, for example, for the estimation of vibration of offshore structures under irregular waves.

The spectral data given in Figs. 7.23a and b are taken from Longoria et al. (1991). As mentioned in the preceding paragraphes, Longoria et al. used a normalized form of a Pierson-Moskowitz spectrum as the control spectrum shape for the flow acceleration spectrum. The frequency in the spectral representation of Fig. 7.23 is normalized by the frequency f_0, the peak frequency in the velocity spectrum S_U.

From Fig. 7.23, the following two points may be noted. First, the in-line force spectrum closely follows the velocity spectrum. Specifically, the peak frequency is equal to that of the velocity spectrum. Second, the lift-force spectrum behaves, however, in an entirely different way; the peak frequency, f/f_0, is 2 (Fig.

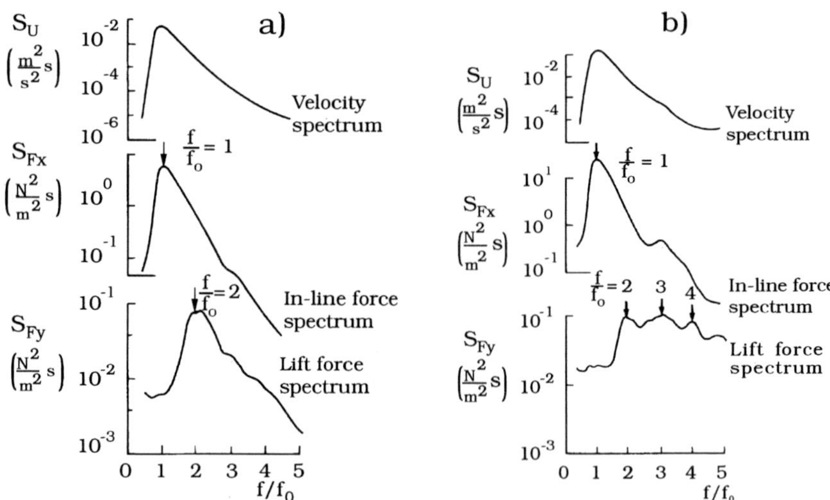

Figure 7.23 Spectral distributions. (a): $KC_r = 6.8$, $Re_r = 8200$, $\beta_r = 1204$, and $q = 0.27$. (b): $KC_r = 11.6$, $Re_r = 14400$, $\beta_r = 1241$, and $q = 0.28$. The data are from random-flow experiments in a water tunnel (Longoria et al., 1991).

7.23a) or larger (Fig. 7.23b), and the spectrum may be a narrow-band spectrum (Fig. 7.23a) or a broad-band spectrum (Fig. 7.23b).

Regarding the shape of the force spectra, Borgman (1967), for a *cylindrical pile* exposed to irregular waves, developed *a linear model* of the wave in-line force on the pile, using the Morison equation with constant drag and inertia coefficients with the values of velocity and acceleration obtained from the linear wave theory. The force spectrum according to Borgman's model may be expressed as in the following

$$S_{Fx}(f) = \frac{8}{\pi} K_d^2 \sigma_U^2 S_U(f) + K_i^2 S_a(f) \tag{7.63}$$

in which K_d and K_i are related to the Morison coefficients C_D and C_M as follows:

$$K_d = \frac{1}{2}\rho C_D D \quad \text{and} \quad K_i = \rho C_M A \tag{7.64}$$

σ_U in Eq. 7.63 is the root-mean-square value of the velocity induced by waves:

$$\sigma_U^2 = \int_0^\infty S_U(f)\,df \qquad (7.65)$$

Subsequently, Borgman related S_{Fx} to the wave spectrum S_η, the spectrum of water surface elevation (see Eq. 7.13), using the linear wave theory:

$$S_U(f) = (2\pi f)^2 \frac{\cosh^2\big(k(h+z)\big)}{\sinh^2(kh)} S_\eta(f) \qquad (7.66)$$

$$S_a(f) = (2\pi f)^4 \frac{\cosh^2\big(k(h+z)\big)}{\sinh^2(kh)} S_\eta(f) \qquad (7.67)$$

in which h is the water depth, z is the vertical coordinate measured from the mean water level upwards, and k is the wave number, which is related to f by the dispersion relation (Appendix III):

$$(2\pi f)^2 = gk\,\tanh(kh)$$

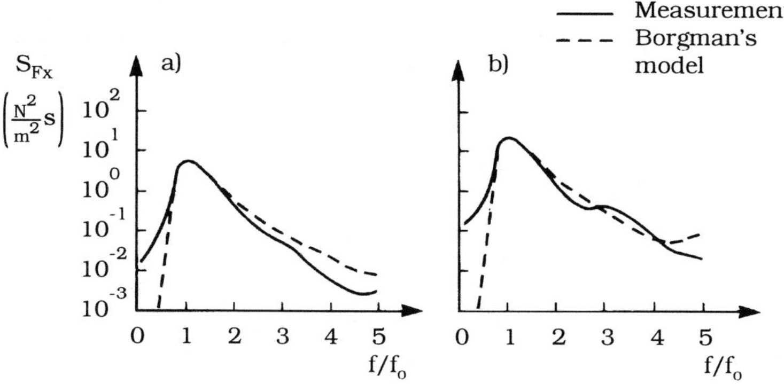

Figure 7.24 Comparison of in-line force spectrum obtained from Borgman's linear model (Eq. 7.63) with measurements of Longoria et al. (1993). Test conditions in (a) and (b) above are the same as in Figs. 7.23a and b, respectively. Taken from Longoria et al. (1993).

Fig. 7.24 compares the spectra obtained from Borgman's linear model in Eq. 7.63 with the measured in-line force spectra in the study of Longoria et al. (1991, 1993) presented in Figs. 7.23a and 7.23b. As is seen, the Borgman model represents the measured spectra quite well in the frequency range where most of the energy is concentrated. Outside this range the Borgman model underestimates the spectrum at the lower end and overestimates it at the upper end of the spectrum. To improve the accuracy of the in-line force spectrum function over all frequencies, Longoria et al. (1993) has developed a non-linear model of the in-line force, using the so-called two-input/single-output model. Apparently, the model has proven to be quite effective in illustrating the contribution of both the inertia and drag components as function of frequency.

Information on other statistical properties of the in-line force such as the probability density function and the autocorrelation function can be found in Borgman (1965 and 1972).

7.2.3 Forces on pipelines in irregular waves

Fig. 7.25 presents the results of the experiments carried out with random oscillatory flows by Bryndum, Jacobsen and Tsahalis (1992). The pipe was a bottom-mounted pipe. The flow in Bryndum et al.'s tests was generated with the carriage technique. The in-line force coefficients were derived from the force time-series data by use of the least-squares-fit method which was applied for the *full length* of the test record, rather than on a cycle-to-cycle basis. The lift-force coefficient, on the other hand, was obtained by using the least-square fit of the measured lift force time series to the time series of the lift force predicted by the following equation

$$F_L(t) = \frac{1}{2}\rho D \, C_L \, U^2(t), \qquad (7.68)$$

for the *full length* of the test record. The Keulegan-Carpenter number was based on the "significant" velocity and the peak period of the velocity spectrum.

Apparently, the difference between the regular-wave results and the irregular-wave ones is not very large as regards the in-line force coefficients. The influence of irregular waves on the lift even appears to be nil (Fig. 7.25c).

We have seen in the preceding sections that, in the case of wall-free cylinder, the force coefficients in irregular waves differ significantly from those in regular waves because the various vortex-shedding regimes which exist for sinusoidal flows are disrupted in irregular waves. However, in the case of bottom-mounted cylinder, the previously mentioned vortex regimes do not exist at all. There is only one single regime (regardless of the KC number) in which a lee-wake vortex is formed behind the pipe for each half period of the motion (Chapter 3, Section 3.4). This flow regime will clearly be there no matter whether the pipe is exposed to regular waves or to irregular waves. Therefore, the force on the pipe will not undergo any significant change when the waves change from regular to irregular.

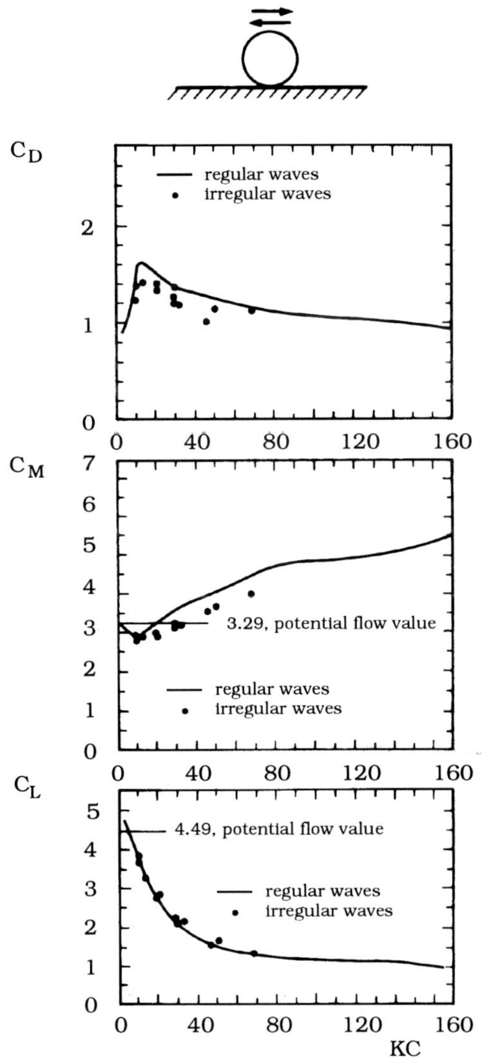

Figure 7.25 Force coefficients versus KC for irregular wave flow for a bottom-mounted cylinder. Re is in the range $0.7 - 2.5 \times 10^5$. The pipe roughness is $k/D = 10^{-3}$. Bryndum et al. (1992).

7.2.4 Forces on vertical cylinders in directional irregular waves

Høgedal, Skourup and Burcharth (1994) (also see Høgedal, 1993) made a systematic experimental investigation of the effect of the wave directionality on the wave forces, local and depth-averaged, on a vertical smooth cylinder.

In the experiments the Reynolds number, Re, and the Keulegan-Carpenter number, KC, were in the following ranges: $1 \cdot 10^4 < Re < 5 \cdot 10^4$ and $2 < KC < 35$. In the 3-D wave field the $\cos^{2s}$ spreading function was applied. The spreading parameter, s, was chosen to be either a constant or a function of frequency. The latter has previously been shown to resemble the directional spreading of waves in the North Sea. The standard deviation of the spreading function, σ_θ, was in the 3-D waves in the range: $22° < \sigma_\theta < 57°$.

Analyses of the measured wave forces, local and depth-integrated, showed a reduction of the extreme resultant and in-line forces in 3-D waves compared to 2-D waves with equal spectral properties, when identical probabilities of non-exceedence were considered. The reduction of the local wave forces strongly depends on the local ratio between the drag and inertia term in Morison's equation and on the degree of directional spreading of the 3-D wave field. In the experiments the resultant wave forces were reduced up to 20% below mean water level (MWL) and up to 50% above MWL; the latter figure is the value of reduction measured just below the wave crest.

The measured reduction of the wave loads agrees with similar experimental results presented by Aage, Jorgensen, Andersen, Dahl and Klinting (1989).

REFERENCES

Aage, C., Jorgensen, P., Andersen, L.W., Dahl, C. and Klinting, P. (1989): Wave loads on a cylinder in 2-D and 3-D deep water waves. Proc. 8th Int. Conf. on Offshore Mechanics and Arctic Engineering, The Hague, 1989, pp 2:175-181.

Abbott, M.B. (1991): Numerical modelling for coastal and ocean engineering. In: Handbook of Coastal and Ocean Engineering, Ed. J.B. Herbich, Vol. 2, Gulf Publishing Company.

Abdalla, S. and Özhan, E. (1993): Third-Generation wind-wave model for use on personal computers. J. Waterway, Port, Coastal and Ocean Eng., ASCE, 119(1):1-14.

Borgman, L.E. (1965): Wave forces on piling for narrow-band spectra. J. Waterways and Harbors Div., ASCE, 91(WW3):65-90.

Borgman, L.E. (1967): Spectral analysis of ocean wave forces on piling. J. Waterways and Harbors Div., ASCE, 93(WW2):129-156.

Borgman, L.E. (1972): Statistical models for ocean waves and wave forces. In: Advances in Hydroscience, Ed. Ven Te Chow, Academic Press, 8:139-181.

Bryndum, M.B., Jacobsen, V. and Tsahalis, D.T. (1992): Hydrodynamic forces on pipelines: Model tests. Trans. ASME, J. Offshore Mech. and Arctic Engrg., 114:231-241.

Chakrabarti, S.K. (1987): Hydrodynamics of Offshore Structures. Computational Mechanics Publications.

DIF (1984): Pile Supported Offshore Steel Structures. Dansk Ingeniørforening's Code of Practice, DS 449.

Goda, Y. (1985): Random Seas and Design of Maritime Structures. University of Tokyo Press.

Hansen, N.-E. O. (1981): Determination of design waves for steel platforms. Progress Report No. 55, Inst. of Hydrodynamics and Hydraulic Engineering, ISVA, Techn. Univ. Denmark, pp. 31.37.

Hasselmann, K. et al. (1973): Measurements of wind-wave growth and swell decay during the joint North Sea wave project (JONSWAP). Deutsches Hydrographisches Institut, Hamburg. Ergänzungsheft zur Deutschen Hydrographischen Zeitschrift, Reihe A (8°), Nr. 12, 1973, 95 p.

Huang, N.E., Chen, D.T., Tung, C.-C. and Smith, J.R. (1972): Interactions between steady non-uniform currents and gravity waves with applications for current measurements. J. Phys. Oceanogr., 2:420-431.

Høgedal, M. (1993): Experimental Study of Wave Forces on Vertical Circular Cylinders in Long and Short Crested Sea. Ph.D.-Thesis, Hydraulics and Coastal Engineering Laboratory Department of Civil Engineering, Aalborg University, Denmark.

Høgedal, M., Skourup, J. and Burcharth, H.F. (1994): Wave forces on a vertical smooth cylinder in directional waves. ISOPE '94, Tokyo.

Isaacson, J. (1988): Wave and current forces on fixed offshore structures. Canadian Journal of Civil Eng., 15:937-947.

Isaacson, M. and MacKenzie, N.G. (1981): Long-term distributions of ocean waves - - A review. J. Waterway, Port, Coastal and Ocean Division, ASCE, 107(WW2):93-109.

Isaacson, M., Baldwin, J. and Niwinski, C. (1991): Estimation of drag and inertia coefficients from random wave data. Trans. of ASME Jour. Offshore Mech. and Arctic Engrg., 113:128-136.

Jothi Shankar, N., Cheong, H.-F., and Subbiah, K. (1987): Forces on a smooth submarine pipeline in random waves – A comparative study. Coastal Engineering, 11:189-218.

Kriebel, D.L. and Dawson, T.H. (1993): Distribution of crest amplitudes in severe seas with breaking. J. Offshore Mechanics and Arctic Engineering, ASME, 115:9-15.

Longoria, R.G., Beaman, J.J. and Miksad, R.W. (1991): An experimental investigation of forces induced on cylinders by random oscillatory flow. Trans. ASME, J. Offshore Mech. and Arctic Engrg., 113:275-285.

Longoria, R.G., Miksad, R.W. and Beaman, J.J. (1993): Frequency domain analysis of in-line forces on circular cylinders in random oscillatory flow. Trans. ASME, J. Offshore Mech. and Arctic Engrg., 115:23-30.

Longuet-Higgins, M.S. (1952): On the statistical distribution of the heights of sea waves. J. of Marine Research, XI(3):245-265.

Longuet-Higgins, M.S. (1975): On the joint distribution of the periods and amplitudes of sea waves. J. Geophys. Res., 80(18):2688-2694.

Longuet-Higgins, M.S. and Stewart, R.W. (1961): The changes in amplitude of short gravity waves on steady non-uniform currents. J. Fluid Mech., 10:529-549.

Muir, L.R. and El-Shaarawi, A.H. (1986): On the calculation of extreme wave heights: A review. Ocean Engineering, 13(1):93-118.

Ochi, M.K. (1981): Stochastic analysis and probabilistic prediction of random seas. Advances in Hydroscience, 13:217-375.

Pierson, W.J. and Moskowitz, L. (1964): A proposed spectral form for fully developed wind seas based on the similarity theory of C.A. Kitaigorodskii. J. Geophys. Res., 69(24):5181-5190.

Press, W.H., Flannery, B.P., Teukolsky, S.A. and Vetterling, W.T. (1989): Numerical Recipes (Fortran Version), Cambridge Univ. Press.

Sand, S.E. (1979): Three-dimensional deterministic structure of ocean waves. Series Paper No. 24, Ph.D.-Thesis, Inst. of Hydrodynamics and Hydraulic Engineering, ISVA, Techn. Univ. Denmark.

Sarpkaya, T. and Isaacson, M. (1981): Mechanics of Wave Forces on Offshore Structures. Van Nostrand Reinhold Company.

Southworth, R.W. (1969): Autocorrelation and spectral analysis. In: Mathematical Methods for Digital Computers, Vol. 1, (Ed. A. Ralston and H.S. Wilf), John Wiley and Sons, Inc., 1960.

Sumer, B.M. and Kozakiewicz, A. (1995): Visualization of flow around cylinders in irregular waves. Int. Journal of Offshore and Polar Engineering, 5(4):270-272. Also see: Proc. 4th Int. Offshore and Polar Engrg. Conf., Osaka, Japan, April 10-15, 1994, 3:413-420.

Tung, C.C. and Huang, N.E. (1973): Statistical properties of wave-current force. Proc. ASCE, J. Waterways, Harbors and Coastal Engineering Division, 99(WW3):341-354.

Wiegel, R.L., Beebe, K.E. and Moon, J. (1957): Ocean wave forces on circular cylindrical piles. ASCE, J. Hydraulics Div., 83(HY2):1199-1-1199-36.

Williamson, C.H.K. (1985): Sinusoidal flow relative to circular cylinders. J. Fluid Mech., 155:141-174.

Chapter 8. Flow-induced vibrations of a free cylinder in steady currents

The description of flow around and forces on fixed structures has been completed in the preceding chapters. The remainder of the book will study flow-induced vibrations of slender structures.

Flow-induced vibrations of structures in general are encountered in various fields of engineering such as aero-space industry, power generation and transmission, civil engineering, wind engineering, ocean engineering and offshore industry. Bridges, tall buildings, smoke stacks may undergo oscillations in a strong wind; ice-coated transmission lines may be subject to large amplitude vibrations in a steady wind; closely packed tubes in heat exchangers move in oval orbits at high flow velocities; suspended spans of pipelines vibrate when exposed to strong current and/or waves, and so on.

In the present treatment, attention will be concentrated mainly on flow-induced vibrations of slender, marine structures. However, quite a substantial amount of the knowledge which is to be reviewed is equally applicable to structures encountered in wind engineering such as smoke stacks, tall buildings, transmission lines, etc. (Chapters 8 and 11).

The information is organized in four main chapters. The present chapter deals with flow-induced vibrations of a free cylinder in steady current, Chapter 9 studies flow-induced vibrations of a free cylinder in waves, while Chapter 10 reviews the knowledge on pipeline vibrations where the effect of close proximity of the bed becomes important, and finally Chapter 11 gives an account of prediction of flow-induced vibrations by mathematical and numerical treatment.

8.1   A summary of solutions to vibration equation

Let us consider Fig. 8.1 which is an idealized description of a vibrating structure; usually the following forces act on the structure:

1) a spring force, $-ky$, in which k is the spring constant and y is the displacement of the structure (from the equilibrium position);
2) a damping force, $c\,\dot{y}$, in which c is the viscous damping coefficient; and
3) a force on the structure, $F(t)$ in which t is the time.

The differential equation of motion of the structure will read as follows

$$m\,\ddot{y}\,(t) + c\,\dot{y}\,(t) + ky(t) - F(t) \tag{8.1}$$

in which m is the total mass of the system.   Dot over the symbols indicates differentiation with respect to time.

To facilitate the following discussion we will consider the total solution to this equation. We begin with the simplest case, namely the case where $c = 0$ and $F = 0$.

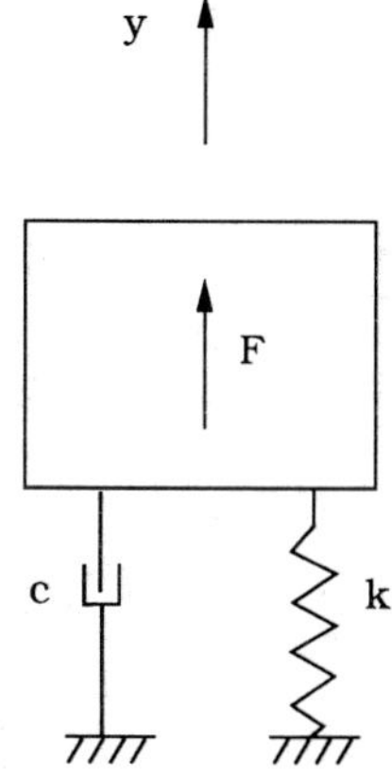

Figure 8.1 Definition sketch. A flexibly-mounted system vibrating in y direction.

8.1.1 Free vibrations without viscous damping

Eq. 8.1 with free vibrations ($F = 0$) in the absence of viscous damping ($c = 0$) will reduce to

$$m \ddot{y}(t) + ky(t) = 0 \qquad (8.2)$$

Because m and k are positive, the solution is

$$y = A_y \cos(\omega_v t) + B_y \sin(\omega_v t) \qquad (8.3)$$

in which ω_v is the angular frequency of the motion,

$$\omega_v = \sqrt{\frac{k}{m}} \qquad (8.4)$$

8.1.2 Free vibrations with viscous damping

In this case the viscous damping is non-zero, therefore Eq. 8.1 with no external force present ($F = 0$; free vibrations) reads

$$m \ddot{y}(t) + c \dot{y}(t) + ky(t) = 0 \qquad (8.5)$$

The trial solution:

$$y = Ce^{rt} \qquad (8.6)$$

and the auxiliary equation, inserting Eq. 8.6 in Eq. 8.5, will be

$$mr^2 + cr + k = 0 \qquad (8.7)$$

The two r values from the preceding equation are determined to be:

$$\left.\begin{array}{c} r_1 \\ r_2 \end{array}\right\} = \frac{1}{2m}\left[-c \pm \sqrt{c^2 - 4\,mk}\,\right] \qquad (8.8)$$

and hence we may take the general solution to Eq. 8.6 as follows

$$y = C_1 e^{r_1 t} + C_2 e^{r_2 t} \qquad (8.9)$$

We examine the solution in the following cases: Case I where $c^2 > 4$ mk and Case II where $c^2 < 4$ mk.

Case I $(c^2 > 4\ \textbf{mk})$. In this case r_1 and r_2 have real values. The constants C_1 and C_2 must be determined from the initial conditions. Let us consider, for example, the following particular case:

$$t = 0: \quad y = A_y \text{ and } \dot{y} = 0 \tag{8.10}$$

From these initial conditions the constants in Eq. 8.9 are found to be as

$$C_1 = -\frac{r_2 A_y}{r_1 - r_2} \quad, \quad C_2 = \frac{r_1 A_y}{r_1 - r_2} \tag{8.11}$$

and for these values the solution (Eq. 8.9) becomes

$$y = \frac{A_y}{r_1 - r_2}(r_1 e^{r_2 t} - r_2 e^{r_1 t}) \tag{8.12}$$

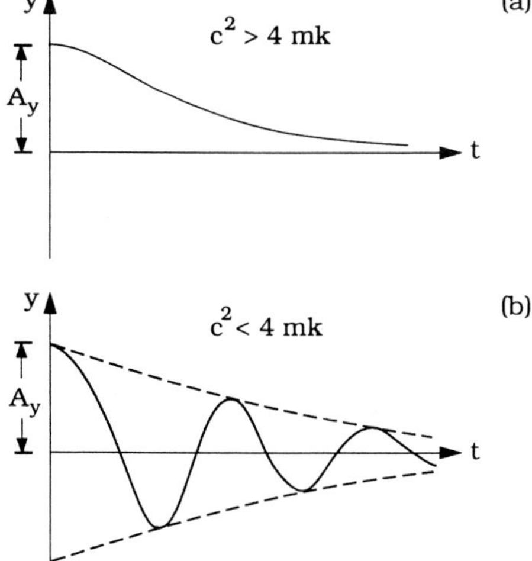

Figure 8.2 Free vibrations (or movement) with viscous damping. (a): Case I $(c^2 > 4\ \text{mk})$ no vibration. The mass creeps back to its equilibrium position. (b): Case II $(c^2 < 4\ \text{mk})$. Damped free vibrations.

Both r_1 and r_2 are negative and r_2 is numerically larger than r_1. Therefore it is readily seen that the solution (Eq. 8.12) is given as that illustrated in Fig. 8.2a: the motion is not a vibration but rather a movement in which the mass, after its initial displacement, gradually retreats towards the equilibrium position. This type of motion is called **aperiodic motion**. Clearly, this case is of little practicle importance as regards the vibrations. One other case of equally little practical importance is when $c^2 = 4\,mk$, where the motion is aperiodic, too. This corresponds to $c = 2\sqrt{mk}$ which is called the **critical damping**.

Case II ($c^2 < 4\,mk$). In this case, the roots r_1 and r_2 are complex:

$$\left.\begin{array}{c} r_1 \\ r_2 \end{array}\right\} = \frac{1}{2m}\left[-c \pm i\sqrt{4mk - c^2}\,\right] \tag{8.13}$$

The real part of the solution (Eq. 8.9) may be written in the following form

$$y = A_y \exp\left(-\frac{c}{2m}t\right)\cos(\omega_{dv}t) \tag{8.14}$$

in which ω_{dv}, the angular frequency, is given by

$$\omega_{dv} = \sqrt{\frac{k}{m} - \left(\frac{c}{2m}\right)^2} \tag{8.15}$$

In Eq. 8.14, A_y is the amplitude of vibrations at time $t = 0$. The solution is illustrated in Fig. 8.2b. As is seen, the vibrations gradually subside with increasing time (damped vibrations).

8.1.3 Forced vibrations with viscous damping

In this case, there exists an external force, $F(t)$, so the differential equation of motion (Eq. 8.1) takes its full form:

$$m\,\ddot{y}\,(t) + c\,\dot{y}\,(t) + k\,y(t) = F(t) \tag{8.16}$$

A particular case of interest with regard to force $F(t)$ is the periodic external force

$$F = F_0 \cos(\omega t) \tag{8.17}$$

in which ω is the angular frequency associated with the periodic force.

A *particular solution* to Eq. 8.16 may be taken as

$$y = C_1 \cos(\omega t) + C_2 \sin(\omega t) \tag{8.18}$$

in which C_1 and C_2 are constants.

Substituting Eq. 8.18 in Eq. 8.16 one gets

$$-\omega^2 C_1 + \frac{c}{m}\omega C_2 + \omega_v^2 C_1 = \frac{F_0}{m} \tag{8.19}$$

and

$$-\omega^2 C_2 - \frac{c}{m}\omega C_1 + \omega_v^2 C_2 = 0 \tag{8.20}$$

in which ω_v is given by Eq. 8.4. C_1 and C_2 are determined from Eqs. 8.19 and 8.20 as

$$C_1 = \frac{(\omega_v^2 - \omega^2)\frac{F_0}{m}}{(\omega_v^2 - \omega^2)^2 + \frac{c^2}{m^2}\omega^2} \tag{8.21}$$

$$C_2 = \frac{\frac{c}{m^2}F_0\omega}{(\omega_v^2 - \omega^2)^2 + \frac{c^2}{m^2}\omega^2} \tag{8.22}$$

The *general solution* of Eq. 8.16 may therefore be written in the following form

$$y - A_y \exp\left(-\frac{c}{2m}t\right)\cos(\omega_{dv}t) + C_1\cos(\omega t) + C_2\sin(\omega t) \tag{8.23}$$

In this equation the first term represents the general solution to the differential equation with the external-force term being zero (Eqs. 8.14 and 8.15). As seen, the contribution of this term to the total solution subsides gradually and the solution asymptotically approaches the particular solution 8.18 (Fig. 8.3):

$$y = C_1\cos(\omega t) + C_2\sin(\omega t) \tag{8.24}$$

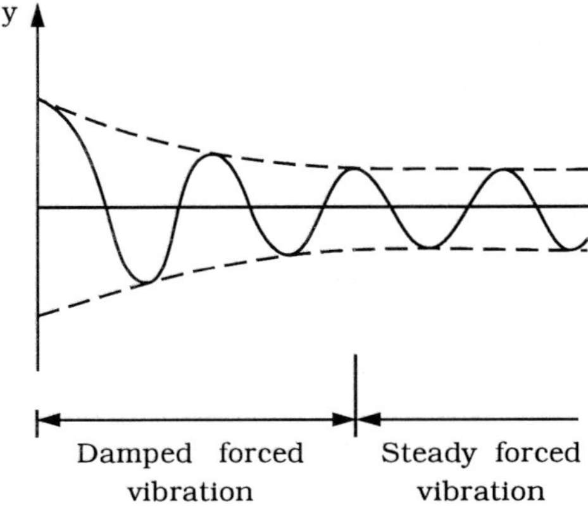

Figure 8.3 Forced vibrations with viscous damping.

This is called **steady forced vibrations**. The solution can be written in the following form

$$y = A \cos(\omega t - \varphi) \tag{8.25}$$

in which

$$A = \sqrt{C_1^2 + C_2^2} = \frac{F_0}{k} \left[\frac{1}{\sqrt{\left(1 - \frac{\omega^2}{\omega_v^2}\right)^2 + \left(\frac{c}{m\omega_v}\right)^2 \frac{\omega^2}{\omega_v^2}}} \right] \tag{8.26}$$

and

$$\varphi = \tan^{-1}\left(\frac{C_2}{C_1}\right) = \tan^{-1}\left[\frac{\left(\frac{c}{m\omega_v}\right)\frac{\omega}{\omega_v}}{1 - \frac{\omega^2}{\omega_v^2}} \right] \tag{8.27}$$

From the solution in Eq. 8.25 it is seen that the steady forced vibration is a simple sinusoidal motion occurring at frequency ω with amplitude A and phase delay φ.

Regarding the amplitude, A, it may be normalized by F_0/k

$$\frac{A}{F_0/k} = \frac{1}{\sqrt{\left(1 - \frac{\omega^2}{\omega_v^2}\right)^2 + \left(\frac{c}{m\omega_v}\right)^2 \frac{\omega^2}{\omega_v^2}}} \tag{8.28}$$

The quantity F_0/k represents the displacement of the mass under static condition. Therefore the normalized amplitude $A/(F_0/k)$ may be interpreted as a *magnification factor*.

Fig. 8.4a illustrates how this quantity varies as function of ω/ω_v, the ratio of the frequency of the external force to the frequency of undamped free vibrations of the system, for various values of parameter $c/(m\omega_v)$. Maximum value of A occurs at

$$\frac{\omega}{\omega_v} = \sqrt{1 - \frac{c^2}{2m^2\omega_v^2}} \tag{8.29}$$

indicated in Fig. 8.4a by a dashed line. As seen the maximum value occurs slightly below resonance.

Since the parameter $c/(m\omega_v)$ usually takes very small values (for which case the maximum value of A occurs very near to resonance), we may take the value of A at resonance as the maximum. This gives the maximum amplitude as

$$A_{\max} \cong \frac{F_0}{k\left(\frac{c}{m\omega_v}\right)} \tag{8.30}$$

Fig. 8.4b, on the other hand, illustrates the variation of φ with ω/ω_v and parameter $c/(m\omega_v)$. While the vibration occurs in phase with the force when

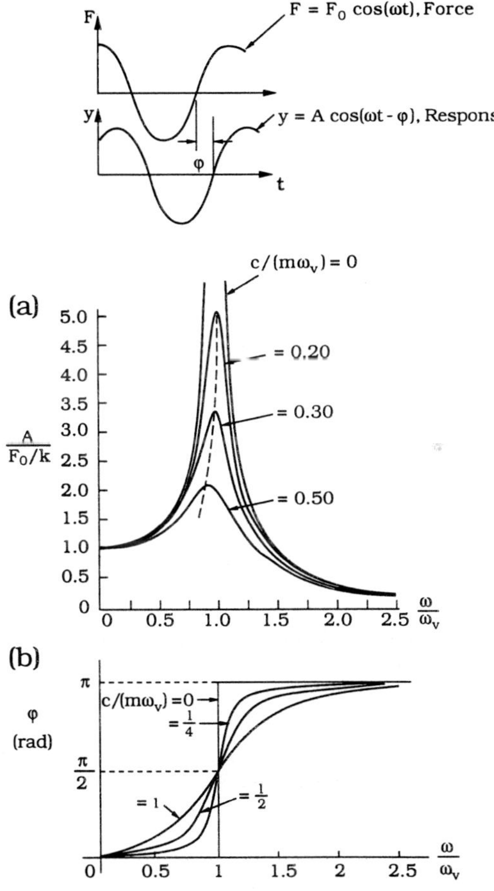

Figure 8.4 *Steady* forced vibrations with viscous damping. Analytical solution given by Eqs. 8.25 - 8.27. $\omega_v = \sqrt{k/m}$.

$\omega/\omega_v \to 0$, the opposite is true when ω/ω_v increases. The motion is $180°$ out of phase for very large values of ω/ω_v.

It is obvious that there will be a delay in the response of the cylinder to the force, as indicated by Fig. 8.4b. This delay, as seen from the figure, is independent of the magnitude of the force, namely F_0, but determined by the forcing frequency, the mass of the system and the structural damping. However, this delay is always $\pi/2$ at the resonance, $\omega/\omega_v = 1$, regardless of the parameter $c/(m\omega_v)$.

8.2 Damping of structures

A vibrating structure dissipates part of its energy into heat. The ability of the structure to dissipate energy is called damping. The role of damping in flow-induced vibrations is that it limits the vibrations (Fig. 8.5).

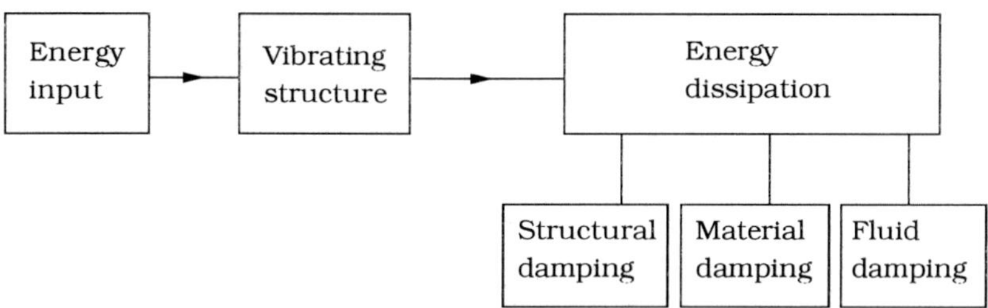

Figure 8.5 Energy input and energy dissipation.

There are three kinds of damping: 1) Structural damping, 2) material damping, and 3) fluid damping. Structural damping is generated by friction, impacting and the rubbing between the parts of a structure. Material damping is generated by the internal energy dissipation of materials (some materials, such as rubber, have very high internal material damping). Fluid-dynamic damping is the result of energy dissipation, as the fluid moves relative to the vibrating structure. In most structures it is the structural and fluid dampings which are dominant, unless the structure is fitted with specially designed material dampers, where also the material damping becomes important.

In the following we shall, for the sake of simplicity, use the term structural damping, referring to the combined effect of the structural damping and the material damping.

8.2.1 Structural damping

In flow-induced vibrations, structural damping and fluid damping are always present side by side. For example, consider a rigid cylinder suspended with springs, resting in still water. When the cylinder is initially displaced to a new position and then released, it will start oscillating. However, the oscillations will eventually subside due to damping. The damping here is caused by the combined action of the structural damping and the fluid damping, and it is theoretically almost impossible to single out the structural damping from the rest of the damping. To isolate the structural damping, we may, however, envisage an idealized situation where the structure is placed in vacuum. In this case, the damping is caused only by the structural damping.

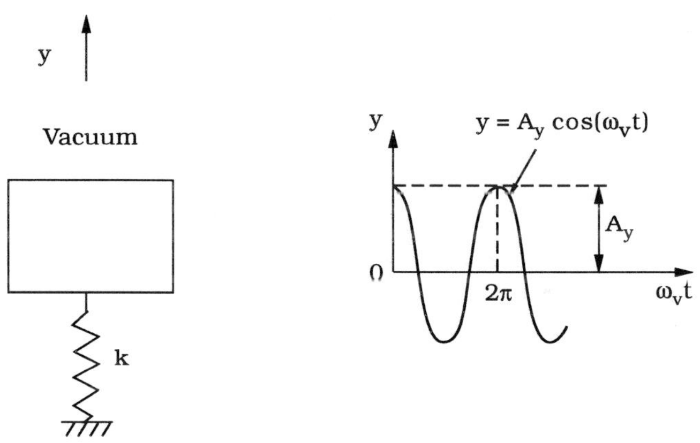

Figure 8.6 Free vibrations in vacuum without damping.

For convenience, let us first consider the simplest case, namely the free vibrations of the structure *in vacuum*, and with *no* damping (Fig. 8.6). In this case, the equation of motion reads

$$m\,\ddot{y} + ky = 0 \qquad (8.31)$$

in which m is the mass per unit span, and k is the spring constant per unit span. The solution to the preceding equation is (see Section 8.1)

$$y = A_y \cos\left(\omega_v t\right) \qquad (8.32)$$

in which ω_v is the angular frequency

$$\omega_v = \sqrt{\frac{k}{m}} \qquad (8.33)$$

i.e. the angular frequency of undamped free vibrations of the system in vacuum.

Next, consider the case in which damping is *included*, Fig. 8.7. Since the structure is placed in vacuum, this damping force is associated with the structural

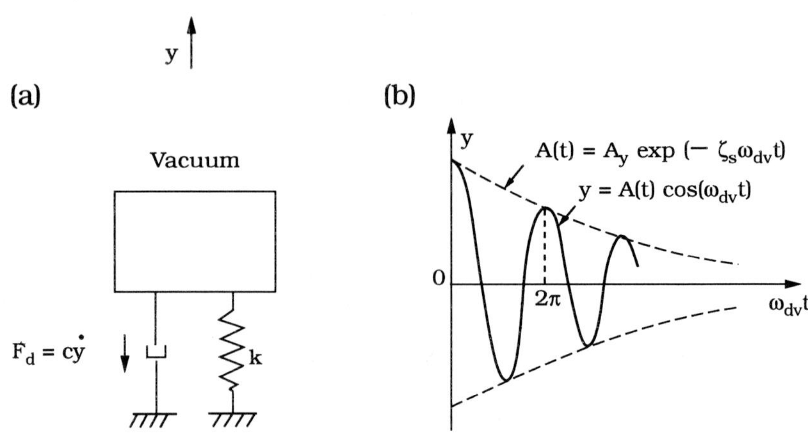

Figure 8.7 Free vibrations with damping in vacuum. To single out the structural damping, the structure is placed in vacuum.

damping alone. Assume that this force is proportional to the velocity of the structure:

$$F_d = c\,\dot{y} \tag{8.34}$$

This model is known as the *linear viscous damper* and proves to be useful in most of the practical cases. In this case, the equation of motion reads

$$m\,\ddot{y} + c\,\dot{y} + ky = 0 \tag{8.35}$$

and the solution becomes (cf. Eqs. 8.14 and 8.15)

$$y = A_y \exp\left(-\frac{c}{2m}t\right)\cos\left(\omega_{dv}t\right) \tag{8.36}$$

where ω_{dv} is the damped angular frequency in vacuum:

$$\omega_{dv} = \sqrt{\frac{k}{m} - \left(\frac{c}{2m}\right)^2} \tag{8.37}$$

For convenience, we replace c, the damping coefficient introduced in Eq. 8.34, with a new quantity ζ_s,

$$\zeta_s = \frac{c}{2m\,\omega_{dv}} \tag{8.38}$$

The latter quantity turns out to be proportional to the energy dissipated by the structural damping, as will be seen later in the section.

In terms of ζ_s, the differential equation and its solution (Eqs. 8.35 - 8.37) may be written as follows

$$m\,\ddot{y} + 2m\,\omega_{dv}\,\zeta_s\,\dot{y} + ky = 0 \tag{8.39}$$

$$y = A_y \exp\left(-\zeta_s\,\omega_{dv}t\right)\cos\left(\omega_{dv}t\right) \tag{8.40}$$

with ω_{dv}

$$\omega_{dv} = \omega_v\left(\frac{1}{1+\zeta_s^2}\right)^{1/2} \tag{8.41}$$

which, in view of $\zeta_s \ll 1$, may be approximated to

$$\omega_{dv} = \omega_v\left(1 - \zeta_s^2\right)^{1/2} \tag{8.42}$$

Note that $\omega_{dv} \simeq \omega_v$, the undamped frequency (Eq. 8.33), since ζ_s^2 is usually small compared with unity.

The quantity ζ_s is called the structural **damping factor**. The energy dissipated in one cycle of vibration is

$$E_d = \int_{one\ cycle} F_d\,dy \tag{8.43}$$

where

$$F_d = c\,\dot{y} \quad \text{or} \quad F_d = 2m\,\zeta_s\,\omega_{dv}\,\dot{y} \tag{8.44}$$

and

$$dy = \dot{y}\,dt \tag{8.45}$$

From Eqs. 8.40, 8.44 and 8.45, considering the amplitude of damped vibration $A(t)$ (see Fig. 8.7) approximately constant during one cycle, one gets

$$E_d = 2\pi m\,\zeta_s\,\omega_{dv}^2 A^2(t) \tag{8.46}$$

On the other hand, the total energy is

$$E_T = \text{Kinetic Energy} + \text{Potential Energy} \tag{8.47}$$

The kinetic energy becomes maximum when the potential energy is zero; thus

$$E_T = (\text{Kinetic Energy})_{\max} = \frac{1}{2}m\,\dot{y}_{\max}^2 \tag{8.48}$$

From Eqs. 8.40 and 8.47

$$E_T = \frac{1}{2} m\, \omega_{dv}^2\, A^2(t), \qquad (8.49)$$

and from Eqs. 8.46 and 8.49

$$\frac{E_d}{E_T} = 4\pi \zeta_s \qquad (8.50)$$

This relation shows that the damping factor ζ_s is proportional to the ratio of the energy dissipated per cycle to the total energy of the structure, thus relating the damping factor to the energy dissipation.

8.2.2 Fluid damping in still fluid

Now, consider the damped, free vibrations of the structure in an otherwise still fluid. The picture will not be drastically different from that in vacuum (Fig. 8.7b). The vibrations will subside with time due to damping. The damping in the present case, however, is caused not only by the structural damping but also by the fluid damping. The specific goal of this section is to make an assessment of the fluid-damping component of the total damping.

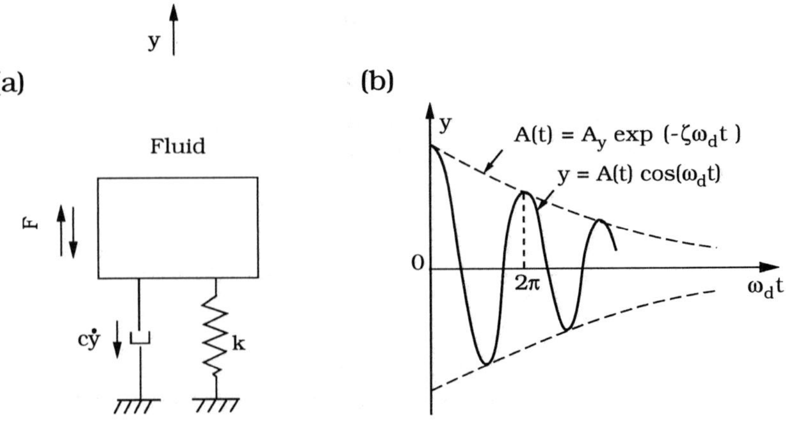

Figure 8.8 Free vibrations with damping in a still fluid (cf. Fig. 8.7).

When the structure undergoes vibrations in an otherwise still fluid, it will be subjected to a hydrodynamic force F (Fig. 8.8). This latter force is actually the Morison force (Chapter 2) on the structure oscillating in the fluid. The equation of motion will be in the form

$$m\,\ddot{y}\,+c\,\dot{y}\,+ky = F \tag{8.51}$$

in which F, the Morison force per unit span, is given by (Eq. 4.30)

$$F = \frac{1}{2}\rho C_D D(-\dot{y})\,|-\dot{y}|\,+\rho C_m A(-\ddot{y}) \tag{8.52}$$

The second term on the right hand-side of the equation, namely $(-\rho C_m A\,\ddot{y})$, may be written in the form $(-m'\,\ddot{y})$ in which m' is the hydrodynamic mass per unit span (Eq. 4.17):

$$m' = \rho C_m A \tag{8.53}$$

Hence, the equation of motion becomes

$$(m + m')\,\ddot{y}\,+c\,\dot{y}\,+\frac{1}{2}\rho C_D D\,|\dot{y}|\dot{y}\,+ky = 0 \tag{8.54}$$

Comparison of the preceding equation with Eq. 8.35 indicates that in the present case: 1) the mass is no longer m but rather $m + m'$, and 2) there is an additional resistance force, namely $(1/2)\rho D\,C_D\,|\dot{y}|\dot{y}$. These changes will obviously affect the total damping.

The solution to Eq. 8.54 may be written in the following form, drawing an analogy between the present case and the vacuum situation (Eq. 8.40)

$$y = A_y\,\exp\!\left(-\zeta\omega_d t\right)\cos(\omega_d t) \tag{8.55}$$

in which ζ is now the total damping factor (comprising the structural damping and the fluid damping), and ω_d is the angular frequency which, in analogy to Eqs. 8.42 and 8.33, must be given by

$$\omega_d = \omega_n(1 - \zeta^2)^{1/2} \tag{8.56}$$

where ω_n is

$$\omega_n = \sqrt{\frac{k}{m + m'}} \tag{8.57}$$

ω_n is called the undamped natural angular frequency. Since ζ is normally small compared with unity, the damped natural angular frequency, ω_d, can be approximated to ω_n, the undamped natural angular frequency:

$$\omega_d = \omega_n(1 - \zeta^2)^{1/2} \cong \omega_n \tag{8.58}$$

The frequency f_n, namely

$$f_n = \frac{\omega_n}{2\pi} = \frac{1}{2\pi}\sqrt{\frac{k}{m+m'}} \qquad (8.59)$$

on the other hand, is called the **undamped natural frequency**, or simply the **natural frequency** of the structure.

Regarding the damping ζ, this quantity is called the equivalent viscous damping factor. It represents the **total damping**, as mentioned earlier, and can be calculated by

$$\zeta = \frac{1}{4\pi}\frac{E_d}{E_T} \qquad (8.60)$$

where E_d is the energy dissipated in one cycle of vibration as defined in Eq. 8.43. The total energy will, in the present case, in analogy to Eq. 8.48, be

$$E_T = \frac{1}{2}(m+m')\,\dot{y}_{\max}^2 \qquad (8.61)$$

Regarding E_d, namely

$$E_d = \int_{\text{one cycle}} F_d\,dy \qquad (8.62)$$

F_d, the total damping force, which opposes the motion of structure needs to be predicted. It is composed of the structural damping force and the fluid damping force, as seen from Eq. 8.54:

$$F_d = c\,\dot{y} + \frac{1}{2}\rho D C_D\,|\dot{y}|\dot{y} \qquad (8.63)$$

Now substituting Eq. 8.63 into Eq. 8.62 along with $dy = \dot{y}\,dt$ gives

$$E_d = \int_{\text{one cycle}} c\,\dot{y}^2\,dt + \int_{\text{one cycle}} \frac{1}{2}\rho D C_D\,|\dot{y}|\dot{y}^2\,dt \qquad (8.64)$$

Inserting Eq. 8.64 into Eq. 8.60 and assuming the damped amplitude $A(t)$ in

$$y = \underbrace{A_y \exp\left(-\zeta\omega_d t\right)}_{A(t)} \cos(\omega_d t) \qquad (8.65)$$

to be approximately constant during one cycle of vibration, gives the following expression for the equivalent viscous damping factor ζ

$$\zeta = \frac{c}{2(m+m')\omega_d} + \frac{\rho D^2}{4\pi(m+m')}\frac{8}{3}C_D\frac{A}{D} \qquad (8.66)$$

The first term on the right hand side of the preceding equation represents the structural-damping component, as it involves c, the structural damping coefficient.

The second term on the other hand represents the fluid damping. Denoting the first term by ζ_s and the second by ζ_f, the total damping is

$$\zeta = \zeta_s + \zeta_f \tag{8.67}$$

in which

$$\zeta_s = \frac{c}{2(m + m')\omega_d} \tag{8.68}$$

and

$$\zeta_f = \frac{\rho D^2}{4\pi(m + m')}\frac{8}{3}C_D\frac{A}{D} \tag{8.69}$$

As seen from the preceding equation, fluid damping in a still fluid is a function of amplitude, the dimension of the structure, the drag coefficient, the hydrodynamic mass and the actual mass of the structure.

Eq 8.69 may be used to predict the fluid damping in a free-decay test. Since the total damping could be predicted from the so-called log decrement (see the next subsection), this would presumably enable the structural damping to be predicted by simply subtracting the fluid-damping component from the total damping. Example 8.1 illustrates this procedure with data obtained in an actual free-decay test.

Measurement of structural damping

It is extremely difficult to estimate the energy dissipation caused by the structural damping (by friction, impacting and the rubbing between the parts of a structure). This is due partly to the uncertainty about the details of the joints and partly to the large numbers of such joints involved. Therefore, testing seems to be the only solution for determining the structural damping of the great majority of structures. This is usually done in air and/or in water.

The most commonly used techniques for measuring damping are: 1) free decay, 2) bandwidth, 3) magnification factor, and 4) response methods. The key procedure in these techniques is basically as follows: 1) apply a known excitation to the structure, 2) record the response of the structure, and 3) find the unknown damping by matching the theoretically predicted response with the test record.

In a free-decay test, the structure is given a certain amount of initial displacement and then released, and the resulting damped vibration of the structure is recorded.

If y_n and y_{n+1} are two consecutive amplitudes in such a record (Fig. 8.9) the ratio of y_n and y_{n+1}, from Eq. 8.55, will be

$$\frac{y_n}{y_{n+1}} = \frac{A_y\exp(-\zeta\omega_d t)}{A_y\exp\{-\zeta\omega_d(t+T)\}} = \exp(\zeta\omega_d T) \tag{8.70}$$

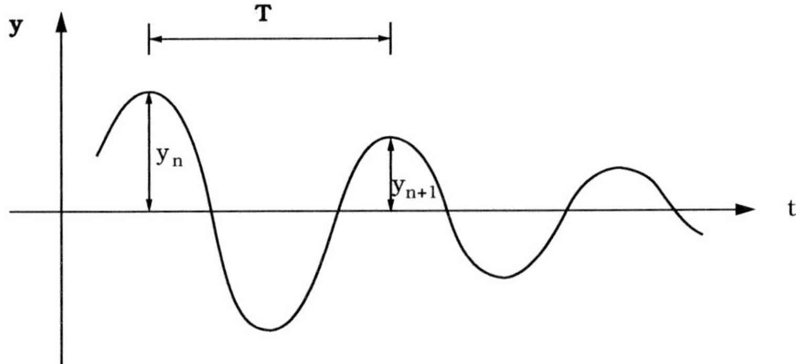

Figure 8.9 Cylinder displacement time series in a free-decay test.

in which T is the period of the vibration (Fig. 8.9):

$$T = \frac{2\pi}{\omega_d} \qquad (8.71)$$

Hence, ζ from the preceding two equations is determined as follows

$$\zeta = \frac{1}{2\pi} \ln \frac{y_n}{y_{n+1}} \qquad (8.72)$$

This equation enables the damping to be calculated from a free-decay test. The quantity $\delta = \ln(y_n/y_{n+1})$ is called the **logarithmic decrement** and is sometimes used to characterize the damping in favour of ζ, which is actually $\zeta = \delta/(2\pi)$.

The following example illustrates how the free-decay technique is used to measure the damping of a flexibly mounted rigid cylinder. Detailed information about the techniques for measurement of damping can be found in the book by Blevins (1977, pp. 232-244).

Example 8.1:   Free decay test to determine structural damping

In an investigation to study vibrations of a spring-supported, rigid cylinder the structural damping of the experimental system, shown in Fig. 8.10, has been determined both in air and in water by free-decay tests. The following paragraphs

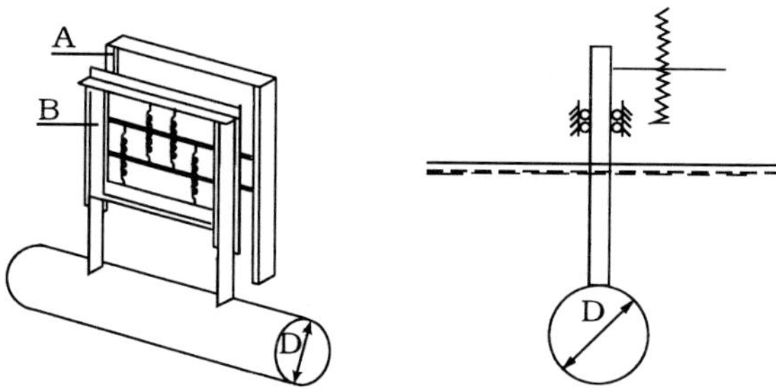

Figure 8.10 Experimental system used in the tests referred to in Example 8.1.

will briefly summarize the results of these tests. The system properties are given in Table 8.1.

In the air experiments, a counter-balance weight, mounted nearly frictionless, was used to eliminate the negative effect of the actual weight of the cylinder through the arrangement shown in Fig. 8.11.

The structural damping was determined directly from Eq. 8.72, namely,

$$\zeta_s = \frac{1}{2\pi} \ln \frac{y_n}{y_{n+1}} \quad , \tag{8.73}$$

considering that the fluid damping in air would be negligible and therefore may be omitted. The quantities y_n and y_{n+1} are two consecutive amplitudes, as indicated in Fig. 8.9.

Table 8.1 System properties for the tests given in Example 8.1.

Cylinder diameter	Cylinder surface roughness	Spring constant	Relative density	Mass ratio
D	k_s	k/ρ	$\rho_{\text{cylinder}}/\rho$	$\dfrac{m + m'}{\rho D^2}$
(cm)	(mm)	(m^2/s^2)		
10.5	5	0.336	1.09	1.6

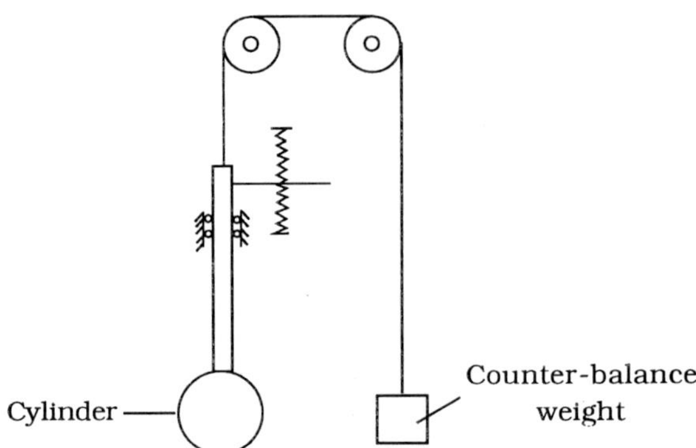

Figure 8.11 Schematic description of the arrangement to measure the structural damping in air.

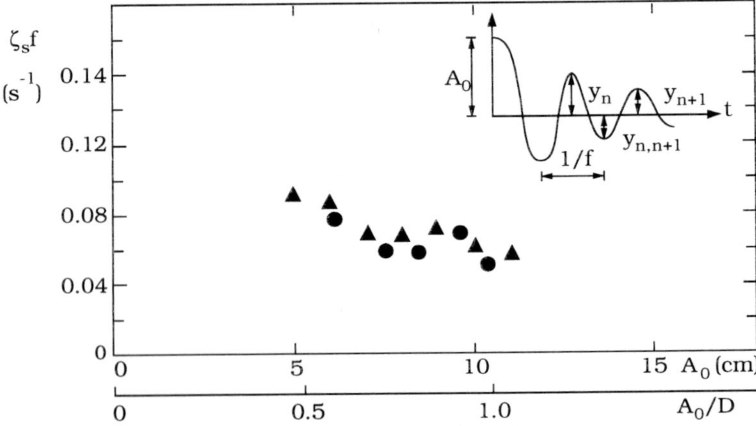

Figure 8.12 Structural damping per unit time, $\zeta_s f$, versus the initial excitation displacement A_0 for the tests given in Example 8.1. Triangles: From air experiments. Circles: From water experiments.

In the water experiments, on the other hand, the structural damping was predicted from

$$\zeta_s = \zeta - \zeta_f \tag{8.74}$$

the total damping, ζ, in the experiments was found from Eq. 8.72, while ζ_f was calculated from Eq. 8.69. A, the amplitude of the cylinder motion for one cycle of the vibration in Eq. 8.69, was calculated from

$$A = \frac{1}{4}\left(y_n + y_{n+1} + 2y_{n,n+1}\right) \tag{8.75}$$

in which $y_{n,n+1}$ is the trough amplitude between the two consecutive crest amplitudes y_n and y_{n+1}. The drag coefficient C_D in Eq. 8.69 was taken from the diagram given by Fredsøe and Justesen (1986, Fig. 7) as a function of Re and KC numbers. The calculations were made for each cycle of the free decay test, in which $KC = 2\pi A/D$ and $Re = \dot{y}_{\max} D/\nu$.

Fig. 8.12 depicts the experimentally determined structural damping (per unit time), namely $\zeta_s f$, as a function of initial displacement A_0. The air and water results are seen to be in general agreement, although there is a slight tendency that the ζ_s values are underpredicted by the water experiments.

8.3 Cross-flow vortex-induced vibrations of a circular cylinder

It has been seen in Chapters 1 and 2 that 1) a cylinder exposed to a steady current experiences vortex shedding if $Re > 40$ and 2) this phenomenon results in periodic variations in the force components on the cylinder; the lift force oscillates at the vortex-shedding frequency, while the drag force oscillates at twice the vortex-shedding frequency.

Now, if the cylinder is a flexibly-mounted cylinder, these forces may induce vibrations of the cylinder. The lift force may induce **cross-flow vibrations**, while the drag force may induce **in-line vibrations** (Fig. 8.13). These vibrations are generally termed the **vortex-induced vibrations**. There exist excellent reviews on the subject by Blevins (1977), King (1977), Sarpkaya (1979), Griffin (1981), Bearman (1984), Chen (1987) and Pantazopoulos (1994).

We shall first focus on cross-flow vibrations and subsequently in Section 8.4 we shall examine the in-line vibrations.

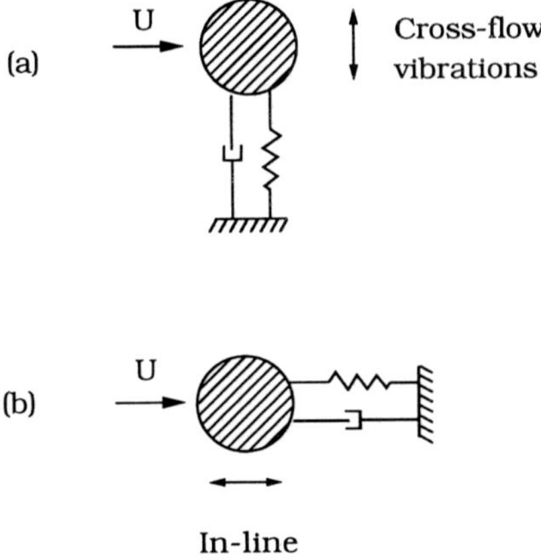

Figure 8.13 Definition sketch.

8.3.1 Feng's experiment

The cross-flow vibrations of a circular cylinder can be best described by reference to the experiment of Feng (1968). The experimental set-up employed by Feng is shown schematically in Fig. 8.14. It is basically a flexibly-mounted circular cylinder with one degree of freedom of movement in the y-direction. The system is exposed to air flow. The flow speed is increased in small increments, starting from zero. To see if there is any hysterisis effect, experiments are repeated also with a decreasing flow speed (again in small increments). For each flow velocity, U, the following quantities are measured: the vortex-shedding frequency, f_v; the vibration frequency, f; the vibration amplitude, A; and the phase angle, i.e. the phase difference between the cylinder vibration and the lift force, φ. The measured quantities are then plotted as a function of normalized velocity $V_r (= U/Df_n)$ (Fig. 8.15). Here, f_n is the natural frequency of the vibrating system.

The diagrams in Fig. 8.15 show the following.

1) As the flow velocity is increased from zero, no vibration is experienced until the velocity reaches a value of about $V_r = 4$. At this point, vibrations begin to emerge. The frequency diagram indicates that these small-amplitude vibrations occur at the natural frequency of the system, namely $f/f_n = 1$, while the vortex shedding (therefore, the oscillation in the lift-force) occurs at the stationary-

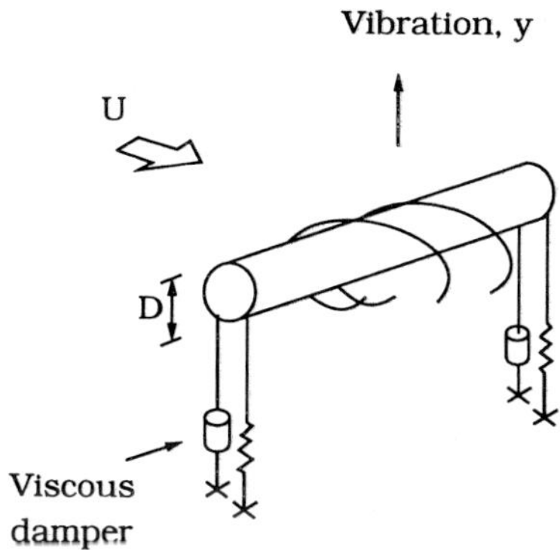

Figure 8.14 Definition sketch.

cylinder Strouhal frequency, namely at

$$\frac{f}{f_n} = St\frac{U}{Df_n} \tag{8.76}$$

with St approximately equal to 0.2. The identity (8.76) is depicted in Fig. 8.15a as a reference line.

2) Fig. 8.15a shows that the vortex-shedding frequency follows the stationary-cylinder Strouhal frequency until the velocity V_r reaches the value of 5. With a further increase in the velocity beyond this point, however, it departs from the Strouhal frequency and begins to follow the natural frequency of the system (the horizontal line $f/f_n = 1$ in Fig. 8.15a). As is seen, this takes place over a rather broad range of V_r, namely over the range $5 < V_r < 7$.

The preceding observation implies that the vortex shedding frequency locks into the natural frequency of the system at $V_r = 5$ and remains locked in until V_r reaches the value of about 7. Therefore it may be concluded that, in this range, the vortex shedding is controlled not by the Strouhal law; rather the cylinder vibration itself has an important influence as well. The flow-visualization work of Williamson and Roshko (1988) clearly shows that the separation vortices are forced to interact by the cylinder vibration, leading to vortex shedding, at a frequency equal to the vibration frequency rather than the exact value of the Strouhal frequency.

This phenomenon is known as the **lock-in** phenomenon. Other terms such as "resonance", "syncronization", "wake capture" are also used in literature to refer to this phenomenon.

Presumably, in the lock-in range, three frequencies, namely the cylinder

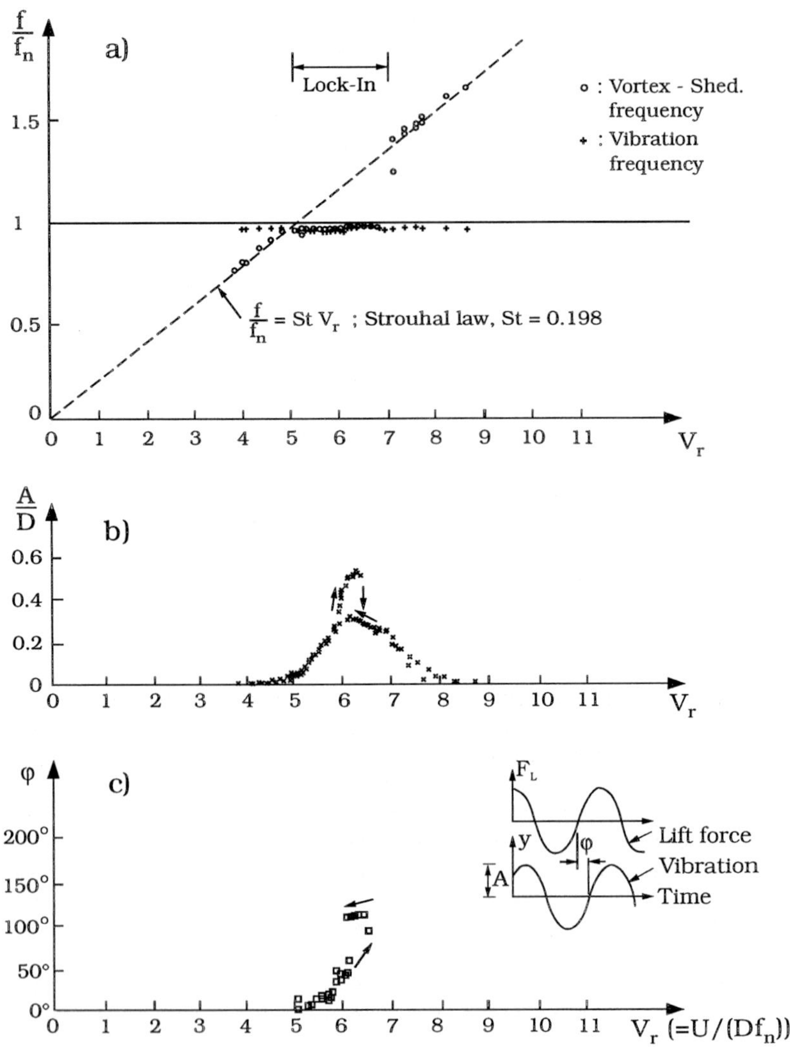

Figure 8.15 Cross-flow response of a flexibly-mounted circular cylinder subject to steady current *in air*. $2m\,\zeta_s/(\rho D^2) = 0.4$, m being mass per unit span. Feng (1968).

vibration frequency, the vortex-shedding frequency and the natural frequency coincide: $f = f_v = f_n$. This means that, in this range, the shedding, therefore the lift force, oscillates in sympathy with the cylinder motion. This obviously results in vibrations with very large amplitudes (Fig. 8.15b).

3) As the flow velocity is increased even further ($V_r \overset{\sim}{>} 7$), the shedding frequency suddenly unlocks from the natural frequency and experiences an abrupt jump, to assume its Strouhal value again. This occurs around $V_r \simeq 7.3$.

The width of the lock-in range in terms of V_r may depend on the vibration amplitude. The larger the vibration amplitude, the broader the lock-in range. This is because larger vibration amplitudes (obtained as a result of smaller structural damping) may require larger values of V_r for the shedding frequency to unlock from the vibration frequency, to restore the Strouhal frequency.

Fig. 8.15a shows that, at $V_r \simeq 7.3$, while the vortex-shedding frequency assumes its Strouhal value, the vibrations still occur at the natural frequency. The consequence of this is a reduction in the vibration amplitude (Fig. 8.15b), since the forcing frequency (namely, the vortex-shedding frequency) is no longer in sympathy with the motion of the cylinder. As the velocity is increased further, the forcing frequency will move even further away from the natural frequency, therefore this effect will be even more pronounced, resulting in even larger reduction in the vibration amplitude, as is clearly seen from Fig. 8.15b. The figure shows that the vibrations completely disappear when V_r reaches the value of about 8.5.

4) Fig. 8.15b indicates that there is a hysteresis effect in the amplitude variation. Likewise, from Fig. 8.15c it is seen that the phase angle experiences the same kind of behaviour.

For convenience, the phase angle and the amplitude variation with respect to V_r are plotted schematically in Fig. 8.16. Also shown in the figure are the sketches that illustrate the vortex-shedding mode experienced during the course of increase or decrease of the velocity. The latter information is due to the works of Williamson and Roshko (1988) and Brika and Laneville (1993). The term "2S" mode means that there are two single vortices shed for each cycle of vibrations. This mode represents the familiar vortex shedding (Section 1.2). The wake in this case will be the Karman street-type wake, as indicated in the figure. The term "2P" mode, on the other hand, indicates that two vortex pairs form in each cycle of vibrations during the course of shedding process, and the sketch depicted in the figure illustrates what the cylinder wake looks like in this case. (Detailed description of vortex motions around a vibrating cylinder will be given in Section 8.3.3).

The mode of vortex shedding undergoes a sudden change, switching from "2S" mode to "2P" mode, when V_r reaches the value of about 6.5. This is due to the fact that the former mode apparently can not be maintained for larger amplitudes, as will be seen later in Section 8.3.3. As a result of this sudden change in the shedding mode, both the amplitude and the phase angle undergo abrupt changes; the amplitude is reduced considerably and the phase angle is increased

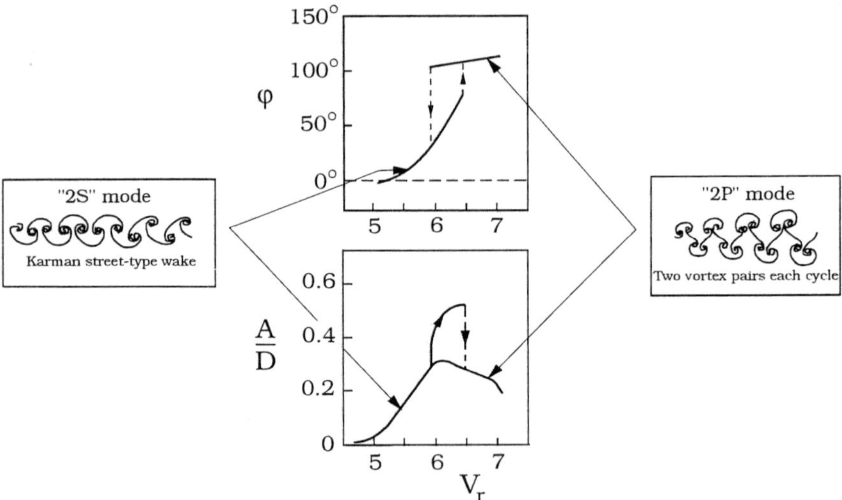

Figure 8.16 Hysteresis effect. The sketches regarding the mode of vortex
shedding are due to Williamson and Roshko (1988).

to a higher value. The experimental work of Brika and Laneville (1993) gives a
detailed account of the hysteresis effect and its relation to the mode of vortex
shedding.

One important implication of the preceding observation in relation to the
change in the mode of vortex shedding is that it limits the amplitude of vibrations.
Although no simultaneous force measurements have been made, the change in the
mode of vortex shedding from "2S" mode to "2P" mode may probably cause
the lift force to experience an abrupt reduction, as suggested by Williamson and
Roshko (1988), with reference to the work of Bishop and Hassan (1964). This
would presumably cause the sudden drop in the amplitude observed in Fig. 8.16,
(see Section 8.3.3 for a detailed discussion).

Example 8.2:   Cross-flow vibrations in water

The response described in the preceding paragraphs is typical for a flexibly-
mounted cylinder exposed to air flow. When the cylinder is exposed to water flow,
however, although the response with respect to amplitude will be more or less the

same as in the case of air flow, the response as regards the frequency will be different. Fig. 8.17 reveals this (cf. Fig. 8.17 and Fig. 8.15a). The vibration frequency in the water case is not equal to the natural frequency of the system in the lock-in range, but rather it increases monotonously with V_r. Fig. 8.18 illustrates schematically the frequency response experienced in the lock-in range in both the air and the water cases. Since the cylinder, in the lock-in range, should respond with a frequency equal to its natural frequency, Fig. 8.18b implies that the natural frequency in the water situation is not maintained at its still-water value f_n, but rather it undergoes a constant increase with increasing V_r. This aspect will be examined in details in the following.

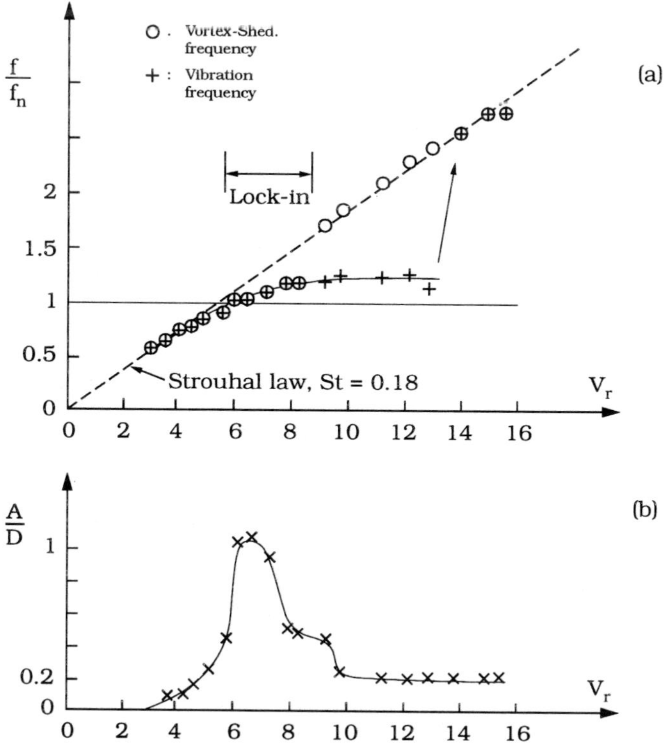

Figure 8.17 Cross-flow response of a flexibly-mounted circular cylinder subject to steady current in water. $(m/\rho D^2) = 5.3$. Anand (1985).

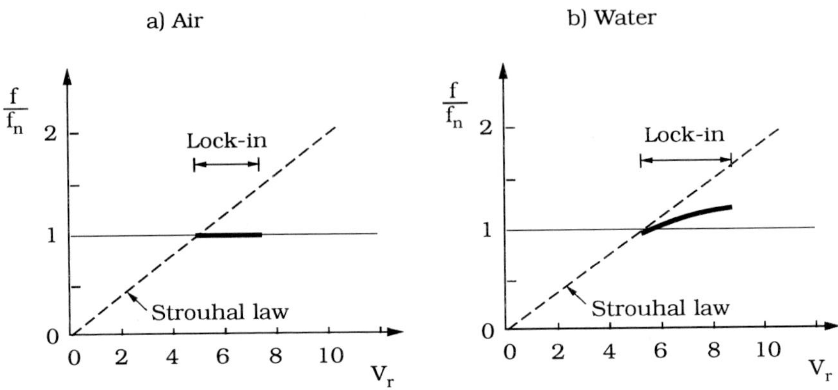

Figure 8.18 Schematic illustrations of frequency response in air and water
with regard to cross-flow vibrations.

The natural frequency, f_n, adopted in Figs. 8.15 and 8.17 as the scaling
parameter, is the frequency measured in still fluid. However, this quantity may
undergo a marked change when the structure is subject to a flow. To recognize
this, consider the definition given in Eq. 8.59:

$$f_n = \frac{1}{2\pi}\sqrt{\frac{k}{m+m'}} \tag{8.77}$$

in which m' is (Eq. 4.17):

$$m' = \rho C_m A = \rho C_m \frac{\pi D^2}{4} \tag{8.78}$$

For small vibrations in still fluid ($KC \lesssim 5$ or alternatively $A/D \lesssim 0.8$), from Figs.
4.10 and 4.11, C_M approches to 2 and therefore $C_m(= C_M - 1)$ approaches to
unity for a circular cylinder. When the cylinder is subject to a current (Fig. 8.19),
however, C_m will no longer be the same as in the case of still fluid. Let us denote
the new hydrodynamic-mass coefficient in the case of current by C_{mc}. Hence, the
new hydrodynamic mass will be

$$m'_c = \rho C_{mc} \frac{\pi D^2}{4} \tag{8.79}$$

and therefore the new natural frequency will be

$$f_{nc} = \frac{1}{2\pi}\sqrt{\frac{k}{m+m'_c}} \tag{8.80}$$

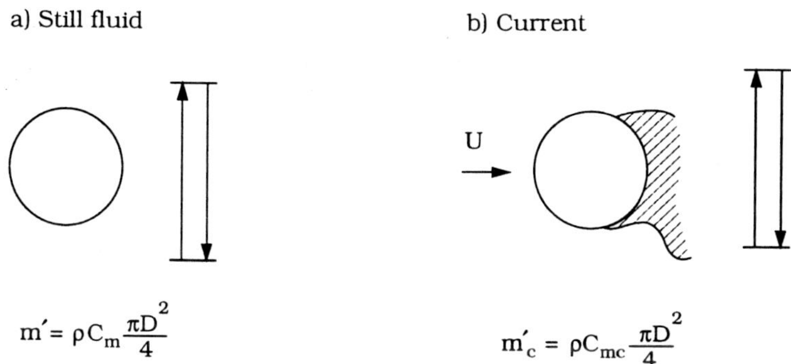

Figure 8.19 Hydrodynamic mass in still fluid and in current.

From Eqs. 8.77 and 8.80, the ratio between the natural frequency in current and that in still fluid is found as

$$\frac{f_{nc}}{f_n} = \left(\frac{m+m'}{m+m'_c}\right)^{1/2}$$
(8.81)

From Eqs. 8.78 and 8.79, this ratio may be re-written as

$$\frac{f_{nc}}{f_n} = \left(\frac{1+C_m/(m/\rho D^2)}{1+C_{mc}/(m/\rho D^2)}\right)^{1/2}$$
(8.82)

C_{mc} values have been measured by Sarpkaya (1978) for a circular cylinder subject to a steady current and oscillating in the cross-flow direction (forced oscillations). Sarpkaya expressed the lift force on the cylinder in terms of the Morison equation and determined the force coefficient through the conventional Fourier analysis. The results of Sarpkaya's study regarding the hydrodynamic-mass coefficient is reproduced in Fig. 8.20. The horizontal axis is the reduced velocity defined by

$$V_r = \frac{U}{Df}$$
(8.83)

in which f is the frequency of the forced vibrations. Although the forced-vibration experiments may not be able to simulate the free, self-excited vibrations such as those depicted in Figs. 8.15 and 8.17 for the full range of V_r, they are, however, reasonable approximations of the self-excited, free vibrations around the lock-in velocity. Therefore, the results of Sarpkaya's experiments may be used to assess the natural frequency. Picking up the values of C_{mc} from the Sarpkaya diagram, and taking C_m to be unity, the ratio f_{nc}/f_n can then be worked out (Fig. 8.21).

Fig. 8.21 shows that the natural frequency of structure remains almost the same as that in still fluid ($f_{nc}/f_n \cong 1$) when $m/(\rho D^2) = 200$, a typical value representing the air situation, while it increases monotonously with V_r for $m/(\rho D^2) = 2.5$ and 5, typical values representing the water situation.

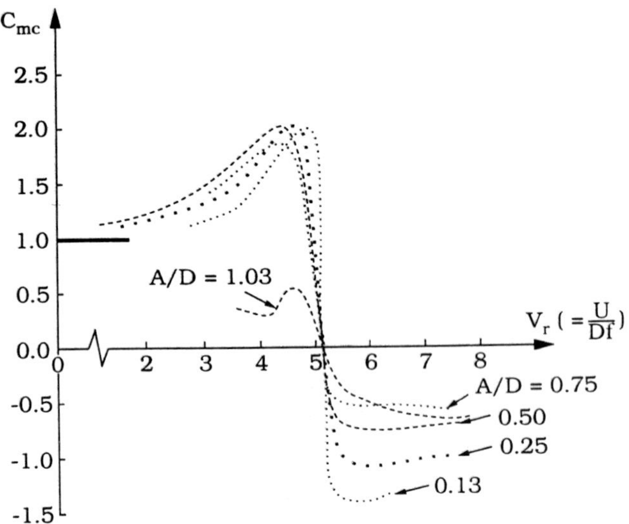

Figure 8.20   Hydrodynamic mass coefficient for a circular cylinder vibrating in the cross-flow direction and subject to a current. The results were obtained from *forced-vibration* experiments.   Sarpkaya (1978).

In the case of $m/(\rho D^2) = 200$, f_{nc}/f_n is approximately equal to unity, simply because both m' and m'_c (Eq. 8.81) are small compared with m, therefore can be neclected, leading to $f_{nc} \cong f_n$. In the case of $m/(\rho D^2) = 2.5$ and 5, the hydrodynamic masses m' and m'_c, are no longer negligible. Therefore, f_{nc} will be different from f_n, the still-water value of the natural frequency.

The preceding analysis, although it is at best suggestive, may help explain why a flexibly-mounted structure in water (Fig. 8.17) responds differently from that in air (Fig. 8.15).

It is interesting to note the following observations made by other researchers. Bearman and Mackwood (1991) carried out experiments with a circular cylinder exposed to an oscillating water flow generated in a U-tube. The mass parameter $m/(\rho D^2)$ was 3.69. In their study for the in-line fixed cylinder the largest peak amplitude occurred not for multiples of f_n/f_w like 2 and 3 (in which f_w = the wave

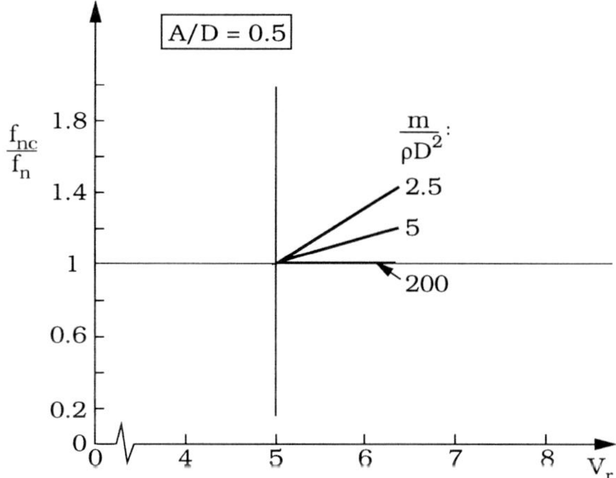

Figure 8.21 Change in natural frequency of structure in current. f_n is the natural frequency corresponding to still fluid, and f_{nc} is that corresponding to the case when the structure is subject to a current. $m/(\rho D^2) = 200$ represents air situation, while $m/(\rho D^2) = 2.5$ and 5 represents water situation for a circular cylinder.

frequency), but for smaller values, 1.79 and 2.72, respectively. This suggests that the actual natural frequency of their system was higher than f_n, in agreement with the preceding analysis. Parallel results were obtained by Kozakiewicz, Sumer and Fredsøe (1994) $(m/(\rho D^2) = 0.91)$. Maull and Kaye (1988), in their experiments where $m/(\rho D^2) \cong 12$, found only a very little shift of resonances of the response of a flexible cylinder in waves. This suggests that an increase in the natural frequency disappears for $m/(\rho D^2)$ greater than about $O(10)$.

Finally, it may be noted that the expression adopted for the natural frequency in Eq. 8.77 is only an approximation, since it represents the undamped natural frequency. The damped natural frequency is actually given by Eq. 8.56, namely $f_n = (1/2\pi)\sqrt{k/(m + m')}(1 - \zeta^2)$ where ζ, the damping factor, includes not only the structural damping but also the fluid damping.

8.3.2 Non-dimensional variables influencing cross-flow vibrations

One way of identifying the non-dimensional variables which govern the cross-flow vibrations of a flexibly-mounted structure subject to a steady current is to consider the full differential equation of motion. The governing parameters may then easily be identified upon the normalization of this equation.

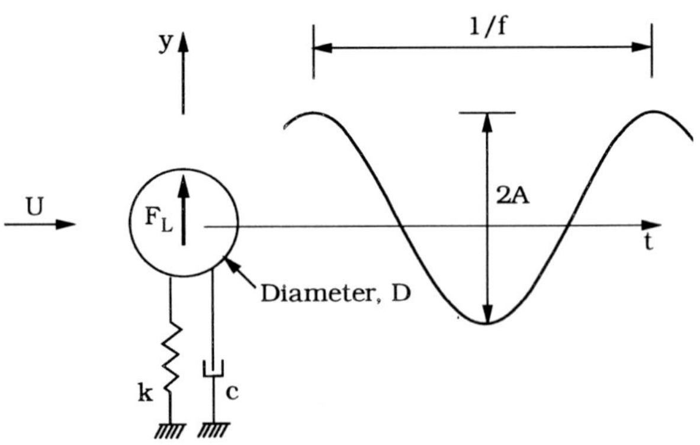

Figure 8.22 Definition sketch for cross-flow vibrations.

This differential equation (in the y direction, Fig. 8.16) is

$$(m + m')\, \ddot{y}\,(t) + c\, \dot{y}\,(t) + k\, y(t) = F_L(t) \tag{8.84}$$

in which F_L is the lift force on the cylinder. For a free cylinder, this lift force oscillates around zero:

$$F_L(t) = \overline{F}_L + F'_L(t) = F'_L(t) \tag{8.85}$$

in which F'_L may be expressed in terms of oscillating force coefficient:

$$F'_L(t) = \frac{1}{2}\rho C'_L(t)DU^2 \tag{8.86}$$

The structural damping c (Eq. 8.68):

$$c = 2(m + m')\omega_d \zeta_s \cong 2(m + m')\omega_n \zeta_s \tag{8.87}$$

in which ω_n is (Eq. 8.57):

$$\omega_n = \frac{2\pi}{f_n} = \sqrt{\frac{k}{m + m'}} \tag{8.88}$$

Furthermore, y and t may be normalized in the following manner:

$$Y = \frac{y}{D} \quad , \quad T = \frac{tU}{D} \tag{8.89}$$

Inserting Eqs. 8.85 - 8.89 into Eq. 8.84 and normalizing the equation gives the following non-dimensional equation:

$$\ddot{Y} + \frac{4\pi\zeta_s}{V_r}\,\dot{Y} + \frac{4\pi^2}{V_r^2}Y = \frac{C_L'(t)}{2M} \tag{8.90}$$

in which

$$M = \frac{m + m'}{\rho D^2} \tag{8.01}$$

Here M is called the **mass ratio**.

Regarding the force coefficient $C_L'(t)$, it may be approximated to

$$\begin{aligned}
C_L'(t) &= \sqrt{2}\,(\overline{C_L'^2})^{1/2}\,\sin(\omega t)\\
&= \sqrt{2}\,(\overline{C_L'^2})^{1/2}\,\sin\left(2\pi\left(\frac{f_v D}{U}\right)T\right)
\end{aligned} \tag{8.92}$$

Simple dimensional considerations suggest that $(\overline{C_L'^2})^{1/2}$ and $f_v D/U$ must be dependent on the following non-dimensional quantities:

$$(\overline{C_L'^2})^2 = f_1\left(\frac{A}{D}\,,\ V_r\,,\ Re\,,\ \frac{k_s}{D}\right) \tag{8.93}$$

$$\frac{f_v D}{U} = f_2\left(\frac{A}{D}\,,\ V_r\,,\ Re\,,\ \frac{k_s}{D}\right) \tag{8.94}$$

Now, returning to Eq. 8.90, and considering the functional dependencies indicated in Eqs. 8.93 and 8.94 along with the expression given in Eq. 8.92, the independent, non-dimensional variables which govern the cross-flow vibrations of a flexibly-mounted circular cylinder may be identified as:

$$V_r\,,\ M\,,\ \zeta_s\,,\ Re\,,\ \frac{k_s}{D} \tag{8.95}$$

Traditionally, ζ_s and M are combined to give a new non-dimensional variable, the so-called **stability parameter**, according to the following definition:

$$K_s = 2M(2\pi\zeta_s) = \frac{2(m + m')(2\pi\zeta_s)}{\rho D^2} \tag{8.96}$$

Therefore, the governing non-dimensional variables will be:

$$V_r \ , \ M \ , \ K_s \ , \ Re \ , \ \frac{k_s}{D} \tag{8.97}$$

If the flow is a sheared flow, obviously the shear effect, plus the level of incoming turbulence will influence the vibrations in addition to the above mentioned parameters. The following paragraphs will summarize the influence of each individual variable on the cross-flow vibrations.

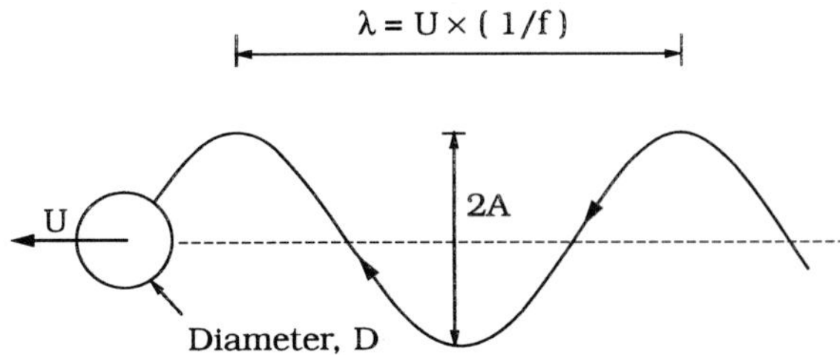

Figure 8.23   Cross-flow vibrations of a cylinder subject to a steady current may also be viewed as the periodic motion of the cylinder towed with a constant velocity in an otherwise still fluid. In this case, the reduced velocity V_r will be identical to λ/D, λ being the wave length of the trajectory of the cylinder.

Effect of reduced velocity.   The role of V_r in relation to cross-flow vibrations has already been discussed extensively in conjunction with Figs. 8.15 and 8.17. The preceding analysis indicates that the variable V_r emerges in the non-dimensional formulation of the problem through two effects: 1) It appears already in Eq. 8.90 in relation to the response of the mechanical system, and 2) it makes its way through the formulation of the force term, namely $C'_L(t)$ in Eq. 8.90 (see Eqs. 8.92 - 8.94). A simple interpretation of V_r with regard to the latter effect would be given by viewing this quantity as the ratio of the wave length of the cylinder trajectory, λ, to the diameter D, λ/D, if the cylinder was towed in still fluid with a constant velocity U having a periodic trajectory as illustrated in Fig. 8.23. In this case, the wave length of the periodic motion of the cylinder will be

$$\lambda = U \times (1/f) \tag{8.98}$$

and therefore λ/D

$$\frac{\lambda}{D} = \frac{U}{fD} = V_r \tag{8.99}$$

Hence, it is obvious that λ/D (or, alternatively, V_r) must play an extremely important role with respect to the vortex motion around (and hence with respect to the lift force on) the cylinder. This would apparently determine the excitation range of the vibrations (Figs. 8.15 and 8.17).

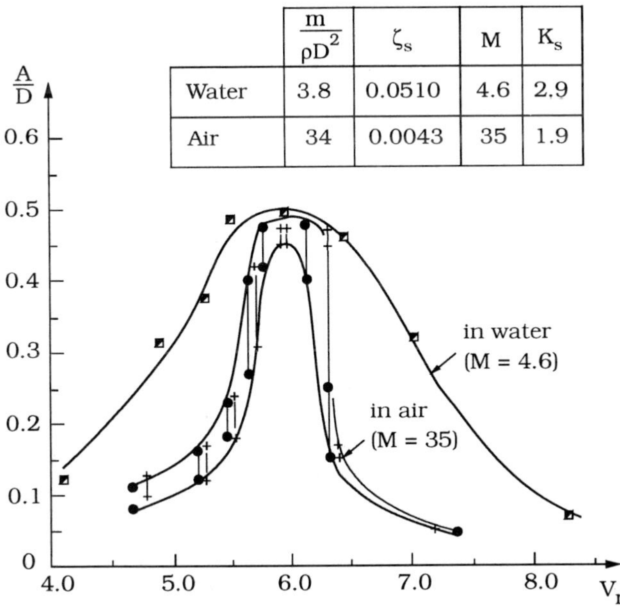

	$\dfrac{m}{\rho D^2}$	ζ_s	M	K_s
Water	3.8	0.0510	4.6	2.9
Air	34	0.0043	35	1.9

Figure 8.24 Effect of mass ratio, M, on cross-flow amplitude response. Griffin (1982).

The presently available data indicate that the excitation range of cross-flow vibrations in terms of V_r extends over $4.75 < V_r < 8$ in air where the maximum amplitude occurs in the range $5.5 < V_r < 6.5$ (see for example Fig. 8.15), (King, 1977). In water, however, the excitation range may cover a significantly broader range such as $3 < V_r < O(10)$ with maximum amplitude occurring in the range $6 < V_r < 8$ (see, for example, Fig. 8.17).

Effect of mass ratio. The mass ratio influences both the frequency response and the amplitude response.

Regarding the frequency response, this has already been discussed in Section 8.3.1, and it has been seen that, for very large values of M such as $O(100)$, the structure responds in the lock-in range with its still-fluid, natural frequency (Fig. 8.15), while for small values of M such as $O(1)$, the structure responds in the lock-in range with a frequency which increases with the reduced velocity (Fig. 8.17).

Regarding the amplitude response, Fig. 8.24 shows the role of M in the amplitude response; in the figure the responses of two systems (with not extremely different K_s but very different M values) are compared. Although the maximum responses in the two cases occur at similar values of V_r, away from the maximum response, however, the amplitudes are apparently quite different. The cylinder with larger M responds over a much narrower range.

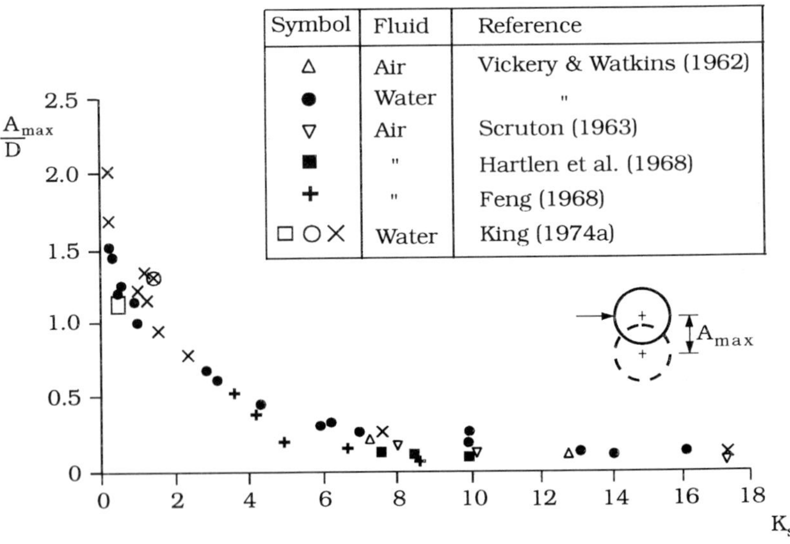

Figure 8.25 Amplitude of response as functions of stability parameter for cross-flow vibrations. Data compiled by King (1974a).

Effect of stability parameter. This is the only variable which comprises the structural damping (Eq. 8.96), and as such it must be expected to influence heavily the maximum amplitude of vibrations. It has been demonstrated that an increase in the stability parameter reduces significantly the maximum amplitude of vibrations (King (1974a) and Griffin (1981)).

The data from wind tunnel and water channel experiments compiled by

King (1974a) (Fig. 8.25) indicate that the cylinder remains virtually stationary for K_s values larger than approximately 18. For very lightly damped systems, on the other hand, the maximum amplitudes increase considerably. However, there seems to be a limit to the maximum amplitudes; no amplitudes larger than $A/D = 2$ have been observed (King, 1974a). The data compiled by Griffin (1981) appear to confirm this result.

Example 8.3: A crude model for maximum vibration amplitude as function of stability parameter

The solution to the differential equation of motion with forced vibrations and viscous damping has been discussed in Section 8.1.3. The solution presented there is obtained in the case when the external force is sinusoidal (Eq. 8.17). The analytical expressions regarding the amplitude of vibrations and the phase angle are given in explicit forms in Eqs. 8.26 and 8.27. The formulation adopted in Section 8.1.3 may be regarded as a crude model for the cross-flow vibrations of a flexibly-mounted cylinder subject to a current, where the sinusoidally varying external force is induced by vortex shedding.

The solution obtained in Section 8.1.3 gives the *maximum* amplitude as (Eq. 8.30):

$$A_{\max} = \frac{F_o}{k(2\zeta_s)} \tag{8.100}$$

in which F_0 is the amplitude of the oscillating lift force. Now, we may approximate F_0 to

$$F_0 = \frac{1}{2}\left(\sqrt{2}\left(\overline{C_L'^2}\right)^{1/2}\right)\rho D U^2 \tag{8.101}$$

since the force is assumed to be varying sinusoidally and therefore $\sqrt{2}(\overline{C_L'^2})^{1/2}$ represents the maximum value of the oscillating external force. Inserting Eq. 8.101 into Eq. 8.100 and normalizing the equation gives

$$\frac{A_{\max}}{D} = \frac{\sqrt{2}}{2\pi}\left(\overline{C_L'^2}\right)^{1/2}\frac{V_r^2}{K_s} \tag{8.102}$$

Now, first, we insert $V_r = 5$ in the preceding equation, since the maximum amplitude occurs at this point (the resonance point). Second, regarding the coefficient $(\overline{C_L'^2})^{1/2}$, although this depends on the parameters, A/D, V_r, Re, and k_s/D (Eqs. 8.92 and 8.93) we may, to a first approximation, adopt the value $(\overline{C_L'^2})^{1/2} \cong 0.3$ obtained for a stationary smooth cylinder, an average value for the subcritical

Reynolds number regime (Fig. 2.8). When these values are inserted in Eq. 8.102, the normalized maximum amplitude is obtained as

$$\frac{A_{\max}}{D} = \frac{1.7}{K_s} \tag{8.103}$$

The preceding equation is plotted in Fig. 8.26 together with the data given in Fig. 8.25. Despite the rather approximate nature of the model, the predicted amplitude of response is in surprisingly good agreement with the data for $K_s \gtrsim 2$, corresponding to $A/D \lesssim 0.8$. For smaller K_s (larger amplitudes), however, the model result begins to deviate from the data, mainly because of the inadequate representation of the forcing term (Eq. 8.101 with $\left(\overline{C_L'^2}\right)^{1/2} \cong 0.3$) in the case of a cylinder vibrating with large amplitudes.

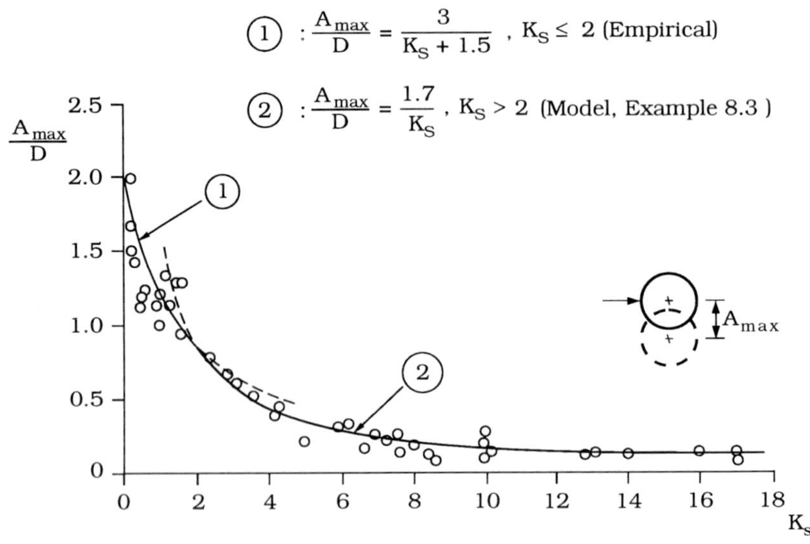

Figure 8.26 Amplitude of response as function of stability parameter for cross-flow vibrations. Comparison of the model result with the data from Fig. 8.25.

For values of K_s smaller than 2, the amplitude data in Fig. 5.26 may be represented by the following empirical equation

$$\frac{A_{\max}}{D} = \frac{3}{K_s + 1.5} \; ; \; K_s \leq 2 \tag{8.104}$$

which, together with the theoretical relation obtained in the preceding paragraphs

$$\frac{A_{\max}}{D} = \frac{1.7}{K_s} \; ; \quad K_s > 2 \; , \tag{8.105}$$

may be taken as design equations with regard to the maximum amplitude.

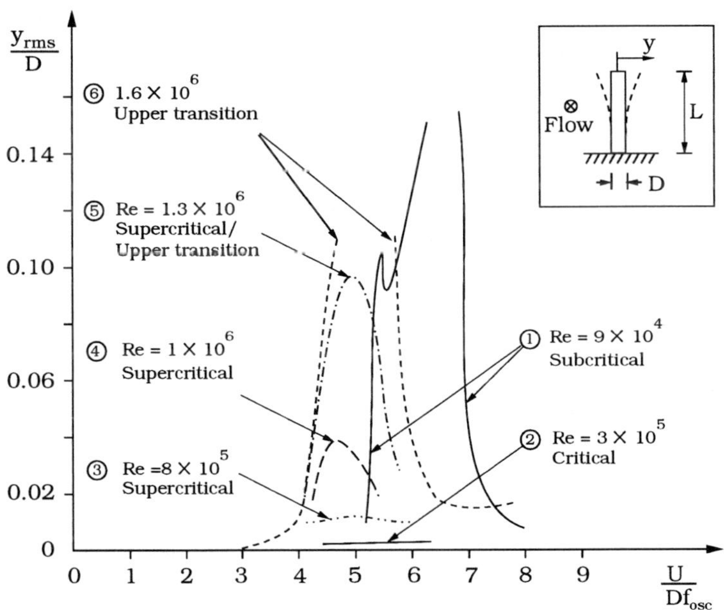

Figure 8.27 Cross-flow amplitude response against V_r with different values of Re. Stack model tested in wind tunnel. The length-to-diameter ratio $L/D = 11.5$. Stability parameter $K_s = 10$. The surface is slightly roughened $k/D = 1.3 \times 10^{-4}$, but it is hydraulically smooth. The model stack is free to move only in the cross-wind direction. Wootton (1969).

Effect of Reynolds number. Wootton (1969) studied the effect of Reynolds number on the cross-flow vibrations. Fig. 8.27, which is reproduced from Wootton's work, illustrates this effect. Note that the frequency appearing in the reduced velocity here is the frequency of oscillations rather than the natural frequency. A circular cylinder stack model was used in Wootton's study. Although the surface of the model was slightly roughened ($k/D = 1.3 \times 10^{-4}$), the surface behaved as a hydraulically smooth surface; the roughness elements were well submerged in the cylinder boundary layer near separation, as k/δ_c is estimated to be about 2% for example for $Re = 3 \times 10^5$ in which δ_c is the cylinder boundary-layer thickness near separation, $\delta_c = 3/\sqrt{Re}$ (Jensen and Sumer, 1986).

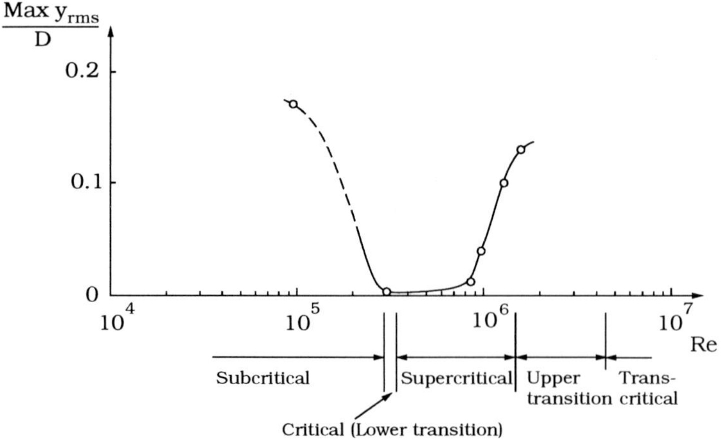

Figure 8.28 Maximum cross-flow amplitude response against Re for the data given in the previous figure. Maximum amplitudes for $Re = 8.3 \times 10^4$ and $Re = 1.7 \times 10^6$ are extrapolated values. For test conditions, see the caption of the previous figure.

Fig. 8.28 presents the maximum amplitudes extracted from Fig. 8.27 plotted as a function of Re. From Figs. 8.27 and 8.28 it is clear that the Reynolds number influences the amplitude response quite significantly. The picture in Fig. 8.28 is rather similar to that of the variation of lift force as function of Re, (Fig. 2.8b). As seen from Fig. 8.28 the amplitude is reduced considerably in the range from $Re \cong 2 \times 10^5$ to $Re \cong 10^6$. This range apparently coincides with the critical and supercritical flow regimes, and it is known that the oscillating lift is reduced tremendously in this range of Re (Fig. 2.8b), meaning that the cross-flow amplitudes should also be reduced. However, as Re is increased ($Re \lesssim 10^6$), with the regular vortex shedding re-established in the upper transition and further the transcritical flow regimes, the lift force recovers (Fig. 2.8b) and therefore the cross-flow vibrations are restored again, as clearly indicated by Fig. 8.28.

To make a rough estimate of the maximum amplitudes experienced in different Reynolds-number regimes, the results of the crude model set in Example 8.3, namely Eqs. 8.104 and 8.105, may be used where the right-hand-side of the equations must be multiplied by the factor, $(\overline{C_L'^2})^{1/2}/0.3$, in which $(\overline{C_L'^2})^{1/2}$ is the r.m.s. value of the lift coefficient. The value of $(\overline{C_L'^2})^{1/2}$ can be picked up from the diagram given in Fig. 2.8.

Regarding the influence of Re on the frequency response, Fig. 8.29 presents the frequency data obtained in Wootton's study (1969). The plotted frequencies are those deduced from the reduced velocity for maximum peak amplitudes.

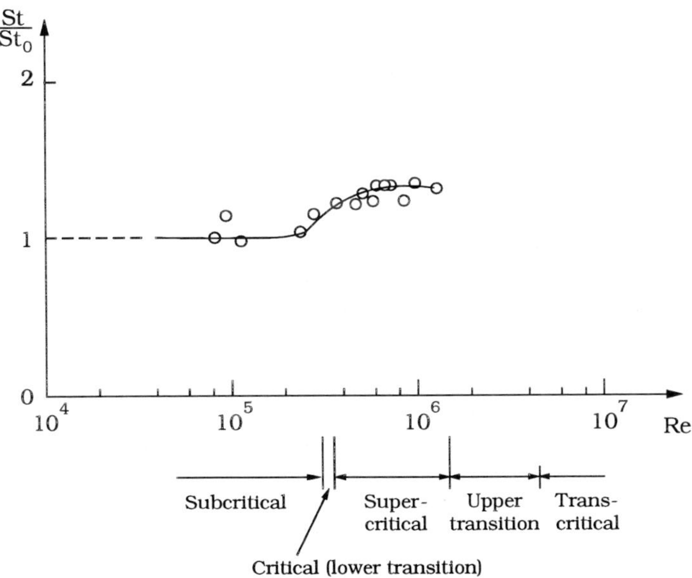

Figure 8.29 Strouhal frequency for vibrating cylinder, deduced from the reduced velocity for maximum peak amplitudes, presented in Fig. 8.27. St_0 is the Strouhal number in the subcritical range. St_0, deduced from the reduced velocity for maximum peak amplitude, is 0.16. For test conditions, see the caption of Fig. 8.27. Wootton (1969).

Clearly, these frequencies may be regarded as Strouhal frequencies, since the vibration frequency and the vortex-shedding frequency coincide when the maximum peak amplitudes are experienced (lock-in). As seen, the variation of Strouhal frequency as function of Re looks rather similar to that for a stationary cylinder (Fig. 2.8c); namely, St increases in the supercritical flow regime in the same manner as in Fig. 2.8c. However, it may be noticed that the increase in Strouhal frequency (by a factor of 1.5) is not as large as that experienced in the case of stationary cylinder where an increase in St by a factor of 2-2.5 is observed, as the flow regime changes from subcritical to supercritical (Fig. 2.8c).

As seen from the preceding discussion, the effect of Re on cross-flow vibrations is very significant. The key point here is that the process of vortex shedding (therefore the vortex-induced lift on the cylinder) changes with the Reynolds number. It is known, however, that the process of vortex shedding is heavily influenced also by several other factors such as the presence of turbulence in the incoming flow (Figs. 1.15 and 2.17) the surface roughness of the cylinder (Fig. 1.12) and the shear in the incoming flow (Figs. 1.17 and 1.18). Likewise, the presence of in-line vibrations should be expected to influence the vortex shedding process, therefore the oscillating lift, on the cylinder. In practice, one or more of these effects will always be present, and therefore the ideal conditions in Wootton's experiment (namely, the low level turbulence, (about 0.5%), the extremely small surface roughness, the shear-free flow, and no in-line movement) can harly be achieved. Hence the Reynolds-number influence, which is related exclusively with a smooth-surface cylinder in a smooth uniform flow, may not be felt as strongly as in Wootton's study (Fig. 8.28).

Fig. 8.30a presents the cross-flow amplitude data obtained in a field study with a full-scale submarine pipeline span in the strongly tidal Severn Estuary (U.K.) (Raven, Stuart and Littlejohn, 1985). The surface of the pipe was hydraulically smooth. The gap between the pipe and the bed was two times the pipe diameter so that the pipe could be regarded as a free cylinder. The data in Fig. 8.30a shows a constant increase in the amplitude, as the velocity increases during the tidal flow. As seen, the Reynolds number influence is not present, since the r.m.s. amplitudes reach as large values as 0.5 times the pipe diameter (cf. Fig. 8.27) in the supercritical range. This behaviour may be due partly to the presence of in-line vibrations and partly to the turbulence in the approach flow. This example substantiates the argument put forward in the preceding paragraph that the Reynolds number effect in a real-life situation may not be very strong (or it may be totally absent), as indicated in Raven et al.'s field experiments. Other field data are available, revealing the presence of large vibrations in the Reynolds number range where the Re-number influence is expected to be present so as to reduce the vibrations; large cross-flow vibrations were measured with full-scale steel marine piles during the construction of Immingham Oil Terminal where the Reynolds number was in the order of magnitude of 10^6 (Sainsbury and King, 1971).

Effect of surface roughness. It is known that the surface roughness influences the vortex shedding and therefore the oscillating lift. As has been seen in Chapter 2, the critical, supercritical, and upper transition flow regimes merge into one single narrow range for rough cylinders ($k_s/D > 3 \times 10^{-3}$) (Fig. 2.12), and the flow regime switches directly from subcritical to transcritical over this narrow range. The latter implies that the extremely small lift oscillations experienced in the case of smooth cylinder in the previously mentioned transitional regimes shown in Fig. 2.8b may not be experienced in the case of rough cylinders. Therefore it may be anticipated that in these transitional regimes, the rough cylinder does not

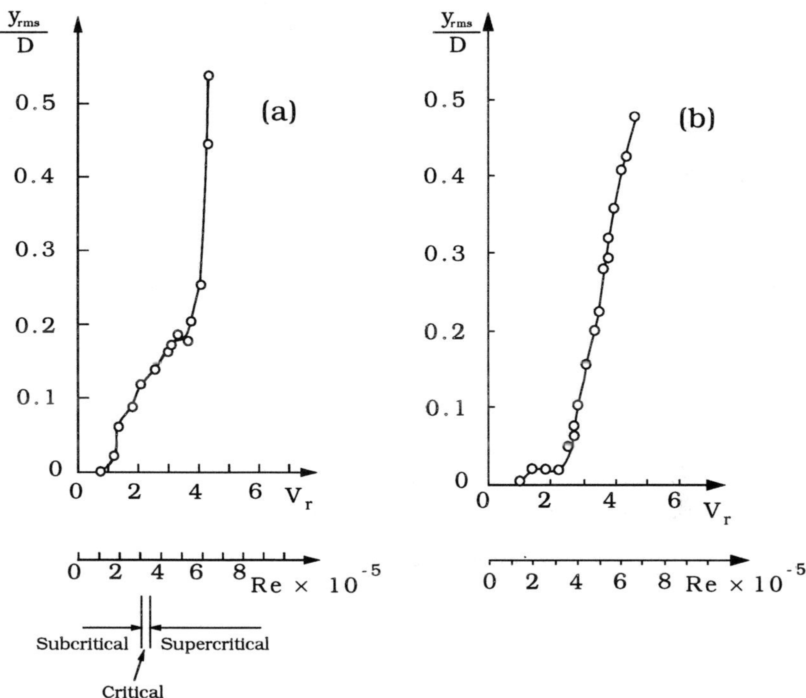

Figure 8.30 Cross-flow amplitude response against V_r at large Re numbers
with a full-scale pipeline of 50.8 cm diameter with 40 m span
length. a: smooth pipe. b: rough pipe ($k/D = 8.5 \times 10^{-3}$).
Raven et al. (1985).

experience the same kind of reduction in its cross-flow amplitude as in the case
of smooth cylinder (cf. Fig. 8.28). This aspect of the problem unfortunately has
not been investigated in a systematic manner. However, the sporadic data available suggest that the magnitude of the response amplitude does not change very
significantly with changing roughness, provided that the roughness is sufficiently
large (larger than approximately 3×10^{-3}).

Fig. 8.30b presents the cross-flow amplitude response obtained with the
same full scale pipeline as in Fig. 8.30a but with a surface roughness of $k/D =
8.5 \times 10^{-3}$. As seen, the change in the roughness does not cause any significant
change in the value of V_r where the onset of vibrations occurs. Neither does it
cause any significant change in the maximum amplitudes of vibrations.

Effect of sheared flow. Humphries and Walker (1987) made a study of

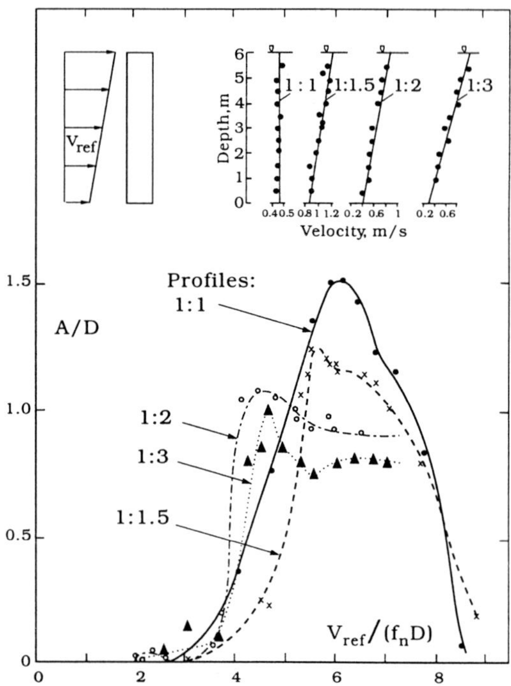

Figure 8.31 The influence of shear on cross-flow vibration amplitude for a circular smooth pipe with $D = 16.8$ cm. Humphries and Walker (1987).

the influence of sheared current on cross-flow vibrations (Fig. 8.31). They found that increasing the slope of the linear shear profile decreased the peak amplitude response but broadened the range of lock-on over which large vibrations occurred. Apart from sporadic indications, partly discussed in the preceding paragraphs, no systematic study is available today investigating the influence of the level of turbulence in the incoming flow.

8.4   In-line vibrations of a circular cylinder

As mentioned earlier, a cylinder subject to a steady current may, due to vortex shedding, experience an oscillating drag force (Fig. 2.3). If the cylinder is a flexibly-mounted cylinder, this oscillating drag force may induce in-line vibrations (Fig. 8.13b).

Observations show that there are three kinds of in-line vibrations. Of the three, two of them occur at small values of the reduced velocity, namely one in the region $1 \lesssim V_r \lesssim 2.5$, the so-called **first instability region**, and the other in the region $2.5 \lesssim V_r \lesssim 4$, the so-called **second instability region**, see Fig. 8.32. The third kind of in-line vibrations (observed for cylinders with two degrees of freedom of movement), on the other hand, occurs at somewhat higher flow velocities, at velocities where the cross-flow vibrations are observed. The following paragraphs give a detailed account of these three kinds of in-line vibrations.

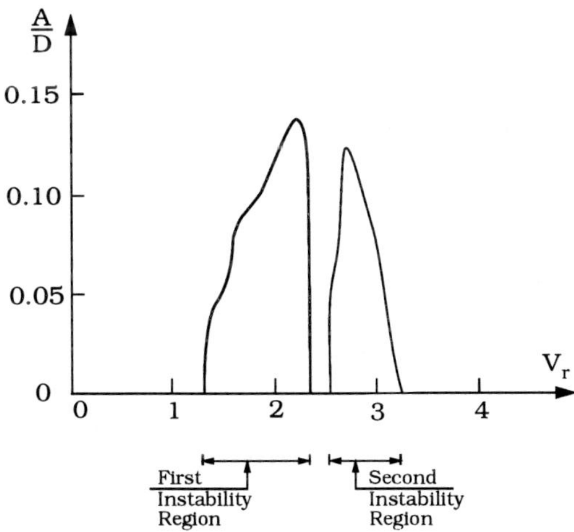

Figure 8.32 In-line vibrations. $Re = 6 \times 10^4$. King (1974b).

First- and second-instability in-line vibrations

The first-instability-region in-line vibrations are caused by the combined action of 1) normal vortex shedding giving rise to two oscillations per shedding, and 2) secondary, symmetric vortex shedding which occurs as a result of in-line motion of the cylinder relative to the fluid (Fig. 8.33). This vortex shedding creates a flow situation where the in-line force ocillates with a frequency, approximately three times the Strouhal frequency:

$$\frac{f_x D}{U} = 3\, St \tag{8.106}$$

where f_x = the frequency of in-line force and St = Strouhal number (Wootton et al., 1974). If this frequency is close to the natural frequency of the system f_n, the cylinder will vibrate in the in-line direction with large amplitudes:

$$f_x = f_n \tag{8.107}$$

From Eqs. 8.106 and 8.107, it is seen that this will occur when the value of reduced velocity V_r becomes

$$V_r = \frac{1}{3\,St} = \frac{1}{3 \times 0.2} \cong 1.7 \tag{8.108}$$

This kind of vibration is what occurs in the first instability region.

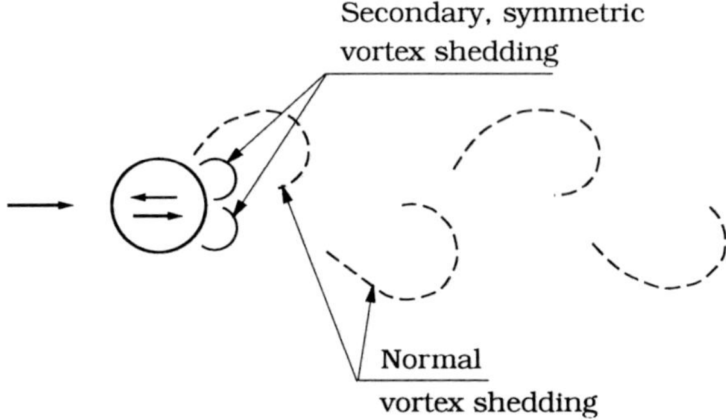

Figure 8.33 Schematic description of vortex shedding pattern in the first-instability region.

As the above mechanism suggests the vibrations must be existent to get the secondary, symmetric vortex shedding. In a smooth flow with no turbulence or any other disturbances, the vibrations may not be excited. Currie and Turnbull's (1987) study is quite indicative in this regard: they observed no in-line vibrations when the turbulence was removed from the flow.

If we gradually increase V_r from 1.7, Eq.8.108 shows that the in-line force frequency $f_x (= 3\,St\,\frac{U}{D})$ will become higher and higher, moving steadily away from f_n, thereby ending the previously mentioned type of vibrations.

As the vibrations stop, the normal vortex shedding will be restored again, and the in-line force will start oscillating with the familiar frequency (Section 2.3):

$$\frac{f_x D}{U} = 2\,St \tag{8.109}$$

If the velocity is increased even further, f_x will increase according to Eq. 8.109, and the large-amplitude in-line vibrations will occur again when the frequency f_x in Eq. 8.109 becomes equal to f_n (second lock-in).

These second large-amplitude in-line vibrations will therefore come into existence when

$$V_r = \frac{1}{2\,St} = \frac{1}{2 \times 0.2} = 2.5 \qquad (8.110)$$

This is termed the second instability.

The actual location of the V_r value at which the maximum lock-in vibrations occur is determined by the St number. The St number itself is dependent upon various factors such as the Reynolds number, pipe roughness, wall proximity, etc. (Chapter 1). Fig. 8.34 clearly reveals this. The Strouhal number for the full scale marine pile (transcritical flow regime) is higher than for the model pile (subcritical flow regime). This means that, according to Eq. 8.108, the response curve should shift to the left in Fig. 8.34 because the St number increases in this flow regime.

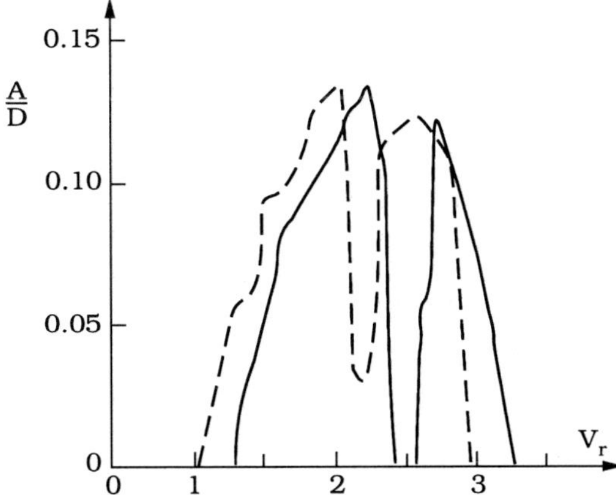

Figure 8.34 Comparison of full-scale marine pile in-line vibrations with small-scale model results. Dashed curve: Full-scale pile with $Re = 6 \times 10^5$. Solid curve: Small-scale model with $Re = 6 \times 10^4$. King (1974b).

Likewise, the V_r range over which the in-line vibrations occur depends on the previously mentioned parameters. King et al. (1973) report that the first instability region covers the range $1.25 < V_r < 2.5$ and the second $2.5 < V_r < 3.8$.

As for the maximum amplitudes attained, the presently available data are not conclusive as to how the first-instability-region vibration amplitude compares with the second-instability-region one. Although the data reproduced in Fig. 8.35 indicate that the second-instability-region amplitude is larger than the first-instability-region one, there is one case in the figure where the opposite is true.

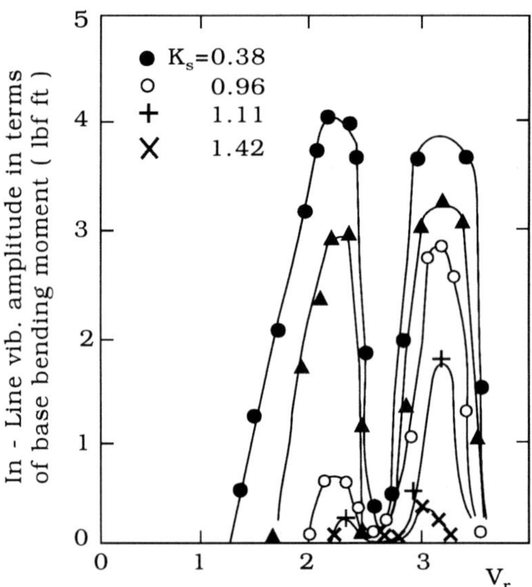

Figure 8.35 Comparison of several first- and second-instability-region in-line vibrations. King et al. (1973).

Fig. 8.36 gives the maximum in-line vibration amplitudes (irrespective of the instability regions) versus the stability parameter, which is reproduced from King (1977). It represents the data compiled from laboratory experiments (King, 1974a) and a full-scale test (Wootton, 1972). King (1977) notes that the results of Wooton's full-scale test fall on the common curve. The latter implies that no Reynolds number effect is experienced. This may be due to several factors, as mentioned previously in conjunction with cross-flow vibrations, such as the presence of turbulence in the flow, the surface roughness of the cylinder and the shear in the incoming flow.

When compared with the cross-flow vibration amplitudes (Fig. 8.25), the in-line vibration amplitudes are seen to be one order of magnitude smaller than the cross-flow vibration amplitudes (Fig. 8.37). This is because the force (and therefore the flow velocity) which is required to initiate the in-line vibrations is

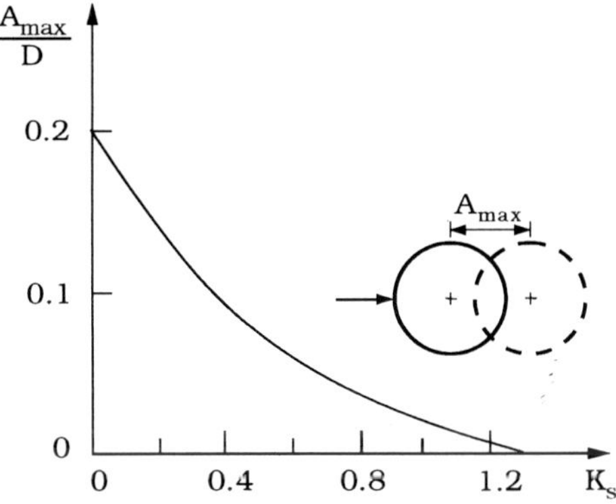

Figure 8.36 Maximum amplitude of in-line vibrations versus the stability parameter. King (1977).

far smaller than that which is required for the cross-flow ones. As an additional effect, the force coefficient (C'_D) is also smaller in the former case than C'_L in the latter (cf. Fig. 2.15).

Third kind in-line vibrations

As mentioned in the beginning of this section, observations reveal that there exists a third type in-line vibrations. These vibrations occur in the region where the cross-flow vibrations take place with a system with two degrees of freedom of movement; see Fig. 8.38, which is reproduced from Tsahalis (1984). Similar behaviour has been observed also by Bryndum et al. (1989).

As seen from the figure, the in-line vibration amplitudes experienced in this region are much larger than in the second instability region. The explanation for this third type of in-line vibrations may be given as follows. From Fig. 8.38 it is seen that the in-line vibrations in this region occur mainly at a frequency which is twice the cross-flow vibration frequency. This implies that the in-line force acting on the cylinder is still oscillating at the frequency

$$\frac{f_x U}{D} = 2\,St \tag{8.111}$$

which is now far from the natural frequency f_n, meaning that the in-line vibrations are now occurring well away from the second-instability-region lock-in point. Yet the vibration amplitudes are considerably larger than those experienced in the second-instability region.

This can be explained by the considerable increase in the in-line force amplitude in this region. The following two effects may be responsible for this increase:

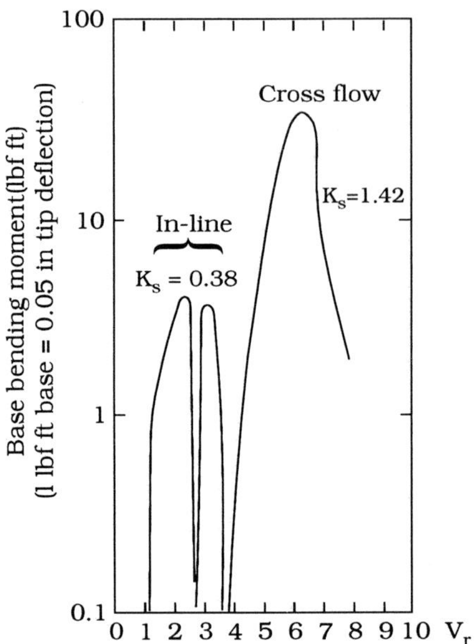

Figure 8.37 Comparison of in-line vibrations with the cross-flow ones. King
et al. (1973).

1) The cylinder now experiences much higher velocities; and more impor-
tantly:

2) When the cylinder is vibrating with large amplitudes in the cross-flow
direction, the strength of the shed vortices will become stronger, and also the vortex shedding itself will occur in a more orderly fashion (larger correlation lengths),
which will altogether lead to a considerable increase in the force coefficient C_D',
see also Fig. 8.47. As seen from this figure, the magnification in the fluctuating
drag force becomes substantial only after the cross-flow amplitude A/D becomes
greater than 0.2-0.3. For this reason, no significant in-line vibrations should be
expected if the cross-flow vibration amplitudes are below that level.

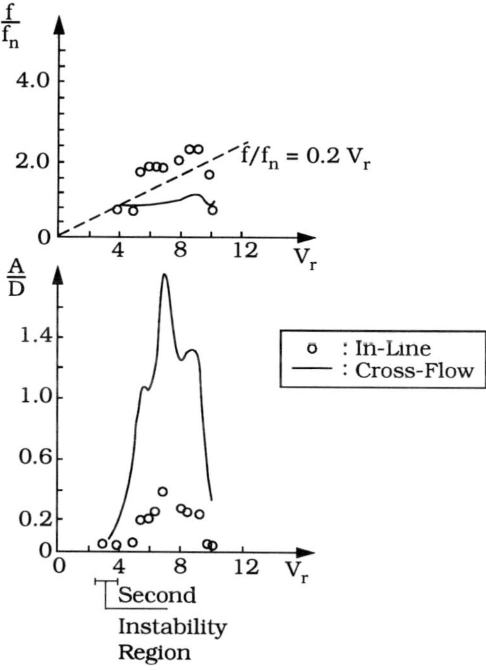

Figure 8.38 In-line and cross-flow vibrations in Tsahalis' (1984) experiments. The cylinder has two-degrees of freedom of movement. $Re = 10^3 - 10^4$. $K_s = 0.5$.

8.5 Flow around and forces on a vibrating cylinder

8.5.1 Cylinder oscillating in cross-flow direction

Flow. In the case of a cylinder oscillating transversely in a steady current (cross-flow oscillations), the relevant parameters to describe the flow are

$$\frac{A}{D} \quad \text{and} \quad V_r \tag{8.112}$$

in addition to the parameters governing the case of a stationary cylinder like Re, k_s/D, etc.. Here V_r is based on the cylinder vibration frequency, f, namely $V_r = U/(Df)$.

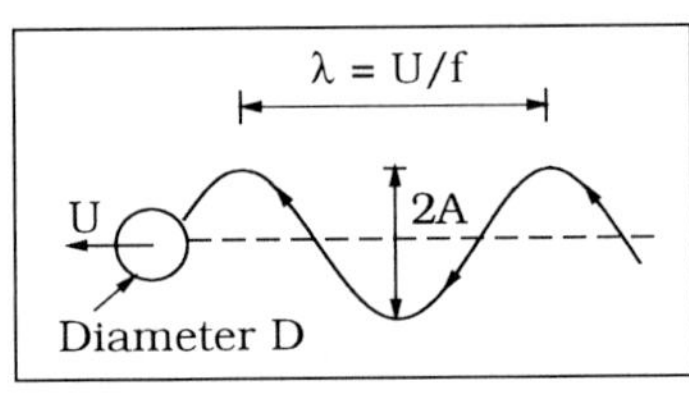

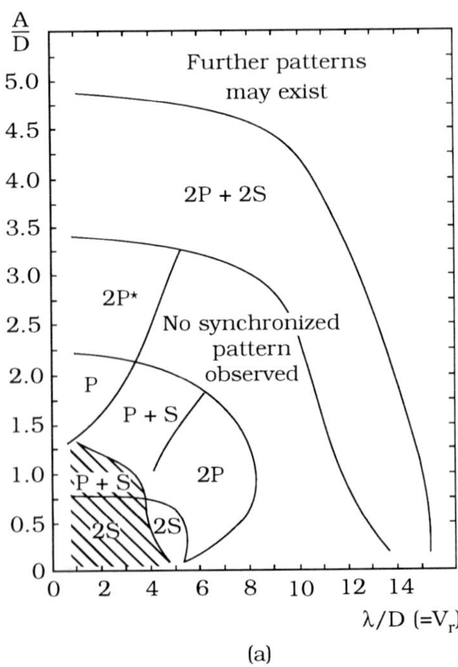

(a)

Figure 8.39a Map of vortex-flow regimes for a cylinder oscillating cross-flow in a steady current. See Fig. 8.39b for the legend. $3 \times 10^2 < Re < 10^3$. Williamson and Roshko (1988).

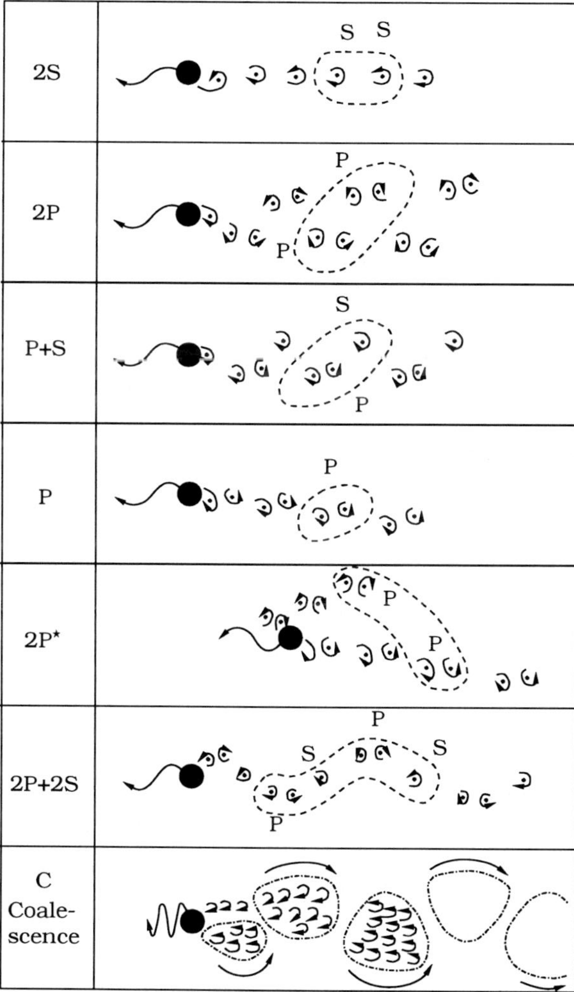

Figure 8.39b Legend for Fig. 8.39a. Sketches of the vortex shedding patterns that are found in the map in Fig. 8.39a. "P" means a vortex pair and "S" means a single vortex, and each pattern is defined by the number of pairs and single vortices formed per cycle. The dashed line encircles the vortices shed in one complete cycle. The wavy arrow at the cylinder indicates its movement relative to the still water. Williamson and Roshko (1988).

The involvement of the parameter A/D is quite straightforward. A simple interpretation of the second parameter, V_r, on the other hand, has been given earlier (Fig. 8.25). Namely, V_r may be viewed as the ratio of the wave length of the cylinder trajectory, λ, to the diameter D, if the cylinder is towed in still fluid with a constant velocity U:

$$V_r \equiv \frac{\lambda}{D} \tag{8.113}$$

Emphasizing the trajectory in the point of view is often useful to better understand the cylinder-vortex interaction, as has been pointed out by Williamson and Roshko (1988). We may therefore write the relevant parameters as

$$\frac{A}{D} \, , \, \frac{\lambda}{D} \tag{8.114}$$

This section will focus on the influence of these two parameters.

Williamson and Roshko (1988) has made an extensive study of flow around a circular cylinder oscillating in a steady current. The cylinder in Williamson and Roshko's experiments was forced to oscillate. They found several flow regimes as a function of A/D and λ/D. Their key diagram, summarizing these flow regimes, is reproduced here in Fig. 8.39a. The legend for the figure is given in Fig. 8.39b. The Reynolds-number range in the experiments of Williamson and Roshko was $3 \times 10^2 < Re < 10^3$. No experimental data exists for higher Re-numbers.

The A/D axis in the $(A/D, \ \lambda/D)$ plane corresponds to the special case of planar oscillatory flow (i.e., zero current velocity). In this case, repeatable vortex-flow regimes have been found for certain ranges of amplitude-to-diameter ratio A/D (or alternatively $KC(= 2\pi A/D)$), as discussed extensively in Sections 3.1 and 3.2. The flow regimes observed for an oscillating cylinder in a steady current must therefore approach asymptotically to the previously mentioned flow regimes observed in the case of planar oscillatory flow, as λ/D tends to zero.

As mentioned earlier, the cylinder in Williamson and Roshko's experiments was forced to oscillate. In this case, depending on the values of A/D and λ/D, the frequency of vortex formation may not be synchronized with the body-motion frequency. The region in which no synchronization has been observed is indicated in Fig. 8.39a.

Figs. 8.40 and 8.41 illustrate how the vortex-flow patterns evolve during the course of one cycle of oscillations for two most important cases (regarding the practical application), namely in the case of "2S" (Fig. 8.40) and "2P" (Fig. 8.41) modes, for values of $\lambda/D = 4.5$ (Fig. 8.40) and 5.5 (Fig. 8.41) for the same value of A/D, namely $A/D = 0.5$. As is seen, the small change in the value of λ/D from 4.5 to 5.5 causes the flow regime to change from one mode ("2S" mode) to another ("2P" mode).

Forces. Fig. 8.42 is a close-up picture of the maps of vortex synchronization regions extracted from Fig. 8.39a.

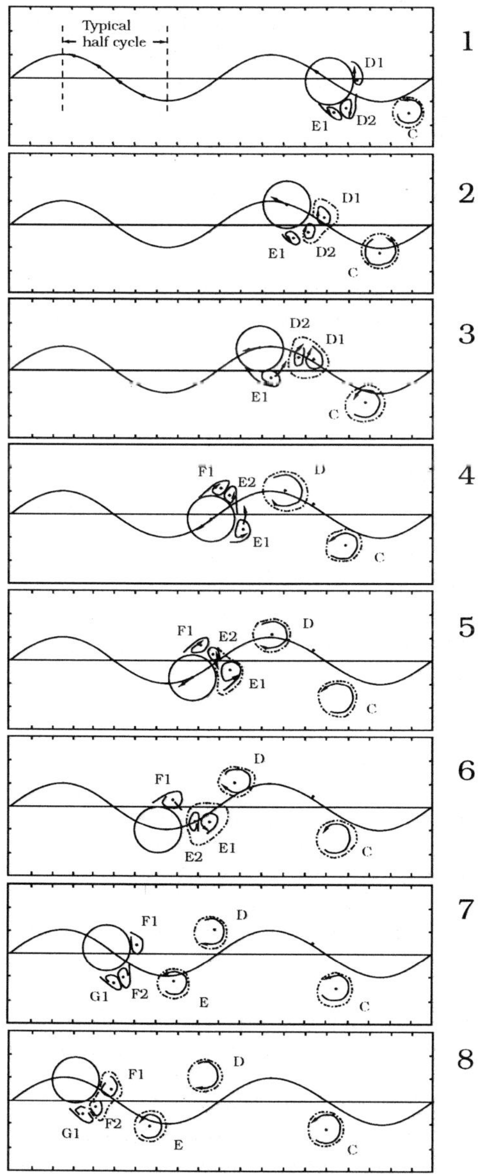

Figure 8.40 "2S" mode. Sketch of vortex motions. $\lambda/D = 4.5$, $A/D = 0.5$, $Re = 392$. Williamson and Roshko (1988).

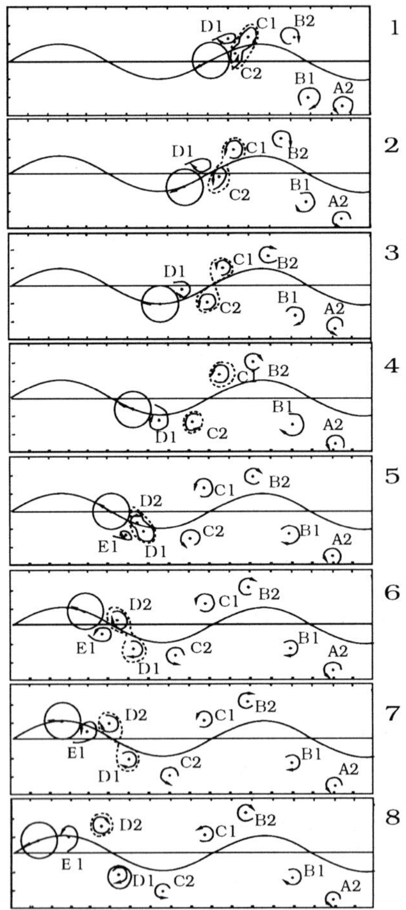

Figure 8.41 "2P" mode. Sketch of vortex motions. $\lambda/D = 5.5$, $A/D = 0.5$, $Re = 392$. Williamson and Roshko (1988).

The implication of Fig. 8.42 is that the hydrodynamic forces on the cylinder may undergo drastic changes if the boundaries between different regions in the plane $(A/D, \lambda/D)$ are crossed. Typical examples of this are given in Figs. 8.43a and b, taken from Bishop and Hassan (1964). Regarding the lift force variation (Fig. 8.43a), the lift force continually increases until λ/D (i.e., the reduced velocity V_r) reaches the value of 5.3. At this point, however, it undergoes a sudden drop.

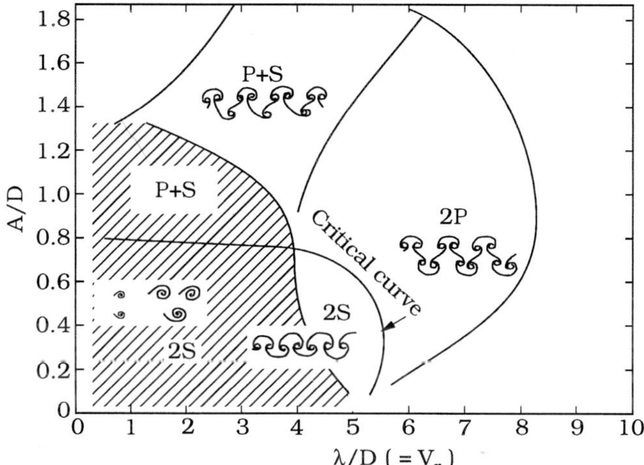

Figure 8.42 Map of vortex synchronization patterns near the fundamental lock-in region. The critical curve marks the transition from one mode of vortex formation to another. Hatched area is where the "coalescence" type shedding (see Fig. 8.39b) occurs. Williamson and Roshko (1988).

This sudden drop in the lift force is explained by the fact that, at this point, the flow regime changes from "2S" mode to "2P" mode (Fig. 8.42). Note the similar jump in the phase depicted in Fig. 8.43a. Work by Zdravkovich (1982), who examined other published visualization results (den Hartog (1934), Meier-Windhorst (1939), Angrilli et al. (1974) and Griffin and Ramberg (1974)), and later by Öngören and Rockwell (1988), revealed that the vortex shedding modes are not the same on the two sides of the phase jump. The subject has been elaborated recently by Williamson and Roshko (1988) and Brika and Laneville (1993).

Regarding the drag force given in Fig. 8.43b, similar behaviour is observed also in this case. Fig. 8.43b further shows a hysteresis effect. The same kind of hysteresis has been reported by Bishop and Hassan (1964) also for the lift force and its phase (not included in Fig. 8.43a). This issue has been discussed previously in conjunction with Feng's experiments in relation to cross-flow vibrations of a circular cylinder (Section 8.3.1 and Fig. 8.16). According to Williamson and Roshko (1988), it is possible that, in a certain range of λ/D (or V_r), either of the two modes, namely the "2S" mode and the "2P" mode, can exist. If this is the case, they argue, then the chosen mode will be dictated by the history of the

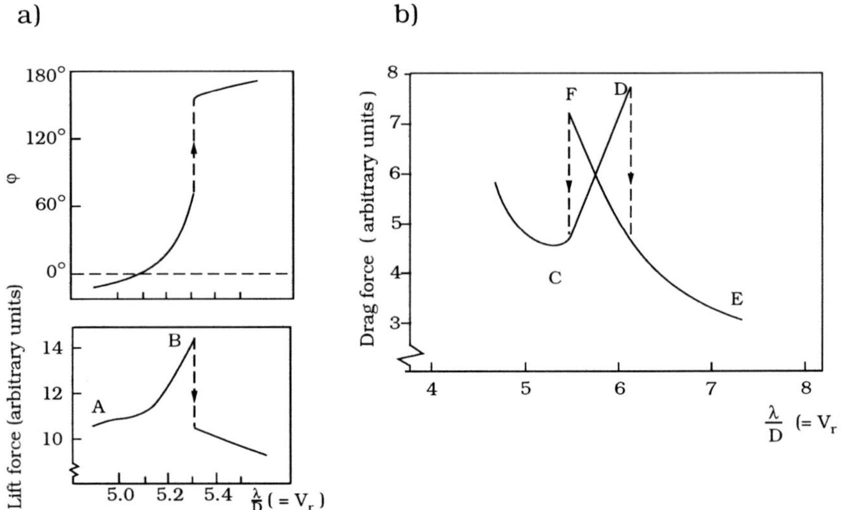

Figure 8.43 a) Variation of lift force and its phase ϕ (ϕ being the phase angle between the lift and the body motion) as λ/D is varied for a cylinder forced to oscillate. $A/D = 0.2$. b) Variation of mean drag. $A/D = 0.3$. Bishop and Hassan (1964).

flow, and this would explain Bishop and Hassan's hysteresis. Obviously, the way in which the quantity λ/D (or V_r) is increased/decreased is important; the history of the flow can be important only in the case where λ/D is increased/decreased at small increments (Brika and Lanewille, 1993).

Figs. 8.43a and b indicate that both the drag and the lift increase considerably in the synchronization range. This increase is due partly to the increase in the spanwise correlation when the cylinder is oscillated, as discussed in Section 1.2.2 (see Fig. 1.28). The continuos increase in the force with λ/D (from A to B in Fig. 8.43a and from C to D and from E to F in Fig. 8.43b), on the other hand, is linked to the relation between the vibration frequency and the frequency of vortex formation. As vibration frequency approaches the frequency of vortex formation (i.e., as one proceeds from point A to B in Fig. 8.43a or from point C to D or from point E to F in Fig. 8.43b), the process of vortex shedding will occur at the same tempo as the cylinder vibrations (synchronization). Hence, the end result will be a substantial enhancement in the force components.

The ratio of the maximum force to the force corresponding to the case of stationary cylinder may be termed the **force amplification factor**. Fig. 8.44

gives the lift-force amplification factor as a function of the amplitude-to-diameter ratio.

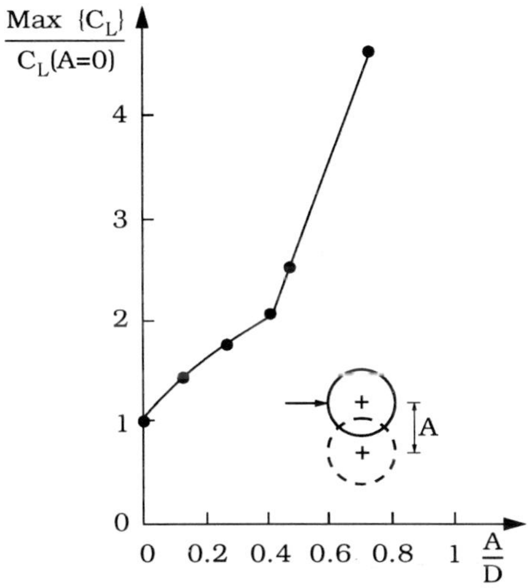

Figure 8.44 Amplification in the lift force for a cylinder vibrating in the cross-flow direction $Re = 6 \times 10^3$. Bishop and Hassan (1964).

This diagram shows that the force amplification increases with increasing amplitude. Bishop and Hassan's (1964) data agree quite well with King's (1974a, also see 1977) predictions made with the help of a linear, mathematical model in the range $A/D \stackrel{\sim}{<} 0.5$, see Fig. 8.45. (King's model is basically similar to the crude model given in Example 8.4, see Eq. 8.102; the lift coefficient is calculated, based on the experimentally obtained amplitudes of flexibly-mounted cylinders through an equation similar to Eq. 8.102). The amplification factor predicted by King's model begins to decrease from $A/D \cong 0.5$ and becomes zero when $A/D = 1.5 - 2$ (Fig. 8.45). This must be linked to the change in the mode of vortex synchronization patterns summarized in Fig. 8.42.

Fig. 8.46 illustrates the amplification in the mean drag obtained by Sarpkaya (1978). It is evident that the in-line force increases with A/D; this may be related to the fact that the cylinder, undergoing cross-flow oscillations, presents a larger projected area to the mean flow. Sarpkaya notes that a calculation based on the steady-flow drag coefficient for a stationary cylinder and the apparent projected area for the in-line force yields

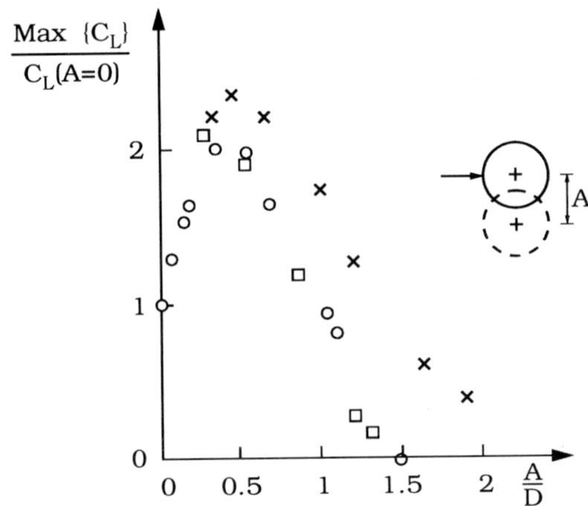

Figure 8.45 Amplification in the lift for a cylinder vibrating in the cross-flow direction. Circles: Vickery and Watkins (1962), $Re = 10^4$. Squares and crosses: King (1974a, 1977) with different cylinder roughness, $Re = 4 \times 10^4$ where lift is not measured directly but predicted from the measured amplitudes of flexibly-mounted cylinders. King (1974a, 1977).

$$\frac{\text{Max}\{C_D\}}{C_D(A=0)} = 1 + 2\frac{A}{D} \tag{8.115}$$

The preceding relation is apparently in very good agreement with the data plotted in Fig. 8.46.

Finally, Fig. 8.47 shows the amplification factor regarding the fluctuating drag force obtained in the study of Bishop and Hassan (1964). The increase in the fluctuating drag may be interpreted in the same way as in the case of lift (Fig. 8.44).

Effect of close proximity of a wall. In the case when the cylinder is oscillating transversely near a wall (the pipeline situation), the presence of the wall will influence the force coefficients. Figs. 8.48 and 8.49 show the results by Sumer, Fredsøe, Jensen and Christiansen (1994)

The definitions of the lift coefficients in the figure are given as

$$F_{yA} = \frac{1}{2}\rho C_{LA} D U^2 \tag{8.116}$$

$$F_{yT} = \frac{1}{2}\rho C_{LT} D U^2 \tag{8.117}$$

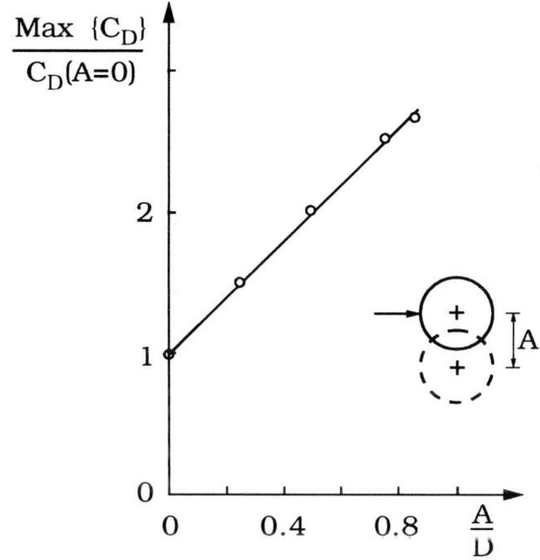

Figure 8.46 Amplification in the mean drag for a cylinder vibrating in the cross-flow direction $Re = 5 \times 10^3 - 2.5 \times 10^4$. Sarpkaya (1978).

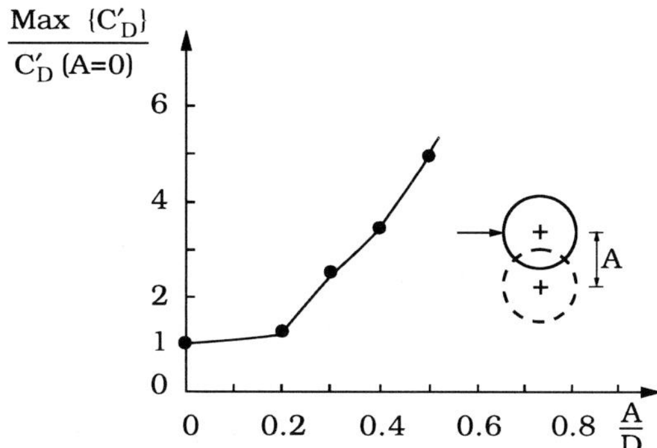

Figure 8.47 Amplification in the fluctuating drag coefficient for a cylinder vibrating in the cross-flow direction. $\left(\overline{C_D'^2}\right)^{1/2}$ is denoted by C_D' in the figure for simplicity. $Re = 1.1 \times 10^4$. Bishop and Hassan (1964).

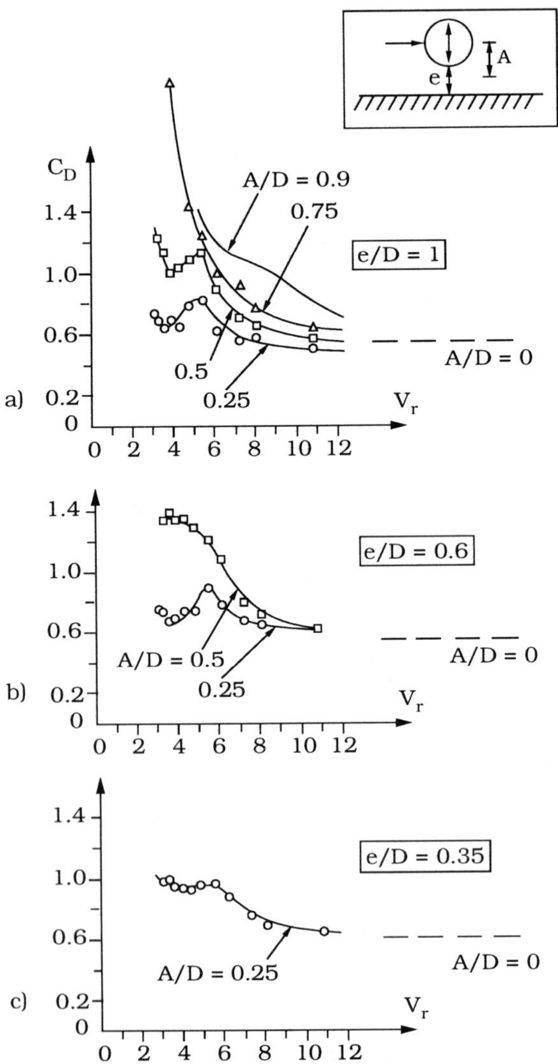

Figure 8.48 Mean in-line force coefficient for a cylinder subject to forced vibrations in the cross-flow direction. Effect of close proximity of a wall. $Re = 6 \times 10^4$. Sumer et al. (1994).

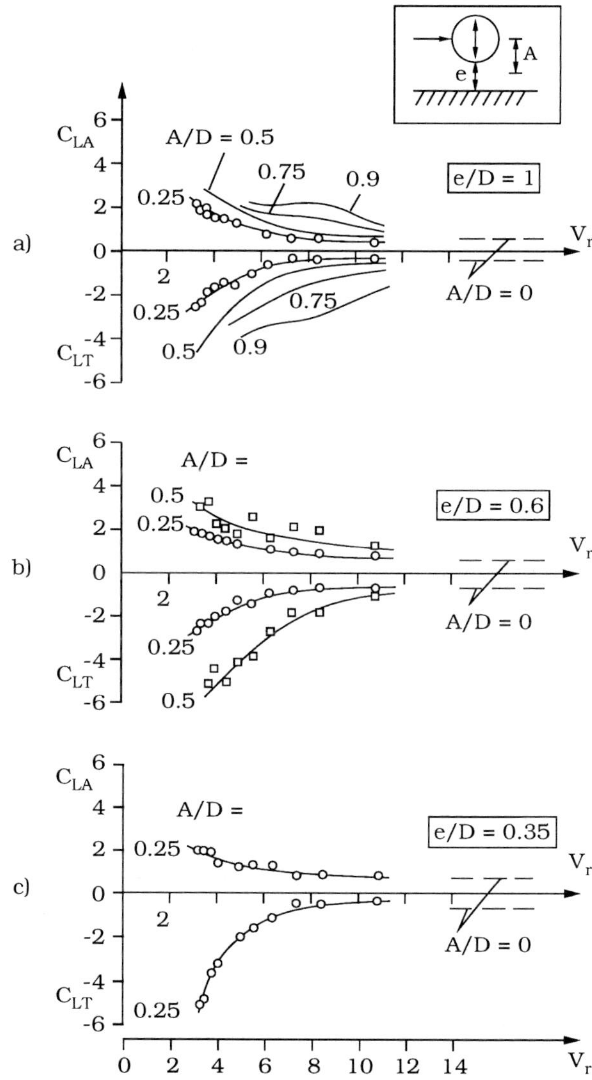

Figure 8.49 Lift force coefficient defined by Eqs. 8.124 and 8.125 for a cylinder subject to cross-flow vibrations in the cross-flow direction. Effect of close proximity of a wall. $Re = 6 \times 10^4$. Sumer et al. (1994).

in which F_{yA} is the maximum value of the lift force away from the wall and F_{yT} that towards the wall.

Fig. 8.49 indicates that the lift increases with increasing amplitude. However, as V_r is increased (i.e. the frequency is decreased), the lift asymptotically goes to its stationary-cylinder value, similar to Fig. 8.48. One may also note from the figure that the cylinder experiences considerable negative lift. This aspect of the problem is common to all types of flow, steady or oscillatory (as demonstrated by Sumer et al.). These large, negative lift forces are experienced at the instants when the cylinder approaches the wall.

8.5.2 Cylinder oscillating in in-line direction

In the case of a flexibly-mounted cylinder oscillating in the in-line direction, a visualization study of the vortex formation mechanism by King et al. (1973) (see also King, 1977) revealed that, in the first instability region, vortices were shed symmetrically, as shown in Fig. 8.50a. (However, the symmetrically-shed vortices adopt the familiar vortex-street configuration within a short distance from the cylinder. This is because the symmetric arrangement of vortices is theoretically unstable; see the concluding remarks in Example 5.3). King et al.'s visualization study furthermore indicated that vortices were shed from alternate sides of the cylinder in the second instability region (Fig. 8.50b).

Similar vortex pattern, namely symmetrical and alternate vortex shedding in the first and second instability region, respectively, were observed in the wake behind the full-scale piles (Wootton et al., 1974).

As far as the fluctuating drag (i.e., the force driving the in-line vibration) is concerned, King (1974a, 1977) gives this as a function of A/D for both the first and the second instability regions, as shown in Fig. 8.51. (The way in which the data are presented in this figure is slightly different from King's presentation). The figure shows the fluctuating drag increasing linearly with amplitude. Unlike the cross-flow vibration results (Fig. 8.45), Fig. 8.51 implies that the amplitudes do not tend to a finite limiting value although, in practice, the maximum amplitude recorded are of the order of 0.2 diameters (Fig. 8.36).

Finally, it may be noted that force measurements have been reported recently by Moe, Holden and Yttervoll (1994) and Sarpkaya (1995) for a cylinder which is oscillating freely in both the in-line and cross-flow directions.

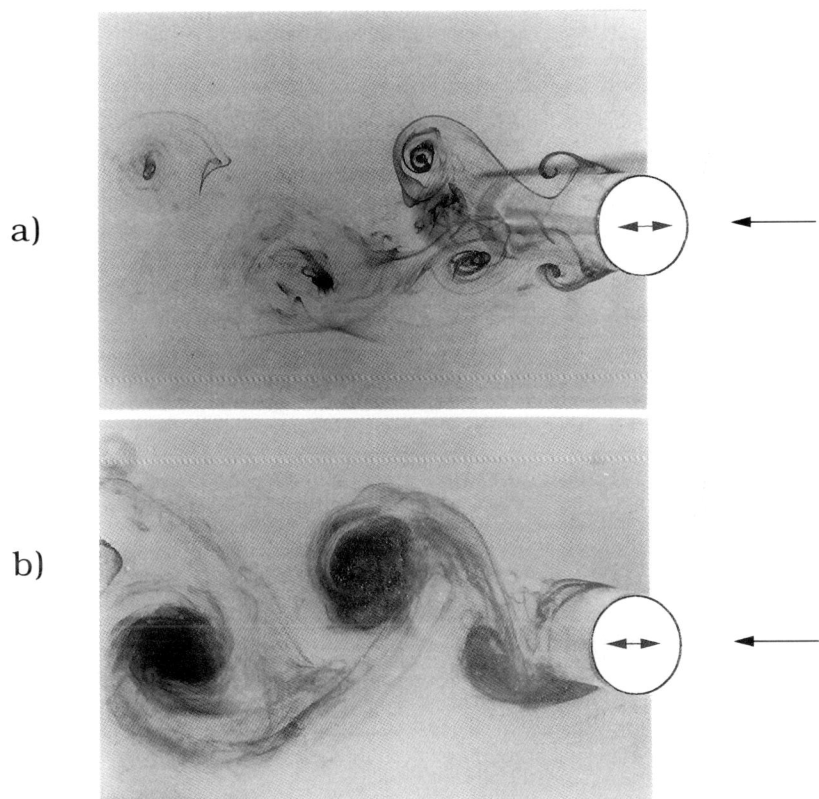

Figure 8.50 (a): Symmetric vortex shedding in the case of a cylinder oscillating in the in-line direction due to self-excited vibrations.
The first instability region. (b): That in the second instability
region. King (1977) with permission – see Credits.

8.6 Galloping

In practice, flow-induced vibrations may be caused by effects other than
vortex shedding. Pipeline vibration in close proximity of the bed (Chapter 10)
and the so-called galloping type vibration are two examples. In the latter, the

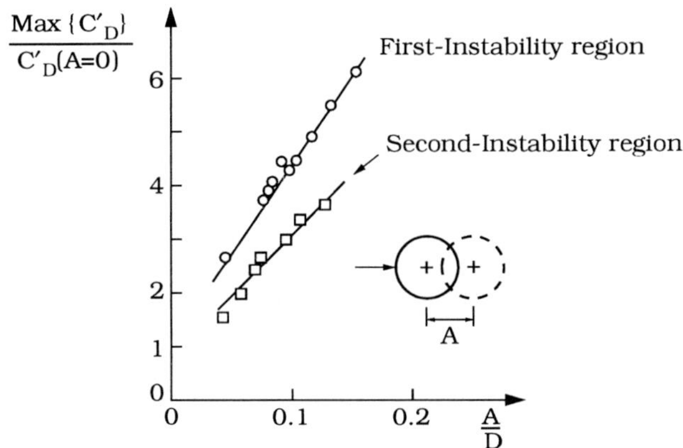

Figure 8.51 Amplification in the fluctuating drag for a cylinder vibrating in the in-line direction. $Re = 4 \times 10^4$. King (1974a, 1977).

body shape, may be such that a small tortional motion of the body causes a flow asymmetry. This, in turn, creates a force which drives the body in direction of its initial motion, resulting in the so-called **galloping instability**. The classic example of galloping is the vibration of ice-coated power lines because the ice cover normally forms an asymmetric (unfavorable) shape. Other examples include vibrations of group of risers or tethers on a tension-leg platform and the vibration of a flowline attached to one leg of an offshore tower in the area of offshore engineering.

In literature, the term *"flutter"* is also used for vibration of bluff bodies, particularly with regard to aircraft vibrations. Although galloping vibrations and aircraft flutter are induced by similar mechanisms there are significant differences between the two. Blevins (1977) makes the following distinction: In aircraft flutter, the aerodynamic forces are often sufficiently large, compared with the weight and inertia of the cross section, to produce large shifts in the natural frequencies. In galloping vibrations, the aerodynamic forces are usually small compared with the massive structures, so shifts in natural frequency are generally very small. In addition, aerodynamic flutter is ordinarily produced by the interaction of a torsion mode and a displacement mode, whereas galloping instabilities often affect only a single mode.

Detailed accounts of the subject, mostly related to the aerodynamic galloping and the galloping encountered in nuclear engineering, have been given by Blevins (1977) and Chen (1987).

Example 8.4: Galloping vibration of a cylinder with a rectangular cross-section

The following example is adopted from Blevins (1977) and deals with a flexibly-mounted cylinder with a rectangular cross-section (Fig. 8.52) exposed to a steady current. When the cylinder is displaced slightly from its equilibrium position, the flow relative to the cylinder at the moment of displacement will be asymmetric, and due to this asymmetry, a lift force will be created (Fig. 2.16) in the direction of the initial displacement (F_y in Fig. 8.52). If this force is sufficiently large, the so-called galloping instability will set in, and the cylinder will begin to vibrate in the y-direction. These vibrations are *galloping* type vibrations. Clearly, the mechanism behind these vibrations is different from that associated with the vortex-induced vibrations. In some cases, galloping vibrations may occur concurrently with the vortex-induced vibrations (see Example 8.5).

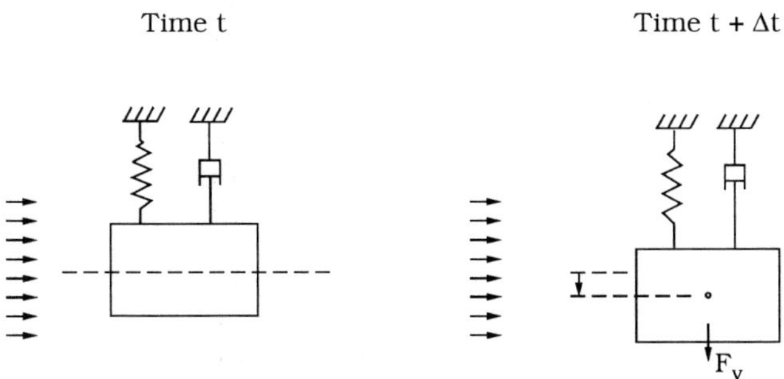

Figure 8.52 Force on a rectangular cross-section induced by a small motion of the cylinder.

The instability and the resulting galloping vibrations of a cylinder with rectangular cross-section may be predicted through the familiar vibration equation:

$$(m + m')\,\ddot{y} + 2\,(m + m')\,\zeta_s\,\omega_n\,\dot{y} + ky = F_y \tag{8.118}$$

Now let us consider the force term in the preceding equation. F_y is (Fig. 8.53)

$$F_y = -F_L \cos(\alpha) - F_D \sin(\alpha) \tag{8.119}$$

in which

$$F_L = \frac{1}{2} C_L \, \rho D \, U_{rel}^2 \tag{8.120}$$

and F_D and F_y are defined similarly. Here U_{rel} is the velocity of fluid relative to the body (Fig. 8.53). Hence,

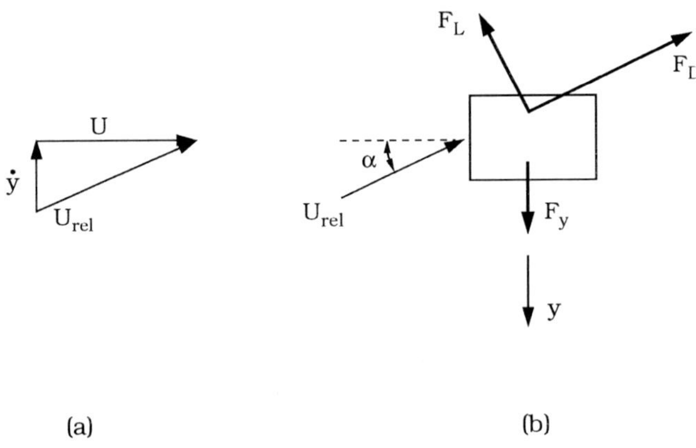

Figure 8.53 (a): Velocity relative to the vibrating cylinder and (b): Forces on the cylinder.

$$C_y = -C_L \cos(\alpha) - C_D \sin(\alpha) \tag{8.121}$$

For small α, the coefficient C_y

$$C_y \cong C_y(\alpha = 0) + \alpha \left(\frac{\partial C_y}{\partial \alpha} \right)_{\alpha=0} \tag{8.122}$$

$C_y(\alpha = 0)$ from Eq. 8.121 is found

$$C_y(\alpha = 0) = -C_L(\alpha = 0) \equiv 0 \quad , \tag{8.123}$$

and $\left(\frac{\partial C_y}{\partial \alpha} \right)_{\alpha=0}$, again from Eq. 8.121

$$\left(\frac{\partial C_y}{\partial \alpha} \right)_{\alpha=0} = -\left(\frac{\partial C_L}{\partial \alpha} \right)_{\alpha=0} - C_D(\alpha = 0) \tag{8.124}$$

Inserting Eqs. 8.123 and 8.124 in Eq. 8.122 gives

$$C_y = -\left[\left(\frac{\partial C_L}{\partial \alpha} \right)_{\alpha=0} - C_D(\alpha = 0) \right] \alpha \tag{8.125}$$

Since

$$\alpha = \frac{\dot{y}}{U} \tag{8.126}$$

then, C_y will be

$$C_y = -\left[\left(\frac{\partial C_L}{\partial \alpha}\right)_{\alpha=0} + C_D(\alpha = 0)\right]\frac{\dot{y}}{U} \tag{8.127}$$

or F_y

$$F_y = -\frac{1}{2}\rho DU\left[\left(\frac{\partial C_L}{\partial \alpha}\right)_{\alpha=0} + C_D(\alpha = 0)\right]\dot{y} \tag{8.128}$$

Substituting Eq. 8.128 into the equation of motion (Eq. 8.118) one gets

$$\ddot{y} + 2\zeta\,\omega_n\,\dot{y} + \omega_n^2 y = 0 \tag{8.129}$$

where ζ, the total damping, is defined by

$$2\zeta\,\omega_n = 2\zeta_s\,\omega_n + \frac{1}{2}\frac{\rho DU}{m+m'}\left[\left(\frac{\partial C_L}{\partial \alpha}\right)_{\alpha=0} + C_D(\alpha = 0)\right] \tag{8.130}$$

and ω_n is the natural angular frequency $= \sqrt{k/(m+m')}$ (Eq. 8.57).

Solution to Eq. 8.129 is:

$$y = A_y \exp(-\zeta\,\omega_d t)\sin(\omega_d t + \phi) \tag{8.131}$$

where ω_d is the damped angular natural frequency:

$$\omega_d = \omega_n(1 - \zeta^2)^{1/2} \tag{8.132}$$

(see Section 8.2.2).

Now, the vibrations grow with respect to time when

$$\zeta < 0 \tag{8.133}$$

From Eqs. 8.130 and 8.133 one obtains:

$$\text{Vibrations grow when } -\left[\left(\frac{\partial C_L}{\partial \alpha}\right)_{\alpha=0} + C_D(\alpha = 0)\right]\frac{U}{Df_n} > 2K_S \tag{8.134}$$

where f_n is the natural frequency of the system ($\equiv \omega_n/2\pi$) and K_S is the stability parameter (Eq. 8.96).

Eq. 8.134 indicates that large amplitude vibrations occur when

$$\frac{U}{Df_n} > \frac{2K_s}{-\left[\left(\frac{\partial C_L}{\partial \alpha}\right)_{\alpha=0} + C_D(\alpha = 0)\right]} \tag{8.135}$$

Table 8.2 (borrowed from Blevins 1977) gives numerical values of the quantity $\left[\left(\frac{\partial C_L}{\partial \alpha}\right)_{\alpha=0} + C_D(\alpha = 0)\right]$ for various cross-sections.

Table 8.2 - The values of $\left[\left(\frac{\partial C_L}{\partial \alpha}\right)_{\alpha=0} + C_D(\alpha = 0)\right]$ for various cross-sections [1]. From Blevins (1977).

Cross Section	$\left(\dfrac{\partial C_L}{\partial \alpha}\right)_{\alpha=0} + C_D(\alpha=0)$	Re
	-2.7	6.6×10^4
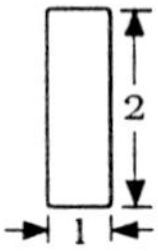	0	6.6×10^4
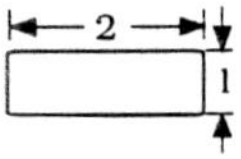	-3.0	3.3×10^4
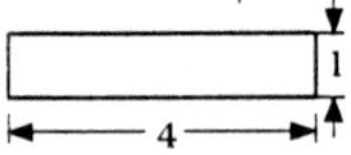	10.0	$2 \times 10^3 - 2 \times 10^4$
	0	6.6×10^4
	0.5	5.1×10^4
	-0.66	7.5×10^4

1) α is in radians; flow is from left to right.

The analysis given in the previous paragraphs assumes that α is small. In the case of unstable vibrations, however, the oscillations build up so that the small

α assumption would no longer be valid. In this case, the response of the system should be calculated, taking into consideration the non-linear effects. In such a situation, the solution to the equation of motion may be sought numerically, using the information given in Fig. 2.16 for C_L and Table I.1, Appendix I for $\overline{C}_D$.

Blevins (1977) gives an analytic treatment of the problem, fitting a cubic curve to the lift-force data. The normalized amplitude of vibrations in Blevins' analysis is obtained as follows

$$\underline{A} = \sqrt{\frac{4}{3a_3}\underline{U}\,(1 - a_1\underline{U})} \tag{8.136}$$

in which $\underline{A}$ is defined by

$$\underline{A} = \frac{A}{D}\frac{\rho D^2}{4\,(\mathrm{m} + \mathrm{m}')}\zeta_s \tag{8.137}$$

and $\underline{U}$, the normalized velocity:

$$\underline{U} = \frac{U}{Df_n}\frac{\rho D^2}{4(\mathrm{m} + \mathrm{m}')\,(2\pi\zeta_s)} \tag{8.138}$$

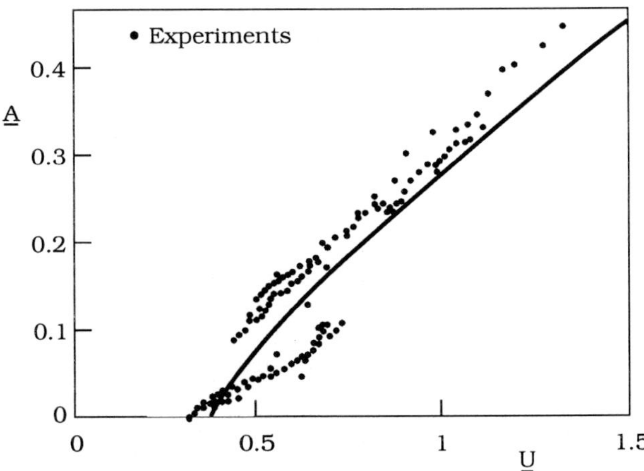

Figure 8.54 Experimental data regarding the response of a square section after Parkinson and Smith (1964) at $Re = 4,000$ to 20,000. Solid curve: Analytical solution (Eq. 8.136). Blevins (1977).

The coefficients a, and a_3 are related to the previously mentioned curve fitting and they are given as a, $= 2.7$ and $a_3 = -31$. The preceding solution is compared with the data from experiments by Parkinson and Smith (1964) (Fig. 8.54); as seen, the agreement is good. Blevins (1977) demonstrates that, with a higher-order curve fit to the measured lift-force data of Fig. 2.16, the hysteresis effect observed in the experiments could be produced.

Example 8.5:   Vibration of two interfering cylinders

Two cylinders arranged in tandem or side by side or in a staggered arrangement may undergo galloping vibrations. The subject has been investigated by Bokaian and Geoola (1984a) for the case when the upstream cylinder is flexibly-mounted and the downstream one is fixed (Fig. 8.55) and, again, by the same authors (1984b) when the downstream cylinder is flexibly-mounted and the upstream one is fixed (Fig. 8.56).

Regarding the former case, while the vibrations are vortex-induced vibrations for the spacings between the cylinders marked "a" in Fig. 8.55, they are galloping vibrations for the area marked "b" in the figure. The latter involves very small spacings between the cylinders such as $\frac{x_0}{D} < O(1.5)$ and $\frac{y_0}{D} < O(0.5)$. For such small spacings, vortex shedding ceases to exist. Therefore the vortex-induced lift disappears. Yet, there will be another kind of lift acting on the flexibly-mounted cylinder. This lift is caused by the asymmetry in the flow due to the proximity effect. Fig. 8.57 illustrates the variation of this lift with respect to y. As the figure implies, a small initial displacement of the upstream cylinder may be enough to trigger the galloping instability and therefore the vibrations, just as in the previous example. These vibrations may be called *proximity-induced galloping vibrations* in the present case.

Fig. 8.56 depicts the vibration regimes regarding the downstream cylinder, the upstream cylinder being fixed. As seen, in this case, the two regimes, namely the vortex-induced vibrations and galloping vibrations may occur concurrently (Fig. 8.56). The galloping here is directly related to the wake of the upstream cylinder, and therefore these vibrations may be called *wake-induced galloping vibrations*. Fig. 5.107 gives an example regarding the cylinder response, illustrating the cylinder response for the case when $y_0 = 0$ for different values of x_0 (the line at the bottom in the regime diagram given in Fig. 8.56). As seen, the vibration regime changes, as x_0 is increased. Of particular interest is the presence of the combined vortex-induced and galloping response for $x_0/D = 1.5$. Also, it may be interesting to note that the frequency response experienced with the galloping vibrations (Fig. 8.58b) is different from that with the vortex-induced vibrations (Fig. 8.58a).

Bokaian and Geoola (1987) have extended their work so as to include the cases where the diameter of the fixed cylinder is much larger than that of the

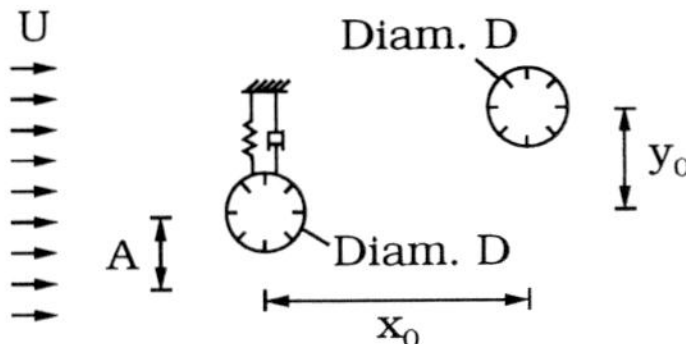

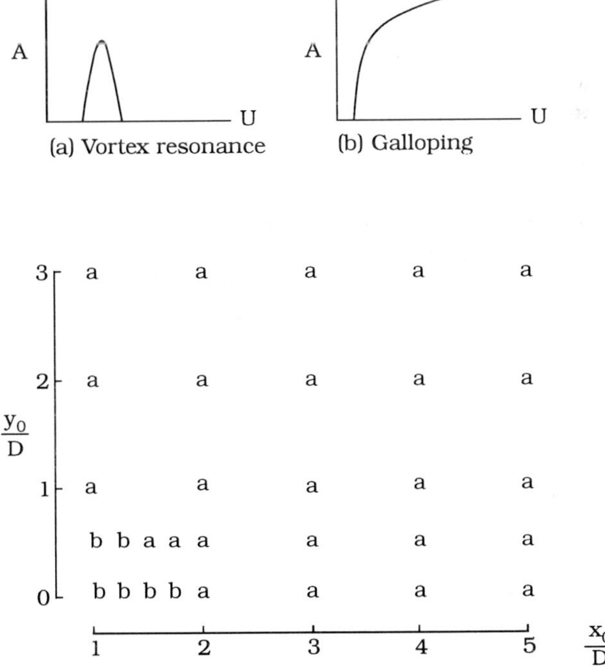

Figure 8.55 Variation of the reduced amplitude against the reduced velocity at various cylinders separations. Downstream cylinder is fixed. Bokaian and Geoola (1984a).

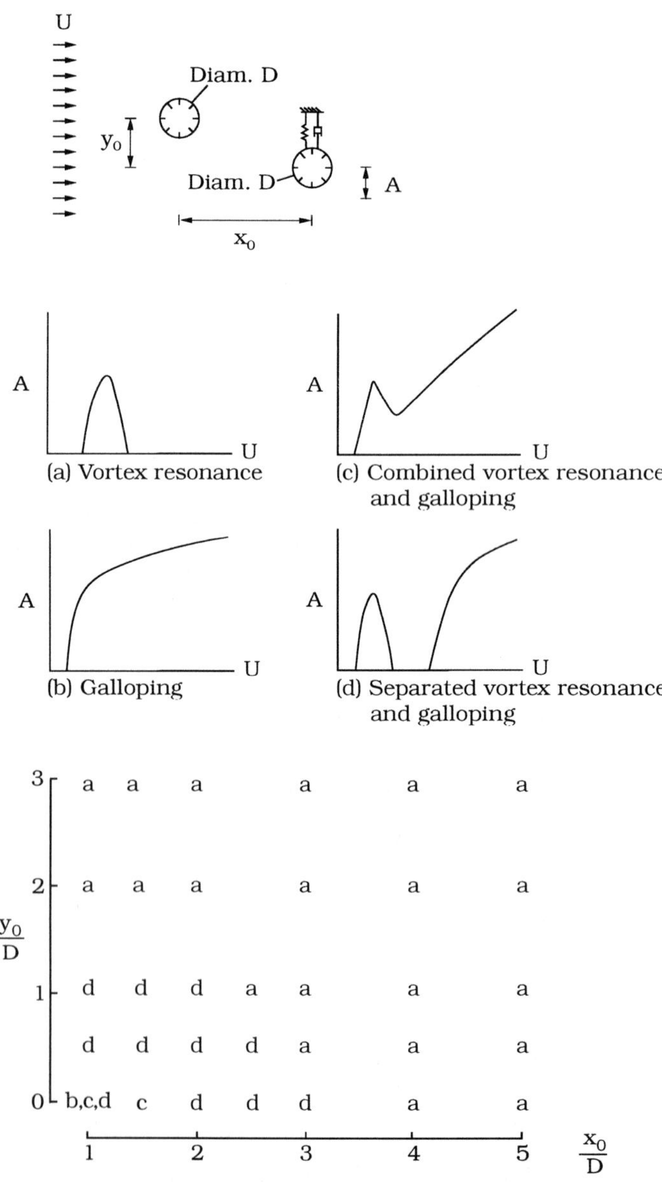

Figure 8.56 Type of instability observed at various cylinder separations. Upstream cylinder is fixed. Bokaian and Geoola (1984b).

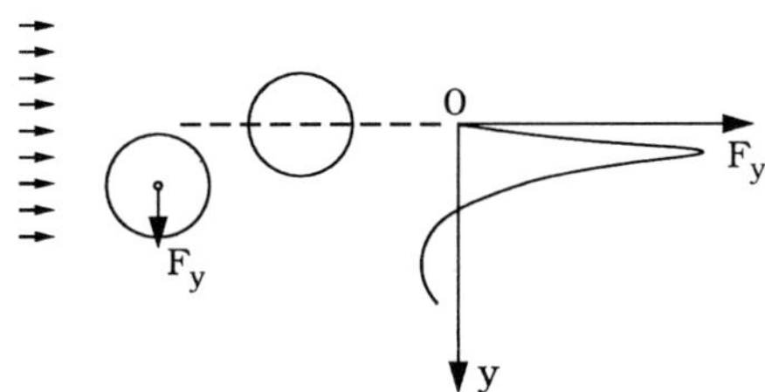

Figure 8.57 Lift force induced by the proximity effected. Adapted from Bokaian and Geoola (1984a).

flexibly-mounted one, considering their application to vibrations of flowlines attached to the leg of a platform.

Finally, it may be mentioned that forces on composite, multiple risers have been measured, as related to vibrations of these structures (Hansen, Jacobsen and Lundgren, 1979), while forces as well as hydroelastic vibrations of such structures have been investigated by Overvik (1982).

8.7 Suppression of vibrations

There are three methods of suppressing vibrations (Hallam et al., 1978):

1. Controlling the reduced velocity;

2. Controlling the mass and damping; and

3. Controlling the vortex shedding.

Controlling the reduced velocity

The structure can be designed such that the critical value of the reduced velocity, $U/(Df_n)$ for the onset of vibrations will never be exceeded. This can be done either increasing the natural frequency of the structure f_n or increasing the diameter D or both. Increasing the natural frequency can be achieved for example by bracing the structure.

In some cases, it may not be possible to maintain the reduced velocity below the critical value. In such cases, it is important to ensure that the vibrations are reduced as much as possible by preventing the coincidence of vortex shedding

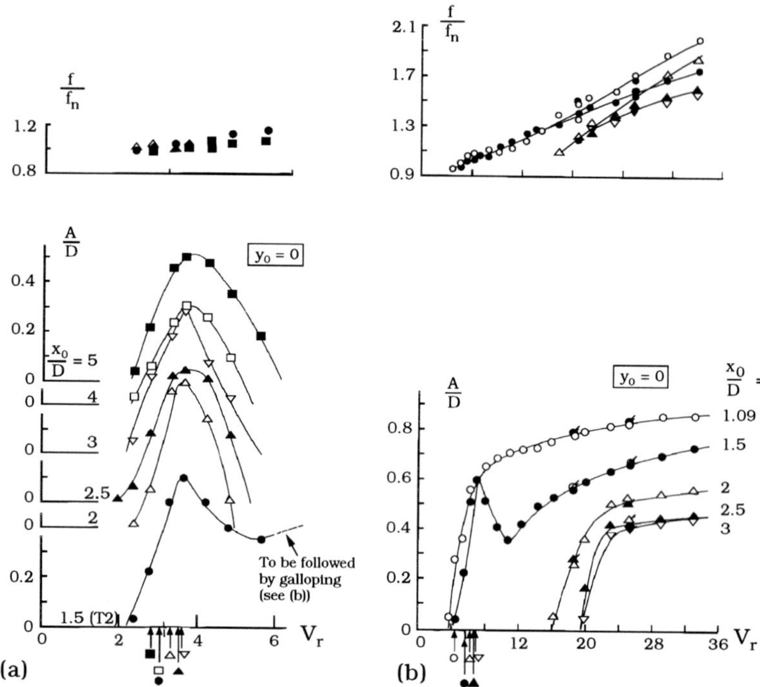

Figure 8.58 Variation of the reduced amplitude and the oscillation-frequency ratio versus the reduced velocity in tandem arrangement: $\bigcirc$, $\bullet\!\!\!\diagup$, $x_0/D = 1.09$; $\bullet$, $\varnothing$, 1.5; $\triangle$, $\blacktriangle\!\!\!\diagup$, 2.0; $\blacktriangle$, $\triangle\!\!\!\diagup$, 2.5; $\triangledown$, $\blacktriangledown$, 3.0 (the first symbol represents the 6.5% turbulence intensity while the second one denotes the 11.9% turbulence level); $\square$, 4.0; $\blacksquare$, 5.0 (6.5% turbulence intensity). Bokaian and Geoola (1984b).

frequency with resonances in the design of the structure. This method normally works for shorter cables or risers. For longer structures, the natural frequencies are densely distributed. Therefore, it may be difficult to avoid resonant vibrations. In such cases, a device called wave absorbing termination, described in Vandiver and Li (1994), may be capable of suppressing the vibrations. The idea behind this device is that when incident vibration waves reach the point of termination of the cable, they are absorbed rather than reflected, and hence the cable behaves with the dynamic properties of an infinitely long structure.

Controlling the mass and damping

The vibrations are virtually eliminated when K_s, the stability parameter, becomes larger than about 18 for cross-flow vibrations (Fig. 8.25) and larger than about 1.2 for in-line vibrations (Fig. 8.36). Therefore, if the structure is designed such that K_s is sufficiently large, then the vibrations will in effect be suppressed. For K_s to have large values (see Eq. 8.102), the mass parameter $m/(\rho D^2)$ and/or the damping ζ_s should be increased.

Caution must be exercised, however, when the increase in the mass is considered. The increase may result in a reduction in the natural frequency of the structure, therefore a decrease in the critical velocity for the onset of vibrations. Hallam et al. (1978) notes that this has happened to some marine structures, with disastrous results.

Controlling the vortex shedding

The idea here is to control the vortex shedding or indeed the flow so that the excitation forces are eliminated or weakened. In case of vortex-induced vibrations, a wide range of controlling devices may be implemented for this. An extensive review of these devices was given by Zdravkovich (1981). Fig. 8.59, reproduced from Zdravkovich, gives a summary of various methods for interfering the vortex shedding mainly in wind engineering. The methods can be grouped into three categories.

The first category devices (Figs. 8.59a.I and a.II) are various types of **surface protrusions**. These can be grouped into two sub-categories; one with the omnidirectional response (Fig. 8.59a.I) and the other with the unidirectional response (Fig. 8.59a.II). The omnidirectional devices are the ones which are not influenced by the direction of the flow. These are basically helical strakes, helical wires, etc. The uni-directional devices, on the other hand, are rectangular fins, straight fins extending along the length of the structure, straight wires extending along the length of the structure, etc.

The second category devices are various types of **shrouds** (Fig. 8.59b). These include perforated shrouds (with square or circular holes), array of rods encircling the structure, fine mesh gauze, etc.

The third category devices are **wake stabilizers** (Fig. 8.59c) such as sawtooth fins, splitter plates, guide plates, etc. Clearly, while the first two category devices act as spoilers to disrupt the boundary layer on the surface of the structure, the third category devices (wake stabilizers) prevents the interaction between the two shear layers, peresumably leading to the complete or partial elimination of vortex shedding.

Some of the previously mentioned devices may not be suited to marine work because they have to be welded to the structure, causing problems with fouling, fabrication and perhaps corrosion. Nevertheless, *strakes, shrouds, and fins and fairings* are the most commonly used methods also in marine environments (Every,

King and Weaver, 1982, discusses these devices in the light of laboratory tests, site investigations and extensive literature surveys). The following paragraphs give a closer account of these devices.

As regards the strakes, apart from the difficulties in installation and handling, the drag coefficient of these devices is rather large, being 1.3 - 1.4. Hallam et al. (1978) gives the optimum strake configuration as that of *three* helically-wound fins of about 10% of the cylinder diameter, with a pitch of 5D. Jones and Lamb (1992) described a detailed model investigation of the use of helical strakes aimed at controlling the level of vibrations of a conductor for an exploration well. Results showed that notable levels of vibration suppression could be achieved with a partial strake coverage over the length of the structure provided that strake was properlly positioned. Reducing the strake coverage also had a beneficial effect on the overall level of drag on the conducter. One last point regarding the helical strakes concerns their application in an environment with marine growth. In the case where removal of marine growth is not contemplated, the fouling of strakes result in a rounded form which is similar to a situation where the structure is attached with helical cables. It is known that the latter has also proven to be quite effective in reducing vortex-induced vibrations.

Perforated shrouds (Fig. 8.59b) have been used in marine environment (King, Prosser and Verley, 1976). King et al. give the optimum shroud geometry as follows: the shroud should have a diameter 20% larger than the cylinder and an open area ratio of 36%, and should extend for 20% of the wetted length. King et al. further reports that a fairly thick layer of marine growth (in the form of crustacea) did not reduce the shroud's effectiveness in full scale marine tests.

Fins and fairings have been successfully used in marine applications. Fins extending along the length of the cylinder (Fig.8.59a.II) were quite effective in reducing the in-line vibrations (King et al., 1976). The latter authors give the dimensions of the fins as follows: 10% of the diameter of the cylinder and fitted over about 20% of the cylinder's length at 45 degrees from the front stagnation point. Note that this device is effective only for unidirectional flows.

A fairing for an exploratory drilling riser was used successfully in a high current region (Grant and Patterson, 1977). Its design was such that it encircled the riser and extended about 2 diameters into the wake of the riser (Fig. 8.60). It was designed to align itself with the current direction similar to a weather-vane. It not only eliminated vortex excited motions of both the riser and its kill and choke lines (Fig. 8.60), but also reduced the hydrodynamic drag.

Stansby, Pinchbeck and Henderson (1986) developed special spoilers which meet the following requirements: 1) effective in currents and waves; 2) attachable to existing structures, by divers if necessary; 3) easily removable; 4) free of marine growth; and 5) inexpensive to produce. The device is supported on a hoop encircling the structure. It is made of strong resilient plastic, polypropelene, with antifouling additives. The hoop diameter is slightly larger than the cylinder diameter so that small relative movement of the hoops prevent marine fouling on the surfaces in contact with each other. The antifouling additive prevents fouling

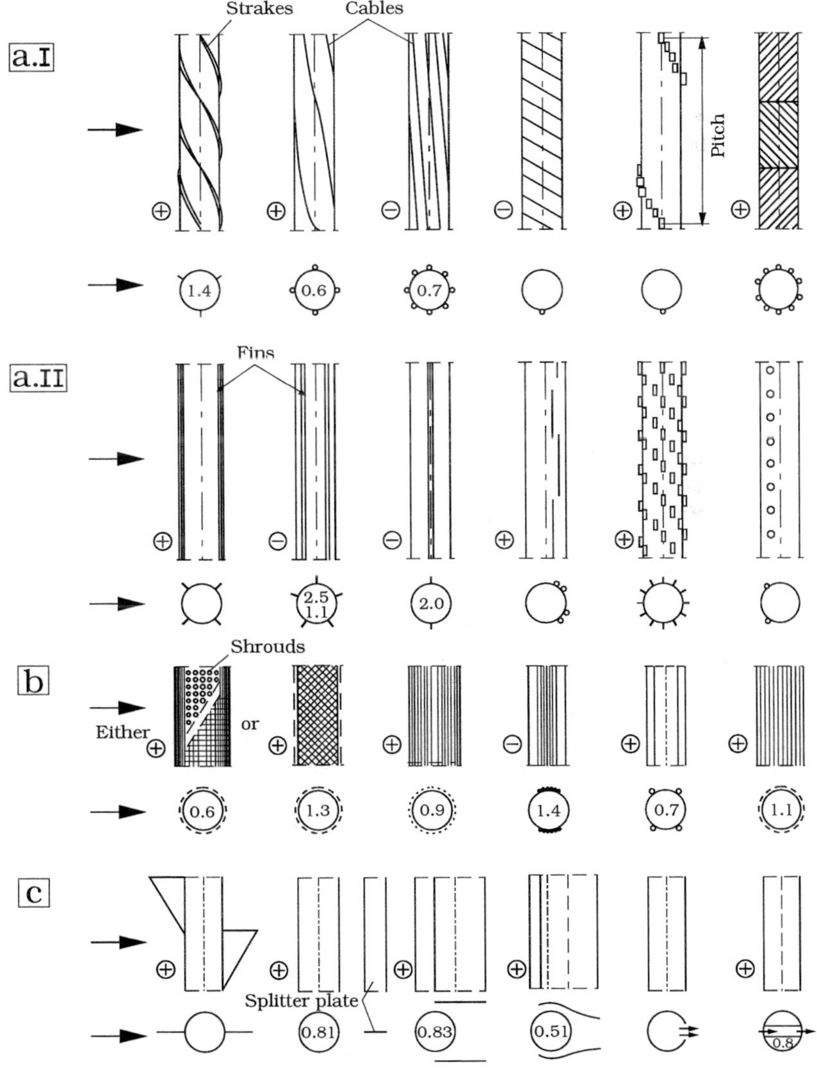

Figure 8.59 Aerodynamic and hydrodynamic means for interfering with vortex shedding: (i) surface protrusions ((a) omnidirectional and (b) unidirectional), (ii) shrouds, (iii) nearwake stabilisers (⊕ effective. ⊖ ineffective). Zdravkovich (1981).

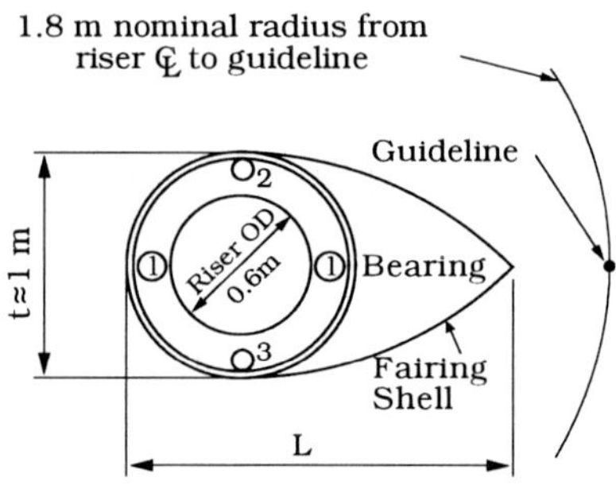

1. Blowout Preventer Control Lines
2. Kill Line
3. Choke Line

Figure 8.60 Riser and fairing geometry used by Grant and Patterson (1977).

on the device itself. Tests made with very lightly damped cylinders equiped with this device indicated that cross-flow vibrations were reduced markedly in steady currents. No tests were carried out for waves. Also tested were various configurations of two cylinder situations. There will be an increase in drag with respect to the plain cylinder situation.

Finally, Moros and Swan (1992) describes a laboratory investigation of a new method where a second phase (air) is introduced in the vicinity of the structure. The bubly plume apparently acts as a wake stabilizer, presumably reducing the amplitude of vibrations. As much as 80% reduction in the amplitude of vibrations was achieved when air was released at $0.5L$ and $0.4L$ (L being the cylinder length). Furthermore, the drag force on the structure is reduced by approximately 10% as compared to a plain cylinder.

Chung, Whitney, Lezius and Conti (1994) investigated the moment and lift on a pipe, straked with power cables arranged in helical form, and a pipe with a perforated shroud with the purpose of 1) finding a means of reducing vortex-induced vibration of a deep-ocean mining pipe 6000 m in length, and also 2) measuring flow-induced torsional moment of the pipe. It was found that these configurations generated the least vortex-shedding intensity, had minimum unsteady lift and the smallest increase in drag compared to a bare pipe. Tests with the straight-down power-cable configuration (cable parallel to the pipe axis) showed, however, significant drag and lift and a larger flow-induced torsional moment, as reported by Chung and Whitney (1993).

REFERENCES

Anand, N.M. (1985): Free span vibrations of submarine pipelines in steady and wave flows. Thesis (Dr. Eng. Degree), Div. of Port and Ocean Engineering, The Univ. of Trondheim, The Norwegian Institute of Technology, Trondheim, Norway.

Angrilli, F., Di Silvio, G. and Zanando, A. (1974): Hydroelasticity study of a circular cylinder in a water stream. In: Flow-Induced Structural Vibrations (ed. E. Naudascher), Berlin Springer-Verlag, pp. 504-512.

Bearman, P.W. (1984): Vortex shedding from oscillating bluff bodies. Annual Rev. Fluid Mech., 16:195-222.

Bearman, P.W. and Mackwood, P.R. (1991): Non-linear vibration characteristics of a cylinder in an oscillating water flow. Proc. 5th Conf. on Flow-Induced Vibrations. Inst. of Mech. Eng., Brighton, U.K., 21-23 May, 1991, pp. 21-31.

Bishop, R.E.D. and Hassan, A.Y. (1964): The Lift and Drag Forces on a Circular Cylinder Oscillating in a Flowing Fluid. Proc. Roy. Soc. London, A 277:51-75.

Blevins, R.D. (1977): Flow-Induced Vibrations. Van Nostrand.

Bokaian, A., Geoola, F. (1984a): Proximity induced galloping of two inferfering circular cylinders. J. Fluid Mech., 146:417-449.

Bokaian, A. and Geoola, F. (1984b): Wake induced galloping of two interfering circular cylinders. J. Fluid Mech., 146:383-415.

Bokaian, A. and Geoola, F. (1987): Flow-induced vibrations of marine risers. Proc. ASCE, J. Waterway, Port, Coastal and Ocean Engineering, 113(1):22-38.

Brika, D. and Laneville, A. (1993): Vortex-induced vibrations of a long flexible circular cylinder. J. Fluid Mech., 250:481-508.

Bryndum, M.B., Bonde, C., Smitt, L.W., Tura, F. and Montesi, M. (1989): Long free spans exposed to current and waves: Model tests. Proc. 21st Annual Offshore Technology Conf. (OTC), Houston, TX, May 1-4, 1989, Paper OTC 6153, pp. 317-328.

Chen, S.-S. (1987): Flow-induced vibration of circular cylindrical structures. Hemisphere Publishing Corporation.

Chung, J.S. and Whitney, A.K. (1994): Flow-induced moment and lift for a circular cylinder with cable attachment. Int. J. of Offshore and Polar Engrg., 3(4):280-287.

Chung, J.S., Whitney, A.K., Lezius, D. and Conti, R. (1994): Flow-induced torsional moment and vortex suppression for a circular cylinder with cables. Proc. 4th Int. Offshore and Polar Engrg. Conf., Osaka, Japan, April 10-15, 1994, III:447-459.

Currie, I.G. and Turnbull, D.H. (1987): Streamwise oscillations of cylinders near the critical Reynolds number. J. Fluids and Structures, 1:185-196.

Den Hartog, J.R. (1934): The vibration problem in engineering. Proc 4th Int. Congress in Appl. Mech., Cambridge, U.K., pp. 34-53.

Every, M.J., King, R. and Weaver, D.S. (1982): Vortex-excited vibrations of cylinders and cables and their suppression. Ocean Engrg., 9(2):135-157.

Feng, C.C. (1968): The measurement of vortex-induced effects on flow past stationary and oscillating circular and D-section cylinders. M.Sc. Thesis. The University of British Columbia, 1968.

Fredsøe, J. and Justesen, P. (1986): Turbulent separation around cylinders in waves. J. Waterway, Port, Coastal and Ocean Engineering., ASCE, 112:217-233.

Grant, R. and Patterson, D. (1977): Riser fairing for reduced drag and vortex suppression. Proc. 9th Annual Offshore Technology Conf., OTC Paper No. 2921, pp. 343-352.

Griffin, O.M. (1981): OTEC cold water pipe design for problems caused by vortex-excited oscillations. Ocean Engineering, 8(2):129-209.

Griffin, O.M. (1982): Flow-Induced Oscillations of OTEC Mooring and Anchoring Cables: State of the Art. Naval Research Laboratory Washington, D.C., Memorandum Report 4766, May 27, 1982.

Griffin, O.M. and Ramberg, S.E. (1974): The vortex street wakes of vibrating cylinders. J. Fluid Mech., 66:553-576.

Hallam, H.G., Heaf, N.J. and Wootton, L.R. (1978): Dynamics of Marine Structures. Construction Industry Research and Information Association (CIRIA) Report UR8, London.

Hansen, N.-E.O., Jacobsen, V. and Lundgren, H. (1979): Hydrodynamic forces on composite risers and individual cylinders. Proc. 11th Annual Offshore Technology Conf. (OTC), Houston, TX, April 30 - May 3, 1979, Vol. III, Paper OTC 3541, pp. 1607-1621.

Hartlen, R.T. et al. (1968): Vortex-excited oscillations of a circular cylinder. U.T.I.A.S. Report UTME-TP-6809, Nov. 1968.

Humphries, J.A. and Walker, D.H. (1987): Vortex excited response of large scale cylinders in sheared flow. Proc. OMAE, Houston, TX, 2:139-147.

Jensen, B.L. and Sumer, B.M. (1986): Boundary layer over a cylinder placed near a wall. Progress Report No. 64, Inst. of Hydrodynamics and Hydraulic Engineering, ISVA, Techn. Univ. Denmark, pp. 31-39.

Jones, G.S. and Lamb, W.S. (1992): The use of helical strakes to suppress vortex induced vibration. BOSS '92, 2:804-835.

King, R., Prosser, M.J. and Johns, D.J. (1973): On vortex excitation of model piles in water. J. Sound and Vibration, 29(2):169-188.

King, R. (1974a): Vortex-excited structural oscillations of a circular cylinder in flowing water. Ph.D. Thesis. Loughborough University of Technology, U.K., July 1974.

King, R. (1974b): Vortex-excited structural oscillations of a circular cylinder in steady currents. 6th Annual Offshore Technology Conf., Paper No. OTC 1948, Houston, TX, May 6-8, 1974, pp. 143-154.

King, R. (1977): A review of vortex shedding research and its application. Ocean Engineering, 4:141-172.

King, R., Prosser, M.J. and Verley, R.L.P. (1976): The suppression of structural vibrations induced by currents and waves. BOSS '76, NTH, Trondheim, 1:263-283.

Kozakiewicz, A., Sumer, B.M. and Fredsøe, J. (1994): Cross-flow vibrations of a cylinder in irregular oscillatory flow. J. Waterway, Port, Coastal and Ocean Engrg., ASCE, 120(6):515-533.

Maull, D.J. and Kaye, D. (1988): Oscillations of a flexible cylinder in waves. Proc 5th Conf. on Behaviour of Offshore Structures, BOSS, Trondheim 1988, 2:535-547.

Meier-Windhorst, A. (1939): Flatterschwingungen von Zylindern in gleich-mässigen Flüssigkeitsstrom. Mitteilungen des Hydraulischen Instituts der Technischen Hochschule, München, Heft 9, pp. 3-39, 1939.

Moe, G., Holden, K. and Yttervoll, P.O. (1994): Motion of Spring Supported Cylinders in Subcritical and Critical Flows. Proc. 4th Offshore and Polar Engineering Conf., Osaka, Japan, 3:468-475.

Moros, A. and Swan, C. (1992): The introduction of a second phase as a mean of reducing vortex induced vibrations. BOSS '92, 2:791-803.

Öngören, A. and Rockwell, D. (1988): Flow structures from an oscillating cylinder. Part 2. Mode competition in the near wake. J. Fluid Mech., 191:225-245.

Overvik, T. (1982): Hydroelastic motion of multiple risers in a steady current. Dr. Eng. Degree Thesis, The Univ. of Trondheim, The Norwegian Inst. of Technology, Trondheim, Norway, August 1982, Vii+173 p.

Pantazopoulos, M.S. (1994): Vortex-induced vibration parameters: Critical Review. Proc. 13th Int. Conf. on Offshore Mechanics and Arctic Engineering, OMAE, 1994, 1:199-255.

Parkinson, G.V. and Smith, J.D. (1964): The square prism as an aeroelastic non-linear oscillator. Quart. J. Mech. Appl. Math. 17:225-239.

Raven, P.W.C., Stuart, R.J. and Littlejohns, P.S. (1985): Full-scale dynamic testing of submarine pipeline spans. 17th Annual OTC in Houston, Texas, May 6-9, 1985, Paper No. 5005, pp. 395-405.

Sainsbury, R.N. and King, D. (1971): The flow-induced oscillation of marine structures. Proc. of Institution of Civil Engineers, London, 49:269-302.

Sarpkaya, T. (1978): Fluid forces on oscillating cylinders. J. Waterway, Port, Coastal and Ocean Div., ASCE, 104(WW3):275-290.

Sarpkaya, T. (1979): Vortex-induced oscillations - A selective review. J. Appl. Mech. Trans. of ASME, 46:241-258.

Sarpkaya, T. (1995): Hydrodynamic damping, flow-induced oscillations and biharmonic response. Trans. ASME, J. Offshore Mech. and Arctic Engineering, 117:232-238.

Scruton, C. (1963): On the wind-excited oscillations of stacks, towers, and masts. Paper No. 16, Proc. Conf. on Wind effects on Buildings and Structures, Teddington, U.K., 2:797-832(836).

Stansby, P.K., Pinchbeck, J.N. and Henderson, T. (1986): Spoilers for the suppression of vortex-induced oscillations (Technical Note). Applied Ocean Research, 8(3):169-173.

Sumer, B.M., Fredsøe, J., Jensen, B.L. and Christiansen, N. (1994): Forces on a vibrating cylinder near a wall in steady and oscillatory flows. J. Waterway, Port, Coastal and Ocean Engineering, ASCE, 120(3):233-250.

Tsahalis, D.T. (1984): Vortex-induced vibrations of a flexible cylinder near a plane boundary exposed to steady and wave-induced currents. Trans. ASME, J. Energy Resources Technology, 106:206-213.

Vandiver, J.K. and Li, L.L. (1994): Suppression of cable vibration by means of wave absorbing terminationns. BOSS '94, 2:633-643.

Vickery, B.J. and Watkins, R.D. (1962): Flow-induced vibrations of cylindrical structures. Proc 1st Australiasian Conf., pp. 213-241.

Williamson, C.H.K. and Roshko, A. (1988): Vortex formation in the wake of an oscillating cylinder. J. of Fluids and Structures, 2:355-381.

Wootton, L.R. (1969): The oscillation of large circular stacks in wind. Proc of Institution of Civil Engineers, London, 43:573-598.

Wootton, L.R. (1972): Oscillations of piles in marine structures. C.I.R.I.A., Report 40.

Wootton, L.R., Warner, M.H. and Cooper, D.H. (1974): Some aspects of the oscillations of full-scale piles. IUTAM-IAHR Symposium, Karlsruhe, Federal Republic of Germany, August 14-16, 1972. The proceedings book (ed. E. Naudascher), pp. 586-601, Springer-Verlag, 1974.

Zdravkovich, M.M. (1981): Review and classification of various aerodynamic and hydrodynamic means for suppressing vortex shedding. J. of Wind Engrg. and Industrial Aerodynamics, 7:145-189.

Zdravkovich, M.M. (1982): Modification of vortex shedding in the synchronization range. ASME, J. of Fluids Engineering, 104:513-517.

Chapter 9. Flow-induced vibrations of a free cylinder in waves

9.1 Introduction

A cylinder subjected to an oscillatory flow experiences periodic forces: the transverse component of the force, i.e., the lift force, oscillates at its fundamental lift frequency, while the in-line component of the force oscillates at the frequency of the oscillatory motion.

Regarding the latter, although the in-line force primarily oscillates at the frequency of oscillatory motion, there will be small periodic fluctuations superimposed on this force; these small fluctuations are induced by vortex motions around the cylinder due to vortex shedding and flow reversals. These small oscillations in the in-line force mostly occur at frequencies significantly higher than the flow frequency.

The cause-and-effect relationships between the forces and the vibrations of a flexible cylinder (Fig. 9.1) may be summarized as in Table 9.1. Note that there may be a significant coupling between the in-line and cross-flow vibrations in the case of a system with two degrees of freedom. This occurs when the cylinder oscillates in the in-line direction in the resonance regime (i.e., $f_w/f_n \cong 1$ in which f_w = the wave frequency); in this case, the amplitudes of the in-line oscillations may become large, and therefore the in-line motion may begin to influence the cross-flow vibrations (Lipsett and Williamson, 1991b). When the wave frequency is outside the resonance range (i.e., $f_w/f_n \neq 1$), however, no significant coupling between the in-line and cross-flow vibrations takes place (Sumer et al., 1989; Maull and Kaye, 1988).

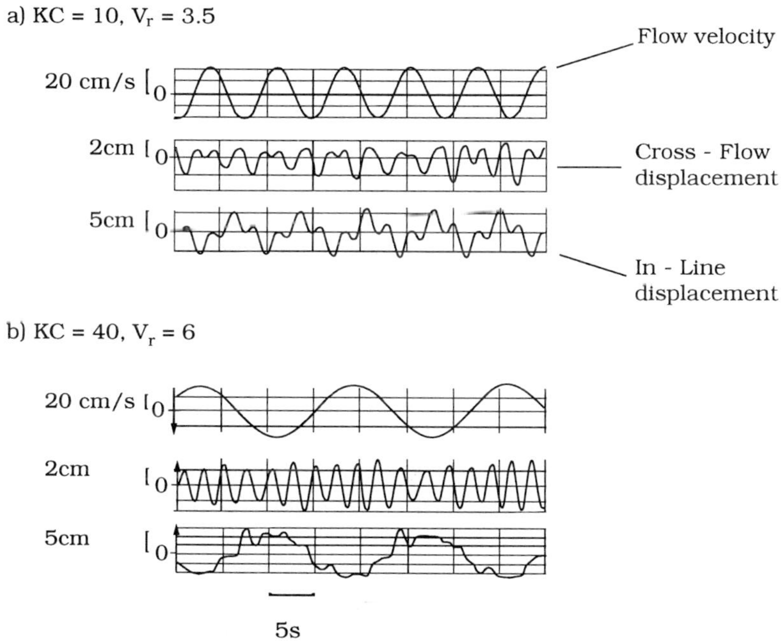

Figure 9.1 Time series of cross-flow and in-line displacements of a flexibly-mounted circular cylinder in oscillatory flow. Sumer et al. (1989).

As seen from Table 9.1, there are two kinds of in-line vibrations: one is caused by the Morison force and the other by the vortex-induced in-line force. To differentiate one from the other, the vibration caused by the Morison force will be called the *in-line oscillatory motion* (or *in-line motion* for short), while the second will be called the *in-line vibration* (Fig. 9.2).

Fig. 9.3 shows the ranges regarding the wave frequency and the fundamental lift frequency encountered in the ocean environment. The fundamental lift frequency, f_L, in Fig. 9.3, is determined from the relation $f_L = N_L f_w$ in which f_w is the wave frequency and N_L is the number of oscillations in the lift force per flow

Table 9.1 Cause and effect relationships between the forces and vibrations.

Force	Frequency of force on stationary cylinder	Vibrations
Lift force	Fundamental lift frequency	Cross-flow vibrations
In-line force (Morison force)	Frequency of oscillatory flow	In-line oscillatory motion
Vortex-induced component of in-line force	Frequencies significantly higher than those of oscillatory flow in most of the cases	In-line vibrations super-imposed on in-line oscillatory motion

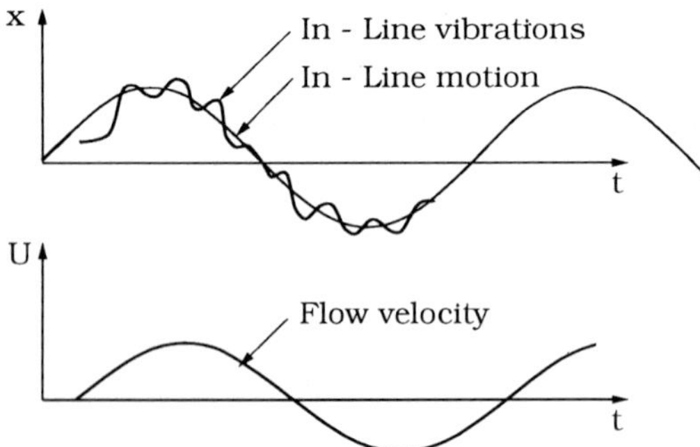

Figure 9.2 Schematic description of two kinds of in-line oscillations of a flexibly-mounted cylinder.

cycle (taken in the range $2 \le N_L \le 20$, Fig. 3.16) (Eq. 3.13). Fig. 9.3 also shows the natural frequency ranges corresponding to both the compliant structures and the fixed structures in the ocean.

It is not surprising to see that the natural frequency, f_n, remains outside the wave-frequency range; this is simply to avoid resonance with regard to in-

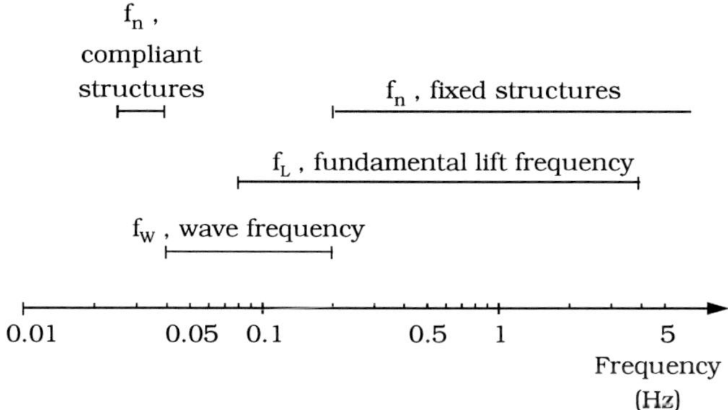

Figure 9.3 Ranges of typical wave and fundamental lift force frequencies compared with that of natural frequency (f_n) of offshore structures.

line motion induced by the Morison force. However, the possibility of the wave frequency coinciding with f_n can not be entirely ruled out, particularly at the two ends of the wave-frequency range. Regarding the fundamental lift frequency, the figure clearly shows that this frequency can coincide with f_n in the case of fixed structures. The latter implies that the frequency of vortex-induced oscillations in the in-line force also can coincide with f_n, since this frequency is in the same order of magnitude as the fundamental lift frequency, as mentioned earlier. So, from the above considerations, it may be concluded that while the compliant structures may undergo in-line oscillatory motion, the fixed structures may undergo all three types of vibrations, the cross-flow vibrations, the in-line oscillatory motion, and in-line vibrations, indicated in Table 9.1.

In the following sections, we shall first focus on cross-flow vibrations, then we shall examine in-line vibrations, and finally, we shall concentrate our attention on the in-line motion of structures.

9.2 Cross-flow vibrations

It has been seen in Chapter 8 that the cross-flow vibrations of a flexibly-mounted cylinder exposed to a steady current are governed mainly by the following non-dimensional parameters (Eq. 8.103):

$$V_r , M , K_s , Re , \frac{k_s}{D}$$

In the case when the cylinder is exposed to an oscillatory flow (Fig. 9.4), similar considerations as in Section 8.3.1 lead to the following non-dimensional variables

$$V_r \ , \ KC \ , \ M \ , \ K_s \ , \ Re \ , \ \frac{k_s}{D} \tag{9.1}$$

in which V_r is the reduced velocity defined by

$$V_r = \frac{U_m}{D f_n} \tag{9.2}$$

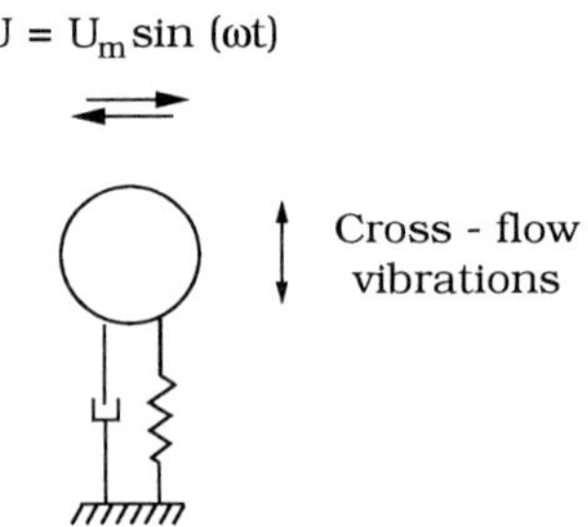

Figure 9.4 Definition sketch regarding cross-flow vibrations in oscillatory flow.

Some investigators prefer to use the frequency ratio f_n/f_w as an independent variable in favour of V_r (Isaacson and Maull (1981), Angrilli and Cossalter (1982), Bearman and Hall (1987), Maull and Kaye (1988) and Bearman and Mackwood (1991)). The two parameters are related, however, by $f_n/f_w = KC/V_r$. One advantage of using V_r instead of f_n/f_w is that it makes it possible for us to reconcile with the case of steady current, as a special case, when $KC \to \infty$.

As seen, in addition to the non-dimensional parameters already known from the steady-current research, there is one new parameter, namely the Keulegan-Carpenter number. This is not an entirely unexpected result, however, since it is known that the KC number is one of the major parameters which govern the lift force on a cylinder exposed to oscillatory flows (Chapter 3).

Cross-flow vibrations of cylinders in waves have been the subject of extensive research in recent years; Zedan and Rajabi (1981), Isaacson and Maull (1981), Angrilli and Cossalter (1982), Verley and Johns (1983), Bearman and Hall (1987), Maull and Kaye (1988), Borthwick and Herbert (1990) and Kaye and Maull (1993) in the case of a vertical cylinder in real waves and Sarpkaya (1979), Sarpkaya and Rajabi (1979), McConnell and Park (1982a and b), Jacobsen, Hansen and Petersen (1985), Bearman and Hall (1987), Sumer and Fredsøe (1988, 1989), Bearman and

Mackwood (1991), Lipsett and Williamson (1991a and b), Graham and Djahan-souzi (1991), Bearman, Lin and Mackwood (1992), Slaouti and Stansby (1992), and Kozakiewicz, Sumer and Fredsøe (1994) in the case of a cylinder exposed to a planar oscillatory flow.

In the following paragraphs, the general features of cross-flow vibrations of a circular cylinder will be described, based mainly on the work of Sumer and Fredsøe (1988).

9.2.1   General features

Figure 9.5 illustrates typical records of cylinder vibration. Figure 9.6, on the other hand, represents the amplitude and frequency data for KC number equal to 10, 20, 30, 40 and 100, including the data corresponding to the current case. The data are plotted in the form f/f_n, f/f_w and $2A/D$ versus the reduced velocity V_r. Here $f =$ the cylinder vibration frequency, $f_n =$ the natural frequency, $f_w =$ the frequency of the oscillatory flow, $2A =$ the double amplitude of cylinder vibration.

In Fig. 9.6, the identity

$$\frac{f}{f_n} = \frac{N}{KC}V_r \tag{9.3}$$

is plotted as a reference line, (the radiating lines issuing from the origin of the $(f/f_n , V_r)$ coordinate system). Here, $N =$ the number of vibrations in one cycle of the oscillatory flow

$$N = \frac{f}{f_w} \tag{9.4}$$

In the same figure, the relation

$$\frac{f}{f_n} = 0.2V_r \tag{9.5}$$

is plotted as a reference line for the current case (cf. Fig. 8.15).

$KC = 20$. This KC number constitutes a good example which enables us to explain distinct features of the cylinder response in oscillatory flows, common to other KC-numbers as well.

1) First, let us focus on the *frequency response*. The question here is: Why does the number of vibrations per cycle jump down to a one-less value at some points, as V_r is increased? This is explained as follows. In the tests, V_r is increased by increasing U_m. However, also the frequency of the flow, f_w, has to be increased parallel to the increase in U_m, to maintain the value of $KC = U_m/(Df_w)$ unchanged.

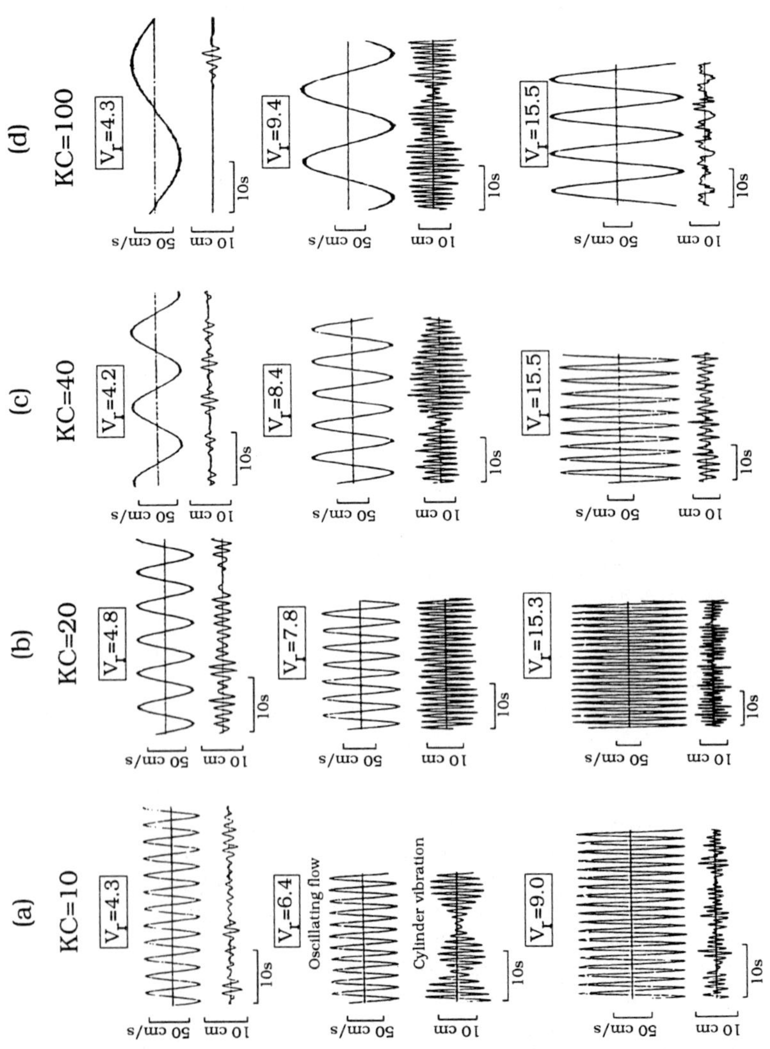

Figure 9.5 Sample records of oscillating-flow velocity and cylinder cross-flow vibration. Sumer and Fredsøe (1988).

Now let us follow the frequency response, as the velocity V_r is increased. The number of vibrations for one cycle of the motion $N(= f/f_w)$ is maintained at 4, as V_r increases; and this appears to be the case until V_r reaches the value of $V_r = 5.5$. At this point, the lock-in occurs where the vibration frequency f becomes approximately equal to the natural frequency of the system; $f \simeq f_n$. Now if V_r is increased further, then there will be a corresponding increase in f_w according to the argument in the preceding paragraph; in that case the ratio f/f_w will become 3 plus some fraction, because f cannot maintain the value 4, since it is locked into f_n:

$$N = \underbrace{\frac{f}{f_w}}_{\text{is increased}} \overset{\overset{\text{remains}}{\text{locked into}}}{\underset{\longrightarrow}{\text{------}\longrightarrow}} = \frac{f_n}{f_w} = 3 + \text{ some fraction} \qquad (9.6)$$

Owing to the very nature of the phenomenon, the number of vibrations per cycle has to have an integer value. Thus, the ratio f/f_w cannot be maintained at the value "3 plus some fraction", it has to drop to the next integer value, which is 3. Therefore, once the lock-in point is reached, some further increase in V_r will lead to a sudden drop in the number of vibrations to a one-less value.

The further drop in the number of vibrations from 3 to 2, which occurs at about $V_r = 9$, can be explained exactly the same way as in the preceding paragraphs.

Once the value $N(= f/f_w) = 2$ has been reached, this value is maintained steadily for further increases in V_r, because $N = 2$ is the absolute minimum for the number of vibrations in one cycle of the oscillating motion.

This behaviour is observed also in the case of $KC = 10$, where N starts with the value 2, and this value is maintained continuously throughout the V_r range scanned in the experiments.

Note that N becomes unity only *(i)* in the case where KC is in the range $4 < KC < 7$, as demonstrated in the following paragraphs; and *(ii)* in the case where f_w overlaps f_n.

One important feature of the frequency response, which can be observed from Fig. 9.6, is that the cylinder oscillates at the lock-in points with a frequency that is slightly higher than f_n, the natural frequency of the cylinder in still water. This is because the natural frequency of the system is slightly increased when the cylinder is exposed to a water flow, as discussed in Example 8.3. Kozakiewicz et al. (1994) give data on the variation of the natural frequency as a function of V_r, obtained from oscillatory-flow experiments (Fig. 9.7).

2) Now let us consider the *amplitude response*. Once the frequency response is explained, it is no longer difficult to explain the amplitude response. In fact, there should be an obvious peak in the amplitude response, whenever the frequency ratio f/f_n in its zigzagging path on the $(f/f_n, V_r)$ plane hits the value

KC=10
KC=20
KC=30
N=2
N=2
N=3
N=4
N=7 N=6 N=5 N=4 N=3 N=2
f/f_n
f/f_w
f/f_w (=N)
2A/D
V_r
a
b
c

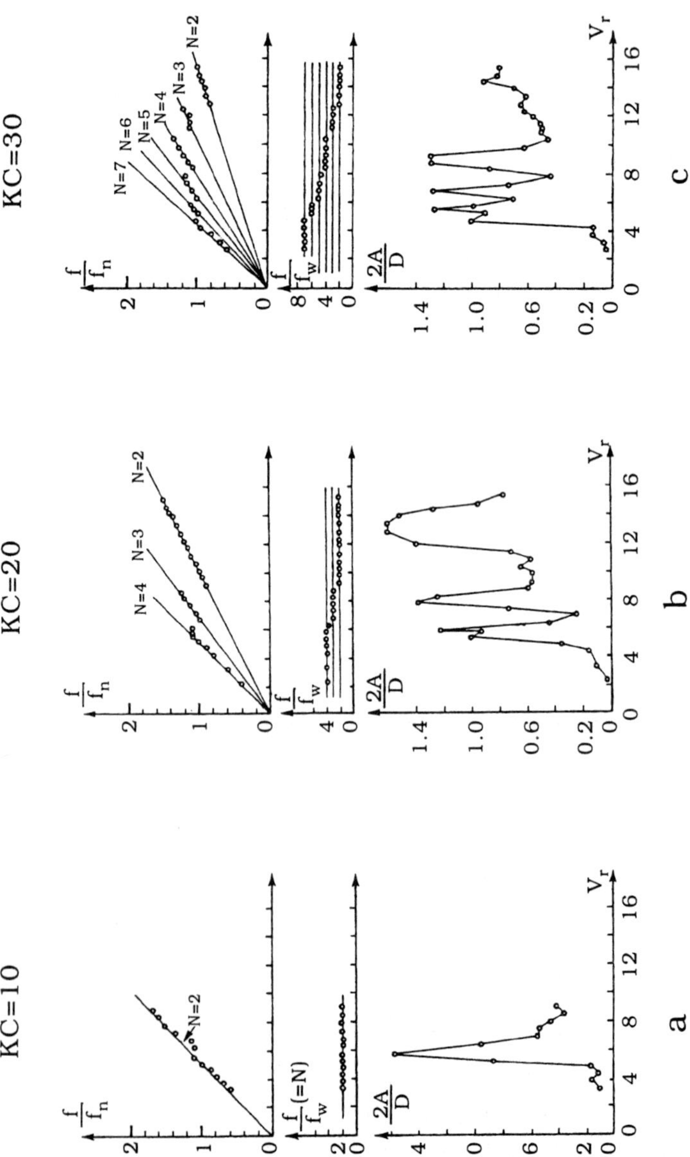

Figure 9.6

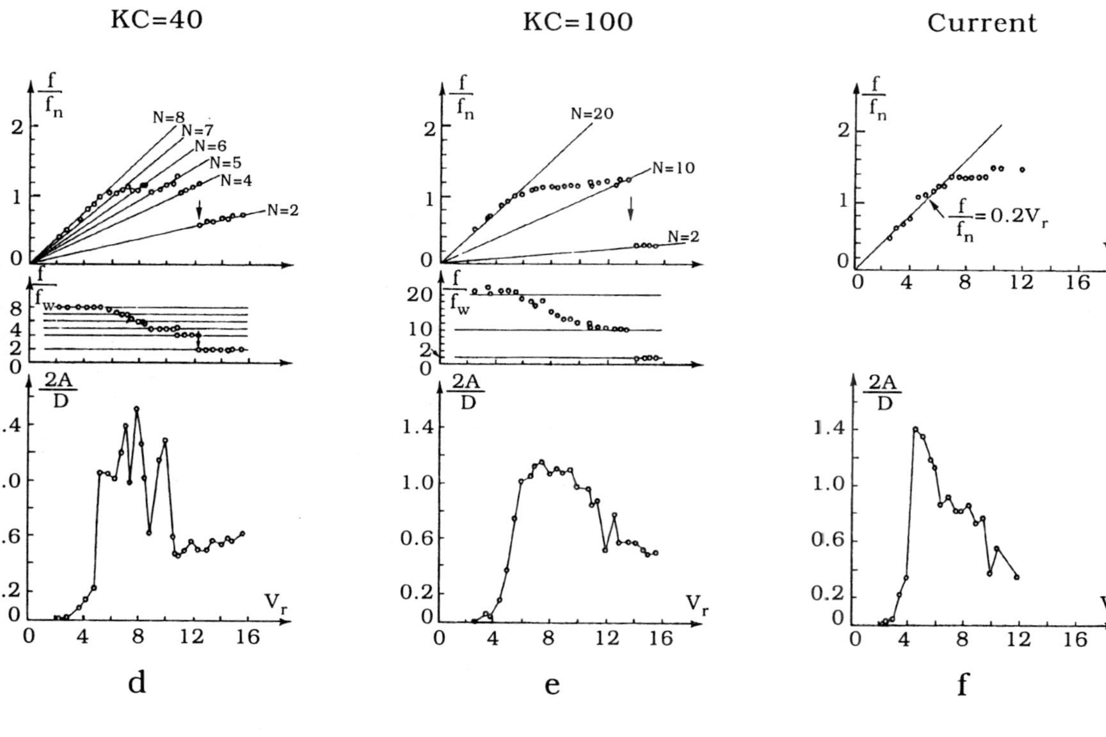

Figure 9.6 Frequency and amplitude response for cross-flow vibrations of a cylinder subject to an oscillatory flow. Radiating lines in f/f_n versus V_r, diagrams: equation 9.3. $M = 1.6$, $K_s = 0.9$, $k/\rho = 0.336 m^2/s^2$. Sumer and Fredsøe (1988).

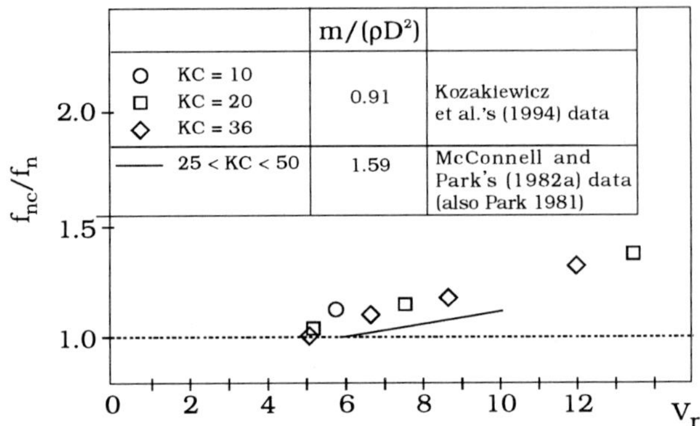

Figure 9.7 Natural frequency of a circular cylinder in oscillatory flow. f_n
is the natural frequency in still water.

of approximately unity (lock-in). Indeed, this is what happens in the amplitude response depicted in Fig. 9.6b.

3) Although no lift frequency measurements have been made in Sumer and Fredsøe's (1988) study, the question how the cylinder-vibration frequency relates to the lift frequency can be discussed in the light of the preceding description.

Sarpkaya (1976) reports that for a stationary cylinder, the fundamental lift frequency for one cycle of the motion is 4, when $KC = 20$ (see Fig. 3.16; see Table 5.2 for a full comparison). On the other hand, the aforementioned measurements show that the number of vibrations for one cycle of the motion is maintained at 4 until the first lock-in point is reached. Therefore, it appears that the cylinder vibration frequency follows the stationary cylinder fundamental lift frequency as V_r is steadily increased from zero up to the first lock-in point.

When the lock-in point is reached, however, the three frequencies, namely the vibration frequency, the lift frequency, and the natural frequency collapse onto one value, where the lift frequency is locked into the natural frequency of the system, just as the vibration frequency is locked into the latter frequency. Then the arguments put forward under the foregoing item 1 for the vibration frequency should equally be applicable to the lift frequency. Thus, the number of oscillations in the lift per flow cycle should be expected to jump down to a one-less value, namely to 3, when the first lock-in point is reached and further to 2, when the second one is reached.

From the foregoing arguments, it can be concluded that the fundamental lift frequency for one cycle of the oscillatory flow is a function not only of the KC number, but also of the reduced velocity V_r in case of a vibrating cylinder.

KC = 30. There is nothing special about the response obtained in this case. Every aspect of the cylinder response can be explained in the same way as in the case of $KC = 20$.

1) The jump in the number of vibrations per cycle down to a one-less value is clearly seen from the frequency response. It occurs five times as V_r is increased from zero up to approximately 16.

2) The distinct feature of the amplitude response, namely the multi-peak behaviour, is very clear from the amplitude diagram.

3) The arguments put forward under Item 3 in the previous section in connection with the lift force are equally applicable here, too. Also, note the good agreement between the number of vibrations per flow cycle (for small values of reduced velocity) and the number of oscillations in the lift force on a stationary cylinder per flow cycle for this KC number in Table 9.2.

Table 9.2 Sarpkaya's (1976) data (see Fig. 3.16) on stationary-cylinder lift-force frequency and Sumer and Fredsøe's (1988) data on cross-flow vibration frequency of a flexibly-mounted cylinder.

Number of cross-flow vibrations of a flexibly-mounted cylinder per flow cycle (for small values of reduced velocity) $N = f/f_w$			Number of oscillations in the lift force on a stationary cylinder per flow cycle $N = f_v/f_w$
Sumer og Fredsøe (1988)			Sarpkaya (1976)
KC	Experiment I	Experiment II	$N = f_L/f_w$
10	2	2	2
20	4	4-5	4
30	7	6-7	6
40	8	8-10	8
60	13	12	10-15
100	21	21	15-?

KC = 40. The comments in the preceding paragraphs also apply to this case. However, there is one aspect of the frequency response which needs an explanation. As is seen from the figure (Fig. 9.6d); the number of vibrations per cycle, N, jumps from 4 directly down to 2. This occurs at about $V_r = 12.5$. The same behaviour is seen even more clearly from the frequency response corresponding to the case $KC = 100$, where N jumps down from 10 to 2. This occurs at about $V_r = 13.5$. This behaviour can be attributed to the boundary layer transition. This aspect of the problem will be studied in detail later in this section in conjunction with the influence of Re number.

$KC = 100$.   Here it appears that the multipeak amplitude response disappears. For such high values of the KC number, the response characteristics should be expected to degenerate into those similar to the one obtained in steady currents. Indeed, there is a good deal of resemblance between the $KC = 100$ and "current" cases in Fig. 9.6 (except of course the V_r range where N becomes equal to 2 in the case of $KC = 100$). When plotted in the form of f/f_w versus V_r diagram, the frequency response obviously does not show any steplike variation like the ones obtained for the previous KC numbers 20, 30, and 40.

Finally, it may be noted that data presented by other investigators (Zedan and Rajabi (1981), McConnell and Park (1982a and b), Bearman and Mackwood (1991), and Kaye and Maull (1993)) verify the above description.

Cylinder response for KC below 7.   It is known that, for a KC number below 7, there is no vortex shedding. Then, at the first glance, it may seem that no cylinder vibration can be obtained when $KC < 7$, since there is no vortex shedding. However, the tests conducted for $KC = 5$ in Sumer and Fredsøe's (1988) study demonstrated that cylinder vibrations with amplitudes as large as $2A/D = 1.4$ very well can be obtained.

Fig. 9.8 shows the amplitude and frequency responses for this KC number. In the figure, the dotted curve represents the vibrations which do not come into existence themselves, but rather are initiated by a large external disturbance where the cylinder is displaced from its equilibrium position a distance of about half a pipe diameter away and then released.

The occurrence of vibrations can be attributed to the lift force originated from the asymmetry in the strength of the two attached vortices, which form behind the cylinder every half cycle of the flow.

This asymmetry occurs once the KC number exceeds the value 4 (Fig. 3.16), giving rise to a lift force at the oscillation frequency of the flow (Williamson, 1985b). The fact that the vibrations in our case occur at the oscillation frequency of the flow ($f/f_w = 1$) confirms the hypothesis that the vibrations are caused by the lift force originated from the asymmetry of the attached vortices.

The onset of vibrations occurs at about $V_r = 5$ for the vibrations which are initiated with a large external disturbance. However, in the absence of such disturbances, the onset velocity can be as high as $V_r = 8$.

As for the amplitude response, when the forcing frequency (i.e., f_w) is near the natural frequency, the amplitude obviously takes very large values. However, it decreases steadily as the forcing frequency moves away from the natural frequency.

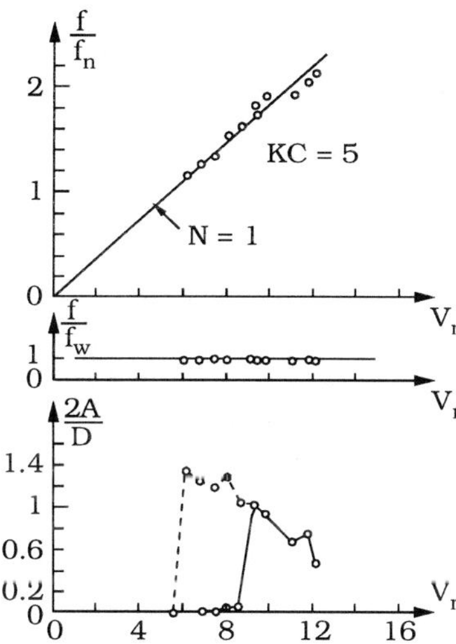

Figure 9.8 Frequency and amplitude of cross-flow vibrations in oscillatory flow. $KC = 5$. Dotted curve: vibrations initiated by external disturbance. Solid curve: vibrations come into existence by themselves. Straight line in f/f_n versus V_r diagram: Eq. 9.3. $M = 1.6$, $K_s = 1.5$, $k/\rho = 0.168 \; m^2/s^2$. Sumer and Fredsøe (1988).

Response with two degrees of freedom of movement. Maull and Kaye (1988) made tests with a cylinder with two kinds of freedom of movement: in one test, the cylinder was restrained in the in-line direction, in the other it was unrestrained. For the range of f_n/f_w tested in the experiments $(f_n/f_w \gtrsim 1.3)$, the cross-flow response of the cylinder in the unrestrained case was practically the same as in the case of restrained in-line only. This is because, in the unrestrained case, the motion of the cylinder relative to the fluid in the in-line direction was relatively small (the amplitudes being about 15-20% of the amplitudes of the fluid motion). However, when the cylinder oscillates in the in-line direction in the resonance regime $(f_n = f_w)$, then the amplitude of the in-line oscillation will become quite large, as will be seen in Section 9.4. Therefore the in-line motion may begin to influence the cross-flow vibrations in this case (Lipsett and Williamson, 1991b).

9.2.2 Effect of mass ratio and stability parameter

Sumer and Fredsøe (1988) studied the effect of the mass ratio and that of the stability parameter on the cross-flow vibrations in oscillatory flow. The trends were found to be similar to those found in the case of steady current (Section 8.3.1), namely 1) the higher the mass ratio, the narrower the response range in V_r, and 2) the smaller the stability parameter, the larger the response amplitude. Fig. 9.9 illustrates these effects for three different KC numbers.

One may also note that, in Fig. 9.6, maximum amplitudes experienced at different KC numbers are not drastically different from that measured in the case of steady current (Fig. 9.6f). Similar observations can be made with the other test series achieved in the studies of Sumer and Fredsøe (1988) and Sumer, Fredsøe and Jacobsen (1986). This suggests that the steady-current data given in Fig. 8.25 regarding the dependence of maximum amplitude on K_s may, to a first approximation, be implemented for the case of waves too. Zedan and Rajabi's (1981) results as regards the maximum amplitude measured in their tests also support the above assessment.

9.2.3 Effect of Reynolds number and surface roughness

Sumer and Fredsøe (1989) made a systematic investigation of the effect of the Reynolds number and the surface roughness on cross-flow vibrations in oscillatory flows. They used three kinds of circular cylinders shown in Table 9.3.

Fig. 9.10 compares the cross-flow response of the three cylinders for $KC = 20$. Let us first consider Fig. 9.10a. It is clear from the figure that the response of the large cylinder is not the same as that of the small cylinder. While the small cylinder vibrates with large amplitudes at the first lock-in point $V_r \cong 5.5$, this is not the case for the large cylinder. This is because the Reynolds number attains its critical value already at this point (i.e., at a value a little larger than 1×10^5), therefore the vortex shedding is "weakened" and presumably the cylinder does not respond in the way as it does in the subcritical flow regime. Although the response amplitudes of the two cylinders are much the same at the second lock-in point, namely at $V_r \cong 8$, large differences are observed for further values of V_r.

Considering also other KC numbers tested in the study, Sumer and Fredsøe (1989) concluded that the vibrations at high Reynolds numbers for a cylinder with a smooth surface can be markedly different from those at low Re numbers (corresponding to the subcritical flow regime). It may be noticed that this conclusion is very much in line with the results obtained in the case of steady current (Figs. 8.27 and 8.28).

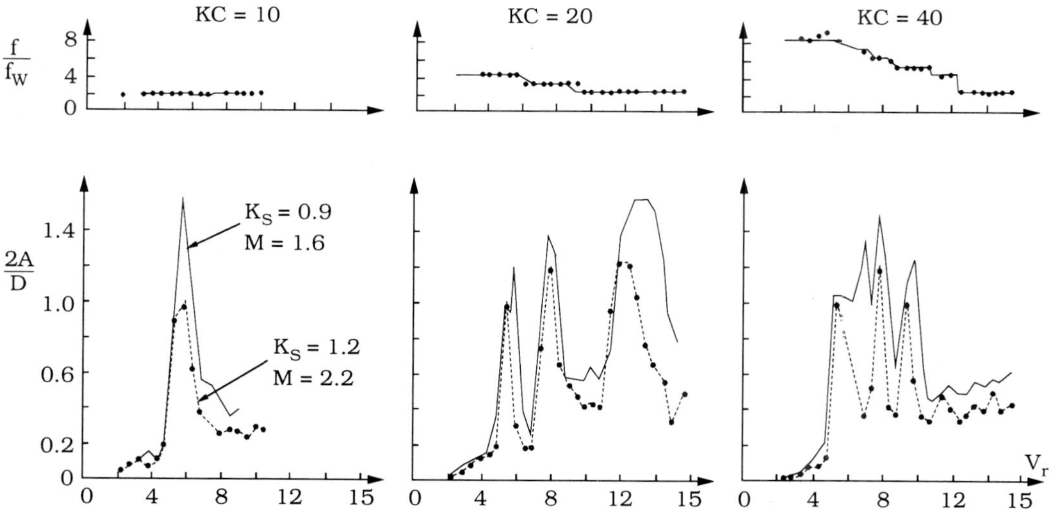

Figure 9.9 Frequency and amplitude response. Effect of stability parameter and mass ratio. Solid curve: $K_s = 0.9$, $M = 1.6$, $k/\rho = 0.336 m^2/s^2$, $f_n = 0.71$ Hz. Circles: $K_s = 1.2$, $M = 2.2$, $k/\rho = 0.336 m^2/s^2$, $f_n = 0.61$ Hz. Sumer and Fredsøe (1988).

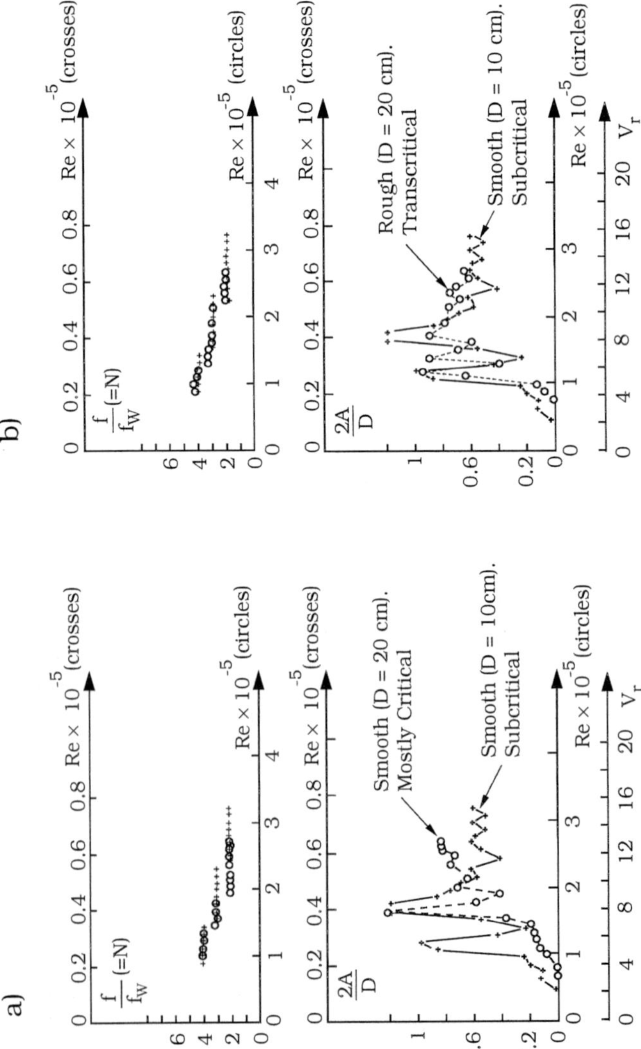

Figure 9.10 Comparison of cross-flow response in Sumer and Fredsøe's (1989) large cylinder experiments in different flow regimes. $KC = 20$. Crosses: $D = 10$ cm, $M = 1.6$, $K_s = 1.7$, $k/\rho = 0.168 m^2/s^2$, $f_n = 0.5$ Hz. Circles: $D = 20$ cm.

Table 9.3 Flow conditions in Sumer and Fredsøe's (1989) experiments. The hydroelastic properties of the three systems are practically the same.

Cylinder surface	Cylinder diameter D(cm)	Surface roughness k_s/D	Range of Re	Approximate flow regimes
Smooth	10	0	$Re \lesssim 1 \times 10^5$	Subcritical
Smooth	20	0	$1 \times 10^5 \lesssim Re \lesssim 3 \times 10^5$	Mostly critical
Rough	20	13×10^{-3}	$1 \times 10^5 \lesssim Re \lesssim 3 \times 10^5$	Transcritical

As for the response of the rough, large cylinder in Sumer and Fredsøe's (1989) study (Fig. 9.10b), it is interesting to note that the response frequency and the response amplitude are not much different in the two cases indicated, namely the case of subcritical flow regime and that of transcritical flow regime (achieved by the rough-wall cylinder).

Similar arguments put forward in conjunction with the effect of surface roughness on cross-flow vibrations in steady current (Section 8.3.1) may be used in the present case, too. Namely, the rough cylinder does not experience the same kind of large change in amplitude response as the smooth cylinder due to the weak presence of transitional flow regimes. Sumer and Fredsøe (1989), considering the results obtained for other KC numbers tested, concluded that if the cylinder is rough, the Re number effect may disappear, depending on the roughness parameter k_s/D. It was found that the Re number effect is practically non-existent for a cylinder with a roughness parameter $k_s/D = 13 \times 10^{-3}$. Available data on stationary cylinders suggest that the Re number effect on the vibration of cylinders practically disappears for $k_s/D \gtrsim 3 \times 10^{-3}$.

An important practical consequence of Sumer and Fredsøe's (1989) large-cylinder experiments concerns the laboratory model study of vibrations of marine risers and pipelines. Normally, the marine growth on such flexible offshore structures satisfies the relation $k_s/D \gtrsim 3 \times 10^{-3}$. The results of Sumer and Fredsøe's study suggests that, in such situations, the model similarity can very well be achieved with a smooth small-scale model cylinder as far as Re number is concerned.

If the marine growth is such that $k_s/D \lesssim 3 \times 10^{-3}$, however, the Re number effect (thus the scale effect) is felt; obviously, the smaller the roughness of the pipe, the more pronounced the Re number effect.

9.2.4 Cross-flow vibrations in irregular waves

A thorough and systematic investigation of cross-flow vibrations of a flexibly-mounted cylinder in irregular flow conditions has been made by Kozakiewicz et al. (1994). To eliminate the additional effects encountered in an actual wave environment, such as wave non-linearity, wave asymmetry, wave drift (and its associated return flow) and orbital flow velocities, Kozakiewicz et al. (1994) preferred to experiment with an irregular *oscillatory flow*. The flow in Kozakiewicz et al.'s study was simulated by the motion of a carriage in an otherwise still water. The cylinder was a hydraulically-smooth cylinder, and it had one degree of freedom of movement, namely in the cross-flow direction. Kozakiewicz et al. conducted experiments also with regular oscillatory flow for reference purposes. Fig. 9.11 compares the time series of flow velocity and the cylinder vibration in regular and irregular oscillatory flow conditions.

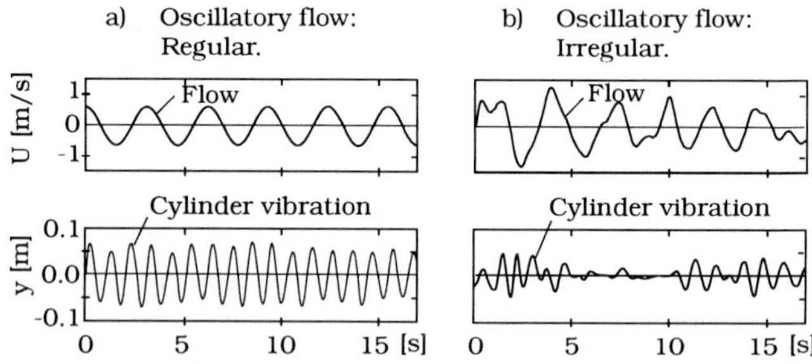

Figure 9.11 Time series of flow velocity (U) and cylinder displacements (y). $KC = 20$, $V_r = 7.6$. Kozakiewicz et al. (1994).

The Keulegan-Carpenter number and the reduced velocity in the irregular oscillatory flow are defined by

$$KC = \frac{\sqrt{2}\sigma_U T_w}{D} \tag{9.7}$$

$$V_r = \frac{\sqrt{2}\sigma_U}{D f_n} \tag{9.8}$$

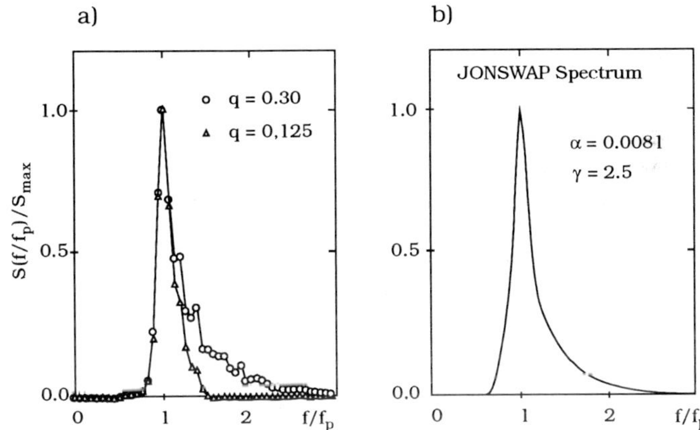

Figure 9.12 Non-dimensional spectra: (a) Velocity control spectra used in Kozakiewicz et al.'s (1994) random oscillatory flow tests; (b) Example of a JONSWAP spectrum.

respectively. Note that $\sqrt{2}\sigma_U$ (σ_U being the r.m.s. value of flow velocity) will reduce to U_m in the case of regular oscillatory flow.

Measured in-situ water elevation spectrum for the North Sea storm conditions was used as the control spectrum to generate carriage control irregular velocity signals in Kozakiewicz et al.'s study. This spectrum shown in Fig. 9.12a with $q = 0.30$ is well described by the JONSWAP wave spectrum with relevant parameters (Section 7.1.1). An example of a normalized JONSWAP spectrum for given parameters is depicted in Fig. 9.12b for comparison. Here, q is the spectral width parameter (Longoria et al., 1991) defined by

$$q = \sqrt{1 - \frac{m_1 m_1}{m_0 m_2}}, \quad m_n = \int_0^\infty f^n S(f) df, \qquad (9.9)$$

where $S(f)$ is a power spectrum, and m_n is the spectral moment of the nth order (Section 7.1.1). For broad-band spectra, q approaches 1, while for narrow-band spectra q is close to 0. For the input velocity power spectrum, the parameter ε, also characterizing the width of the power spectrum (Chapter 7, Eq. 7.20), was calculated to be 0.59.

Fig. 9.13 compares the power spectra related to the cylinder response. It is clear that, in contrast to an extremely narrow band spectrum function of vibrations (Fig. 9.13a, the bottom spectrum) in the regular oscillatory-flow situation, the spectrum in the case of irregular oscillatory flow (Fig. 9.13b, the bottom spectrum)

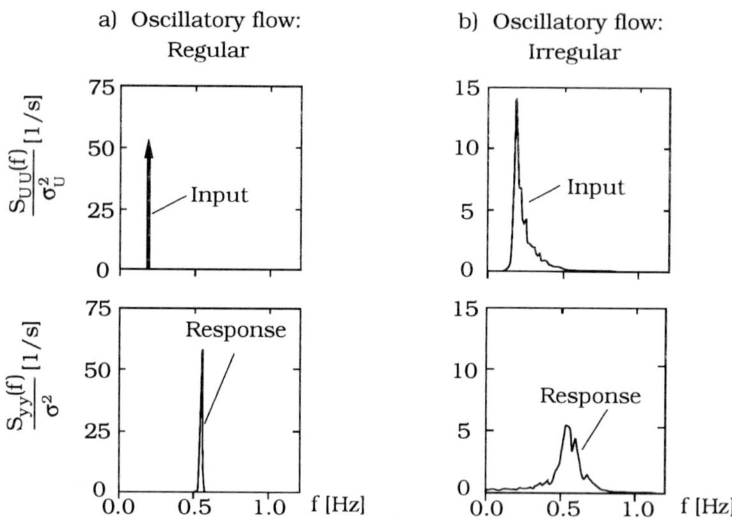

Figure 9.13 Normalized input velocity power spectra $(S_{UU}(f))$ and normalized response power spectra $(S_{yy}(f))$: (a) $KC = 20$, $V_r = 7.6$, $q = 0$; (b) $KC = 20$, $V_r = 7.6$, $q = 0.3$. Kozakiewicz et al. (1994).

is definitely a broad-band spectrum, as anticipated.

Fig. 9.14 compares the results regarding the frequency and amplitude response of the cylinder in the regular and irregular oscillatory flows, for one of the three KC numbers tested, namely for $KC = 20$. In the figure, f is the frequency of vibrations, corresponding to the peak frequency of the response spectrum, while $2A/D$ is the mean peak-to-peak amplitude of cylinder oscillations. The quantity σ, on the other hand, is the r.m.s. value of cylinder displacement from its mean.

Amplitude response. First of all, the amplitude response does not reveal the multipeak behaviour of the regular-flow case, being nearly constant in the present case over the large portion of the V_r-axis.

Second, vibration amplitudes are considerably smaller over the part of that V_r-axis ($V_r > 5$ in Fig. 9.14) that comprises the lock-in ranges for regular oscillatory flow. This is due to the continuously changing nature of the forcing, which occasionally results in periods of small cylinder activity (Fig. 9.11b). As will be seen later, the response amplitude systematically decreases with increasing width of the spectrum.

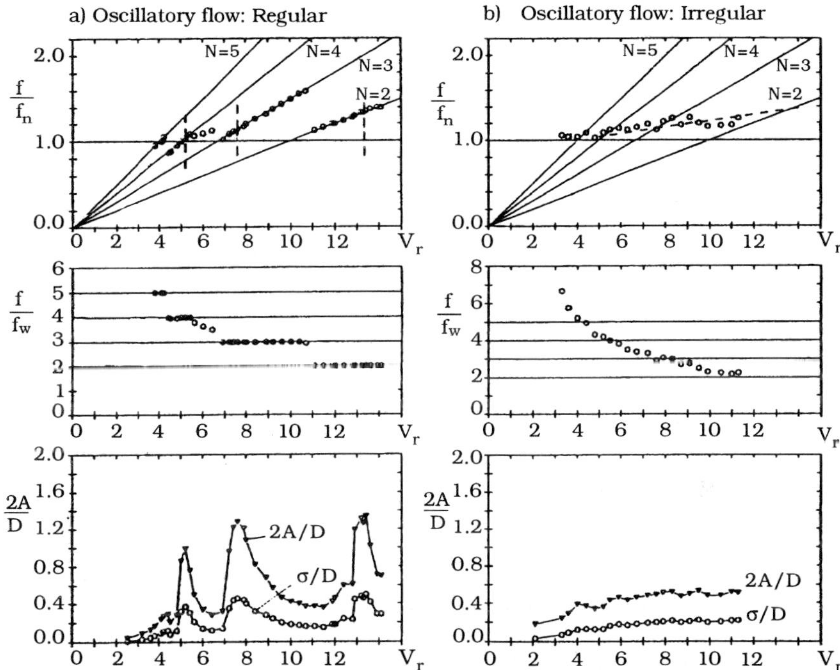

Figure 9.14 Frequency and amplitude response of a flexibly-mounted cylinder. (a): Regular oscillatory flow, $KC = 20$. (b): Irregular oscillatory flow for a broad-band input velocity spectrum ($q = 0.30$ or alternatively $\varepsilon = 0.59$), $KC_r = 20$. $M = 1.7$, $K_s = 2.1$, $k = 295.4 \ N/m$, $f_n = 0.48$ Hz.

Third, vibration amplitudes are larger than in regular oscillatory flow for small reduced velocities $V_r < 5$ and they are initiated earlier at $V_r \approx 2$. This is connected with the broad frequency band of the input-velocity power spectrum, which can result in high instantaneous KC numbers that in turn incite a larger than expected lift in a subsequent cycle.

Finally, it should be emphasized that the aforementioned comparison is based on the information obtained with an experimental setup, where the in-line motion was suppressed. Also, the results refer to the mean amplitudes averaged over the whole length of the test, including the resting periods that the cylinder could experience in a typical irregular-oscillatory-flow situation.

Frequency response. The frequency response of the cylinder in the case of irregular oscillatory flow (Fig. 9.14) differs considerably from that in the regular-oscillatory-flow situation. The characteristic zigzagging behaviour of f/f_n completely disappears. While the lock-in occurs at discrete V_r values in regular oscillatory flows at $V_r = 5.2$, 7.6 and 13.5, it apparently occurs over the whole range of V_r in the case of irregular oscillatory flows. Indeed, Fig. 9.14 indicates that the frequency f/f_n appears to follow the best straight line for the points in Fig. 9.7, representing the ratio of the natural frequency in oscillatory flow, f_{nc}, to the natural frequency in still water, f_n, versus V_r (this line is shown as dashed lines in the uppermost graph of Fig. 9.14). In other words, in the case of irregular waves, the system is selective in the sense that, when it oscillates (Fig. 9.11b), it oscillates with its natural frequency, regardless of the value V_r experienced. This is not an entirely unexpected result, since over a wide variety of frequencies in the input-velocity spectrum at any value of V_r, there are always components at frequencies close to the natural frequency which presumably excite the system. It is remarkable, however, that the response frequency in irregular waves can well be approximated by the relationship for the natural frequency obtained from regular-wave tests, shown in Fig. 9.7.

Finally, we may note that the vibration time series in irregular oscillatory flows may contain resting periods, as seen in Fig. 9.11b. So caution must be exercised considering the fact that the lock-in referred to in the preceding paragraph does not occur continuously, but rather, intermittently.

Effect of spectrum width. Kozakiewicz et al. (1994) repeated their tests with an input velocity power spectrum with $q = 0.125$ (Fig. 9.12a), a narrow-band spectrum, to examine the effect of spectrum width on the vibratory response of the cylinder.

Regarding the frequency response, the way in which the frequency changes with V_r was found to be not totally the same as in the case of broad-band input spectrum; it resembled partially the frequency response observed under regular oscillatory flow conditions and partially that observed under irregular oscillatory flow conditions with broad-band spectrum. The explanation of this fact is that, for a narrow-band spectrum, the regular-oscillatory-flow vortex shedding regimes occur more frequently, that is, in longer intervals than for the broad-band case, as was shown by Sumer and Kozakiewicz (1995). A consequence of this is a narrower spectrum of cylinder displacements (lower values of q_c) when compared to the results for $q = 0.30$.

It is remarkable, however, from Kozakiewicz et al.'s (1994) study that the amplitude response of the cylinder for this narrow-band input spectrum resembles quite closely that observed for the broad-band input spectrum.

The maximum amplitudes of cylinder vibrations, $A_{\max}$, are presented in Fig. 9.15 as a function of q, the spectral width parameter. It is evident that the increase in the width of the input velocity spectrum results in a systematic decrease of oscillations. The figure indicates that, for $q = 0.3$, a value characterizing the

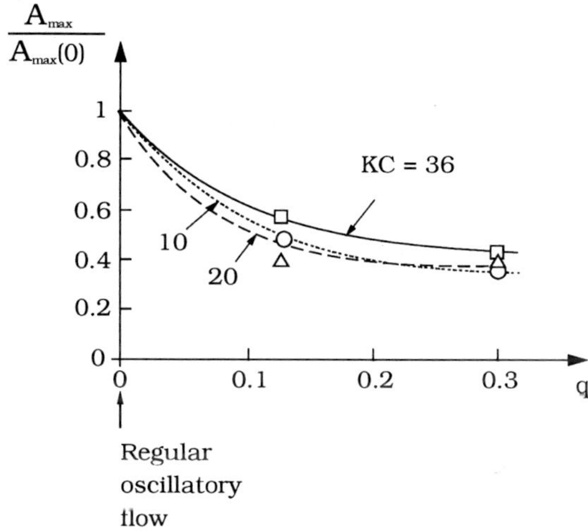

Figure 9.15 Maximum amplitude as a function of q, the parameter characterizing the width of spectrum function of irregular oscillatory
flow. $A_{max}(0)$ is the maximum amplitude in the case of regular
oscillatory flow. Kozakiewicz et al. (1994).

JONSWAP spectrum (see Fig. 9.13), the maximum amplitudes are reduced by about 60% with respect to the values experienced in the case of regular waves. This is a very significant reduction in the vibration amplitude.

9.3 In-line vibrations.

There are two kinds of in-line motion: 1) Periodic in-line movement, which occurs at the wave frequency and is caused by the total in-line (Morison) force; and 2) High-frequency in-line vibrations, which are induced by effects such as vortex shedding. These latter vibrations are superimposed on the wave-induced in-line movement, as is seen from Fig. 9.1. These small-amplitude, high frequency in-line vibrations have not been investigated extensively, therefore our knowledge on the subject is very limited. One reason behind this may be that their effect is normally overshadowed by the low-frequency, large-amplitude in-line movement. However, considering the relatively high-frequency oscillations associated with these vibrations, they may contribute to the total fatigue damage fairly significantly.

The mechanism behind such vibrations must be closely associated with 1) vortex shedding (as in the case of steady currents; see Section 8.3.2) and also 2) with the motion of vortices around the cylinder caused by flow reversals. If the cylinder has two degrees of freedom of movement, on the other hand, the presence of cross-flow vibrations will also be important.

These vibrations have been reported by Jacobsen et al. (1985) in relation to physical-model tests of a mono-tower platform exposed to waves, and by Sumer et al. (1989) in relation to an experimental study with the aim of determining the hydrostatic vibrations of marine pipelines exposed to waves. In the latter study, tests were also carried out for a pipe which was placed away from the sea-bottom with a gap-to-diameter ratio $e/D = 2$. Clearly, the cylinder in these latter tests may be regarded practically as a wall-free cylinder (the time series presented previously in Fig. 9.1 are taken from these tests).

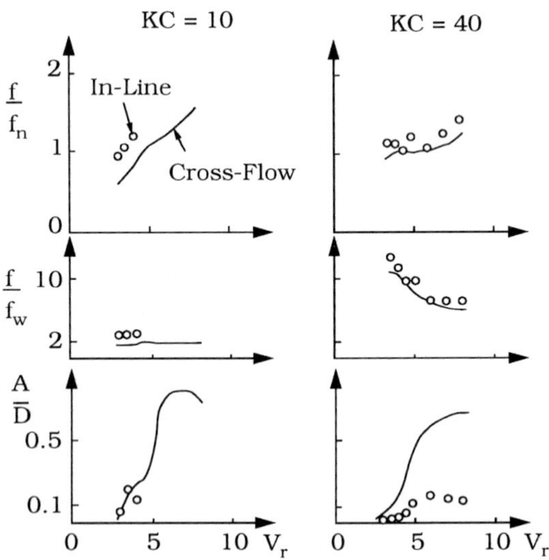

Figure 9.16 Amplitude and frequency of high-frequency in-line vibrations in oscillatory flows. The cylinder has two degrees of freedom of movement. Also shown in the figure are the amplitude and frequency of cross-flow vibrations measured in the tests. $M = 1.9$, $K_s = 0.1$, $f_n = 0.4$ Hz, $D = 15$ cm, $k_s/D = 4 \times 10^{-3}$. Sumer et al. (1989).

Fig. 9.16 depicts the measured in-line vibrations of the test cylinder in the previously mentioned tests. The cylinder was a flexibly mounted-rigid cylinder with two degrees of freedom of movement. Also shown in the figure are the measured cross-flow vibrations of the cylinder. It appears that the number of in-line vibrations, namely $f/f_w(=N)$, is one more than the number of cross-flow vibrations occurring in one wave cycle. For example for $KC = 10$, N for the in-line vibrations is 3, while it is 2 for the cross-flow vibrations.

This result is not in accord with the picture given in Section 8.3.2 with reference to in-line vibrations in steady currents. Apparently, the measured in-line vibrations depicted in Fig. 9.16 are the so-called third kind in-line vibrations described in conjunction with in-line vibrations in steady currents (Section 8.3.2).

The latter vibrations occur generally at a frequency twice the frequency of cross-flow vibrations, whereas, in the present case, the frequency of in-line vibrations is well below this frequency. In fact, it is only moderately higher than the frequency of cross-flow vibrations, f, in the case of $KC = 10$ and slightly higher than f in the case of $KC = 40$. No clear explanation has been offered for this kind of behaviour. Fig. 9.16 indicates that the number of in-line vibrations per flow cycle, f/f_w, is in most of the cases one larger than that of cross-flow vibrations. Sumer et al. (1989) attributes this increase in the number of in-line vibrations to the flow reversals.

9.4   In-line oscillatory motion

As noted earlier, a flexibly-mounted cylinder may undergo in-line oscillations induced by the total in-line force (Table 9.1). This type of movement has been called the in-line motion (Section 9.1), to differentiate it from the vortex-induced in-line vibrations. These oscillations will occur at the frequency of the oscillatory motion f_w, because the force itself (the Morison force) oscillates at f_w.

The main objective regarding this type of oscillations is to predict the response of the flexibly-mounted cylinder when the structure is exposed to waves. In practice, this may be achieved by the application of the equation of motion in the form

$$m\,\ddot{x} + c\,\dot{x} + kx = F(t) \tag{9.10}$$

where x is the in-line displacement of the structure, and $F(t)$ is the total in-line force on the structure per unit span, which, for a cylindrical element, may be written as

$$F(t) = \frac{1}{2}\rho C_D D(U - \dot{x})|U - \dot{x}| + \rho C_m \frac{\pi D^2}{4}\frac{d}{dt}(U - \dot{x}) + \rho\frac{\pi D^2}{4}\dot{U} \tag{9.11}$$

This is the Morison formulation of the force on a non-stationary cylinder (see Section 4, Eq. 4.30).

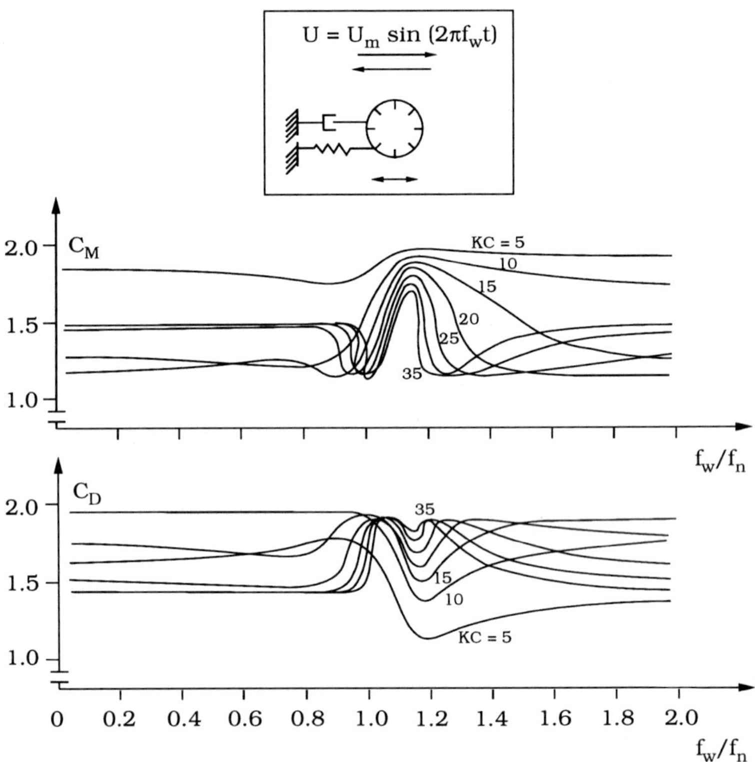

Figure 9.17 Variation of drag (C_D) and inertia coefficient $(C_M = 1 + C_m)$ as function of oscillatory flow frequency for a flexibly-mounted cylinder exposed to an oscillatory flow and undergoing in-line oscillations induced by Morison force. Prediction due to Williamson's semi-empirical model. $m/(\rho D^2) = 5.11$; $\zeta_s = 0.02$; and $\beta(= D^2/(\nu T_w)) = 730$. Williamson (1985a).

To predict the response of the structure, the preceding set of equations (Eqs. 9.10 and 9.11) must be integrated numerically (Williamson (1985a) and Bearman et al. (1992)). Regarding the force coefficients C_D and C_m, the usual view is that these coefficients should be taken as corresponding to stationary-cylinder values. However, whether C_m and C_D may be assumed constant (equal to their stationary-cylinder values) for a complete range of oscillation amplitudes has not been investigated extensively. It may be expected that there is a feedback between the response and the fluid force (represented by C_D and C_m); the response in Eq. 9.10 is determined by the force coefficients, and the force coefficients themselves must depend on the response (or, to be exact, on the motion of water relative to the structure). Williamson (1985a) studied this aspect of the problem by using a semi-empirical model where the force coefficients were chosen from fixed cylinder data, yet they were selected to be those corresponding to the relative amplitude of fluid motion rather than the absolute fluid motion. Williamson's predictions indicated that C_D and C_m show much fluctuation at the resonance point and beyond for a range of wave frequency satisfying $f_w/f_n > 1$. The results of Williamson's (1985a) prediction are shown in Fig. 9.17 where C_M is the inertia coefficient defined in the usual way, $C_M = 1 + C_m$ (Eq. 4.28).

From the results, Williamson concluded that, near resonance, it may be advisable to take account of the change in the force coefficients due to the response of the structure.

Fig. 9.18 illustrates how the amplitude of the Morison-induced in-line oscillations varies with respect to the wave frequency predicted by Williamson (1985a). In the solution, the previously mentioned change in the force coefficient due to the response of the structure was considered. In the same figure (Fig. 9.18b), also the variation of relative amplitude is included. The figure indicates that the cylinder in the neighbourhood of the lock-in point follows very closely the fluid. It may be noticed that the value of $\beta(= Re/KC)$ taken in the calculations regarding the predictions presented in Fig. 9.18, namely $\beta = 730$, is rather small.

This is because Williamson maintained β at this value, identical to that experienced in his experiments, to facilitate comparison. Obviously, β values encountered in practice are an order of magnitude (or more) larger than this. Besides, the surface roughness may be present.

So, caution must be exercised when the diagrams in Fig. 9.18 are to be used. However, similar predictions can be made fairly easily for any set of input parameter values, as described in the preceding paragraphs.

9.5   Flow around and forces on a vibrating cylinder

Flow around and forces on a cylinder exposed to an oscillatory flow and undergoing vibrations are dependent on the vibration amplitude (normalized by the cylinder diameter) and the vibration frequency (or alternatively, the reduced velocity, in the normalized form), in addition to the usual parameters, such as the Re number, the KC number, etc., cf. Section 8.5.1. Although quite a substantial amount of knowledge has accumulated on this subject in recent years in the case of steady currents (see Section 8.5), our knowledge is very limited in the case of waves. Kozakiewicz, Sumer, Fredsøe and Hansen (1996) have studied the case where the cylinder has one degree of freedom of movement, namely in the cross-

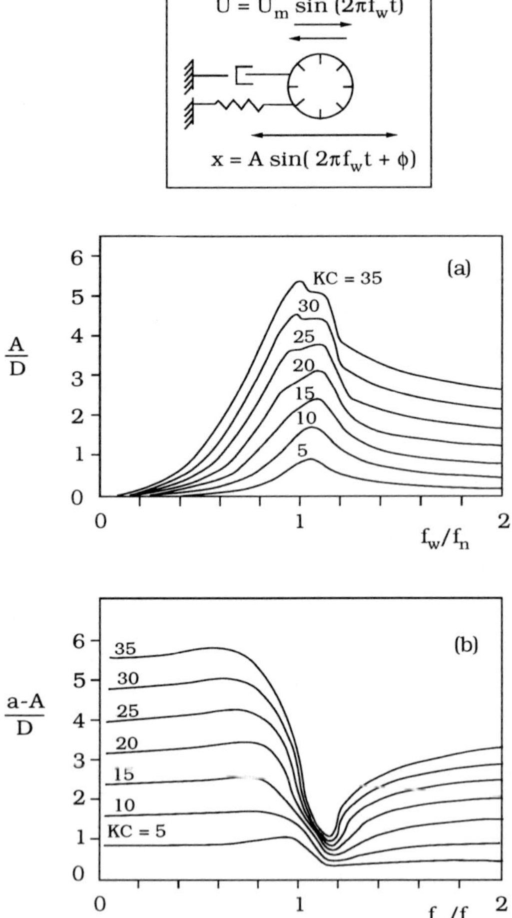

Figure 9.18 Response amplitude of a flexibly-mounted cylinder exposed to an oscillatory flow and undergoing in-line oscillatory motion induced by Morison force. Prediction due to Williamson's (1985a) semi-empirical model for the following input parameters. $m/(\rho D^2) = 5.11$; $\zeta_s = 0.02$; and $\beta(= D^2/(\nu T_w)) = 730$. (a): Amplitude of absolute motion of the cylinder A/D. (b): Relative amplitude $(a - A)/D$ where a is the amplitude of the fluid motion.

flow direction. They conducted two kinds of studies: 1) flow-visualization tests, carried out for two KC numbers, $KC = 10$ and 20, and for the ranges of reduced velocity $V_r = 4$-7 and 5-8, respectively; and 2) a numerical simulation using the discrete vortex model to complement the experiments (see Chapter 5 for a detailed account of the latter model).

Fig. 9.19 compares the vortex flow over one cycle of the motion obtained for the vibrating cylinder case (Fig. 9.19b) with that for the stationary cylinder situation (Fig. 9.19a). The KC number of the tests is $KC = 10$, the Re number being $Re \simeq 10^3$. The vibrating cylinder corresponds to the lock-in situation with $V_r = 6.3$, $f/f_w = 2$ and $A/D \cong 0.8$.

The trajectory of the cylinder is shown in Fig. 9.19b in the upper left frame. It is evident from the figure that, in the vibrating cylinder case, an additional vortex is generated each time before the flow reverses (Vortices b' and A'). Hence there are four vortices generated per flow cycle instead of 2. Analysis of the flow pictures showed that these additional vortices merged with the newly generated ones; Vortex b' merges with Vortex A, and Vortex A' merges with Vortex B. The overall effect is to generate a transverse vortex street, similar to that experienced in the case of a fixed cylinder (cf. Fig. 3.6a).

The generation of additional vortices is the key feature of vortex flows around a vibrating cylinder, common to all other cases tested in Kozakiewicz et al.'s study (provided that the vibration amplitude is sufficiently large). This is linked to the increased length of the cylinder trajectory in the vibrating-cylinder case.

No force measurements were made in Kozakiewicz et al.'s (1996) study. However, the process was simulated numerically, as mentioned earlier, and the forces could be obtained from this simulation.

It may be noted that the numerical-simulation results revealed all the vortex-flow patterns observed in the flow-visualization study. Kozakiewicz et al. (1996, Fig. 5) give an example where the flow pictures obtained from the numerical simulation were compared with those obtained from the flow-visualization experiment.

Fig. 9.20 displays the time series of the lift force obtained in the numerical simulation. The lift force variation over one flow period for the case of fixed cylinder (Fig. 9.20a) agrees quite well with Williamson's (1985b) experiments (cf. Fig. 3.11).

Comparison of Figs. 9.20a and 9.20b indicates the following.

First, the frequency of the lift force generally does not change in the case of vibrating cylinder, namely $f_L = 2f_w$.

Second, when closely inspected, it is found that the kinks in the lift force traces (such as $2'$ in Fig. 9.20b) are caused by the previously mentioned additional vortices generated prior to the flow reversals (e.g., Vortex B' at $t = 7T/8$ in Fig. 9.19b).

Third, the peaks marked 2 and 4 are caused by the return of the vortex, shed previously, just after the flow reversal, in much the same way as in the case

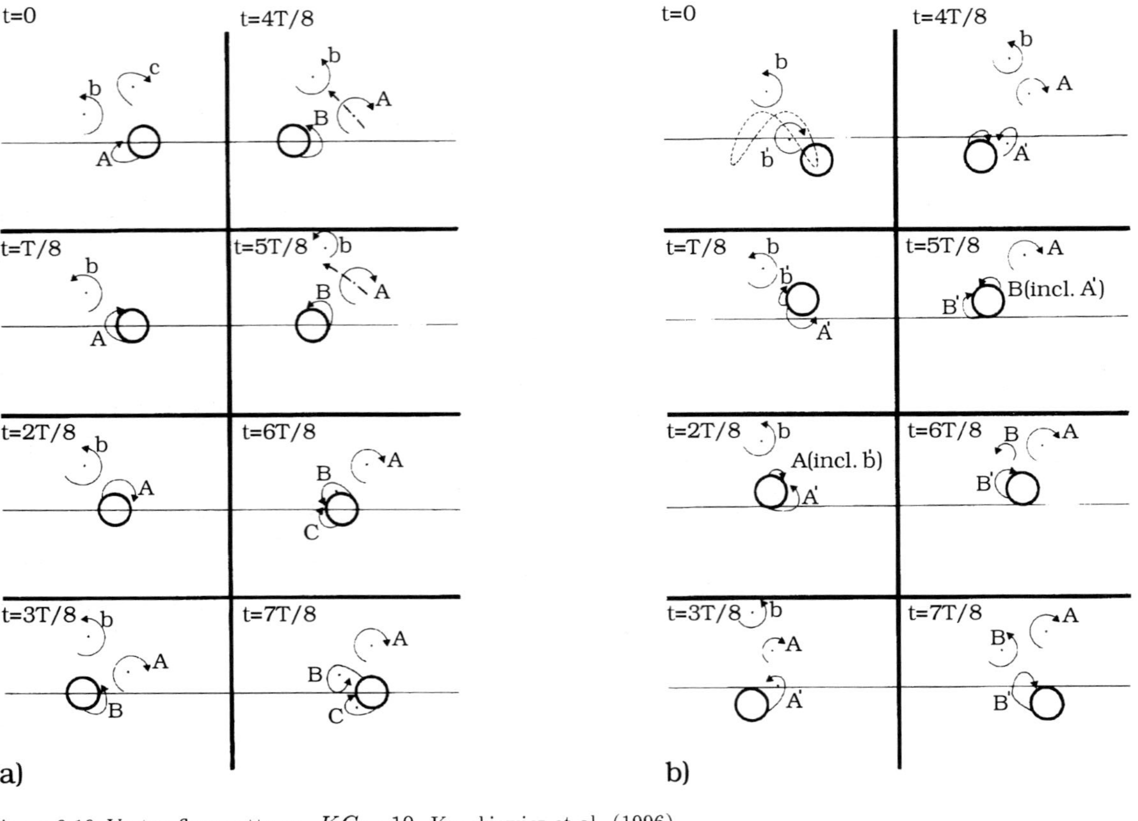

Figure 9.19 Vortex flow patterns. $KC = 10$. Kozakiewicz et al. (1996).

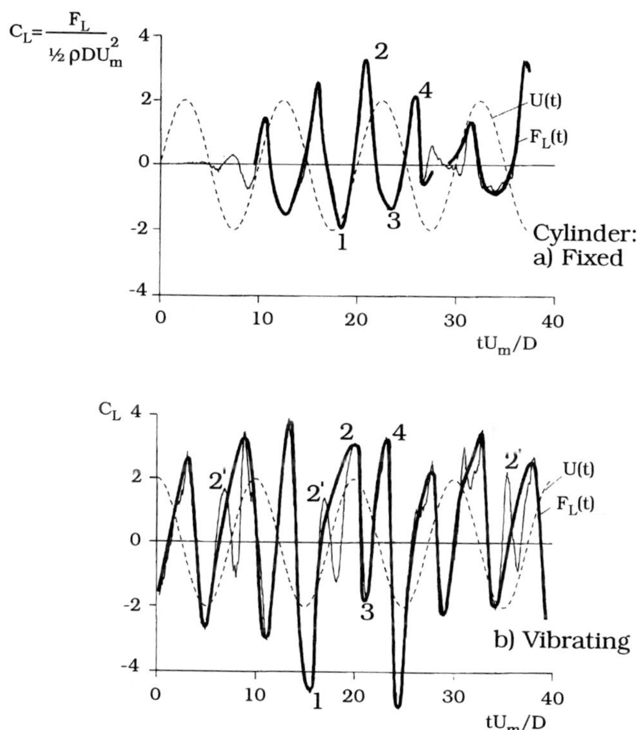

Figure 9.20 Time series of lift force. $KC = 10$. Numerical simulation. Kozakiewicz et al. (1996).

of fixed cylinder (Fig. 3.11): the peak marked 2 is caused by the return of Vortex b at $t = 2T/8$ in Fig. 9.19b, while the peak marked 4 is caused by the return of Vortex A at $t = 4T/8 - 5T/8$ in Fig. 9.19b; however, the latter peak is enhanced by the formation of the additionally generated vortex, Vortex A' (see $t = 4T/8$ in Fig. 9.19b).

Fourth, there appears quite a substantial amount of increase in the magnitude of the lift force when the cylinder is vibrating. This is due to the increase in the strength of the vortices in the vibrating-cylinder-case. Kozakiewicz et al.'s (1996) numerical results showed an increase in $C_{L\,max}$ by a factor of 1.7 when the cylinder was vibrating with an amplitude of $A/D \cong 0.8$, for both $KC = 10$ and $KC = 20$ tested in the study. These results appear to be in good agreement with the experimental results of Sumer, Fredsøe, Jensen and Christiansen (1994).

Finally, Kozakiewicz et al. (1996) observed in their numerical experiments that, similar to the lift force, the in-line force also underwent substantial changes when the cylinder was vibrating. The effect of vibrations was a) to increase the magnitude of the force (a factor of as much as $2 - 2.5$ increase); and b) to superimpose high frequency fluctuations on the in-line force. The frequency of these fluctuations was twice the lift frequency.

REFERENCES

Angrilli, F. and Cossalter, V. (1982): Transverse oscillations of a vertical pile in waves. Trans. of ASME, Journal of Fluids Engrg., 104:46-53.

Bearman, P.W. and Hall, P.F. (1987): Dynamic response of circular cylinders in oscillatory flow and waves. Proc. Int. Conf. on Flow Induced Vibrations, organized by BHRA, Bowness-on-Windermere, England, May 12-14, 1987, pp. 183-190.

Bearman, P.W. and Mackwood, P.R. (1991): Non-linear vibration characteristics of a cylinder in an oscillating water flow. Proc. 5th Conf. on Flow Induced Vibrations, Inst. of Mech. Eng., Brighton, U.K., May 21-23, 1991, pp. 21-31.

Bearman, P.W., Lin, Y.W. and Mackwood, P.R. (1992): Measurement and prediction of response of circular cylinders in oscillating flow. Proc. of the Behaviour of Offshore Structures (BOSS 92) Conf., July 7-10, 1992, London, U.K., pp. 297-307.

Borthwick, A.G.L. and Herbert, D.M. (1990): Resonant and non-resonant behaviour of a flexibly mounted cylinder in waves. J. Fluids and Structures, 4:495-518.

Graham, J.M.R. and Djahansouzi, B. (1991): A computational model of wave induced response of a compliant cylinder. Proc 5th Conf. on Flow Induced Vibrations, Inst. of Mech. Eng., Brighton, U.K., May 21-23, 1991, pp. 333-341.

Isaacson, M. and Maull, D.J. (1981): Dynamic response of vertical piles in waves. Int. Symposium on Hydrodyn. in Ocean Engineering. The Norwegian Inst. of Technology, pp. 887-903.

Jacobsen, V., Hansen, N.-E. O. and Petersen, M.J. (1985): Dynamic response of mono-tower platform to waves and currents. Proc. 17th Annual Offshore Technology Conf., OTC Paper No. 5031, Houston, TX.

Kaye, D. and Maull, D.J. (1993): The response of a vertical cylinder in waves. J. Fluids and Structures, 7:867-896.

Kozakiewicz, A., Sumer, B.M. and Fredsøe, J. (1994): Cross-flow vibrations of cylinders in irregular oscillatory flow. ASCE, J. Waterway, Port, Coastal and Ocean Engineering, 120(6):515-534.

Kozakiewicz, A., Sumer, B.M., Fredsøe, J. and Hansen, E.A. (1996): Vortex regimes around a freely-vibrating cylinder in oscillatory flow. Proc. 6th Int. Offshore and Polar Engrg. Conf., Los Angeles, USA, May 25-30, 1996, 3:490-498.

Lipsett, A.W. and Williamson, I.D. (1991a): Modelling the response of flexibly mounted cylinder in oscillatory flow. Proc. 1st Int. Offshore and Polar Engrg. Conf., Edinburgh, U.K., August 11-16, 1991, 3:370-377.

Lipsett, A.W. and Williamson, I.D. (1991b): Two-dimensional response of a flexibly mounted cylinder in oscillatory flow. Proc. 10th Int. Conf. Offshore Mech. and Arctic Eng., I-A, ASME, pp 187-194.

Longoria, R.G., Beaman, J.J. and Miksad, R.W. (1991): An experimental investigation of forces induced on cylinders by random oscillatory flow. J. Offshore Mechanics and Arctic Engineering, 113:275-285.

Maull, D.J. and Kaye, D. (1988): Oscillations of a flexible cylinder in waves. Proc. Int. Conf. on Behaviour of Offshore Structures. Tapir Publications, Trondheim, Norway, pp. 535-549.

McConnell, K.G. and Park, Y.-Y. (1982a): The frequency components of fluid-lift forces acting on a cylinder oscillating in still water. Experimental Mech., 22(6):216-222.

McConnell, K.G. and Park, Y.S. (1982b): The response and the lift-force analysis of an elastically-mounted cylinder oscillating in still water. BOSS, 2:671-680.

Park, Y. (1981): The response and the lift force analysis of a cylinder oscillating in still water. Ph.D.-thesis, Department of Engineering Science and Mechanics, Iowa State University, Ames, Iowa.

Sarpkaya, T. (1976): In-line and transverse forces on smooth and sand-roughened cylinders in oscillatory flow at high Reynolds numbers. Naval Postgraduate School Technical Report No. NPS-69SL76062, Monterey, CA.

Sarpkaya, T. (1979): Lateral oscillations of smooth and sand-roughened cylinders in harmonic flow. In: Mechanics of Wave Induced Forces on Cylinders (Ed. T.L. Shaw), Pitman Advanced Publishing Program, pp. 421-436.

Sarpkaya, T. and Rajabi, F. (1979): Dynamic response of piles to vortex shedding in oscillating flows. Proc. 11th Annual Offshore Technology Conf., April 30 - May 3, 1979, OTC 3647, pp. 2523-2528.

Slaouti, A. and Stansby, P.K. (1992): Response of a circular cylinder in regular and random oscillatory flow at KC = 10. Proc. of the Behaviour of Offshore Structures (BOSS 92) Conf., July 7-10, 1992, London, U.K., pp. 308-321.

Sumer, B.M. and Fredsøe, J. (1988): Transverse vibrations of an elastically mounted cylinder exposed to an oscillating flow. J. Offshore Mechanics and Arctic Engineering, ASME, 110:387-394.

Sumer, B.M. and Fredsøe, J. (1989): Effect of Reynolds number on vibration of cylinders. J. Offshore Mechanics and Arctic Engineering, ASME, 111:131-137.

Sumer, B.M. and Kozakiewicz, A. (1995): Visualization of flow around cylinders in irregular waves. Int. Journal of Offshore and Polar Engineering, 5(4):270-272. Also see: Proc. 4th Int. Offshore and Polar Engineering Conf., Osaka, Japan, April 10-15, 1994, 3:413-420.

Sumer, B.M., Fredsøe, J. and Jacobsen, V. (1986): Transverse vibrations of a pipeline exposed to waves. Proc. 5th Symposium on Offshore Mechanics and Arctic Engineering, Tokyo, 1986, 3:588-596.

Sumer, B.M., Fredsøe, J., Gravesen, H. and Bruschi, R. (1989): Response of marine pipelines in scour trenches. ASCE, J. Waterway, Port, Coastal and Ocean Engineering, July 1989, 115(4):477-496.

Sumer, B.M., Fredsøe, J., Jensen, B.L. and Christiansen, N. (1994): Forces on vibrating cylinder near wall in current and waves. ASCE, J. Waterway, Port, Coastal and Ocean Engineering, 120(3):233-250.

Verley, R.L.P. and Johns, D.J. (1983): Oscillations of cylinders in waves and currents. Proc. 3rd Conf. on Behaviour of Offshore Structures (BOSS), 2:690-701.

Williamson, C.H.K. (1985a): In-line response of a cylinder in oscillatory flow. Applied Ocean Research, 7(2):97-106.

Williamson, C.H.K. (1985b): Sinusoidal flow relative to circular cylinders. J. Fluid Mech., 155:141-174.

Zedan, M.F. and Rajabi, F. (1981): Lift forces on cylinders undergoing hydroelastic oscillations in waves and two-dimensional harmonic flow. Proc. Int. Symposium on Hydrodynamics in Ocean Engineering. The Norwegian Institute of Technology, pp. 239-262.

Chapter 10. Vibrations of marine pipelines

When pipelines are not buried, unsupported pipeline spans may exist in most locations. When spans develop due to scour, they may change location from time to time, while in the case of pipeline crossings or seabed unevenness (where the bed is non-erodible), the locations are fixed. Fig. 10.1 depicts various scenarios related to pipeline spans. The span length can easily attain values as much as 100 times the pipeline diameter, with a clearance from the sea bottom which may be in the range from practically nil to more than 2-3 times the pipeline diameter (Fig. 10.2). When exposed to flow action such a pipeline span may undergo flow-induced vibrations (Fig. 10.3).

There have been several incidents in the past with pipelines floating to the surface because losing their protective concrete coatings as a result of flow-induced vibrations. In one incident, for example, in mid-September 1975, part of the Cormorant-Sullom Voe oil line off the Shetlands in the North Sea (a 36 inch diameter trunk) surfaced after it lost some 60% of its concrete coating over a short section (Offshore Engineer, 1984).

The flow-induced vibrations are important also for the fatigue life of pipelines. It is known that the damage associated with the fatigue life of a pipe undergoing vibrations is proportional to the product of $A^4 f$ in which A is the amplitude and f the frequency of vibrations (Tsahalis, 1983). Small amplitude vibrations with high frequencies may not be detrimental in the short run; they may, however, have serious consequences in the long run as regards the fatigue life of the pipeline.

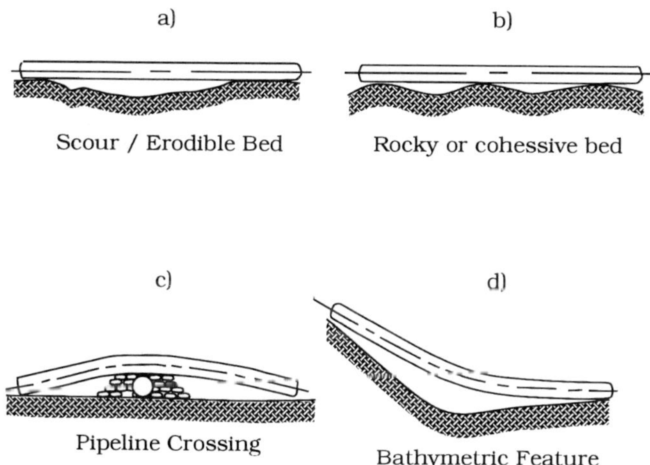

Figure 10.1 Various scenarios related to suspended spans of pipelines. Adapted from Orgill, Barbas, Crossley and Carter (1992).

10.1 Cross-flow vibrations of pipelines

10.1.1 Cross-flow vibrations of pipelines in steady current

The physics behind vibrations of pipelines (i.e., a cylinder placed near a plane boundary) is quite different from that of a free cylinder. For a free cylinder placed in steady flow, the vibrations are caused by regular vortex shedding, the frequency being determined by the Strouhal number, as seen in Chapter 8. This vortex shedding takes place even in the case of the presence of a boundary if the distance e between the wall and the cylinder is larger than about $0.3\,D$, D being the pipe diameter (see Section 1.2.1).

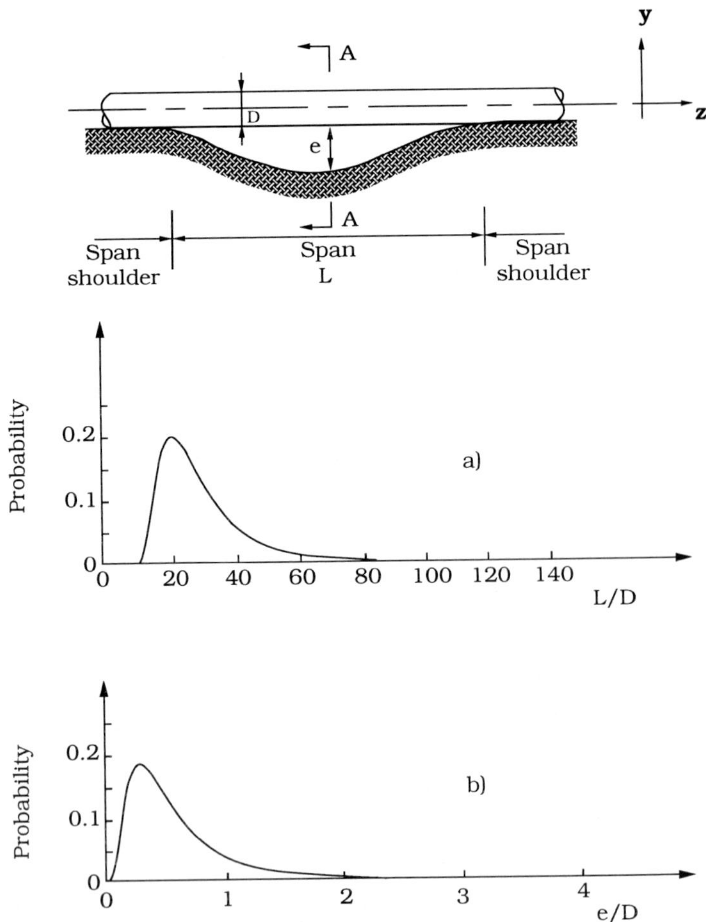

Figure 10.2 Examples of a) distribution of span length-to-diameter ratio
and b) distribution of gap-to-diameter ratio. Reproduced from
Orgill et al. (1992). The data are site-specific.

Closer to the bed, regular vortex shedding is partly suppressed (Fig. 1.21).
However, vibrations still take place, as observed by Jacobsen, Bryndum, Nielsen
and Fines (1984a,b), who studied a pipe suspended in a spring system with gap
ratios equal to zero, 0.5, and larger. Tsahalis and Jones (1981) studied the vibrations of a pipe fixed by simple supports at each end for gap ratios equal to one and

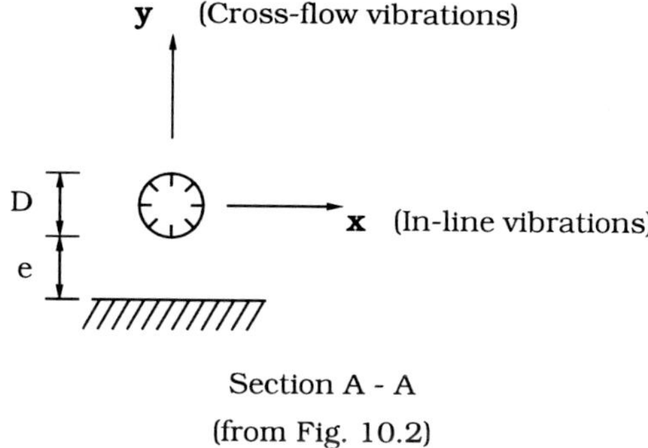

Figure 10.3 Definition sketch of cross-flow and in-line vibrations.

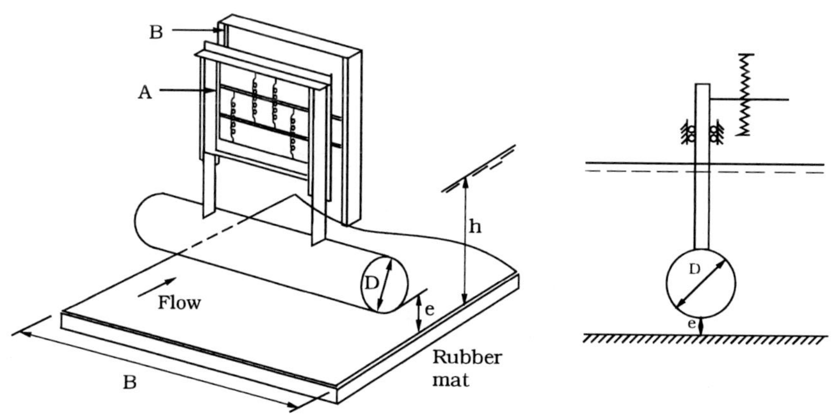

Figure 10.4 Sketch of experimental setup in Fredsøe et al.'s (1985) study.
$B = 2$ m, $D = 0.089$ m, $h = 0.45$ m.

larger. They found that while the frequency of lock-in was in accordance with the
Strouhal number for a free cylinder, the amplitude was decreased for gap ratios
even larger than one.

Tsahalis and Jones (1981) as well as Jacobsen et al. (1984a,b) allowed the pipe to move in two directions. However, the investigations indicate much larger amplitudes in the cross-flow direction than in the in-line direction (about a factor 10). For this reason, the cross-flow vibrations of pipelines may be investigated by a system where the model pipe is allowed to move in the cross-flow direction, according to one degree of freedom. Fredsøe, Sumer, Andersen and Hansen (1985) made an investigation of cross-flow vibrations of a cylinder placed very close to a plane bed, employing such a system. Fredsøe et al.'s work shed light onto the understanding of the influence of bed proximity on the behaviour of the pipe as summarized in the following paragraphs.

Fig. 10.4 illustrates the spring-mounted system employed in the study of Fredsøe et al. (1985). The study reveals the following features:

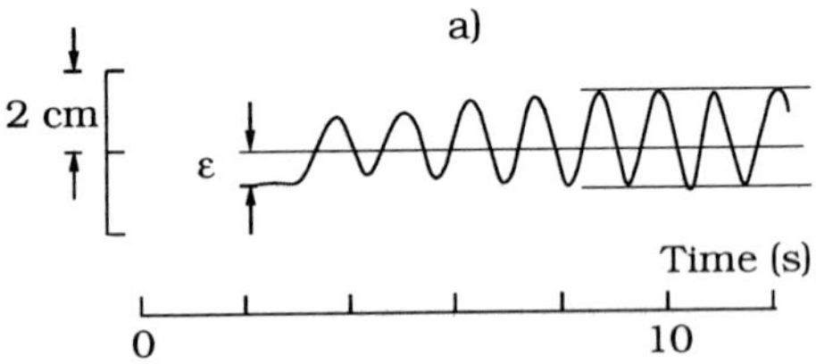

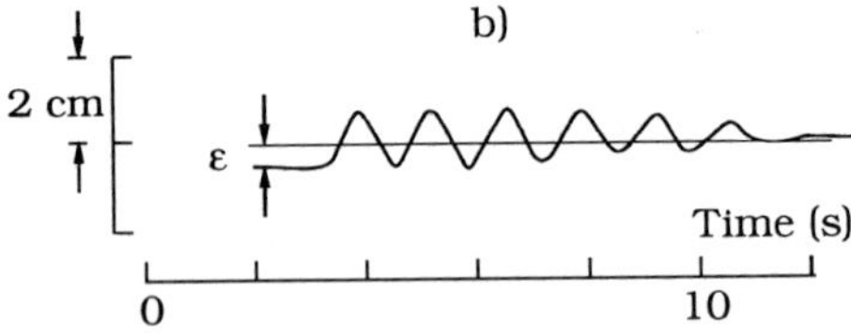

Figure 10.5 (a) Build-up of vibrations, the cylinder being displaced more than ε; (b) vibrations damped out with time, the cylinder being displaced less than ε. Cylinder is very close to bottom ($e/D \sim 0.1 - 0.2$). Steady current. Fredsøe et al. (1985).

1) When the cylinder is well away from the bottom (e/D $\sim$ 1). The cylinder remains motionless for small values of the flow speed. As the flow speed is increased in small increments, a point is reached where the cylinder begins to vibrate. It is observed that both the amplitudes and the frequencies from this point gradually increase as the flow velocity increases. When the flow velocity reaches a certain value, the amplitude of the vibrations increases tremendously, reaching values which are in the same order of magnitude as the pipe diameter due to lock-in. The behaviour is quite similar to the case of a wall-free cylinder (Section 8.3.1, Example 8.2).

2) When the cylinder is very close to the bottom ($e/D \sim 0.1 - 0.2$).
At small velocities, vibrations do not occur unless the cylinder is initially displaced
more than a certain critical distance ε away from its equilibrium position: if the
cylinder is displaced more than ε, vibrations build up immediately and the cylinder
begins to oscillate, see Fig. 10.5a.

If the cylinder is displaced less than ε, the vibrations created "externally"
die out, see Fig. 10.5b. ε depends on the flow velocity as well as on the gap ratio:
for large flow velocities, ε approaches zero as described later.

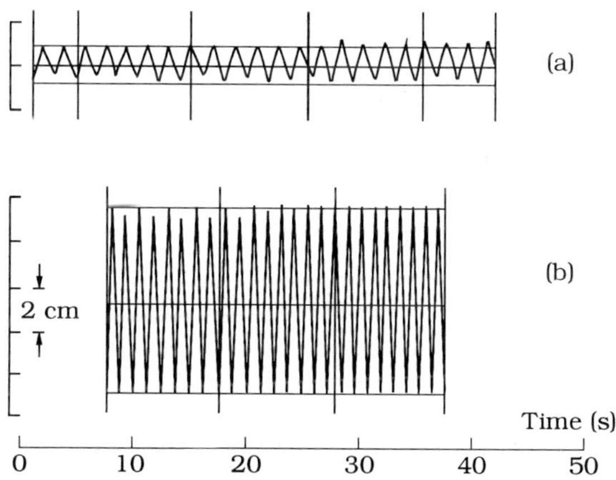

Figure 10.6 Vibrations for $e/D = 0.4$ and $V_r = 3$, (a) no initial displacement, (b) with initial displacement. Steady current. Fredsøe
et al. (1985).

3) When the gap ratio is in the interval 0.3-0.7. Figures 10.6a and
b show the interesting transition case where two different kinds of behaviour of
the pipe exist for exactly the same hydraulic parameters and gap ratio. Fig.
10.6a shows a vortex-excited vibration with a fairly small amplitude and a non-
dimensional frequency fD/U_a equal to 0.22, U_a being the undisturbed velocity in
the mean position of the cylinder away from the bottom.

In Fig. 10.6b, the pipe is originally pressed towards the bottom and then
released. In this case, the pipe keeps the much higher amplitude of vibration.
Furthermore, the frequency is changed so the non-dimensional frequency fD/U_a
becomes 0.35. As in the previous case (2), the displacement ε must have a certain
value at small flow velocities in order to create the large vibrations.

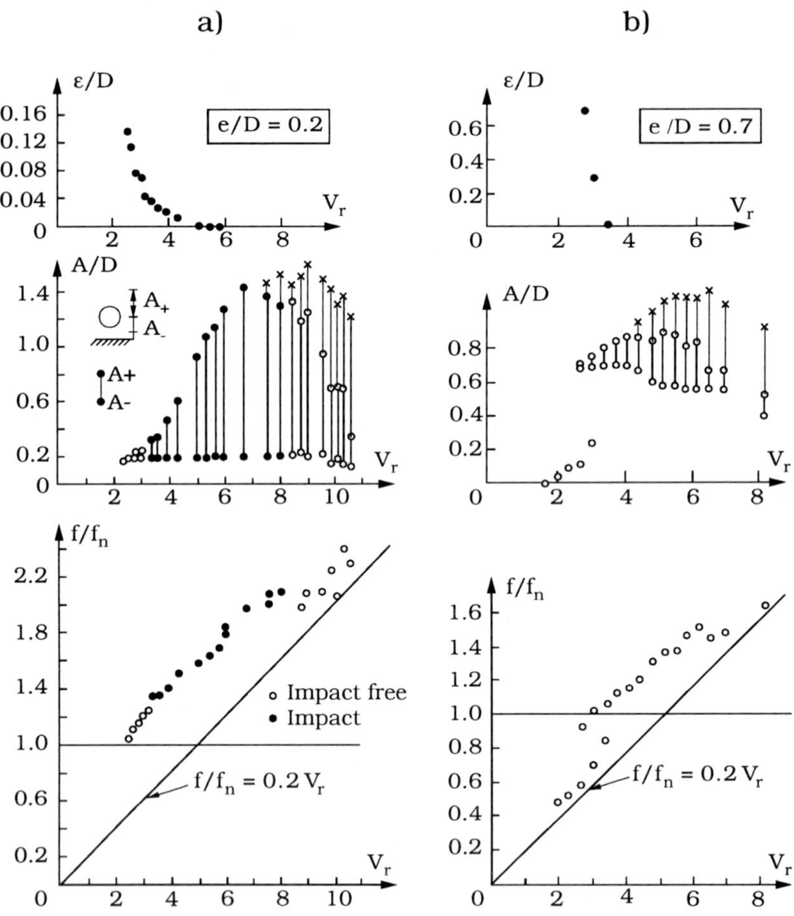

Figure 10.7 Variation in frequencies and amplitudes with reduced velocity. Steady current. a) $\frac{\varepsilon}{D} = 0.2$, b) $\frac{\varepsilon}{D} = 0.7$. $\overset{\circ}{\underset{\circ}{\text{O}}}$ = average value of amplitudes, $\overset{\times}{\underset{\times}{\text{I}}}$ = maximum amplitudes. Fredsøe et al. (1985).

Quantitative description

Fig. 10.7 represents the amplitude and frequency data for two different gap ratios:

In Fig. 10.7b, the gap ratio is so large (0.7) that vortex shedding occurs and may be the driving mechanism of cylinder vibrations. The data are plotted in the familiar form, f/f_n and A/D versus the reduced velocity $V_r = U_a/(Df_n)$ where U_a is the undisturbed velocity in the mean position of the cylinder away from the bottom.

In Fig. 10.7 also the familiar identity

$$\frac{f}{f_n} = 0.2\frac{U_a}{Df_n} = 0.2V_r \tag{10.1}$$

is plotted as a reference line, representing the stationary free cylinder shedding frequency for a Strouhal number equal to 0.2, which is in agreement with a Reynolds number around 5×10^4.

It is seen from Fig. 10.7b that, at small values of V_r, the frequency follows the Strouhal frequency (Eq. 10.1). For V_r slightly larger than 3, the frequency f/f_n increases to become larger than one, and for larger values of V_r up to about 8, the frequencies increase and are significantly larger than those representing the vortex-shedding frequency, Eq. 10.1. For V_r larger than 8, the frequencies again approach the stationary cylinder vortex-shedding frequency.

The amplitudes caused by vortex shedding at $V_r < 3$ are quite small, and a sudden increase in the amplitude is observed to take place at $V_r \sim 3$. However, even for smaller values of V_r, large amplitude vibrations can be observed if a small initial displacement ε is introduced, as described in the previous section. The amplitude and frequency for these movements smoothly follow the amplitudes and frequencies observed at larger values of V_r. In Fig. 10.7b also the variation in ε decreases with increasing values of V_r.

The near wall case depicted in Fig. 10.7a shows that, at small V_r-values, the frequency does not follow the stationary cylinder vortex-shedding frequency. Vibrations without any forced displacement ε do not in fact take place before $V_r = 5.0$.

In Fig. 10.8, the variation in f/f_n and A/D with V_r is depicted for all measured values of the gap ratio. The dashed lines in Fig. 10.8 show the results where an initial displacement is needed. The following tendencies can be concluded from Fig. 10.8: 1) The frequencies do not follow the stationary cylinder vortex-shedding frequency except at very low (< 3) and very high (> 8) values of V_r. 2) The frequency increases with decreasing gap ratio.

It is also seen that maximum amplitude occurs at larger values of V_r for small gap ratios. The smaller the gap ratio, the larger is the maximum amplitude.

Tsahalis and Jones (1981) reported that the amplitude decreases as the gap ratio decreases at gap ratios larger than one. The results shown in Fig. 10.8 confirm this finding, as there is a persistent decrease in the amplitude, when the gap ratio changes from 1.7 to 0.90.

Fig. 10.9a shows the variation in ε/D with reduced velocity for different gap ratios, while Fig. 10.9b displays the borderlines between three different vibration regimes on the e/D and V_r plane. As seen from the latter figure, for the vibrations

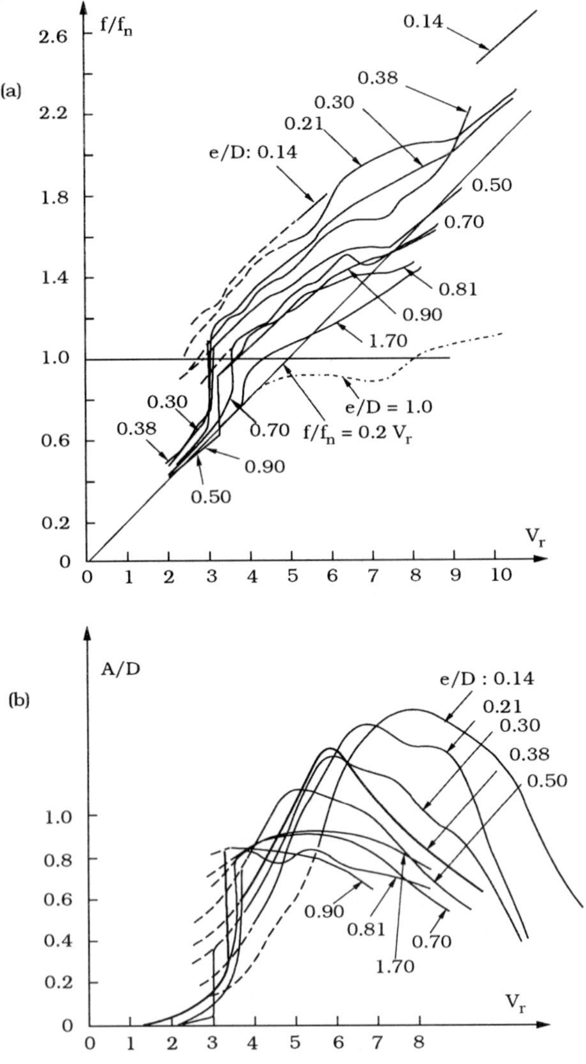

Figure 10.8 Frequency and amplitude response of vibration of a circular
cylinder placed near a plane wall. Dashed lines: an initial
displacement is needed to excite vibrations. Steady current.
$K_s = 1.5$. Dashed-dotted line: Tsahalis and Jones (1981).
Reproduced from Fredsøe et al. (1985).

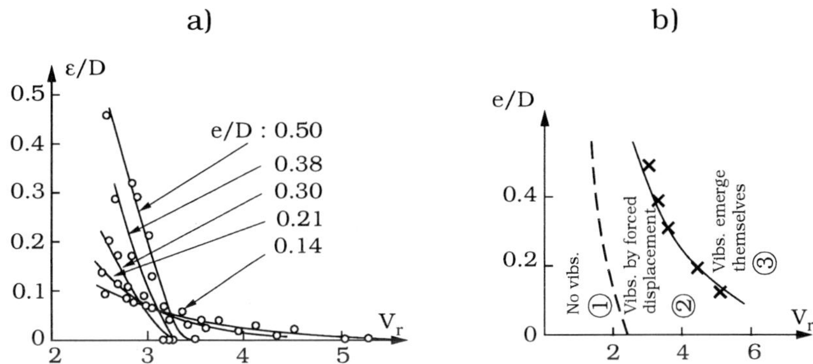

Figure 10.9 (a): Minimum initial displacement, ε, needed to initiate cross-flow vibrations for various values of gap-to-diameter ratio as function of reduced velocity. (b): Various vibration regimes. Steady current. Fredsøe et al. (1985).

to be excited themselves, one needs to move to higher and higher velocities with decreasing clearance between the pipe and the bottom. For the gap ratio $e/D = 0.1$, for example, vibrations emerge themselves without forced initial displacement only after $V_r = 5$. However, in a real-life situation, imperfections in the flow (such as turbulence, etc.) may provide the forced initial displacement to excite the vibrations; in this case, the onset of vibrations may occur at much lower values of V_r such as 1.5-2, as indicated by Fig. 10.9b.

The physical mechanism behind the vibrations

As pointed out, the frequency of the vibrations differs significantly from the frequency of vortex shedding for stationary pipes with small gap ratio. This may to a large extent be explained by the influence of impact. Heavy impact may cut parts of the movement away, resulting in an increase in f. An example of the motion including heavy impact is shown in Fig. 10.10. The vibrations in Fig. 10.10a occur with impact, while those in Fig. 10.10b are impact-free vibrations. It is obvious that the frequency will increase if part of the motion is cut by the impact. However, by considering impact-free vibrations for a near-wall pipeline, which exist for small and large values of V_r (see Fig. 10.7), it is seen that the same increase in f from the stationary cylinder vortex-shedding frequency exists in these experiments. This suggests that the vibrations are not only caused by regular vortex shedding, but are to a certain extent self-excited. This means that close to the wall transfer of energy from the flow to the pipe takes place. It is well known that self-excited vibrations may occur if an originally resting structure being moved in one direction causes a force from the surrounding flow in phase with and in the same direction as the velocity of the structure.

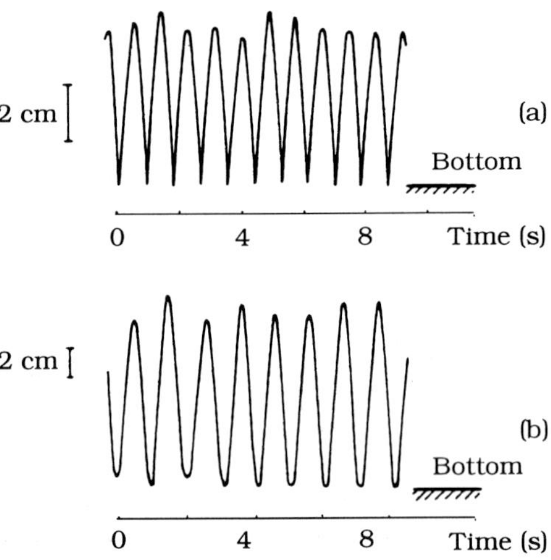

Figure 10.10 (a): Impact run. $\frac{e}{D} = 0.2$; $V_r = 4.3$; $f/f_n = 1.5$ $A_+/D = 0.50$; $A_-/D = 0.2$. (b): Impact-free run for the same reduced velocity. $\frac{e}{D} = 0.7$; $f/f_n = 1.2$; $A_+/D = 0.9$; $A_-/D = 0.65$. Steady current. Fredsøe et al. (1985).

In the present case, such a force can be explained as follows: in the steady case the water discharge below the pipe Q_1 is smaller than above the pipe. If the pipe now suddenly is moved to a new position II (Fig. 10.11), the instantaneous bottom velocity will be smaller at this higher position II than that according to the equilibrium flow situation. In the higher position II, the water discharge Q_2 below the pipe is larger than Q_1 in the equilibrium flow situation, but it takes some time for the flow below the pipe to be accelerated from Q_1 to Q_2. This is because the downstream wake does not immediately adjust its position to the downstream part of the cylinder when the cylinder is vibrating. Hence, just after the change in the position, the instantaneous bottom velocity U_i is smaller than the equilibrium velocity U_e. This means that the bottom pressure just after the pipe has been moved from position I to position II will be larger than in the equilibrium state, (Bernouilli effect), so an excessive force besides the usual lift force will act in the upwards direction on the pipe, the pipe being moved in the upwards direction.

The foregoing argument strongly depends on the presence of the wall as no difference exists between Q_1 and Q_2 in the case of a free cylinder.

Based on the aforementioned description of the vibration process, Hansen, Madsen and Fredsøe (1986) worked out a stability analysis. In this analysis, the growth of a given initial periodic disturbance was investigated by using the familiar vibration equation (Eq. 8.1). The force term was formulated in accordance with the description given in the preceding paragraphs, incorporating Fredsøe and Hansen's (1987) modified potential theory approach. The results indicated that the stability analysis was able to predict the boundary between Regions (1) + (2) and (3) in Fig. 10.9 rather satisfactorily. Also, the frequency of vibrations could be predicted satisfactorily, in agreement with the experiments.

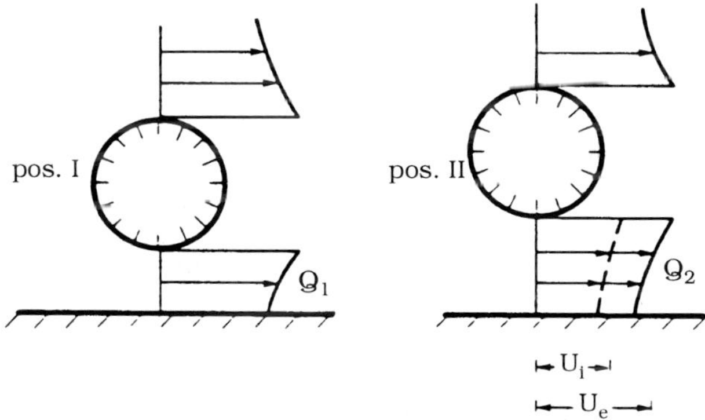

Figure 10.11 Changes in near-pipe flow velocity when the pipe is suddenly lifted a certain distance.

10.1.2 Cross-flow vibrations of pipelines in waves

Important contributions to this topic have been made by Tsahalis (1984, 1985) and Jacobsen et al. (1984a,b). In the work of Tsahalis, model tests were conducted in a wave tank to study the effect of combined steady current and wave action and the proximity of the sea bottom on the vibrations of a flexible pipe for the clearance interval $0 < e/D < 1$. In Jacobsen et al.'s study, similar tests were conducted with a flexibly-mounted rigid cylinder under steady currents, regular and irregular waves and also waves superimposed on steady current.

Subsequently, Tsahalis' and Jacobsen et al.'s experiments have been complemented by Sumer, Fredsøe and Jacobsen's (1986) study. Sumer et al. investigated

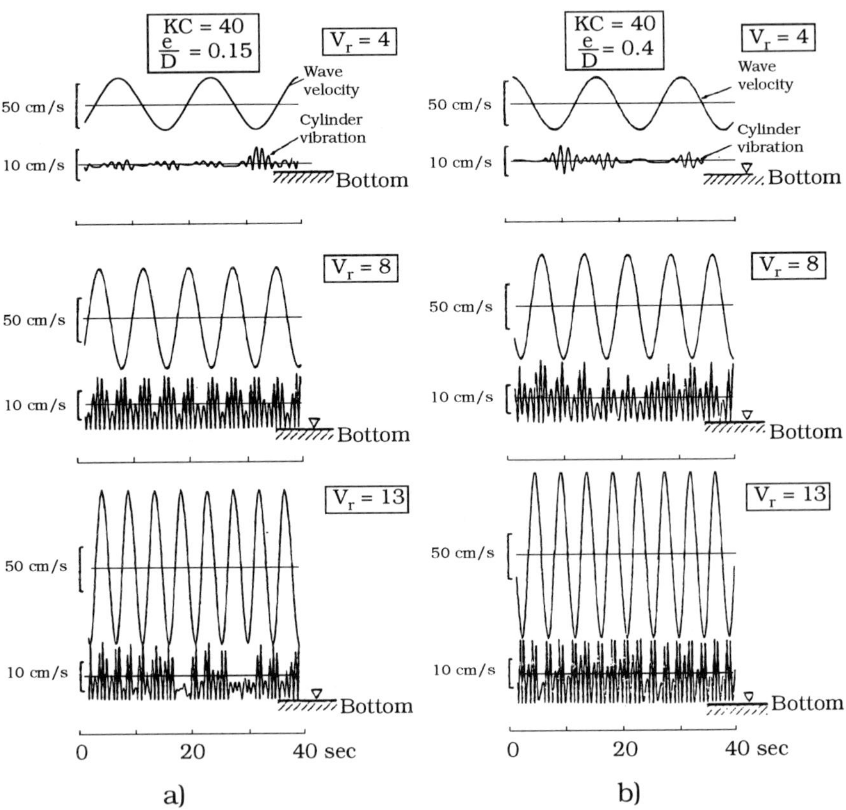

Figure 10.12 Sample records of wave velocity and cylinder vibration near the bottom. $KC = 40$. (a): $e/D = 0.15$ and (b): $e/D = 0.4$. Sumer et al. (1986).

the transverse vibration of a flexibly-mounted, rigid near-wall cylinder exposed to a planar oscillatory flow.

Fig. 10.12 depicts sample vibration records along with the corresponding wave velocity in the study by Sumer et al. (1986). Clearly, the response is rather different from that obtained for a free cylinder (cf. Fig. 9.5c). The influence of the bottom is quite marked, particularly for large-amplitude vibrations, as expected.

Fig. 10.13 represents the mean-gap data plotted as a function of V_r and

KC where $\bar{e}$ = the mean gap, see Fig. 10.13 for the definition sketch. Sumer et al.'s experiments indicated that $\bar{e} = e$ for the case when $e/D = 1$.

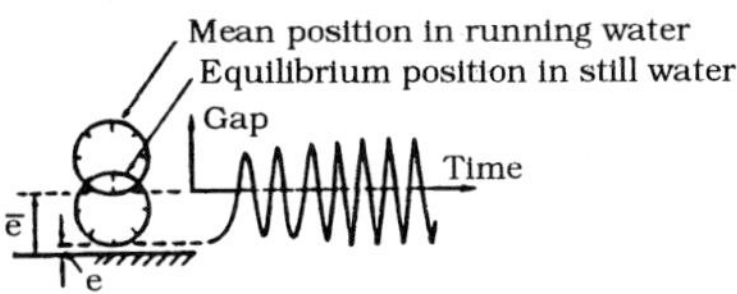

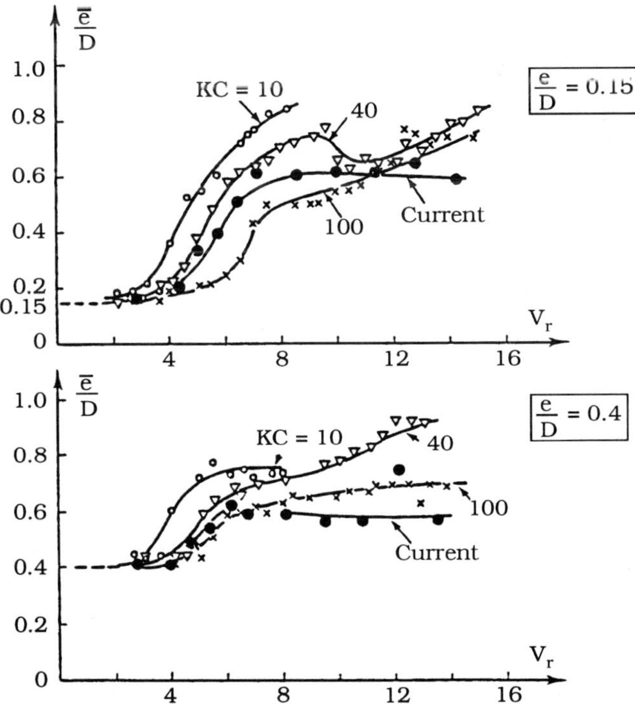

Figure 10.13 Mean gap versus reduced velocity. Waves. Sumer et al. (1986).

Fig. 10.13 shows that, for $e/D = 0.15$ and 0.4, $\bar{e}$ is very close to the corresponding e value for small V_r values, but it increases as V_r increases, which means that, on average, the cylinder is repelled from the bottom. This is due to the presence of a lift force directed away from the wall (cf. Section 2.7).

Fig. 10.13 also shows that the smaller the KC number, the larger the mean gap ratio. This is in accordance with the significant increase in the lift force coefficient found when KC is decreased for bottom-mounted pipes ($e/D = 0$), as reported by Sarpkaya and Rajabi (1979), Bryndum et al. (1983) and Jacobsen, Bryndum and Fredsøe (1984) and also for pipes placed close to the bottom, as reported by Sumer, Jensen and Fredsøe (1991) (see Fig. 4.39).

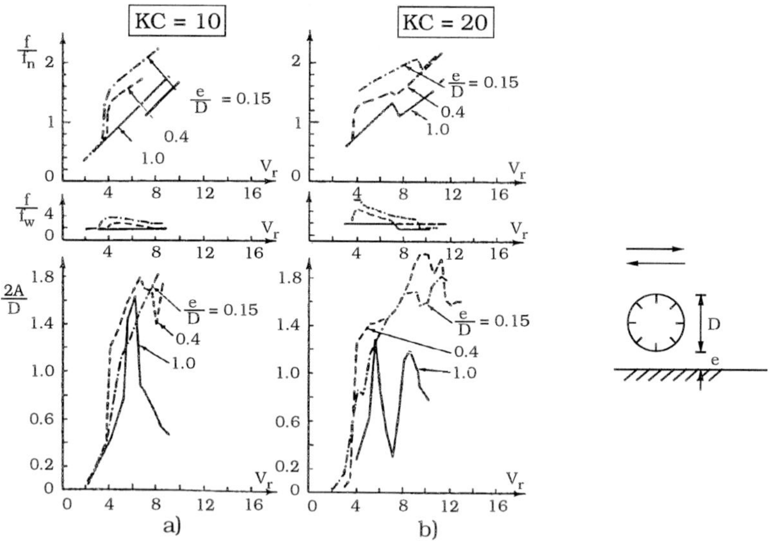

Figure 10.14a,b Frequency and amplitude for three different gap ratios. (a): $KC = 10$, (b): $KC = 20$. $M = 1.6$, $K_s = 1.5$ and $k/\rho = 0.336 \ \mathrm{m^2/s}$. Sumer et al. (1986).

Amplitude and frequency response. The amplitude and frequency responses obtained for $e/D = 0.15$ and 0.4 in Sumer et al.'s (1986) study are plotted in Fig. 10.14 along with the ones obtained for $e/D = 1$. e/D in these plots is the gap ratio corresponding to the equilibrium position in still water as depicted in the sketch in Fig. 10.13. The following conclusions can be drawn from Fig. 10.14:

1) The results obtained for the case of $e/D = 1$ are not drastically different

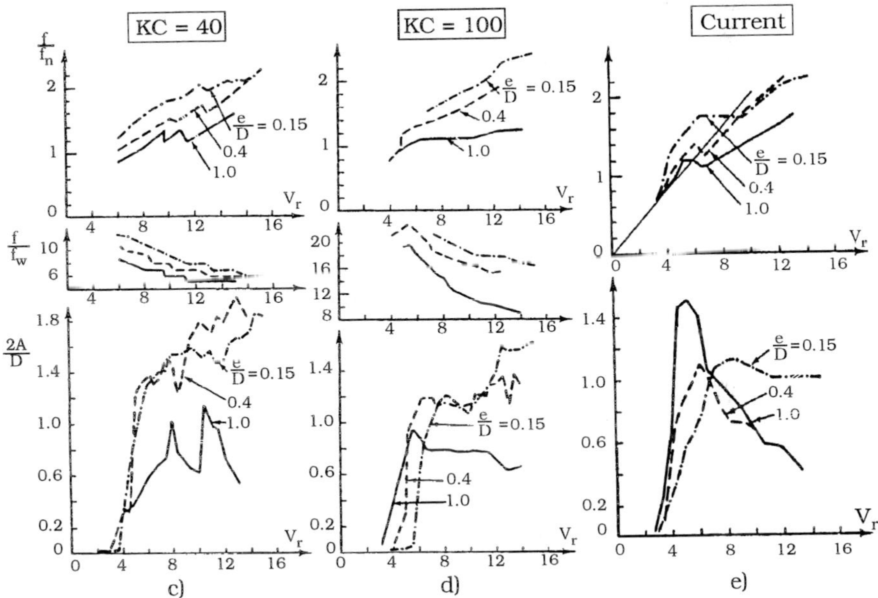

Figure 10.14c,d,e Frequency and amplitude for three different gap ratios. (c): $KC = 40$, (d): $KC = 100$ and (e): $KC = \infty$ (current). $M = 1.6$, $K_s = 1.5$ and $k/\rho = 0.336$ m^2/s. Sumer et al. (1986).

from those obtained for a free cylinder (cf. Fig. 9.6).

2) The frequencies in the case of $e/D = 0.15$ and 0.4 generally do not follow the fundamental lift frequencies (represented here by the frequency results of the case $e/D = 1$ for small values of V_r). The smaller the gap ratio, the higher the frequency.

3) The amplitudes significantly increase as the gap ratio decreases. (Yet, the foregoing generalization does not seem to be valid for the current case for small values of V_r).

4) The amplitude response curves appear to be S-shaped lines (as opposed

to the ones obtained for $e/D = 1$), resembling the "galloping" type instability mechanism (Section 8.6). It should be noted that this type of mechanism results in self-induced vibrations, as described in the previous chapter for the case of current.

Examination of the vibration and wave velocity records (Fig. 10.12) reveals that generally the vibration amplitude is significantly reduced at times when the wave velocity is passing its zero-crossings. This usually leads to a small, impact-free amplitude as can be seen for example from the records $V_r = 8$ and $V_r = 13$ in Fig. 10.12. This impact-free amplitude is followed by a relatively large one as the acceleration stage of the wave is underway. Since there is no impact effect in that latter event, it is obvious that the cylinder is repelled back into the flow by the lift force (of the kind described in the previous section) near the boundary. This lift force may be enhanced by the impact if the cylinder hits the rigid boundary. This phenomenon is an entirely different mechanism compared with the vortex-induced vibration in the case when $e/D = 1$. It is therefore obvious that the frequency response of the near-wall cases, $e/D = 0.15$ and 0.4, should be different from the one obtained for the case when $e/D = 1$. This lift force mechanism has been suggested and verified by experiments by Fredsøe et al. (1985) for a vibrating pipe in the presence of currents (see the previous section). In fact, it is not surprising that the frequency response shown in Fig. 10.14 quite well resembles the general pattern of the frequency response obtained in the work of Fredsøe et al. (1985) which is reproduced in Fig. 10.8. Jacobsen et al. (1984b) suggested a similar lift-force effect when explaining their $e/D = 0$ vibration test results.

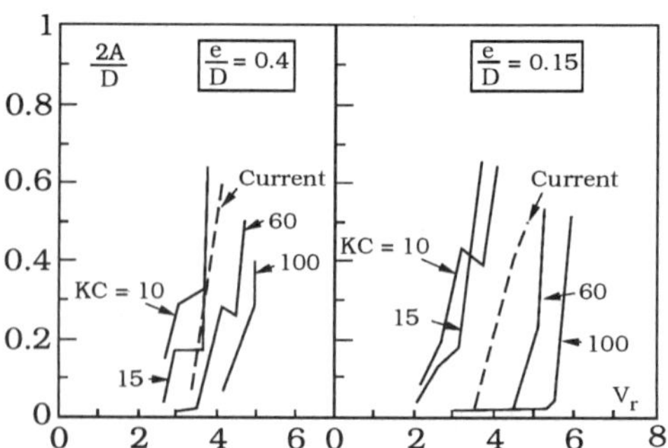

Figure 10.15 Onset of cross-flow vibrations for $e/D = 0.4$ and 0.15. Waves. $M = 1.6$, $K_s = 1.5$ and $k/\rho = 0.336$ m^2/s^2. Sumer et al. (1986).

Fig. 10.15 illustrates the effect of close proximity to the wall upon the onset of vibrations. The onset of vibrations appears to occur at values of V_r

smaller than that in currents for small values of KC number only, as opposed to the result obtained for $e/D = 1$ where the onset of vibrations in waves is observed to occur always earlier, irrespectively of the KC number (Sumer et al. 1986).

Although no clear explanation has been found for the fact that the onset of vibration for large values of KC occurs later in relation to the current case, the earlier initiation of vibration for small KC numbers is attributed to the fact that a near-wall cylinder experiences very large lift forces for small values of the KC number.

When dealing with small gap ratios, the influences of the approaching *bed boundary layer* upon the results should be discussed. The thickness of the wave boundary layer, δ, which is defined as the boundary layer thickness occurring when $U = U_m$ has been calculated by Sumer et al. according to Fredsøe (1984). The results indicated that δ/D is in the range from 0.05 to 0.3 - 0.4. The effect of shear in the incoming flow on the lift forces has been explained in Section 2.7. The latter showed that the shear in the incoming flow has a considerable effect at very small gap ratios (below 0.1). Since the boundary layer thickness in the work of Sumer et al. is less than or in the same order of magnitude as the gaps employed in the study, the results can be considered free from the boundary layer effect.

Sumer et al.'s (1986) work further showed that the effect of increased pipe specific gravity combined with an accompanying increase in the stability parameter was to generally decrease the vibration amplitude. This effect was found to be more pronounced for larger gap ratios. It was also found that this effect caused the reduced velocity for the onset of vibrations to increase.

10.2   In-line vibrations and in-line motions of pipelines

Pipelines exposed to waves may undergo both the wave-induced in-line vibrations and the wave-induced oscillatory motion. Regarding the wave-induced in-line vibrations, even the case of a free cylinder has not been treated very extensively. The major difficulty in the analysis, as pointed out in Section 9.3, is that these high-frequency, small-amplitude vibrations are normally overshadowed by the presence of large-amplitude oscillatory motion of the pipeline induced by the total in-line force (the Morison force). Sumer, Fredsøe, Gravesen and Bruschi (1989) made an experimental investigation of these vibrations in the case of a pipeline placed in the vicinity of a scoured trench, studying the influence of the trench hole on the vibrations (Section 10.4).

As for the Morison-force induced in-line motion of pipelines, the response of the pipeline can be predicted quite easily by application of Eqs. 9.10 and 9.11, similar to the case of a free cylinder (Section 9.4). The force coefficients C_D and $C_m (= C_M - 1)$ must be inserted into the equations this time, using the values given for the case of a stationary pipe near the bottom (Sarpkaya, 1976), see Figs. 4.38,

4.40, 4.41, 4.43, and the same calculation procedure as summarized in Section 9.3 (where the change in the force coefficients due to the response of the pipeline is taken into consideration) must be implemented. The latter may be important, particularly in the resonance region.

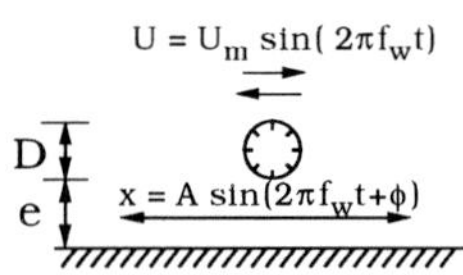

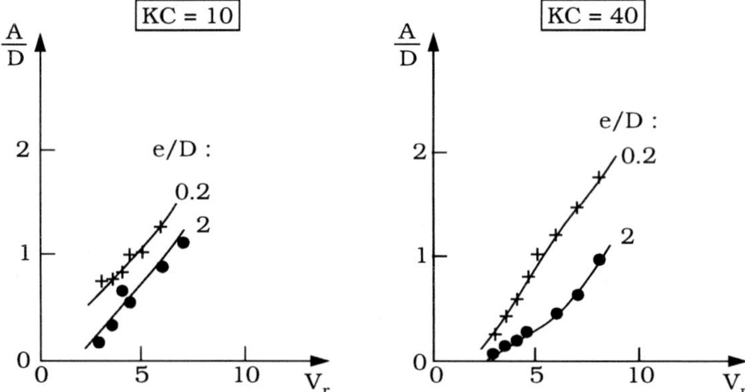

Figure 10.16 Amplitude of a pipeline exposed to waves and undergoing in-line oscillatory motion induced by Morison force. Experiments $D = 15$ cm. Surface roughness $k_s/D = 4 \times 10^{-3}$. $Re = 2 - 7 \times 10^4$. Stability parameter $K_s = 0.1$. Specific gravity of pipe $s = \rho_{\mathrm{pipe}}/\rho = 1.4$ ($M = 1.9$ for wall-freee pipe). Pipe has two degrees of freedom of movement with $f_n = 0.4$ Hz. Sumer et al. (1989).

On the experimental side, Sumer et al. (1989) and Bryndum et al. (1989) report laboratory measurements of in-line motion of pipelines. Fig. 10.16 depicts the results obtained in Sumer et al.'s (1989) study, illustrating the effect of a close proximity of the sea bed on the pipeline's in-line motion. There is a clear increase in the amplitude of the motion very near the bed ($e/D = 0.2$). The results of Bryndum et al.'s (1989) study show the same trend. This increase can be explained by the measured increase in the force coefficients with decreasing gap ratio (see Fig. 4.38); the closer to the bed, the larger the in-line force, therefore the larger the amplitude of the response of the pipeline.

10.3 Effect of Reynolds number

Table 10.1 summarizes the test conditions of the previous work. As is seen, most of the experiments have been conducted in the subcritical flow regime. There are only two investigations (namely Bruschi et al., and Raven et al.) studying pipeline vibrations in currents at high Re numbers, and there is only one study, namely Sumer and Fredsøe (1989), which investigates vibrations at high Re numbers in waves. However, in the latter work, the test pipe was a wall-free cylinder.

It is evident that research work studying pipeline vibrations at high Reynolds numbers is not very extensive. However, based on 1) the knowledge of flow around and forces on pipelines and also 2) of vibrations of free cylinders at high Reynolds numbers, an assessment may be made of the behaviour of pipeline vibration at high Reynolds numbers. The following paragraphs will summarize this assessment.

Steady current

Cross-flow vibrations. It has been seen that a pipeline can undergo two kinds of cross-flow vibrations depending on the value of the clearance between the pipe and the bed (Section 10.1.1): 1) The vortex-shedding induced vibrations, which occur when $e/D \overset{\sim}{>} 0.25$, and 2) The self-excited vibrations which occur when $e/D \overset{\sim}{<} 0.25$.

1) When $e/D \overset{\sim}{>} 0.25$, the regime of water flow around the pipeline should be practically the same as in the case of a free cylinder (see Fig. 1.1). Therefore, marked changes in the response pattern of a pipeline should be expected to occur at high Re numbers if the pipeline-surface roughness k_s/D is less than approximately 3×10^{-3} in accordance with the extensive information given in Section 8.3.2 in conjunction with the effect of Re and the surface roughness on cross-flow vibrations.

If the pipeline-surface roughness k_s/D exceeds 3×10^{-3}, practically no change in the pipeline response pattern should be expected at high Re numbers.

2) When $e/D \overset{\sim}{<} 0.25$, the pipeline vibration is not caused by vortex shedding, but there is an additional contribution from a dynamic lift because of the presence of the bottom, as described in Section 10.1.1. No study is available, though, investigating the scale effects for near-wall pipelines.

Remarks about the works available in the literature. Bruschi et al. (1982) (Table 10.1) state that, in the critical regime at lower velocities, the amplitude of vibrations of wall-free pipes was markedly reduced. This reveals the results of the extensive study of Wootton (1969) on the effect of Re on vibrations of large stacks in wind (see Figs. 8.27 and 8.28).

Table 10.1 Summary of test conditions of pipeline-vibration studies.

Author	Set-up	Gap-ratio e/D	Current	Waves KC	Pipe surface roughness k_s/D	Re number	Flow regime
Tsahalis & Jones (1981)		1 - ∞	Steady	-	Smooth	$2\text{-}8 \times 10^3$	Subcritical
Bruschi et al. (1982)		∞	Tidal (Field) "	-	0.5×10^{-3} 0.5×10^{-3}	? $<2.2 \times 10^5$	Subcritical
Jacobsen et al. (1984a, b)	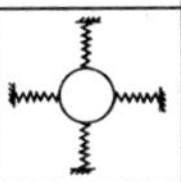	0;0.5;1	Steady	30 - 120	10×10^{-3}	$0.5\text{-}1.7 \times 10^5$	Subcritical
Tsahalis (1985)		1 - ∞	Steady	5 - 25 (Superimposed on current)	Smooth	$<2 \times 10^4$	Subcritical
Fredsøe et al. (1985)	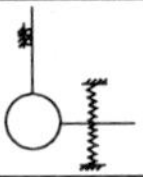	0.1 - 1.7	Steady	-	Smooth	$0.1\text{-}0.6 \times 10^5$	Subcritical
Tørum & Anand (1985)	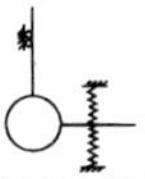	0.5-3	Steady	-	Smooth	$0.7 - 4 \times 10^4$	Subcritical
Raven et al. (1985)		0.5-2	Tidal (Field) "	- -	1×10^{-3} 8.5×10^{-3}	$0.5\text{-}1.7 \times 10^5$ "	Subcritical Critical Supercritical
Sumer et al. (1986)	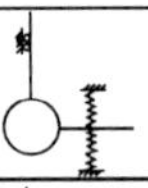	0.15; 0.4;1	Steady	10 - 100	Smooth	$0.2\text{-}1 \times 10^5$	Subcritical
Sumer & Fredsøe (1989)	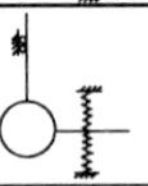	∞	- - -	10 - 100 " "	Smooth Smooth 13×10^{-3}	$0.2\text{-}0.8 \times 10^5$ $1\text{-}4 \times 10^5$ $1\text{-}4 \times 10^5$	Subcritical Subcritical Critical Supercritical
Kristiansen (1988)	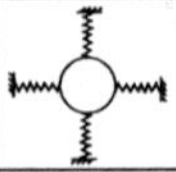	initially nil, on a sandy bed	Steady	-	Smooth	$1\text{-}2.5 \times 10^4$	Subcritical
Mao (1986) Sumer et al. (1988)	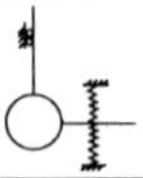	in and outside of a trench hole	"	-	"	$1.5\text{-}7 \times 10^4$	"
Bryndum et al. (1989)	Two models: 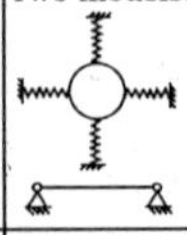	Flat bed tests : 0.2-2 Also: in a trench hole	Steady	Pure Waves: 5-100 Also: Combined Waves and current	$1\text{-}20 \times 10^{-3}$	1.5×10^5 5×10^4	Mostly subcritical
Sumer et al. (1989)	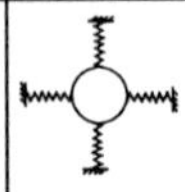	In and outside of a trench hole. Supp.tests with flat bed: 0.2; 2	-	10; 40	10×10^{-3}	$2\text{-}7 \times 10^4$	Mostly subcritical

Raven et al.'s (1985) study concerns full-scale testing of a pipeline span in actual tidal flow situations with a range of clearance $e/D = 0.5 - 2$. The Re numbers of the tests practically cover the critical and part of the post-critical regimes. Raven et al.'s measurements indicated that the r.m.s. amplitudes of cross-flow vibrations reached quite large values in the range of Re number where a reduction in the vibration amplitudes was expected. This indicates that the Reynolds number effect in real-life situations may not be significantly strong due to the high level of turbulence in the incoming flow, the surface roughness caused by marine growth, the shear in the incoming flow, and the presence of in-line movement of the pipeline.

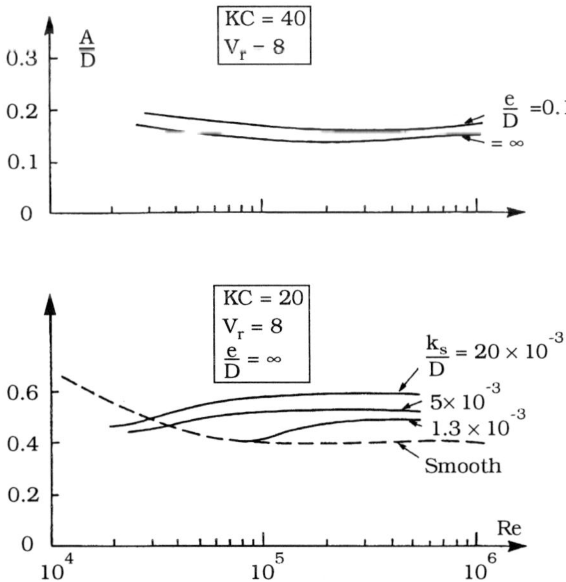

Figure 10.17 In-line vibration amplitude as function of Re number. The structural damping ζ_s is taken to be 0.05 and the specific gravity of the pipe $s = 3$. Jensen (1987).

In-line vibrations. The in-line vibrations in steady currents usually are one order of magnitude smaller than the cross-flow vibrations (see Section 8.4). Like the cross-flow vibrations, there are two contributions to the in-line vibrations, again depending on the clearance between the pipe and the bed: the vortex shedding is solely responsible for the vibrations for large gaps, while the dynamic lift gives an additional component for small gaps.

Table 10.2 Degree of change in the pipeline vibration at high Re numbers (as compared to the vibration obtained in subcritical Re numbers, namely $Re < 3 \times 10^5$).

1) Steady currents

			Flow Regimes	
			Critical, Supercritical and Upper Transition $3\times10^5 < Re < 4\times10^6$	Transcritical $Re > 4\times10^6$
Gap Ratio: $\frac{e}{D} > 0.25$	Cross-Flow Component of Vibration	Pipe Smooth	Marked	Practically None
		Pipe Rough $k_s/D < 3\times10^{-3}$	Moderate	"
		Pipe Rough $k_s/D > 3\times10^{-3}$	Practically None	"
	In-Line Component of Vibration	Pipe Smooth	Marked	"
		Pipe Rough $k_s/D < 3\times10^{-3}$	Moderate	"
		Pipe Rough $k_s/D > 3\times10^{-3}$	Practically None	"
Gap Ratio: $\frac{e}{D} < 0.25$	Cross-Flow Component of Vibration	Pipe Smooth	Moderate	
		Pipe Rough	From Moderate to Practically None	
	In-Line Component of Vibration	Pipe Smooth	Moderate	
		Pipe Rough	From Moderate to Practically None	

Table 10.2 Continued

2) **Waves**

Gap Ratio			Flow Regimes	
			Critical, Supercritical and Upper Transition $3\times10^5 < Re < 4\times10^6$	Transcritical $Re > 4\times10^6$
Gap Ratio: $\dfrac{e}{D} > \dfrac{e_{cr}}{D}$ (see Fig.3.25 for e_{cr}/D)	Cross-Flow Component of Vibration	Pipe Smooth	Marked	Practically None
		Pipe Rough $k_s/D < 3\times10^{-3}$	Moderate	"
		Pipe Rough $k_s/D > 3\times10^{-3}$	Practically None	"
	In-Line Oscillatory Motion Induced by Morison Force	Pipe Smooth	Change only in Amplitude. From Moderate to Practically None	
		Pipe Rough	"	
Gap Ratio: $\dfrac{e}{D} < \dfrac{e_{cr}}{D}$ (see Fig.3.25 for e_{cr}/D)	Cross-Flow Component of Vibration	Pipe Smooth	Moderate	
		Pipe Rough	From Moderate to Practically None	
	In-Line Oscillatory Motion Induced by Morison Force	Pipe Smooth	Change only in Amplitude. From Moderate to Practically None	
		Pipe Rough	"	

The vibrations should obviously be expected to undergo marked changes at high Re numbers due to the change in the flow regime at such numbers if the pipe has a surface roughness k_s/D less than 3×10^{-3}.

If the pipeline-surface roughness k_s/D exceeds 3×10^{-3}, practically no change in the pipeline response should be expected to occur at high Re numbers.

Waves

Cross-flow vibrations. Similar to the case of steady currents, a pipeline may undergo two kinds of cross-flow vibrations when it is exposed to waves, depending on the value of the clearance between the pipe and the bed: 1) The vortex-shedding induced vibrations for large values of e/D, and 2) The self-excited vibrations for small values of e/D.

In the case of waves, the limiting value of e/D below which vortex shedding is surpressed actually depends on the KC number (Fig. 3.25), $e_{cr}/D = f(KC)$. For $KC = 10$, for example, e_{cr}/D can be as small as 0.1, while it increases with increasing KC and attains its asymptotic value, namely 0.25 as $KC \to \infty$.

In the case when $e/D > e_{cr}/D$ the regimes of flow around the pipeline should be expected to be practically the same as in the case of a wall-free cylinder with possibly the critical Re numbers having values relatively lower than in the case of steady flows (cf. Figs. 1.9 and 3.16). It should be noted that the Re number here is based on the velocity of water particles *relative* to the pipeline.

Marked changes in the pipeline's response pattern at high Re numbers may be expected if the surface roughness of the pipeline k_s/D is less than 3×10^{-3}, as described in Section 9.2.3 for the case of a free cylinder.

If the roughness of the pipeline surface k_s/D exceeds 3×10^{-3}, no significant change in the response patterns of the pipeline should be expected on similar grounds given in Section 9.2.3.

In-line oscillatory motion. Things completely change when the in-line oscillatory motion of pipelines is considered. The pipeline in this case moves periodically under the action of the total in-line force at the wave frequency. This contribution is very large, while the contribution to the in-line vibration pattern from the vortex shedding is negligible and is disregarded in the following discussion. Therefore, the question here is how the total in-line force behaves at high Re numbers. A first indication of this behaviour can be obtained by application of Eqs. 9.10 and 9.11 where the measured C_D- and C_M-values obtained by for example Sarpkaya (1977) must be inserted in the Morison equation. Fig. 10.17 shows such a calculation, borrowed from Jensen (1987). However, it must be pointed out that C_M- and C_D-values are obtained for a fixed pipe (see the detailed discussion in Section 9.4). Also, note that especially the presence of cross-flow vibrations could change the C_M- and C_D-values considerably.

Table 10.2 summarizes the changes which should be expected in the response pattern of pipelines at high Re numbers as compared with that obtained in the subcritical flow regime.

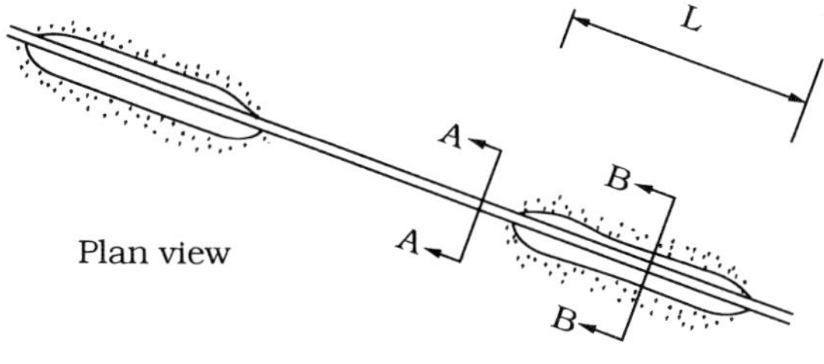

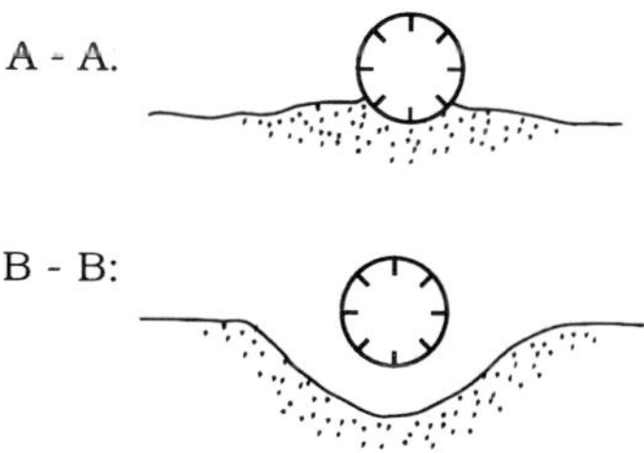

Figure 10.18 Scour holes in plan view and characteristic cross-sections.

10.4 Effect of scoured trench

When pipelines are not buried, spans may develop in most locations due to scour, as mentioned earlier (Fig. 10.18). As the scour spreads along the length of a span, the scoured trench may become sufficiently long, enabling the suspended pipe to sag into its naturally created trench hole. (Leeuwenstein (1985), Bruschi, Cimbali, Leopardi and Vincenzi (1986), Sumer and Fredsøe (1992)). For engineering applications, therefore, it is important to know the vibration pattern of a pipeline that more or less sags into the scoured trench. It is evident that the vibration pattern will be different due to the presence of the trench. This aspect of the problem has been investigated by various researchers. Sumer, Mao and Fredsøe (1988) and Kristiansen (1988) (see also Kristiansen and Tørum, 1989) studied the vibrations of pipelines placed near or on a sandy bottom in steady currents, while

Sumer et al. (1989) and Bryndum et al. (1989) studied the vibrations of a pipeline placed in the vicinity of a frozen model trench in the case of waves.

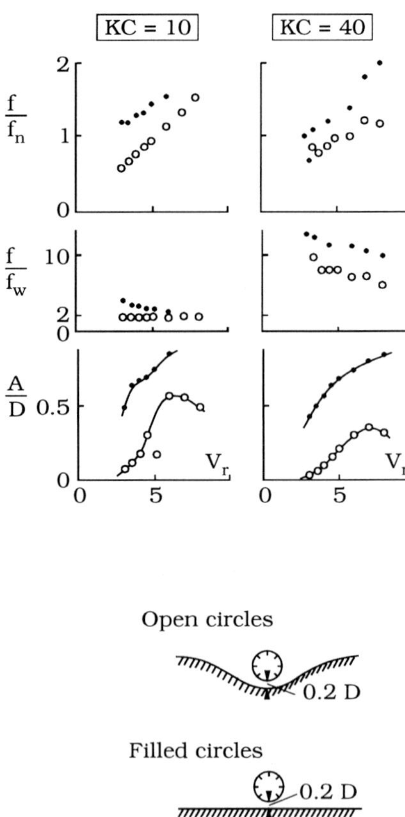

Figure 10.19 Effect of scoured trench on cross-flow vibrations. Hydroelastic properties of the system are the same as in Fig. 10.16. Sumer et al. (1989).

Figs. 10.19 and 10.20 show the results of Sumer et al.'s (1989) study where the trench data are compared with the data obtained from a flat bed. The following conclusions are straightforward from the figure: when the pipe is placed in a trench, the amplitudes and the frequencies are greatly reduced. This can be linked to the fact that the pipe in the trench is well protected and experiences relatively lower

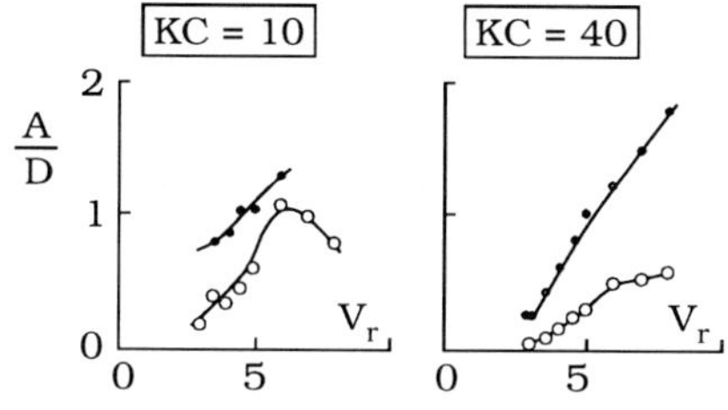

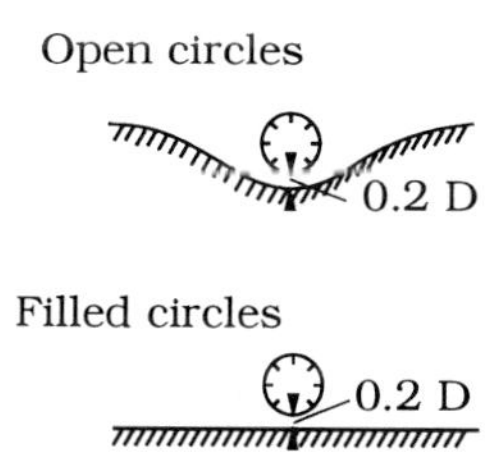

Figure 10.20 Effect of scoured trench on in-line oscillatory motion. Hydroelastic properties of the system are the same as in Fig. 10.16.
Sumer et al. (1989).

velocities. The study further revealed that when the clearance is $2D$, the presence
of the trench below the pipe is practically not felt, and the response is much the
same as that obtained with a flat bed. These effects can be seen very clearly from
the trajectory pictures presented in Fig. 10.21. The preceding conclusions are
valid for the in-line vibrations too.

10.5 Vibrations of pipelines in irregular waves

Cross-flow vibrations

Fig. 10.22 gives two sample vibration records together with the corresponding flow-velocity trace obtained in the work of Sumer et al. (1989), illustrating the
effect of close proximity of the bed (in the present case, in the form of a scoured
trench with a depth equal to the pipe diameter D). Fig. 10.23, on the other hand,
compares the cross-flow vibration data with their regular-wave counterparts.

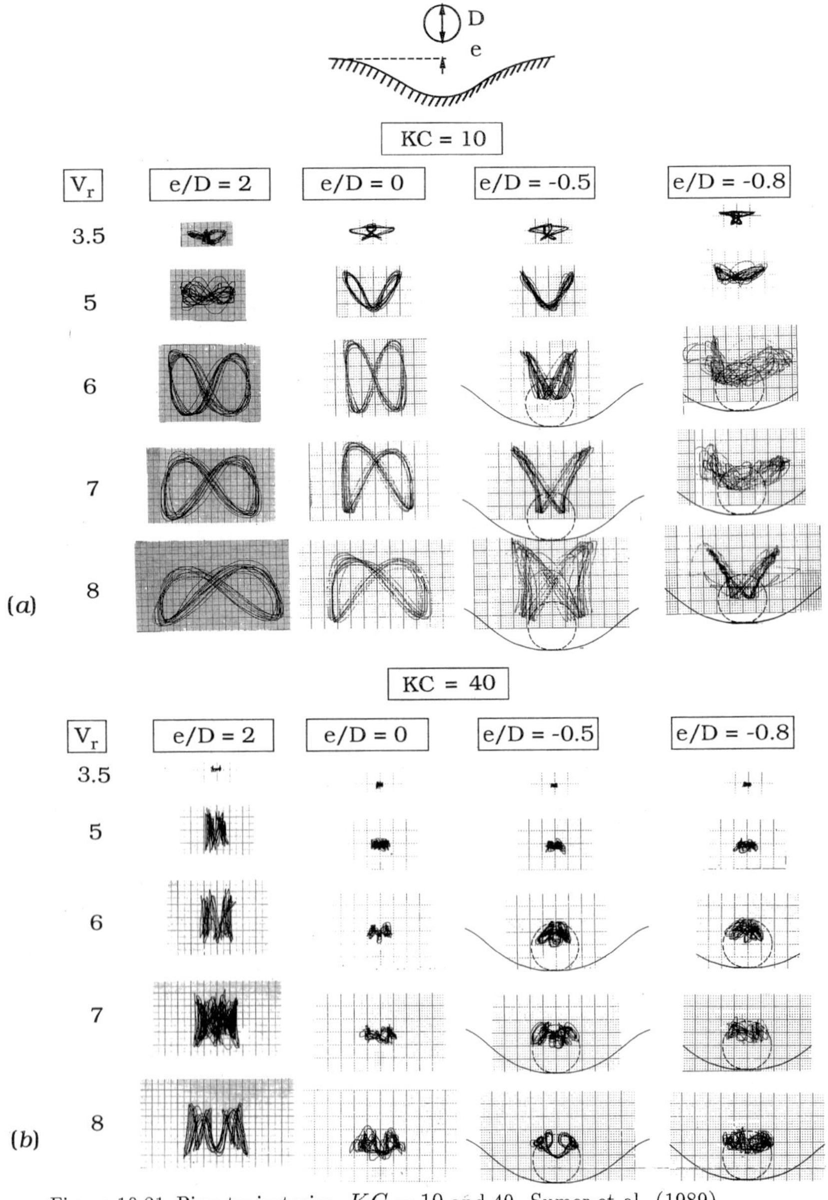

Figure 10.21 Pipe trajectories, $KC = 10$ and 40. Sumer et al. (1989).

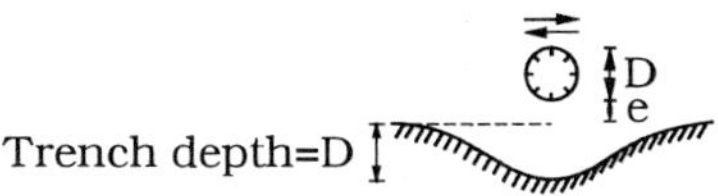

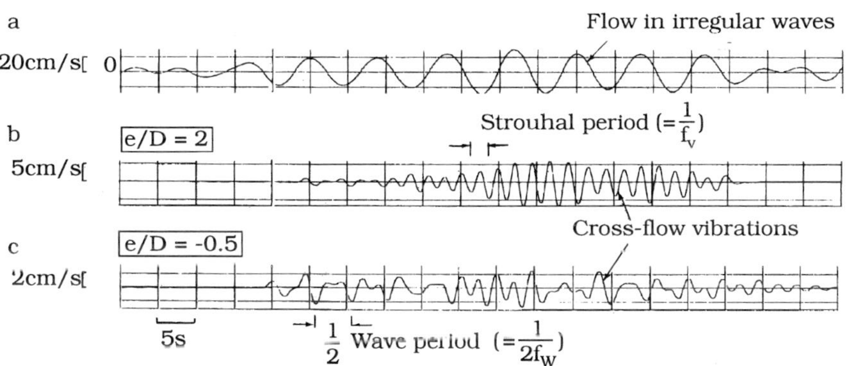

Figure 10.22 Sample records of cross-flow vibrations for $e/D = 2$ and -0.5 obtained under exactly the same irregular-wave conditions: $KC = 10$ and $V_r = 3$. Sumer et al. (1989).

The tests conducted for eight different values of e/D ranging from -0.8 to 2 with a trench depth equal to D indicated that the effect of irregular waves on the amplitude response is such that the maximum amplitudes generally appear to be slightly larger than in the regular waves. This is seen from Fig. 10.23a for the e/D values indicated in the figure.

As for the frequency response, the vibration frequency was found to differ markedly from that in regular waves for values of clearance $e/D > 0$, while it remains practically unchanged for $e/D < 0$ (Fig. 10.23a). This behaviour is directly related to the two different mechanisms driving the vibrations in these two e/D regimes.

For $e/D > 0$, the vibrations are driven by vortex shedding. In irregular waves, the system is selective as far as its vibration frequency is concerned, as has already been seen in Section 9.2.4 in conjunction with the response of free cylinders in irregular waves. As is seen from Fig. 10.22b, the pipe simply starts to vibrate at f_n, when the vortex shedding frequency ($f_v = St\ U_m/D$) comes close to the frequency f_n in the process of time evolution of the flow-velocity amplitude. This explains why the frequency response in irregular waves is different from that in regular waves when $e/D > 0$.

For $e/D < 0$, however, things change. The vortex shedding is suppressed,

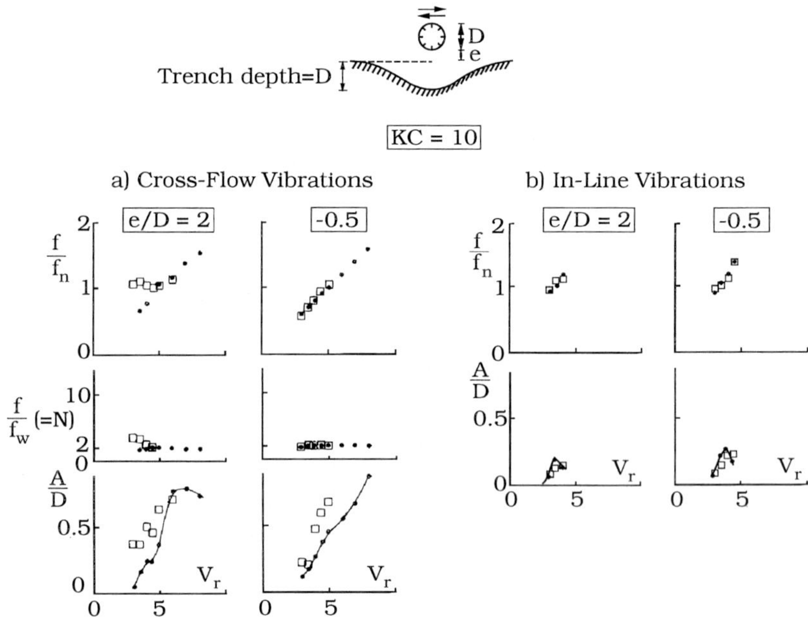

Figure 10.23 Frequency and amplitude response in regular waves and irregular waves, $KC = 10$: (a) Cross-flow vibrations; (b) In-line
vibrations (filled circles = regular waves; squares = irregular
waves). Hydroelastic properties of the system are the same as
in Fig. 10.16. Sumer et al. (1989).

and the pipe is exerted by a lift force oscillating at twice the wave frequency, since
the pipe is now under the influence of close proximity of the bed. Thus, the pipe
should vibrate with this forcing frequency, namely at $f/f_w(= N) = 2$, irrespective
of whether the pipe is exposed to regular waves or to irregular waves. Sumer et
al. (1989) note that the preceding arguments hold true also for $KC = 40$.

Observations in connection with transverse vibrations of pipelines in irregular waves were first made by Jacobsen et al. (1984b). They observed that long
resting periods were interrupted by intervals during which the vibrations built up
and then died out, as groups of large waves passed the pipe. This is a characteristic
feature of the vibrations in irregular waves.

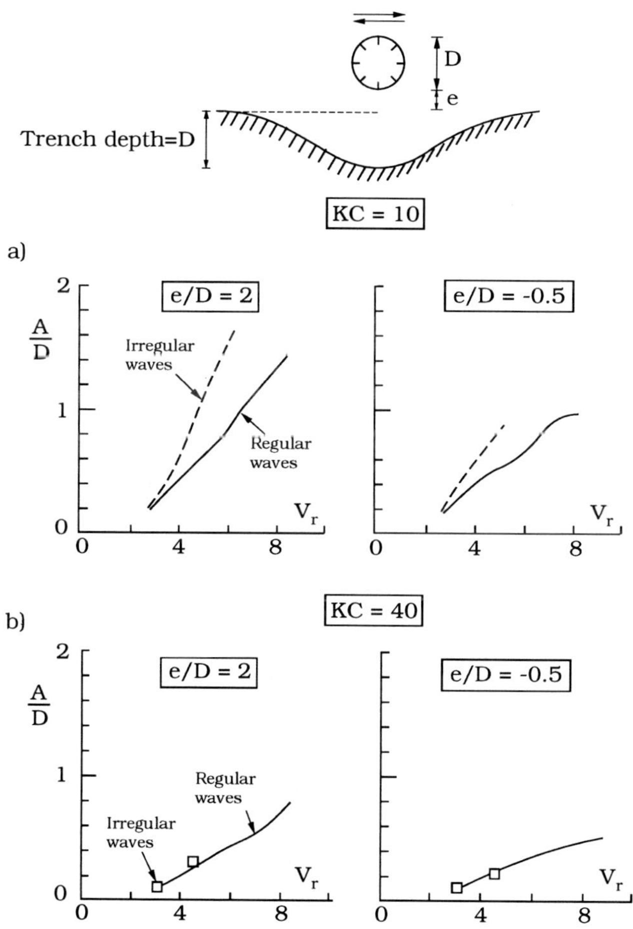

Figure 10.24 Amplitude of in-line motion in regular waves and irregular waves. Hydroelastic properties of the system are the same as in Fig. 10.16. Sumer et al. (1989).

In-line vibrations and in-line movement

Fig. 10.23b shows the in-line vibration data together with their regular wave counterparts for $KC = 10$. Practically no difference exists between the two cases.

Fig. 10.24 compares the amplitude data corresponding to the wave-induced in-line movement of the pipe in irregular waves with those of regular waves. The amplitudes in the irregular waves for $KC = 10$ (Fig. 10.24a) are larger than in the regular waves (see also Fig. 10.25). This is because, as the waves evolve in time, the temporal value of the velocity amplitude can easily reach values high enough to give rise to resonance in-line movements of the pipe. This effect is not pronounced for $KC = 40$ (Fig. 10.24b), where the resonance point is at

$$V_r = \frac{U_m}{D f_n} = \frac{U_m}{D f_w} = 40 \qquad (10.2)$$

since the resonance occurs at $f = f_w$.

10.6   Effect of angle of attack

Bryndum et al.'s (1989) experiments with a long flexible pipe exposed to waves and also to combined waves and current indicate that the response, both in the cross-flow direction and in the in-line direction (the Morison-induced induced in-line motion) is independent of the incident angle. This is in agreement with King's (1977) results in conjunction with a free circular cylinder exposed to steady currents at different incident angles.

Bryndum et al. attribute the observed behaviour of the response to the so-called *cross-flow principle*, namely that the hydrodynamic forces are independent of the incident angle as long as the forces are expressed in terms of the component of the flow perpendicular to the pipe axis (see Section 2.6). However, the observed behaviour may change for very small values of the angle of attack. See Section 4.5 for a detailed discussion regarding the forces on a cylinder placed near a plane wall.

10.7   Forces on a vibrating pipeline

Flow around and forces on a fixed cylinder placed near a wall have been investigated extensively in the last decade or so (an extensive list of references has been given in Section 2.7 and in Section 4.7, covering the cases of steady current and waves, respectively). In the case when the cylinder vibrates in a direction perpendicular to the flow, however, the forces on the cylinder will be influenced by the vibrations of the cylinder. Therefore, the force coefficients will be a function of not only the parameter e/D, but also the parameters characterizing the cylinder vibrations such as the vibration amplitude and the vibration frequency.

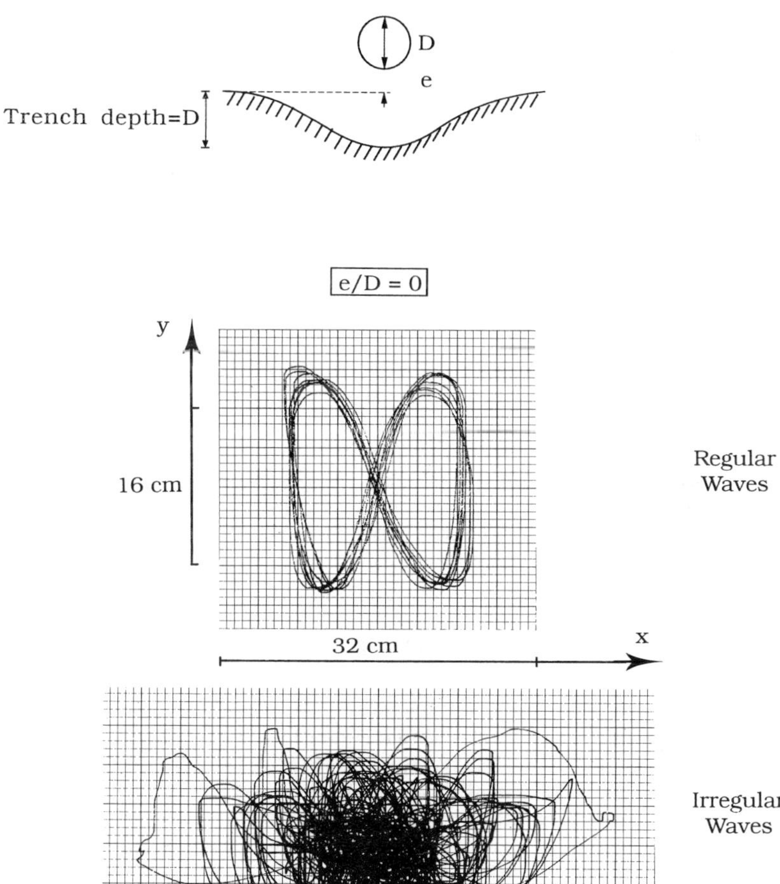

Figure 10.25 Pipe trajectories in regular and irregular waves. Sumer et al. (1989).

In the past, forces on a wall-free cylinder oscillating in the cross-flow direction and subject to a steady current have been measured by several investigators (Bishop and Hassan (1964), Sarpkaya (1978, 1982) and Moe and Wu (1990); see Section 8.5).

Jensen, Sumer and Fredsøe (1992) and Sumer, Fredsøe, Jensen and Christiansen (1994a) have extended the existing work on forces on a vibrating, wall-free cylinder, subject to steady currents, to the case of a vibrating near-wall cylinder. In Jensen et al.'s work, the cylinder was subject to a steady current, while in Sumer et al.'s study both the steady-current and the wave situations have been investigated.

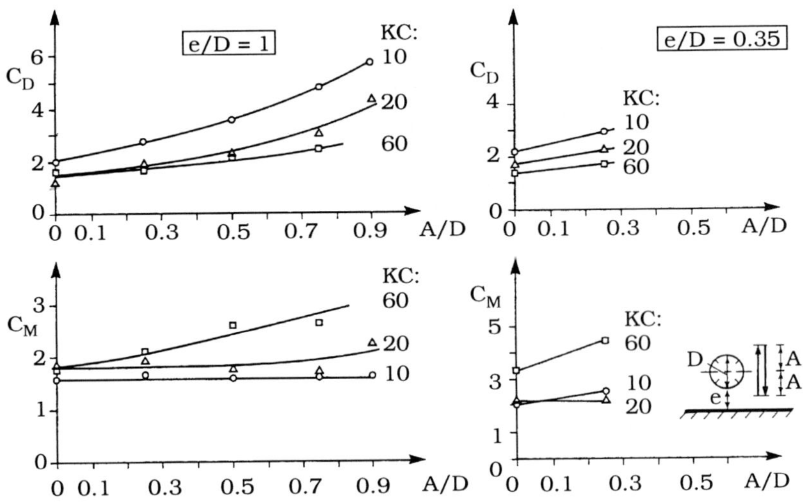

Figure 10.26 In-line force coefficients as function of the normalized amplitude of cross-flow cylinder vibrations A/D, and the flow KC number in the case of forced vibrations. The frequency of cross-flow vibrations is selected such that the reduced velocity in all cases is $V_r = U_m/(Df_y) = 5$. Sumer et al. (1994a).

In a previous study, Sumer et al. (1989) investigated vibrations of and forces on a freely-vibrating pipe placed in the vicinity of a scoured trench and exposed to sinusoidal and random oscillatory flows. The force coefficients were determined for two values of the Keulegan-Carpenter number, namely $KC = 10$ and 40. The main difference between the study by Sumer et al. (1989) and their recent study (1994a) is that the vibrations in the former study were self-excited, therefore they could not be controlled externally. This prevented a systematic investigation of the effect of vibrations on the forces. This shortcoming is avoided in the later study with a system, where the cylinder was vibrated by a hydraulic piston. Figs. 10.26 – 10.29 show the force coefficients as function of the vibration amplitude (Figs. 10.26 and 10.27) and the vibration frequency (Figs. 10.28 and 10.29). Regarding

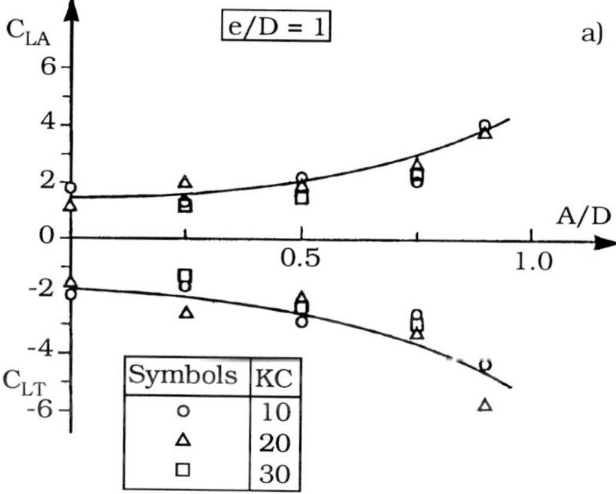

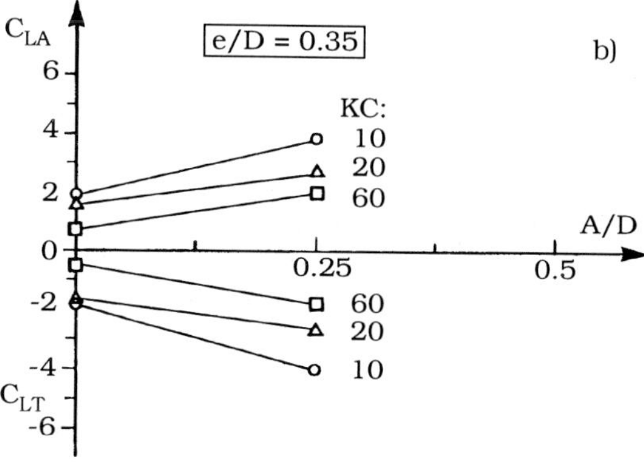

Figure 10.27 Lift force coefficients as function of the normalized amplitude of cross-flow cylinder vibrations, A/D, and the flow KC number in the case of forced vibrations. The frequency of cross-flow vibrations is selected such that the reduced velocity in all cases is $V_r = U_m/(Df_y) = 5$. Sumer et al. (1994a).

the C_D and C_M coefficients, the results of Sumer et al. agree well with the results of Bearman's (1988) study of forces on a flexible, vertical cylinder in waves where KC ranged from 3 to about 20.

Spanwise correlation may be important when the forces on a cylinder are considered. Research has shown that the vibrations of the cylinder have a significant effect on correlations; the correlation for a transversely vibrating cylinder is increased with increasing amplitudes of vibrations (Section 1.2.2 and Section 3.5).

In a recent study (Kozakiewicz, Sumer and Fredsøe, 1992), spanwise correlation measurements have been made for a stationary and transversely vibrating (forced vibrations) cylinder placed near a wall and exposed to oscillatory flows. The correlation was calculated, based on pressure measurements on the surface of the cylinder. Spanwise correlation measurements have been made also for an elastically-mounted cylinder exposed to an oscillatory flow and undergoing self-induced vibration in the transverse direction (Sumer, Fredsøe and Jensen, 1994b). See Section 3.5 for a detailed discussion of the subject.

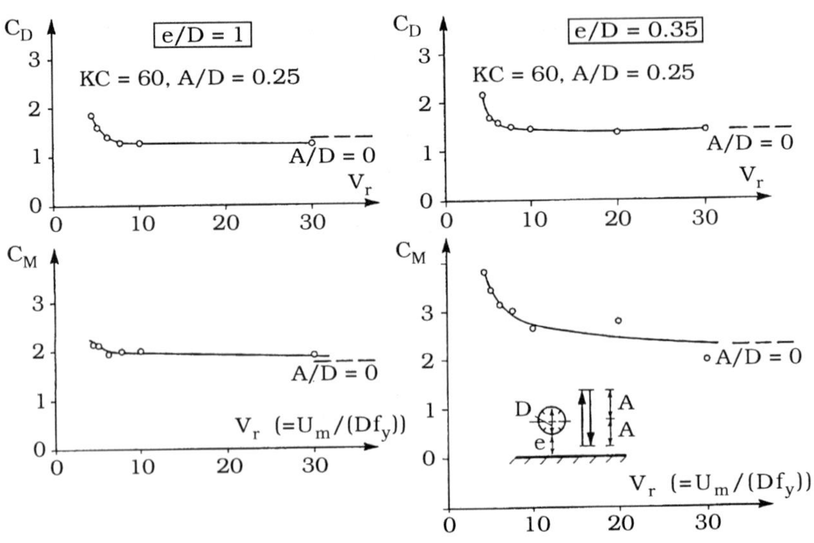

Figure 10.28 In-line force coefficients as function of the normalized frequency of cylinder cross-flow vibrations $V_r = U_m/(Df_y)$. Forced vibrations. $KC = 60$. $A/D = 0.25$. Sumer et al. (1994a).

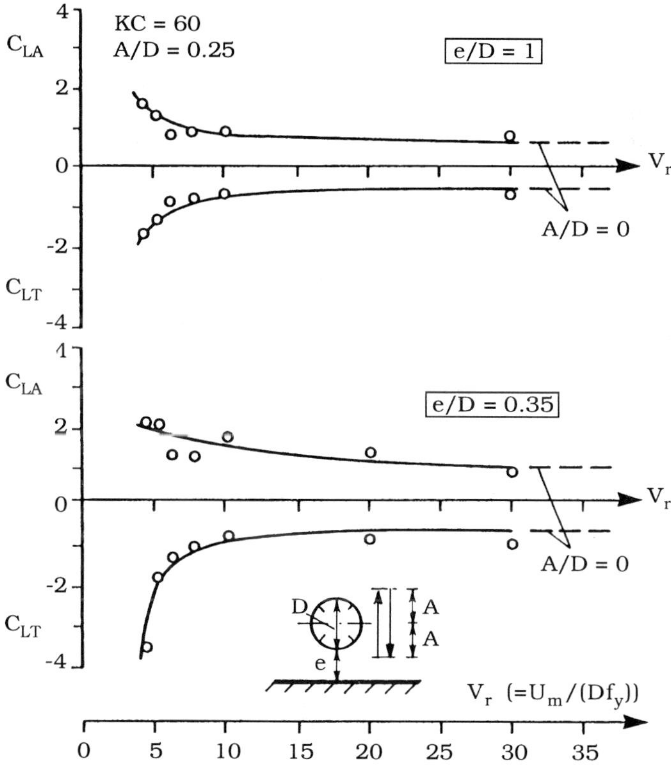

Figure 10.29 Lift-force coefficients as function of the normalized frequency of cylinder cross-flow vibrations $V_r = U_m/(Df_y)$. Forced vibrations. $KC = 60$. $A/D = 0.25$. Sumer et al. (1994a).

REFERENCES

Bearman, P.W. (1988): Wave loading experiments on circular cylinders at large scale. Proc. Int. Conf. on Behaviour of Offshore Struct. (BOSS '88), Trondheim, Norway, June 2, 471-487.

Bishop, R.E.D. and Hassan, A.Y. (1964): The Lift and Drag Forces on a Circular Cylinder Oscillating in a Flowing Fluid. Proc. Roy. Soc. London, A 277, 51-75.

Bruschi, R.M., Buresti, G., Castoldi, A. and Miliavacca, E. (1982): Vortex shedding oscillations for submarine pipelines: comparison between full-scale experiments and analytical models. 14th Annual OTC in Houston, Texas, May 3-6, 1982, Paper No. 4232, 2:21-36.

Bruschi, R., Cimbali, W., Leopardi, G. and Vincenzi, M. (1986): Scour induced free span analysis. Proc. 5th Int. Offshore Mechanics and Arctic Engineering Symposium, April 13-18, 1986, 3:656-669.

Bryndum, M.B., Jacobsen, V. and Brand, L.P. (1983): Hydrodynamic forces from wave and current loads on marine pipelines. Proc. 15th Annual Offshore Technology Conf., Houston, TX, May 2-5, 1983, OTC Paper 4454, pp. 95-102.

Bryndum, M.B., Bonde, C., Smitt, L.W., Tura, F. and Montesi, M. (1989): Long free spans exposed to current and waves: Model tests. Proc. 21st Annual Offshore Technology Conf., Houston, TX, May 1-4, 1989, OTC Paper 6153, pp. 317-328.

Fredsøe, J. (1984): Turbulent boundary layer in wave-current motion. J. Hyd. Engrg., ASCE, 110(8):1103-1120.

Fredsøe, J. and Hansen, E.A. (1987): Lift forces on pipelines in steady flow. J. Waterway, Port, Coastal Ocean Engrg., ASCE, 113:139-155.

Fredsøe, J., Sumer, B.M., Andersen, J. and Hansen E.A. (1985): Transverse vibrations of a cylinder very close to a plane wall. Proc. 4th Int. Offshore Mechanics and Arctic Engineering. Symposium, Dallas, TX, Feb. 17-21, 1985, Vol. I, p. 601-609. Also in J. of Offshore Mechanics and Arctic Engineering, 109(1):52-60, 1987.

Hansen, E.A., Madsen, P.A. and Fredsøe, J. (1986): Self-excited vibrations of pipelines. The Danish Center for Applied Mathematics, DCAMM, The Technical University of Denmark, Report No. 335, 44 p., October 1986.

Jacobsen, V., Bryndum, M.B., Nielsen, R. and Fines, S. (1984a): Vibrations of Offshore Pipelines Exposed to Current and Wave Action. 3rd Int. Symposium on Offshore Mechanics and Arctic Engineering, New Orleans, LA, Feb. 12-16, 1984.

Jacobsen, V., Bryndum, M.B., Nielsen, R. and Fines, S. (1984b): Cross-flow vibrations of a pipe close to a rigid boundary. Trans. ASME, Journal of Energy Resources Technology, Dec. 1984, Vol. 106, p. 451-457.

Jacobsen, V., Bryndum, M.B. and Fredsøe, J. (1984): Determination of flow kinematics close to marine pipelines and their use in stability calculations. Proc. 16th Annual Offshore Technology Conf., Paper No. OTC 4833, pp. 481-492.

Jensen, B.L. (1987): Effect of Reynolds number on in-line vibrations of pipelines. Progress Report No. 65, Inst. of Hydrodynamics and Hydraulic Engineering, ISVA, Techn. Univ. Denmark, pp. 21-30.

Jensen, B.L., Sumer, B.M. and Fredsøe, J. (1993): Forces on a pipeline oscillating in transverse direction in steady current. Proc. 3rd Int. Offshore and Polar Engineering Conf. Singapore, June 6-11, 1993, 3:424-430.

King, R. (1977): Vortex-excited oscillation of yawed circular cylinders. J. Fluids Engrg., 99:495-502.

Kozakiewicz, A., Sumer, B.M. and Fredsøe, J. (1992): Spanwise correlation on a vibrating cylinder near a wall in oscillatory flows. J. Fluids and Structures, 6:371-392.

Kristiansen, Ø. (1988): Current induced vibrations and scour of pipelines on a sandy bottom. Thesis presented to the University of Trondheim, Trondheim, Norway, in partial fulfillment of the requirements of the degree of Doctor of Philosophy.

Kristiansen, Ø. and Tørum, A. (1989): Interaction between current induced vibrations and scour of pipelines on a sandy bottom. Proc. 8th Int. Conf. on Offshore Mechanics and Arctic Engineering. The Hague, The Netherlands, March 19-23, 1989, 5:167-174.

Leeuwenstein, W. (1985): Natural self-burial of submarine pipelines. MaTS - Stability of pipelines, scour and sedimentation. Coastal Engineering Group, Dept. of Civil Engrg., Delft Univ. of Technology, Delft, The Netherlands.

Mao, Y. (1986): The interaction between a pipeline and an erodible bed. Series Paper No. 39, Inst. of Hydrodynamics and Hydraulic Engineering, ISVA, Techn. Univ. Denmark.

Moe, G. and Wu, Z. (1990): The lift force on a vibrating cylinder in a current. J. Offshore Mechanics and Arctic Engineering, 112:297-303.

Offshore Engineer (1984): North Sea lifelines. April issue, pp. 112-127.

Orgill, G., Barbas, S.T., Crossley, C.W. and Carter, L.W. (1992): Current practice in determining allowable pipeline free spans. Proc. 11th Offshore Mechanics and Arctic Engineering Conf., June 7-11, 1992, Calgary, Canada, Pipeline Technology, Vol. 5-A, pp. 139-145.

Raven, P.W.C., Stuart, R.J. and Littlejohns, P.S. (1985): Full-scale dynamic testing of submarine pipeline spans. 17th Annual OTC, Houston, TX, May 6-9, 1985, Paper No. 5005, pp. 395-405.

Sarpkaya, T. (1976): Forces on cylinders near a plane boundary in a sinusoidally oscillating fluid. J. Fluids Engineering, pp. 499-505.

Sarpkaya, T. (1977): In-line and transverse forces on cylinders near a wall in oscillatory flow at high Reynolds numbers. Proc. 9th Annual Offshore Technology Conf., Paper No. OTC 2898, pp. 161-166.

Sarpkaya, T. (1978): Fluid forces on oscillating cylinders. ASCE, J. Waterways, Port, Coastal and Ocean Division, 104(WW3):275-290.

Sarpkaya, T. (1982): Flow induced vibration of roughened cylinders. Proc. Int. Conf. on Flow Induced Vibrations in Fluid Engineering, Reading, England (organized by BHRA Fluid Engrg.), Cranfield, UK, Sept. 14-16, 1982, pp. 131-139.

Sarpkaya, T. and Rajabi, F. (1979): Dynamic response of piles to vortex shedding in oscillating flows. Proc. 11th Annual Offshore Technology Conf., April 30 - May 3, 1979, OTC 3647, pp. 2523-2528.

Sumer, B.M. and Fredsøe, J. (1989): Effect of Reynolds number on vibrations of cylinders. J. Offshore Mechanics and Arctic Engineering, 111:131-137.

Sumer, B.M. and Fredsøe, J. (1992): A review of wave/current induced scour around pipelines. Proc. 23rd Int. Conf. on Coastal Engineering, October 4-9, 1992, Venice, Italy, Chapter 217, 3:2839-2852.

Sumer, B.M., Fredsøe, J. and Jacobsen, V. (1986): Transverse vibrations of a pipeline exposed to waves. Proc 5th OMAE Symposium, Tokyo, Japan, 3:588-596.

Sumer, B.M., Mao, Y. and Fredsøe, J. (1988): Interaction between vibrating pipe and erodible bed. J. Waterway, Port, Coastal and Ocean Engineering, ASCE, 114(1):81-92.

Sumer, B.M., Fredsøe, J., Gravesen, H. and Bruschi, R. (1989): Response of marine pipelines in scour trenches. J. Waterways, Port, Coastal and Ocean Engineering, ASCE, 115(4):477-496.

Sumer, B.M., Jensen, B.L. and Fredsøe, J. (1991): Effect of a plane boundary on oscillatory flow around a circular cylinder. J. Fluid Mech., 225:271-300.

Sumer, B.M., Fredsøe, J., Jensen, B.L. and Christiansen, N. (1994a): Forces on a vibrating cylinder near a wall in steady and oscillatory flows. J. Waterway, Port, Coastal and Ocean Engineering, ASCE, 120(3):233-250.

Sumer, B.M., Fredsøe, J. and Jensen, K. (1994b): A note on spanwise correlation on a freely vibrating cylinder in oscillatory flow. J. Fluids and Structures, 8:231-238.

Tsahalis, D.T. (1983): The effect of seabottom proximity of the vortex-induced vibrations and fatique life of offshore pipelines. J. of Energy Resources Technology, Dec. 1983, 105:464-468.

Tsahalis, D.T. (1984): Vortex-induced vibrations of a flexible cylinder near a plane boundary exposed to steady and wave-induced currents. J. of Energy Resources Technology, June 1984, 106:206-213.

Tsahalis, D.T. (1985): Vortex-induced vibrations due to steady and wave-induced currents of a flexible cylinder near a plane boundary. Proc. 4th Int. Offshore Mechanics and Arctic Engineering Symposium, Dallas, TX, Feb. 17-21, 1985, 1:618-628.

Tsahalis, D.T. and Jones, W.T. (1981): Vortex-induced vibrations of a flexible cylinder near a plane boundary in steady flow. Proc. 13th Annual Offshore Technology Conf., Paper No. OTC 3991, 1:367-381.

Tørum, A. and Anand, N.M. (1985): Free span vibrations of submarine pipelines in steady flows. Effect of free-stream turbulence on mean drag coefficients. J. Energy Resources Technology, Dec. 1985, 107:415-420.

Wootton, L.R. (1969): The oscillation of large circular stacks in wind. Proc. of Institution of Civil Engineers, London, 43:573-598.

Chapter 11. Mathematical modelling of flow-induced vibrations

The equation of motion of a flexibly-mounted structure forms the basis of prediction of flow-induced vibrations. For a system with one degree of freedom of movement, for example, this equation reads (Chapter 8):

$$m\,\ddot{y}\,(t) +\ c\,\dot{y}\,(t) +\ k\,y(t) =\ F(t) \tag{11.1}$$

m being the total mass of the system including the hydrodynamic mass.

The term on the right side of the equation, $F(t)$, represents the force induced by vortex shedding or galloping, or any other effect which causes vibrations. For a vibrating system, there is a feed-back between the motion, $y(t)$, and the force $F(t)$. The major problem encountered in the mathematical and numerical treatment of vibrations is the correct representation of $F(t)$. There are two approaches in this regard. In the first approach, the force term is modelled by a simple expression such as in Eq. 8.17 (where the force term is approximated to a sine, or a cosine, function with a given frequency and an amplitude) while, in the other, the force term is calculated through the hydrodynamic equations (i.e., by solving the flow equations in the form of N.S. equation or by application of the vortex methods, etc.). The models involved in the former approach will be called the *simple models* while those involved in the latter approach will be called the *flow-field models*.

This chapter will review these two approaches with regard to their applications to the case of steady current (Section 11.1) and to the case of waves (Section 11.2). There are also *integrated, general models* used in the offshore engineering practice, which accommodate all kinds of flow environment such as steady currents, sheared currents, waves and their combinations. The remainder of the present chapter (Section 11.3) will briefly describe the underlying principles of this approach by reference to the model developed by Hansen (1982).

11.1 The steady-current case

11.1.1 Simple models

The model given in Eq. 8.17, namely,

$$F(t) = F_0 \cos(\omega t) \tag{11.2}$$

with F_0 expressed as

$$F_0 = \frac{1}{2}\left(\sqrt{2}(\overline{C_L'^2})^{1/2}\right)\rho D U^2 \tag{11.3}$$

(Eq. 8.101) may probably be the simplest model under this category for *cross-flow, vortex-induced vibrations*. As is seen in Example 8.3, this model provides a good agreement with the experimental data with regard to the maximum amplitudes. However, the model prediction begins to diverge from the experimental data for $K_s \stackrel{\sim}{<} 2$, corresponding to $A/D \stackrel{\sim}{>} 0.8$ (Fig. 8.26).

In the past, there have been several attempts to model $F(t)$ in the case of cross-flow, vortex-induced vibrations in a more proper way. Hartlen and Currie (1970) (and later Sarpkaya, 1978) modelled $F(t)$ in the form

$$\frac{F(t)}{\frac{1}{2}\rho D U^2} = C_L = C_{Lm}\sin(\omega t) - C_{Ld}\cos(\omega t) \tag{11.4}$$

in which the velocity of the cylinder in the cross-flow direction is given by $\dot{y} = (\dot{y})_{\mathrm{max}}\cos(\omega t)$. (Sarpkaya (1978) demonstrated that the first term on the right side of Eq. 11.4 represents the inertia component and the second term the drag component of the lift force on the cylinder).

Various authors used the values of C_{Lm} and C_{Ld}, determined from the *forced* vibration experiments as functions of vibration amplitude and frequency, to predict the response of a freely vibrating cylinder, where Eq. 11.1 has been solved numerically (e.g. Sarpkaya (1978), Staubli (1983)). Fig. 11.1 displays a comparison between the predicted response (Staubli, 1983) and the corresponding experimental data depicted earlier in Fig. 8.15.

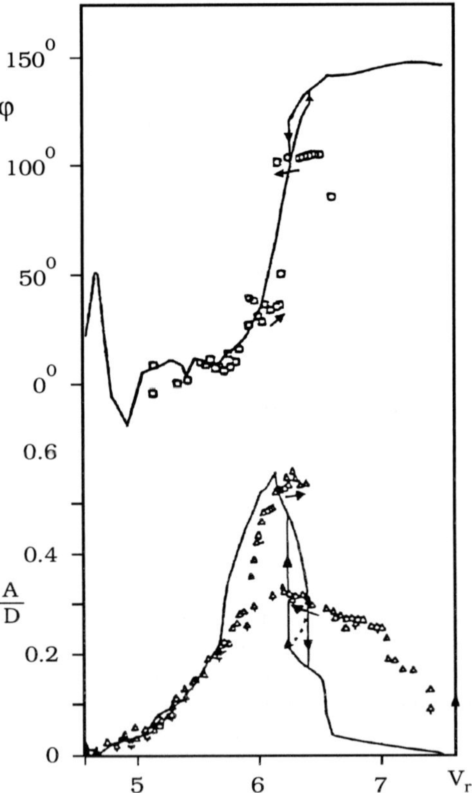

Figure 11.1 Response of cylinder undergoing free, cross-flow vibrations in steady current. Data points: Feng's (1968) experiments (Fig. 8.15). Solid lines: predictions by Staubli's (1983) model.

Hartlen and Currie (1970) took the approach to construct a model in which the lift coefficient, C_L, in Eq. 11.4 is derived from an equivalent oscillator. This model has become known as the **wake oscillator model**. The form of the equation for C_L is assumed in the form

$$\ddot{C}_L + (\text{damping term}) + (2\pi f_v)^2 C_L = (\text{forcing term}) \qquad (11.5)$$

in which f_v = the stationary-cylinder vortex-shedding frequency, and the coefficient $(2\pi f_v)^2$ ensures agreement with the stationary-cylinder Strouhal relation.

The damping term in the preceding equation is related to a linear combination of $\dot{C}_L$ and $(\dot{C}_L)^3$ while the forcing term is related to $\dot{y}$. The coupled equations, Eqs. 11.1 and 11.5 are then solved to obtain C_L and y in the form

$$y = A \, \sin{(\omega t)} \ , \text{ and} \qquad\qquad (11.6)$$

$$C_L = \hat{C}_L \, \sin{(\omega t + \varphi)} \qquad\qquad (11.7)$$

The model involves one experimentally determined lift coefficient and two "tuning" coefficients. These coefficients are selected to give a satisfactory fit to the observed phenomena. It turns out that the model equations have simulated most of the physical phenomena, except the hysteresis effect observed in Fig. 8.15.

Hartlen and Currie's basic formulation has been elaborated on by several researchers (see reviews by Parkinson (1974), Sarpkaya (1979) and Bearman (1984)). The model has been extended by Currie and Turnbull (1987) to the case of in-line vibrations.

11.1.2   Flow-field models

A true description of $F(t)$ can be achieved only by the solution of the flow equations. This may be accomplished either by the direct solution of the N.-S. equations (Section 5.1) or by the application of the vortex methods (Section 5.2). Therefore, the equation of motion of the body, Eq. 11.1, can be coupled with the flow equations, and the solution to the whole system of equations can be sought numerically. This approach has been adopted by several researchers in recent years. The following paragraphs will give a detailed account of this approach.

Cross-flow vibrations. The equation of motion which is to be solved is Eq. 11.1. This equation is, through the force term F, coupled with the flow equations, namely the vorticity-transport equation (Eq. 5.48) and the Poisson equation (Eq. 5.49). In the numerical solution of the coupled equations, the following procedure is followed, to advance the solution from time t to time $t + \delta t$:

1. At the beginning, all the flow quantities (corresponding to time t) are available in the computer memory.

2. The force, F, is found from the aforementioned flow quantities.

3. Under the calculated force, a new value of y is calculated from a finite-difference approximation of Eq. 11.1.

4. Given the new value of $\dot{y}$, a new flow field is calculated through the numerical solution of the flow equations either by the direct solution of the N.-S.

equations or by a vortex method. These are the new flow quantities which are saved in the computer memory. Then the steps from Step 2 to Step 4 are repeated to advance the solution from time $t + \delta t$ to $t + 2\delta t$. This procedure is repeated until the vibrations attain a state of equilibrium.

Regarding Step 4, the usual practice is to solve the flow equations in a stationary coordinate system fixed on the cylinder with the incident velocity input adjusted to take account of the cylinder motion. The force found from the flow quantities determined in the aforementioned fixed coordinate system includes also the "Froude-Krylov" force, namely $\rho(\frac{\pi D^2}{4})\,\ddot{y}$. Clearly, this force must be subtracted from the predicted total force when calculating the force F in Step 2.

A numerical solution for vortex-excited oscillations of a circular cylinder has been obtained by Anagnostopoulos (1994) where the N.-S. equations were solved in the manner described in the preceding paragraphs. The solution was obtained in the Reynolds number range $100 < Re < 140$, to ensure a laminnar, two-dimensional flow. The results regarding the response characteristics were found to be in good agreement with the corresponding results of an experimental study (Anagnostopoulos and Bearman, 1992) (Fig. 11.2). Figs. 11.3–11.5 show the force coefficients obtained in the study of Anagnostopoulos (1994). As seen, the N.-S. solution indicates the familiar amplification in the force coefficients experienced in the syncronization range (cf. Fig. 8.43a).

An interesting point with regard to the results presented in Figs. 11.2–11.5 is that, in contrast to the previous data (Figs. 8.15 and 8.17), the increase in the amplitude occurs quite abruptly at the lower end of the lock-in range. Whether or not this is due to a sudden change in the mode of vortex shedding is not clear, as there is no flow visualization available from the obtained solution.

As mentioned earlier, the flow field in the numerical prediction of cross-flow vibrations may be obtained also through a vortex method. Sarpkaya and Schoaff (1979) were the first to calculate the flow field in conjunction with the prediction of cross-flow vibrations through a discrete vortex model based on potential flow and boundary-layer interaction.

In the calculation of the flow field, the shear layers were rediscretized. Also, the circulation was reduced. This is because the actual flow is a 3-D flow; the circulation needs to be reduced in the application of a 2-D model, to account for the effect of three dimensionality. Fig. 11.6 presents the response characteristics obtained by Sarpkaya and Schoaff's vortex method. Sarpkaya and Schoaff reported that no hysteresis effect was found.

A method similar to that of Sarpkaya and Schoaff (1979) has been used by Kawai (1990) to study free oscillations of a circular cylinder with splitter plate. Yeung and Vaidhyanathan (1993), on the other hand, used the random vortex method (see Section 5.2) in combination with a complex-variable boundary-integral formulation to investigate vortex-induced oscillations of a circular cylinder.

The methods described in the preceding paragraphs may be used to conduct numerical *forced-vibration* experiments. Such numerical experiments were

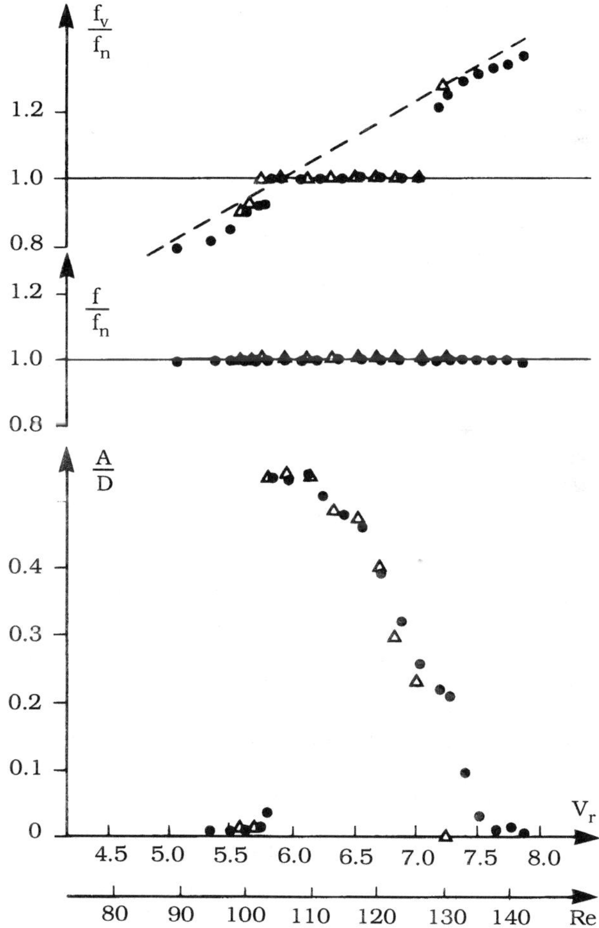

Figure 11.2 Comparison of N.-S. solution of cylinder vibration with the experiments. Circles: N.-S. solution (Anagnostopoulos, 1994). Triangles: Experiments (Anagnostopoulos and Bearman, - 1992). f_v: Vortex-shedding frequency, f: Cylinder frequency. Dashed line: $St = 0.212(1 - 21.2/Re)$ proposed by Roshko (1953). $m/(\rho D^2) = 117$ and $\zeta_s = 0.0012$.

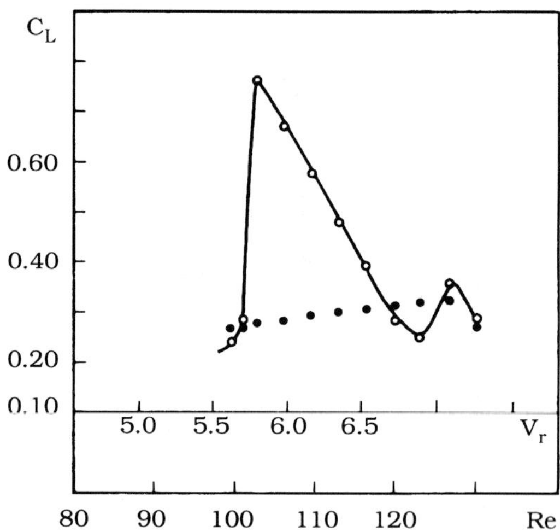

Figure 11.3 Amplitude of lift coefficient. N.-S. solution. •, fixed cylinder;
o, oscillating cylinder. Anagnostopoulos (1994).

undertaken by Hurlbut, Spaulding and White (1982), Lecointe and Piquet (1989), Chang and Sa (1992) and Li, Sun and Roux (1992) (with the direct numerical solution of the N.-S. equations) and by Yeung and Vaidhyanathan (1993) (with the random vortex method). Meneghini and Bearman (1993) simulated the flow, using the discrete vortex method, incorporating viscous diffusion, for $Re = 200$. The latter authors were able to demonstrate that the mode of shedding is different for amplitudes above about 0.6D (cf. Section 8.3.3).

Two-degrees-of-freedom vibrations. In this case, the equations of motion of the vibrating structure will read

$$m\,\ddot{x} + c\,\dot{x} + kx = F_x \tag{11.8}$$

$$m\,\ddot{y} + c\,\dot{y} + ky = F_y \tag{11.9}$$

(Note that the damping and spring constants (c and k, respectively) in the x and y directions may be different). The procedure for the numerical solution of these equations is exactly the same as described for the case of one-degree-of-freedom systems in the preceding section. Namely, to advance the solution from t to $t + \delta t$, the hydrodynamic quantities (and therefore the forces, F_x and F_y) stored in the computer memory are used to calculate the displacements x and y corresponding

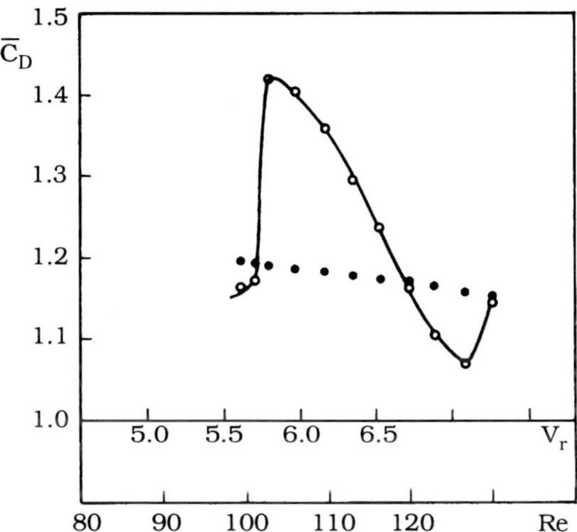

Figure 11.4 Mean drag coefficient. N.-S. solution. •, fixed cylinder; o, oscillating cylinder. Anagnostopoulos (1994).

to time $t + \delta t$, and subsequently the new values of x and y are used to calculate a new flow field (i.e., the flow field corresponding to time $t + \delta t$). Slaouti and Stansby (1994) used the discrete vortex method to determine the flow field. The Re-number range in Slaouti and Stansby's study was 100–200 and the V_r range 2–12.

11.2   The wave case

As seen in Chapter 9, there are three kinds of vibrations of a flexibly-mounted structure in oscillatory flows (Table 9.1): the cross-flow vibrations, the in-line vibrations, and the in-line oscillatory motion.

These vibrations can be determined formally by solving the equations of motion of the structure, namely Eq. 11.1 in the case of a one-degree-of-freedom-of-movement system (cross-flow vibrations or in-line vibrations or in-line motion) and Eqs. 11.8 and 11.9 in the case when the cross-flow vibrations and the in-line vibrations/motion are present concurrently.

Various authors have developed *simple models* where the force terms are modelled in a fashion similar to the case of steady current (Section 11.1.1). Of

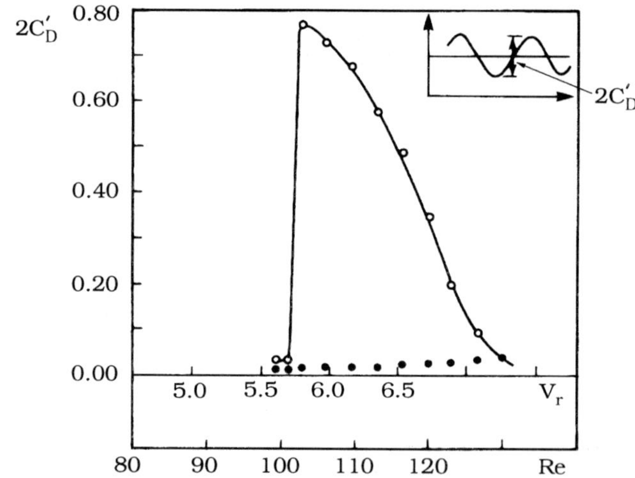

Figure 11.5 Fluctuating drag coefficient (peak-to-peak). N.-S. solution. •, fixed cylinder; ○, oscillating cylinder. Anagnostopoulos (1994).

particular interest is the modelling of the in-line forces. The common practice is to model the in-line force in terms of the Morison equation, as discussed in Section 9.4. Since the Morison equation is not able to resolve the high-frequency, small-amplitude, vortex-induced oscillations in the in-line force, the predicted oscillations from the numerical solution of Eq. 11.1 (or Eqs. 11.8 and 11.9) will represent only the oscillatory in-line motion of the structure. Laya, Connor and Sunder (1984), Williamson (1985) and Bearman et al. (1992) predicted the oscillatory in-line motion of structures by modelling the force term by the Morison approximation for one-degree-of-freedom motion (Section 9.4).

The previously mentioned models have been extended by Lipsett and Williamson (1991 and 1994) to the case of two-degrees-of-freedom systems subject to oscillatory flows. Lipsett and Williamson basically considered the equations of motion in two directions (Eqs. 11.8 and 11.9) with the in-line force, F_x, modelled by the Morison equation (Eq. 4.30) and the lift force, F_y, modelled by

$$F(t) = \frac{1}{2}\rho U^2 D \, C_L \sin(\omega_v t) \tag{11.10}$$

which is originally proposed by McConnell and Park (1982b). Here U is the instantaneous velocity, $U = U_m \sin(\omega t)$, rather than U_m, and ω_v is the angular frequency of vortex-shedding. The latter authors also considered two other models in their study. In one of the models, the velocity U in Eq. 11.10 was replaced by

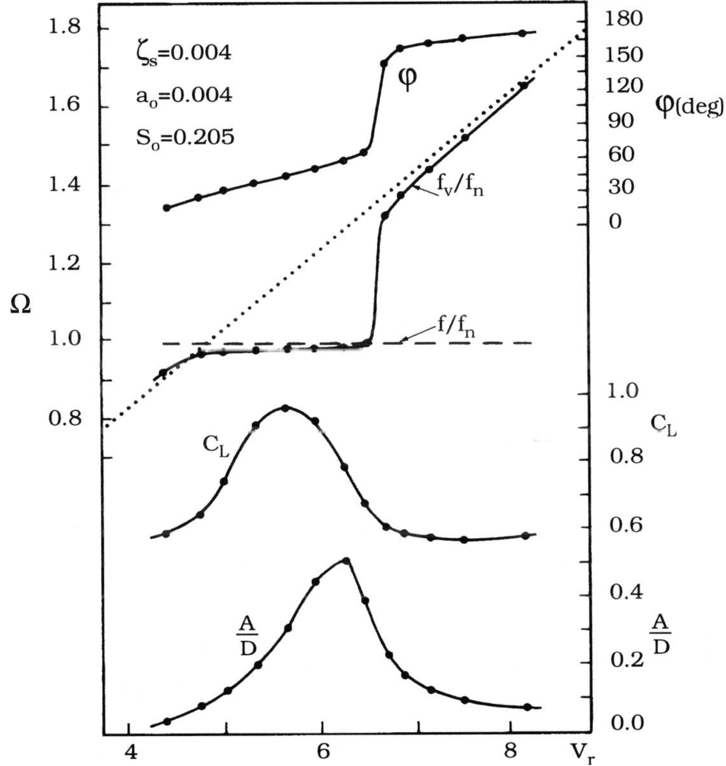

Figure 11.6 Response characteristics of a freely oscillating cylinder as predicted by the discrete vortex model. Dotted line: $f/f_n = (St)V_r$. Sarpkaya and Schoaff (1979).

U_m. The equations of motion (Eqs. 11.8 and 11.9) in this model and in the previous one were uncoupled. In the third model, however, the equations of motion were coupled by considering the in-line and lift force components in directions parallel and perpendicular to the direction of instantaneous relative velocity between the cylinder and the flow. Fig. 11.7 compares the numerically predicted cylinder trajectory from the coupled model with that obtained in the experiments.

Regarding the *flow-field models*, the idea is, as in the case of a steady current, to determine the force on the structure either by the direct numerical solution of the N.-S. equations or by a vortex method.

The equations to be solved will be Eqs. 11.8 and 11.9. The incident flow

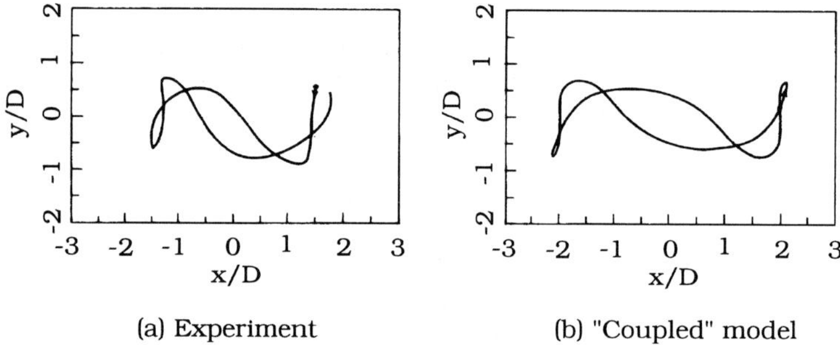

Figure 11.7 Comparison of the trajectory of a flexibly-mounted circular cylinder predicted by the "coupled" model of Lipsett and Williamson (1991) with that obtained by experiments by the same authors. $KC = 19.7$.

will be different, however, in the present case, in that the constant flow in the case of steady flow will be replaced by an oscillatory flow in the present situation.

Graham and Djahansouzi (1991a,b) have used a vortex method to simulate the flow past a circular cylinder in a planar oscillatory flow under conditions, first of a fixed cylinder and, secondly by an elastically mounted cylinder. They presented the results for the range of KC numbers up to 12, and for two ratios of f_n/f_w, namely 2 and 6 in which $f_n/f_w = KC/V_r$. The computations were carried out in the range of Re $1 - 2 \times 10^3$. The in-line force coefficients were determined and compared with their counterparts in the case of stationary cylinder.

Kozakiewicz, Sumer, Fredsøe and Hansen (1996) have used the discrete vortex model to predict the flow around a cylinder vibrating in the cross-flow direction and subjected to a planar oscillatory flow. Two KC numbers were tested, $KC = 10$ and $KC = 20$. The experimentally obtained cylinder trajectories were the input of the calculations. The main aim of the study was to obtain the flow around and the forces on the cylinder. The results were found to be in satisfactory agreement with the experiments (see Section 9.5 for a detailed account of the study).

11.3 Integrated models

Besides the models already described in the preceding sections, there are integrated models used in offshore-engineering practice to predict vibrations of slender structures. These models need to be rather general and to accommodate all kinds of flow environments such as steady currents, sheared currents, waves (2D and 3D; regular and irregular), and their combinations. One such model has been developed by Hansen (1982) (also see Nedergaard, Bendiksen and Andreasen, 1994, and Nedergaard, Hansen and Fines, 1994). The following paragraphs will describe the basic principles of this model.

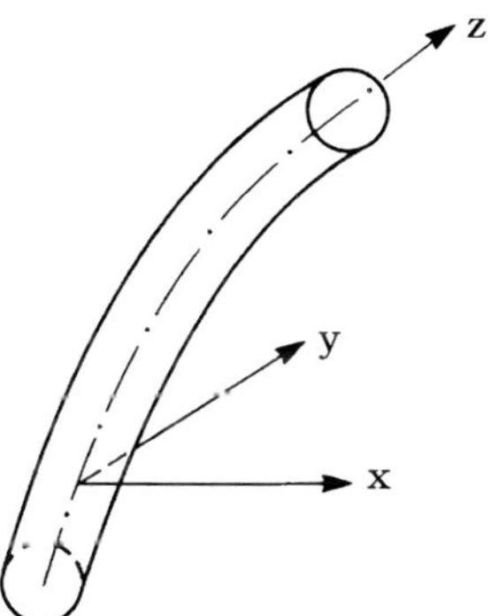

Figure 11.8 Definition sketch.

The vibratory response of the structure is calculated by the equation of motion in x- and y-directions (Fig. 11.8):

$$EI(z)\frac{\partial^4 x}{\partial z^4} - N(z)\frac{\partial^2 x}{\partial z^2} + c\,(z)\frac{\partial x}{\partial t} +$$

$$\quad (11.11)$$

$$+\, m(z)\frac{\partial^2 x}{\partial t^2} = F_x(z,t)$$

and a similar expression for the y-direction. Here, E is the elasticity modulus, I the inertia moment, N the tension, c the structural damping, m the mass per unit length, and F_x the total fluid force in the x-direction. Assuming that the motion can be divided into a forced motion and a "dynamic" motion,

$$x(t,z) = x_f(t,z) + x_d(t,z) \qquad (11.12)$$

(and a similar equation for the y-direction), and expressing the dynamic part by a sum of orthogonal eigenfunctions

$$x_d(t,z) = \sum_{i=1}^{N} X_i(t)\,\psi_i(z) \qquad (11.13)$$

(a similar equation for the y-direction), the following equation is obtained for the undamped eigenvalue solution

$$\omega_0^2 \int_0^L m(z)\psi_i^2(z)dz =$$

$$= \int_0^L \left[EI(z)\psi_i(z)\psi_i^{IV}(z) - N(z)\psi_i(z)\psi_i''(z) \right]\, dz \tag{11.14}$$

in which L is the length of the structure in consideration. From Eqs. 11.11 and 11.12, and integrating along the length of the structure and further utilizing Eqs. 11.13 and 11.14, the following expression is obtained for the damped vibrations:

$$\ddot{X}_i + 2\beta\,\omega_{0i}\,\dot{X}_i + \omega_{0i}^2\,X_i =$$

$$= \frac{1}{m_i} \int_0^L \left[F_X(z,t) - m(z)\,\frac{\partial^2 x_f}{\partial t^2} \right]\psi_i(z)\,dz \tag{11.15}$$

(a similar equation for the y-direction). This equation is the counterpart of Eq. 11.1.

The solution to the preceding equation is obtained in the following integral form:

$$X_i(t) = \frac{1}{m_i\omega_i} \int_0^t \exp\left\{ - \beta_i\,\omega_0(t - z) \times \right.$$

$$\tag{11.16}$$

$$\times \sin\left[\omega_i(t - \tau)\right]\overline{F}_{ix}(\tau)d\tau$$

(and a similar equation for the y-direction) in which

$$\omega_i^2 = \omega_{0i}^2(1 - \beta_i)^2 \tag{11.17}$$

$$\overline{F}_{ix}(t) = \int_0^L \left[F_x(z,t) - m(z)\frac{\partial^2 x_f}{\partial t^2} \right]\psi_i(z)dz$$

Regarding the hydrodynamic load term, the Morison force is considered in the two directions. In addition to that, the vortex-induced lift force (normal to the direction of the instantaneous flow relative to the structure) is considered as in the following

$$F_L = \frac{1}{2}\rho DC_L U^2(z,t)\sin(\omega_{St}t) \tag{11.18}$$

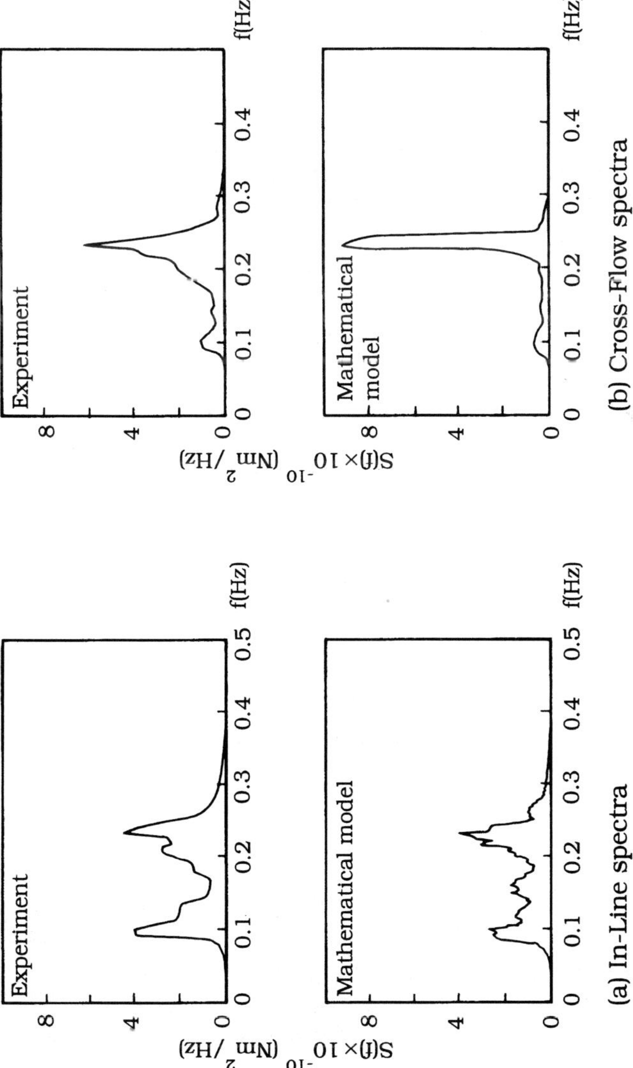

Figure 11.9 a) In-line spectra, measured and calculated. b) Cross-flow spectra, measured and calculated. Nedergaard et al. (1994a).

in which ω_{St} is the Strouhal frequency. According to the model, the forcing is present in "cells" along the length of the structure; however, only when there is lock-in with a structural eigenmode, vibratory response in the cross-flow direction (normal to the direction of the instantaneous flow relative to the structure) will develop. The model takes account of spanwise correlation in a semi-empirical fashion.

Fig. 11.9 shows comparison of model results with the field experiments obtained in a study for predicting the response of a drilling conductor (with a diameter of about 75 cm) installed in the North Sea where the water depth was 93 m. In the figures, the spectra for the bending stress at mudline in the wave direction are depicted.

Similar models have been developed by various authors. The following works can be mentioned in this regard: Iwan (1981), Lyons and Patel (1986), Rajabi, Zedan and Mangiavacchi (1984), Vandiver (1985), Kim, Vandiver and Holler (1985) and Dong and Lou (1991). For the work done before 1982, the review paper by Griffin and Ramberg (1982) can be consulted.

REFERENCES

Anagnostopoulos, P. (1994): Numerical investigation of response and wake characteristics of a vortex-excited cylinder in a uniform stream. Journal of Fluids and Structures, 8:367-390.

Anagnostopoulos, P. and Bearman, P.W. (1992): Response characteristics of a vortex-excited cylinder at low Reynolds numbers. Journal of Fluids and Structures, 6:39-50.

Bearman, P.W. (1984): Vortex shedding from oscillating bluff bodies. Ann. Rev. Fluid Mech., 16:195-222.

Bearman, P.W., Graham, J.M.R. and Obasaju, E.D. (1984): A model equation for the transverse forces on cylinders in oscillatory flow. Appl. Ocean Res., 6:166-172.

Chang, K.-S. and Sa, J.-Y. (1992): Patterns of vortex shedding from an oscillating circular cylinder. AIAA Journal, 30(5):1331-1336.

Currie, I.G. and Turnbull, D.H. (1987): Streamwise oscillations of cylinders near the critical Reynolds number. Journal of Fluids and Structures, 1:185-196.

Dong, Y. and Lou, J.Y.K. (1991): Vortex-induced nonlinear oscillation of tension leg platform tethers. Ocean Engrg., 18(5):451-464.

Feng, C.C. (1968): The Measurement of Vortex Induced Effects on Flow Past Stationary and Oscillating Circular and D-Secton Cylinders. M.Sc. Thesis, Univ. British Columbia.

Graham, J.M.R. and Djahansouzi, B. (1991a): Computation of vortex shedding from rigid and compliant cylinders in waves. Proc. 1st Int. Offshore and Polar Engrg. Conf., ISOPE, Edinburgh, UK, Aug. 11-16, 1991, 3:504-508.

Graham, J.M.R. and Djahansouzi, B. (1991b): A computational model of wave induced response of a compliant cylinder. Proc. 5th Conf. on Flow Induced Vibrations, Brighton, UK, May 21-23, 1991, pp. 333-341.

Griffin, O.M. and Ramberg, S.E. (1982): Some recent studies of vortex shedding with application to marine tubulars and risers. J. of Energy Resources, 104:2-13.

Hansen, N. E.O. (1982): Vibrations to pipe arrays in waves. Proc. of BOSS '82, Boston, Aug. 1982, 2:641-650.

Hartlen, R.T., Currie, I.G. (1970): Lift-oscillator model of vortex-induced vibrations. ASCE, J. Eng. Mech. Div., 96:577-591.

Hurlbut, S.E., Spaulding, M.L. and White, F.M. (1982): Numerical solution for laminar two-dimensional flow about a cylinder oscillating in a uniform stream. Journal of Fluids Engrg., 104:214-222.

Iwan, W.D. (1981): The vortex-induced oscillation of non-uniform structural systems. J. Sound Vibration, 79:291-301.

Kawai, H. (1990): A discrete vortex analysis of flow around a vibrating cylinder with splitter plate. Journal of Wind Engineering and Industrial Aerodynamics, 35:259-273.

Kim, Y.-H., Vandiver, J.K. and Holler, R. (1985): Vortex-induced vibration and drag coefficients of long cables subjected to sheared flows. Proc. 4th Int. Offshore Mech. and Arctic Engrg., OMAE, Symposium, Vol. 1, ASME, Dallas, TX, 1985, 1:584-592.

Kozakiewicz, A., Sumer, B.M., Fredsøe, J. and Hansen, E.A. (1996): Vortex regimes around a freely-vibrating cylinder in oscillatory flow. Proc. 6th Int. Offshore and Polar Engrg. Conf., Los Angeles, USA, May 25-30, 1996, 3:490-498.

Laya, E.J., Connor, J.J. and Sunder, S.S. (1984): Hydrodynamic forces on flexible offshore structures. J. Eng. Mech., ASCE, 110(3):433-448.

Lecointe, Y. and Piquet, J. (1989): Flow structure in the wake of an oscillating cylinder. Journal of Fluids Engrg., 111:139-148.

Li, J., Sun, J. and Roux, B. (1992): Numerical study of an oscillating cylinder in uniform flow and in the wake of an upstream cylinder. J. Fluid Mech., 237:457-478.

Lipsett, A.W. and Williamson, I.D. (1991): Modelling the response of flexibly mounted cylinders in oscillatory flow. Proc. 1st Int. Offshore and Polar Engrg. Conf., ISOPE, Edinburgh, UK, Aug. 11-16, 1991, 3:370-377.

Lipsett, A.W. and Williamson, I.D. (1994): Response of a cylinder in oscillatory flow. Journal of Fluids and Structures, 8:681-709.

Lyons, G.J. and Patel. M.H. (1986): A prediction technique for vortex induced transverse response of marine risers and tethers. J. Sound Vibration, 111:467-487.

McConnell, K.G. and Park, Y.S. (1982b): The frequency component of fluid-lift forces acting on a cylinder oscillating in still water. Experimental Mechanics, 22(6):216-222.

Meneghini, J.R. and Bearman, P.W. (1993): Numerical simulation of high amplitude oscillatory-flow about a circular cylinder using a discrete vortex method. AIAA Shear Flow Conf., July 6-9, 1993, Orlando, FL, Paper AIAA 93-3288.

Nedergaard, H., Bendiksen, E. and Andreasen, K.K. (1994a): Response analysis of slender drilling conductors. Proc. of Int. Conf. on Hydroelasticity in Marine Technology, Trondheim, Norway, May 25-27, 1994, pp. 47-54.

Nedergaard, H., Hansen, N.-E.O. and Fines, S. (1994b): Response of free hanging tethers. Proc. of Behaviour of Offshore Structures Conf., BOSS '94, Massachusetts Inst. of Technology, MA, July 12-15, 1994, 2:315-326.

Parkinson, G.V. (1974): Mathematical models of flow-induced vibrations. In: Flow Induced Structural Vibrations, ed. E. Naudascher, pp. 81-127, Berlin: Springer.

Rajabi, F., Zedan, M.F. and Mangiavacchi, A. (1984): Vortex shedding induced dynamic response of marine risers. J. of Energy Resources, 106:214-221.

Roshko, A. (1953): On the development of turbulent wakes from vortex streets. NACA TN 21913.

Sarpkaya, T. (1978): Fluid forces on oscillating cylinders. J. Waterways, Port, Coastal and Ocean Div, ASCE, 104(WW3):275-290.

Sarpkaya, T. (1979): Vortex-Induced Oscillations – A Selective Review. J. Appl. Mech., ASME Trans, 46:241-258.

Sarpkaya, T. and Schoaff, R.L. (1979): A discrete vortex analysis of flow about stationary and transversely oscillating circular cylinders. Tech. Rep. NPS-69SL79011, Naval Postgrad. Sch., Monterey, CA. The results regarding the discrete vortex analysis were summarized in: Sarpkaya, T., Schoaff, R.L. (1979): Inviscid model of two-dimensional vortex shedding by a circular cylinder. AIAA J., 17:1193-1200.

Slaouti, A. and Stansby, P.K. (1994): Forced oscillation and dynamic response of a circular cylinder in a current - - Investigation by the vortex method. 7th Int. Conf. on Behaviour of Offshore Structures, BOSS '94, Ed. C. Chryssostomidis, Pergamon Press., 2:645-654.

Staubli, T. (1983): Calculation of the vibration of an elastically mounted cylinder using experimental data from forced oscillation. Journal of Fluids Engineering, 105:225-229.

Vandiver, J.K. (1985): The prediction of lock-in vibration on flexible cylinders in a sheared flow. Proc. 17th Annual Offshore Technology Conf., OTC, Houston, TX, May 6-9, 1985, Paper 5006, pp. 405-412.

Verley, R.I.P. (1980): Oscillations of cylinders in waves and currents. Ph.D. Thesis, Loughborough Univ., 1980.

Williamson, C.H.K. (1985): In-line response of a cylinder in oscillatory flow. Applied Ocean Res., 7(2):97-106.

Yeung, R.W. and Vaidhyanathan, M. (1993): Flow past oscillating cylinders. Journal of Offshore Mechanics and Arctic Engineering, 115:197-205.

Appendix I. Force coefficients for various cross-sectional shapes

Table I.1 Force coefficients (compiled by Hallam, Heaf and Wootton (1977)). Notes: 1) Figures in brackets are estimates. 2) Figures with asterisks should be reduced by rounding corners. For elliptic cross-sectional shape, refer to, for example, Hoerner (1965) and Modi, Wiland, Dikshit and Yokomizo (1992).

Shape	Flow direction	$\overline{C}_D$	$\left(\overline{C'^2_D}\right)^{1/2}$	$\left(\overline{C'^2_L}\right)^{1/2}$	Remarks
Circle	-	Fig. 2.7 also Fig. 2.8a	Fig. 2.15a	Fig. 2.15b also Fig. 2.8b	Reynolds number dependent
Square	→ □	2.0*	0.15	0.4	1) Reynolds number independent
"	→ ◇	1.6*	(0.1)	(0.3)	2) Steady lift possible with other directions
Equilateral triangle	→ ◁	1.3*	(0.1)	0.05	"
"	→ ▷	1.8*	(0.15)	(0.5)	"
Rhendex pile	→ ⬡	1.3	(0.1)	(0.4)	"
"	→ ⬡	0.8	(0.1)	(0.8)	"
Octagon (eight sides)	Not critical	1.4	(0.2)	(0.3)	"
Duodecagon (twelve sides)	Not critical	1.1	(0.1)	(0.2)	"

Table I.2a Drag coefficient for different profiles. C_D has been solved into
C_n and C_t and is related to the length a and not the effective
front area. Taken from Danish Society of Engineers (1984).

α	C_n	C_t	C_n	C_t	C_n	C_t
degrees						
0	+1.9	+0.95	+1.8	+1.8	+1.75	+0.1
45	+1.8	+0.8	+2.1	+1.8	+0.85	+0.85
90	+2.0	+1.7	-1.9	-1.0	+0.1	+1.75
135	-1.8	-0.1	-2.0	+0.3	-0.75	+0.75
180	-2.0	+0.1	-1.4	-1.4	-1.75	-0.1

α	C_n	C_t	C_n	C_t	C_n	C_t
degrees						
0	+1.4	0	+2.05	0	+1.6	0
45	+1.2	+1.6	+1.95	+0.6	+1.5	+1.5
90	0	+2.2	+0.5	+0.9	0	+1.9

Table I.2b Drag coefficient for different profiles. C_D has been solved into C_n and C_t and is related to the length a and not the effective front area. Taken from Danish Society of Engineers (1984).

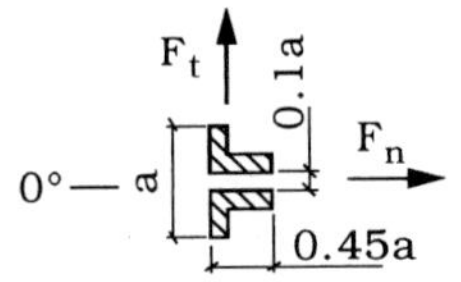

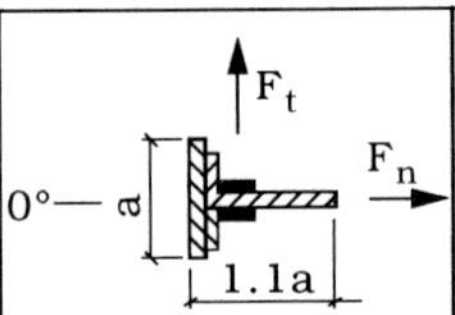

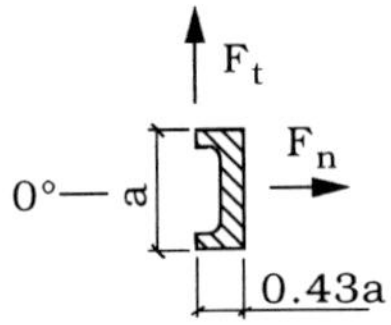

	(0.45a profile)		(1.1a profile)		(0.43a profile)	
α	C_n	C_t	C_n	C_t	C_n	C_t
degrees						
0	+1.6	0	+2.0	0	+2.05	0
45	+1.5	-0.1	+1.2	+0.9	+1.85	+0.6
90	-0.95	+0.7	-1.6	+2.15	+0	+0.6
135	-0.5	+1.05	-1.1	+2.4	-1.6	+0.4
180	+1.5	0	-1.7	±2.1	-1.8	0

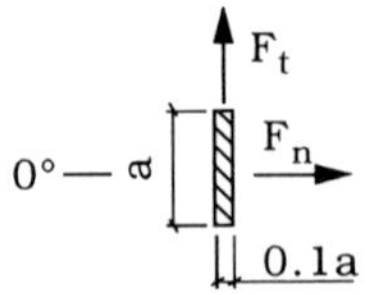

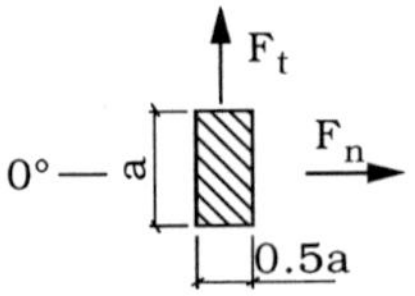

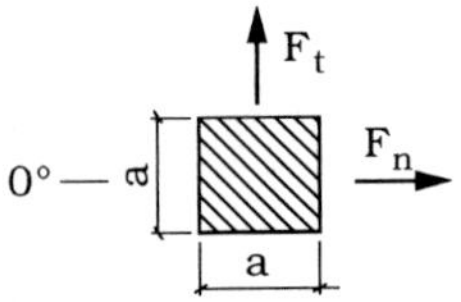

	(0.1a profile)		(0.5a profile)		(a profile)	
α	C_n	C_t	C_n	C_t	C_n	C_t
degrees						
0	+2.0	0	+2.1	0	+2.0	0
45	+1.8	+0.1	+1.4	+0.7	+1.55	+1.55
90	0	+0.1	0	+0.75	0	+2.0

Appendix II. Hydrodynamic-mass coefficients for two- and three-dimensional bodies

Table II.1 Hydrodynamic-mass coefficient C_m for two-dimensional bodies (infinitely long cylinder). $m' = \rho C_m A$ where $m' =$ the hydrodynamic-mass per unit length of the cylinder. Compiled by Danish Society of Engineers (1984).

Section through body	Direction of motion	$\dfrac{a}{b}$	C_m	A
 (circle, 2a)	↕		1.0	πa^2
 (vertical ellipse, 2a)	↕		1.0	πa^2
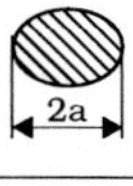 (horizontal ellipse, 2a)	↕		1.0	πa^2
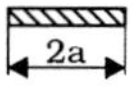 (flat plate, 2a)	↕		1.0	πa^2
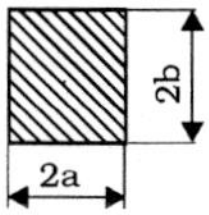 (rectangle, 2a × 2b)	↕	∞	1.00	πa^2
		10.0	1.14	
		5.0	1.21	
		2.0	1.36	
		1.0	1.51	
		0.5	1.70	
		0.2	1.98	
		0.1	2.23	

Table II.2 Hydrodynamic-mass coefficient C_m for three-dimensional bodies. $m' = \rho C_m V$. Compiled by Danish Society of Engineers (1984).

Shape	Direction of motion	$\frac{a}{b}$	C_m	V
Circular disc			0.64	$\frac{4}{3}\pi a^3$
Elliptical disc		1.0	0.64	
		1.5	0.76	
		2.0	0.83	
		3.0	0.90	$\frac{4}{3}\pi ab^2$
		5.0	0.95	
		10.0	0.98	
		∞	1.00	
Rectangular plate		1.0	0.58	
		1.5	0.69	
		2.0	0.76	
		3.0	0.83	$2\pi ab^2$
		5.0	0.90	
		10.0	0.95	
		∞	1.00	
Triangular plate			$\frac{1}{\pi}(\tan\theta)^{\frac{3}{2}}$	$\frac{1}{3}a^3$
Sphere			0.5	$\frac{4}{3}\pi a^3$

Shape	Direction of motion	$\frac{a}{b}$	axial	lateral	V
Ellipsoid		1.0	0.50	0.50	
		1.5	0.30	0.62	
		2.0	0.21	0.70	
		3.0	0.12	0.80	$\frac{4}{3}\pi ab^2$
		5.0	0.06	0.89	
		10.0	0.02	0.96	
		∞	0.00	1.00	

Shape	Direction of motion	$\frac{a}{b}$	C_m	V
Rectangular prism		0.5	1.30	
		1.0	0.68	
		1.5	0.47	
		2.0	0.36	ab^2
		3.0	0.24	
		5.0	0.15	
		10.0	0.08	
		∞	0.00	

Appendix III. Small amplitude, linear waves

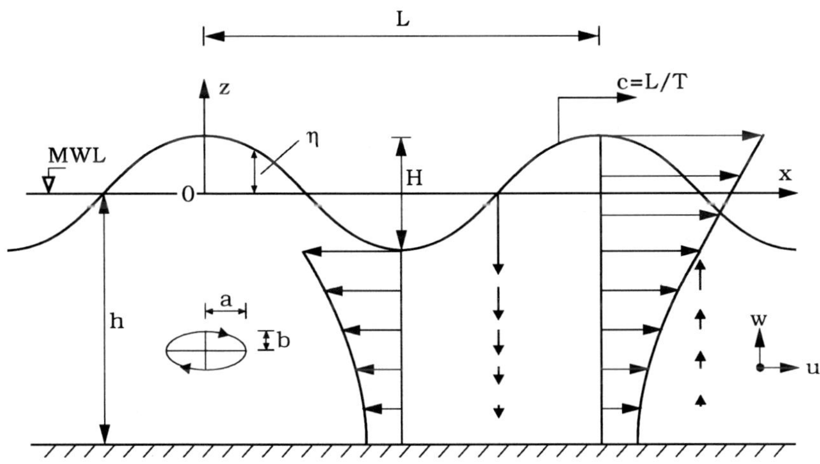

Figure III.1 Description sketch.

Basic equation: $\qquad\qquad \nabla^2\phi = \phi_{xx} + \phi_{zz} = 0$ (III.1)

Bed boundary condition: $\quad w = \phi_z = 0 \ \text{ at } \ z = -h$ (III.2)

Kinematic, free-surface
boundary condition: $\qquad \left(\frac{\partial\phi}{\partial z}\right)_{z=0} = \frac{\partial\eta}{\partial t}$ (III.3)

Dynamic, free surface
boundary condition: $\qquad \left(\frac{\partial\phi}{\partial t} + g\eta\right)_{z=0} = C(t)$ (III.4)

Water surface elevation: $\quad \eta = \frac{H}{2}\cos\left(\omega t - kx\right)$ (III.5)

Potential function: $\qquad \phi = -\frac{Hc}{2}\frac{\cosh\left(k(z+h)\right)}{\sinh(kh)}\sin\left(\omega t - kx\right)$ (III.6)

Wave celerity:

$$c = L/T = \omega/k \tag{III.7}$$
$k = 2\pi/L = \text{ wave number}$
$\omega = 2\pi/T = 2\pi f = \text{ angular wave frequency}$
$f(= 1/T) \text{ being the wave frequency}$

Dispersion relation:

$$\omega^2 = gk\tanh(kh) \tag{III.8}$$

or

$$(2\pi f)^2 = gk\tanh(kh) \tag{III.9}$$
g being the acceleration due to gravity

Horizontal particle velocity:

$$u = \phi_x = \frac{\pi H}{T}\frac{\cosh(k(z+h))}{\sinh(kh)}\cos(\omega t - kx) \tag{III.10}$$

or

$$= \frac{gkH}{4\pi f}\frac{\cosh(k(z+h))}{\cosh(kh)}\cos(\omega t - kx) \tag{III.11}$$

Vertical particle velocity:

$$w = \phi_z = -\frac{\pi H}{T}\frac{\sinh(k(z+h))}{\sinh(kh)}\sin(\omega t - kx) \tag{III.12}$$

or

$$= \frac{gkH}{4\pi f}\frac{\sinh(k(z+h))}{\sinh(kh)}\sin(\omega t - kx) \tag{III.13}$$

Horizontal amplitude of particle motion:

$$a = \frac{H}{2}\frac{\cosh(k(z+h))}{\sinh(kh)} \tag{III.14}$$

Vertical amplitude of particle motion:

$$b = \frac{H}{2}\frac{\sinh(k(z+h))}{\sinh(kh)} \tag{III.15}$$

Pressure:

$$\frac{p}{\rho} = \underset{\text{Hydrostatic}}{-gz} - \underset{\text{Excess}}{\phi_t} \tag{III.16}$$
$$\underset{\text{pressure}}{} \qquad \underset{\text{pressure}}{}$$
ρ being the density of water

Excess pressure

$$\frac{p^+}{\rho} = -\phi_t = g\frac{H}{2}\frac{\cosh(k(z+h))}{\cosh(kh)}\cos(\omega t - kx) \tag{III.17}$$

Wave energy per unit area

$$E = \frac{1}{L}\left(2\int_0^L (\rho g \eta dx)\frac{\eta}{2}\right) \tag{III.18}$$

$$= \frac{1}{16}\rho g H^2$$

REFERENCES FOR APPENDICES

Danish Society of Engineers (1984): Pile-Supported Offshore Steel Structures. Code of Practice. First edition in Danish, April 1983, Dansk Standard DS 449. English translation edition, September 1984.

Hallam, M.G., Heaf, N.J. and Wootton, L.R. (1977): Dynamics of Marine Structures. CIRIA Underwater Engineering Group, Report UR8, Atkins Research and Development, London, U.K.

Hoerner, S.F. (1965): Fluid-Dynamic Drag. Published by the Author, pp. 3-11.

Modi, V.J., Wiland, E., Dikshit, A.K. and Yokomizo, T. (1992): On the fluid dynamics of elliptic cylinders. Proc. 2nd Int. Offshore and Polar Engrg. Conf., San Francisco, USA, 3:595-614.

Author Index

Subject Index